Pvt. Ltd.

Optoelectronics and Systems

by

Aruna Tripathy

Department of Instrumentation and Electronics Engineering,
College of Engineering and Technology,
Techno Campus, Ghatikia,
Bhubaneswar-751 003

2017

Studium Press (India) Pvt. Ltd.

Optoelectronics and Systems

ISBN: 978-93-80012-87-2

Published by:

Studium Press (India) Pvt. Ltd.
4735/22, 2nd Floor, Prakash Deep Building
(Near Delhi Medical Association),
Ansari Road, Darya Ganj, New Delhi-110 002
Tel.: + 91-11-43240200-15 (15 lines); Fax: 91-11-43240215
E-mail: studiumpress@gmail.com

Printed at:

xxxxxxxxx

About the Author

Aruna Tripathy

Aruna Tripathy obtained her Bachelor's degree in Engineering in Electronics and Telecommunication from the then University College of Engineering, Burla (now known as Veer Surendra Sai University of Technology, VSSUT) in 1990. Immediately after, she worked in a core telecom company manufacturing telephones and exchanges for about four years. She obtained her Master's degree and Ph.D. from IIT, Kharagpur. She has been in the teaching profession thereafter for nearly twenty three years in various capacities. She has taught UG courses on a number of subjects including the optoelectronics and PG courses on mainly communication related subjects. She has a number of publications in international journals, conferences. Her current interests include various physical layer aspects of wireless communication and optical character recognition.

Acknowledgement

A project on writing a text book requires inputs, support and help of myriad nature from a multitude of sources. At the outset, I would like to thank Prof. Ranjan Gangopadhaya, Professor, Department of Electronics and Electrical Communication Engineering, IIT, Kharagpur for having introduced me to the world of optical communication systems and optical fibres during my PG studies at IIT, Kharagpur. The class notes prepared under his tutelage have gone a long way to write a lot of material presented in this book.

The UG students of 2010-2014 batch deserve special mention here as they had requested me to teach the subject of "OptoElectronic Devices and Instrumentation" that led to the discovery of the difficulties faced by them in learning this subject without a proper background on Electromagnetic Field Theory. Thanks go to all my students who studied this subject and provided the right stimuli to go for such a book.

Special thanks go to my parent department of Instrumentation and Electronics Engineering, CET, Bhubaneswar for providing the necessary infrastructure and support in making this book a reality. Heartfelt thanks are extended to Dr. Umesh Chandra Pati, Associate Professor, Department of Electronics and Telecommunication Engineering, NIT, Rourkela for making some valuable suggestions and criticisms to improve the quality of the book.

My sincere thanks go to Shri S.K. Acharya, DGM (C&I) Project, and Shri L. Jena, AGM (E&I), both from NALCO, for having reviewed some portions of the book and making useful suggestions.

I will fail in my duties if heartfelt thanks are not extended to my husband Dr. R.R. Dash, Professor, Mechanical Engineering, CET, Bhubaneswar and my son Satwik Dash, himself a good blogger, for providing the right ambience at home without any grumbling over anything.

I will fail in my duties in not acknowledging the publisher. My sincere and earnest thanks to Dr.J.N. Govil and his team at Studium Press for making all steps possible and taking all pains for this book to see the light of the day.

Finally, thanks to all the researchers and people pursuing a passion in the ever enigmatic and challenging field of optics whose work has made this subject so fascinating and inspiring.

Preface

The book is an outcome of teaching the subject of "OptoElectronics and Instrumentation" at graduate level for a couple of years. A strong need was felt to address the issues raised by the students which could have become possible in the form of a new book. Though excellent text books and reference books are available written by experts, for a beginner a different book was required. That has been the motivation for this book. This has been written keeping in mind particularly the prescribed syllabus of the Biju Patnaik University of Technology (BPUT), Odisha. However, it can also be used as a text book/reference book by other universities/ institutes providing subjects on related topics. Material presented in the book may be used, with the appropriate combination of chapters for both UG and PG levels.

Chapter 1 gives an introductory idea of systems built around optical components. Optical sources coupled to detectors through optical fibres are said to constitute a system. The first laser was demonstrated in 1960. A major portion of the optical components is the optical fibre that has been elaborated subsequently. The first low loss optical fibre was realized in 1970. Communication systems and instrumentation systems are introduced. The idea of low loss optical fibres in a certain part of the electromagnetic spectrum has been emphasized.

The systems introduced in chapter 1 rely upon waves propagating through the optical fibres. This requires a good understanding of the electric and magnetic fields. Understanding the electromagnetic waves require a background on Maxwell's equations. Therefore, an attempt has been made to discuss the equations both in the differential as well as the integral forms. Efforts have been made to convince the reader the generation of a wave like phenomena arising out of the solutions to these equations. Efforts have been elaborated to correlate mathematical equations to physical phenomena. This chapter assumes an introductory background on the three widely used coordinate systems and vector algebra.. A couple of numerical examples are provided to help the reader analyzing a given field.

Fields responsible for generating a wave are elaborated in chapter 3. Wave propagation in various media are discussed in detail with the appropriate mathematical expressions. Wave phenomena like reflection, refraction and total internal reflection are described. Boundary value problems are important in visualizing the behaviour of waves at the interface of two different media. This is essential for understanding (i) the mechanism of wave propagation in the core which is made from a dielectric material, (ii) behaviour of the waves at the core-cladding interface of an optical fibre and (iii) air (or any other medium) to fibre boundary. Polarization of waves as regards the perpendicular and parallel cases is also derived for various dielectric and magnetic media. The concept of numerical aperture of a fibre is introduced to aid the understanding of the amount of light coupled into the fibre.

As we said in chapter 1, identification of the right kind of optical source is a very important design issue in building the various systems. The literature on the light emitting diodes and laser diodes is very rich. Therefore, attempt has been made to discuss the sources with the necessary details while keeping the mathematics at an acceptable level. This chapter assumes a background of elementary semiconductor physics on the part of the reader. Several important source parameters are presented.

Fibres basically are of two types; the step index and the graded index fibre. Waves propagate in different paths or modes in these two types. A given application requires a certain type of fibre in preference to the other. Hence, the wave equations along with their solutions are presented in chapters 5 and 6 for the two types of fibres. Detailed derivations of the propagating modes not usually found in other recommended text books are provided. These chapters assume a thorough background on cylindrical coordinate system and its associated geometry on the part of the student. Power flowing in the core is derived.

An optical pulse launched onto an optical fibre undergoes a number of distortions and impairments which prevent data transfer reliably beyond a certain rate over a given fibre link. It is therefore important to identify the causes of these impairments and use a given fibre for a certain maximum data rate. Chapter 7 addresses these issues in the form of various pulse dispersion mechanisms.

We have discussed wave phenomena like reflection and refraction previously. Besides these two, two other important phenomena like interference and diffraction manifested due to the wave nature of light requires treatment as these form the underlying principles of many an optical systems. Hence these have been discussed in chapter 8.

An optical link from the transmitter to the receiver is made possible through the use of a number of optical components besides the optical fibres. Two fibres need to be joined at some point thereby requiring Splicers. Power need to be combined or split from a given point along the link necessitating the inclusion of directional couplers. Optocouplers provide electrical isolation between a source and a load besides eliminating the need of a common ground. Similarly other components find their application in making a fully operational link. The various components form the subject matter of chapter 9.

Having discussed the basic two types of optical sources and optical fibres in the previous chapters, we present the optical detectors in chapter 10. The basic two types of photodetectors are discussed here along with the important parameters of interest. Attempt has been made to keep the underlying mathematics to a minimum.

Once we have the optical source, the fibre, the various components and the optical detectors besides an understanding of various light phenomena, we are ready to develop systems using these ingredients. Optical sensors form a very important class of sensors that find numerous applications in hostile, hazardous and electromagnetic interference susceptible environments. These are constructed based on various properties of light. The discussion of optical sensors is provided in chapter 11.

Next we discuss optical communication systems. The fundamental limitations, issues and types are discussed in chapter 12. Performance index of such type of receivers is derived here. Coherent optical communication receivers form the subject matter of chapter 13. Various types of such receivers are discussed along with their performance indices.

Having gained some insight into optical systems realized in the form of sensors and communication systems, we discuss optical networks in chapter 14. The optical networks continue to be an active field of research and practical systems are being deployed in our country these days in order to meet the rapidly increasing demand for very high speed and reliable data connectivity. With this view, an attempt has been made to cover some important issues of an optical network with emphasis on the ISO-OSI model of network architecture.

The author would feel greatly rewarded if this book is able to ignite a spark of interest amongst the readers for the ever evolving, ever expanding beautiful field of optical sensors, communications and related topics. Every effort has been made to make this book free from errors. However, in spite of the best of efforts, errors might have crept into it unintentionally. Any positive comment in this regard is most welcome that would improve the quality of the book.

Author

Table of Contents

1

Fibre Optics: An Introduction

1.1. INTRODUCTION TO OPTICAL FIBRE COMMUNICATION

Traditional communication systems have been designed and are in operation using electrical signals that come from a host of sources such as speech, music, video on the analog domain or digital data on the digital front. These physical signals after suitable conversion by means of the appropriate transducers to their electrical equivalents are transmitted *via* an electrical carrier using either metallic media or free space. A block diagram of such a system is shown in Fig. 1.1. A system consists of a message source that provides the physical signal for transmission from a source to a destination/sink, a transmitter, the channel through which it is transmitted, a corresponding receiver and the final information sink. The signals typically are analog and nonelectrical in nature. This necessitates the use of an appropriate transducer to generate electrical signals (not shown in this figure). The transmitter provides the necessary signal conditioning and processing so that the characteristics of the signal and that of the channel match approximately. The channel introduces impairments to the signal in various forms that change the signal amplitude and phase in an unpredictable manner. The impairments caused by a channel are very specific to it. The receiver carries out appropriate signal processing to retrieve the original information contained in the impaired signal thus arriving at it before it is sent to the information sink. Systems employing the electrical signals to transmit and subsequently receive information are built around frequencies in the range of a few GHz on the maximum limit. The requirement to send and receive tens of thousands of subscriber's data through one medium cannot be effectively carried out by these electrically suitable physical media. Optical fibres offer a solution to this as they carry information through light waves. The credit of carrying out information transmission and reception goes to Hopkin and Kapany in the UK and Van Heel in Holland. A concise frequency allotment for different services as specified by the electromagnetic spectrum is presented in Table 1.1.

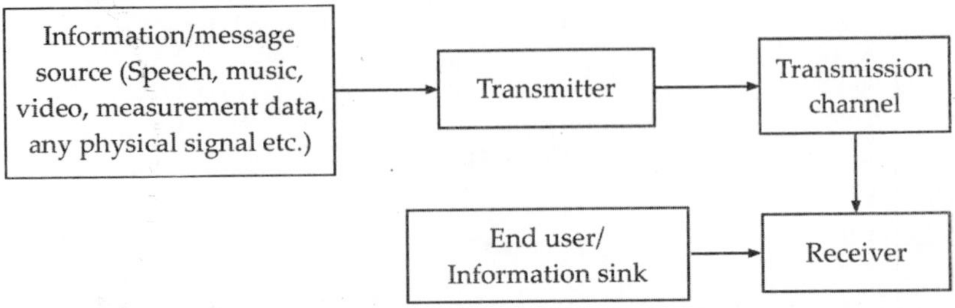

Fig. 1.1: Basic elements of a typical communication system

Communications offered by the optical fibres fall into wideband communications. Very large quantities of data can be transmitted over long distances through the optical fibres. The optical fibres may be thought of as the light equivalent of microwave waveguides with the additional advantage of a very large bandwidth. One optical fibre can carry a data that is about 10^5 times the maximum offered by a conventional microwave link. We can multiplex millions of television channels onto one optical fibre. The other advantages of an optical fibre relative to other guiding media are discussed later.

Physically an optical fibre is a very thin and flexible medium, having a cylindrical shape consisting of three sections: i) the core ii) the cladding and the iii) jacket. The core is the innermost section and is made of glass or plastic. This is the actual fibre and is responsible for guiding the light waves very similar to the microwaves being guided by the waveguides. The core is surrounded by a cladding, a glass or plastic coating which has optical properties very different from that of the core. The outer section is called the jacket that is made of plastic or polymer and other materials and is provided for protection against moisture, absorption, crushing and other-environmental hazards. The core acts like a continuous layer of two parallel mirrors. A signal is first encoded into a light beam, which is then passed in between the two boundaries and propagated as a result of multiple internal reflections.

1.2. ELECTROMAGNETIC SPECTRUM

It is clear from Table 1.1 that we are referring to the last entry of the table; the optical frequencies in the range of 10^5 GHz to 10^6 GHz (10^{14} Hz to 10^{15} Hz). As we may see, the frequencies are extremely high necessitating the development of appropriate technologies for the feasibility of a system designed around optic phenomena.

Table 1.1: Allocation of frequencies for existing applications

Sl. no.	*Frequency range*	*Nomenclature*	*Application/Usage*
1	30 Hz – 300 Hz	Extremely Low Frequencies (ELF)	Underwater Communication
2	300 Hz – 3.0 KHz	Voice Frequency (VF)	Telephone
3	3.0 KHz – 30 KHz	Very Low Frequencies (VLF)	Navigation
4	30 KHz – 300 KHz	Low Frequency (LF)	Radio navigation
5	300 KHz – 3 MHz	Medium Frequencies (MF)	AM radio broadcasting
6	3 MHz – 30 MHz	High Frequencies (HF)	AM, Amateur radio, mobile
7	30 MHz – 300 MHz	Very High Frequencies (VHF)	TV, FM, mobile communications
8	300 MHz – 3 GHz	Ultra High Frequencies (UHF)	TV, radar, satellite communications
9	3 GHz – 30 GHz	Super High Frequencies (SHF)	Terrestrial microwave and satellite communications
10	10^5 GHz – 10^6 GHz	Optical Frequencies	Optical communication √(We are here)

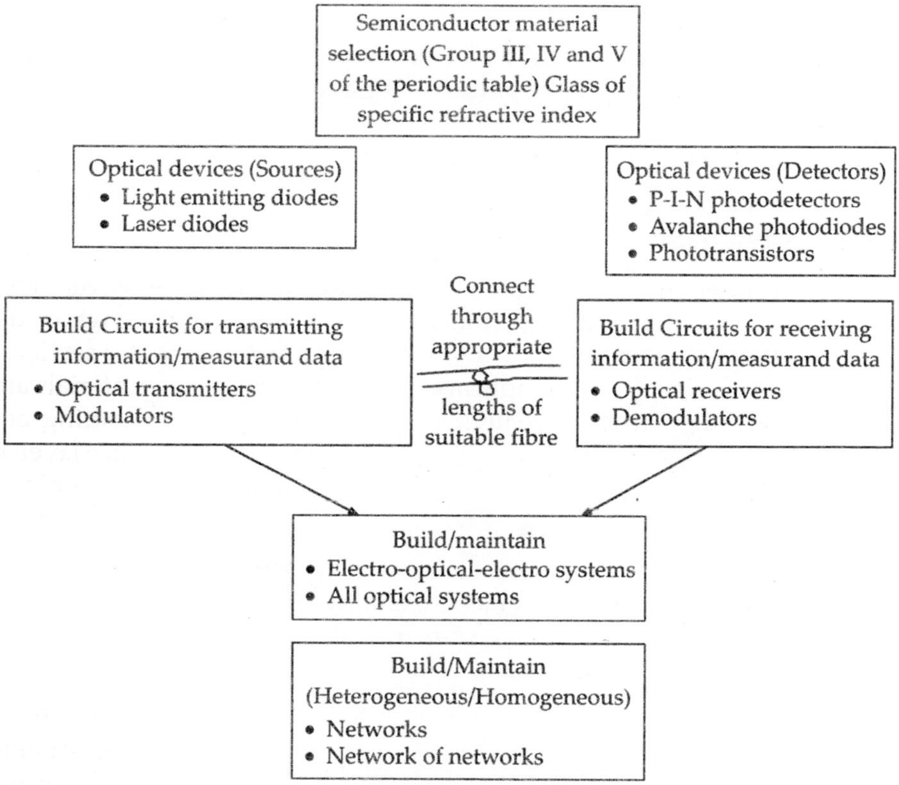

Fig. 1.2: Hierarchy of opto electronics system development

To realize the potential of an optical fiber, it is important to have suitable optical sources which will generate light of an appropriate wavelength that needs to be coupled to the fiber by well designed means while keeping signal leakage to a minimum. In this regard, light emitting diodes and laser diodes act as suitable optical sources. We may view them as devices providing electro optic conversion. Circuits designed to tap the light from the sources need to carry information in a manner appropriate to the particular application. Similarly, at the receiving end, we have to place suitable devices that would provide the reverse process of opto electric conversion (given a light beam, a current or a voltage is generated) , build suitable receiver circuits to process the electrical output thus generated and retrieve information or perform action depending on the measurand sensed if we are building an optical fiber based instrumentation system. All this calls for careful selection of material in the first place. A hierarchical structure may be constructed as shown in Fig. 1.2.

1.3. FEATURES OF OPTICAL FIBRE COMMUNICATION

It is now obvious that the optical fibre is the physical transmission channel between the transmitter and the receiver that accepts input in the form of light pulses and outputs light pulses that have been modified as a result of a number of fibre characteristics. We take a look at the potential benefits and the shortcomings of a typical fibre next.

1.3.1. Advantages

- The attenuation of the light waves through optical fibres is markedly lower than that of the electromagnetic waves through coaxial cables or twisted pair and is constant over a very wide range. Thus transmission of data *via* optical signals within wide range of distances is possible without repeaters.
- Smaller size and lighter weight: Optical fibres are considerably thinner than coaxial cable or bundled twisted-pair cable. So they occupy much less space.
- Electromagnetic isolation: Electromagnetic waves generated from electrical disturbances or electrical noises do not interfere with light signals. As a result, the system is not vulnerable to interference, impulse noise, or cross-talk.
- No physical electrical connection is required between the transmitter and the receiver.
- The fibre is reliable. This is because it can better withstand environmental conditions, such as pollution, radiation and salt produces no corrosion. Moreover, it is nominally affected

by the nuclear radiation. Its life is longer as compared to a copper wire.

- Almost there is no cross-talk in optical fibres and hence transmission is more secure and private, as it is very difficult to tap into a fibre.
- Greater bandwidth: Bandwidth of the optical fibre is higher than that of the equivalent wire transmission line. This is due to the fact that information is carried out by an optical carrier that occur in the THz frequency range. A bandwidth as nominal as 10% of the carrier still gives a very large value that amount to 10^4 GHz – 10^5 GHz. This makes it possible to send millions of speech and television channels together using a single optical fibre through suitable multiplexing and switching technologies.
- As fibres are very good dielectrics, isolating coating is not required.
- Data rate is much higher in a fibre and hence much more information can be carried by each fibre than by equivalent copper cables. As for example, at the immensely high frequencies of an optical fibre, data rates of 2 Gbps over tens of kilometres have been demonstrated. This may be compared with the practical maximum of hundreds of Mbps over about 1 km for coaxial cable and just a few MBps over 1 km for twisted wire.
- The cost per channel is lower than that of an equivalent wire cable system. It is expected that in the near future the optical fibre communication system will be more economical than any other communication system.
- Due to the non-inductive and non-conductive nature of a fibre, there is no radiation and interference on other circuits and systems.
- Greater repeater spacing: Fewer repeaters indicate lower cost and fewer sources of error. It has been observed that a fibre transmission system can achieve a data rate of 5 Gbps over a distance of 111 km. without repeaters, whereas coaxial and twisted-pair systems generally have repeaters every few kilometres.
- The raw material is available in plenty.

1.3.2. Disadvantages

- The fibre is brittle.
- Coupling of light from a given source into the core of the fibre is a difficult task as the core size is very small. Careful design considerations are to me made for coupling power into the fibre in an efficient manner.
- The fibre must be made from pure glass. Any contaminant present in the material of the core would lead to signal loss and other undesired effects.

- The electro optic conversion at the transmitter and the inverse opto electronic conversion at the receiver results in some signal loss and requires careful selection of sources and detectors.
- The signal is subject to a number of impairments specific to the fibre type which means that identical kind of signal processing cannot be applied to systems using different types of fibres.
- Design and practical realization of optical components is very challenging and still under active research.

1.4. PROPAGATION OF LIGHT WAVES THROUGH THE OPTICAL FIBRE

If light wave enters at one end of a fibre in proper conditions, most of it is propagated down the length of the fibre and comes out from the other end of the fibre. There may be some loss due to a small fractional leakage through the side walls of the fibre. This type of fibre is called light-guide or sometimes light-pipe. The reason of confining the light beam inside the fibre is the total internal reflection of light waves by the inside surface of the fibre. Fibre obeys laws of reflection and refraction of light waves.

The light which enters at one end of a fibre at a slight angle to the axis of the fibre follows a zigzag path due to a series of reflections down the length of the fibre.

The central core of a fibre consists of a glass core with a certain refractive index n_1 and totally enclosed by a glass cladding having refractive index n_2. The value of n_1 is greater than n_2 so as to facilitate the mechanism of total internal reflection. This is desired in a fibre as we want all of the light waves to be confined to the core and none to be lost in order to transmit the maximum amount of optical power. A longitudinal cross section of a fibre is shown in Fig. 1.3. Any light wave, which travels

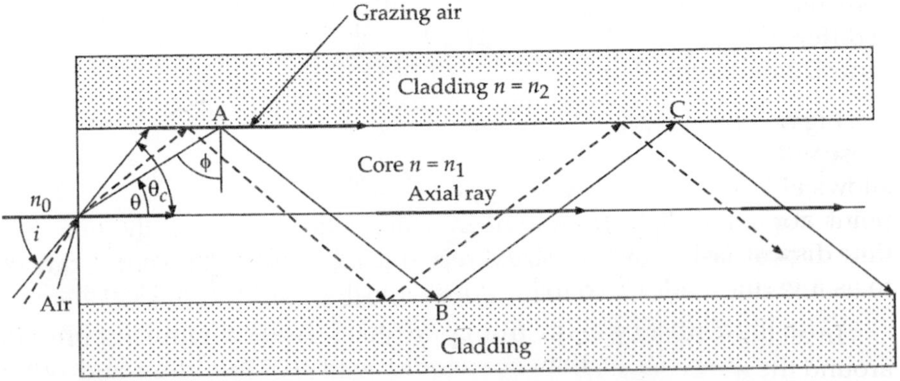

Fig. 1.3: Light wave propagation inside an optical fibre

along the core and meets the cladding at the critical angle of incidence θ_c will be totally reflected. This reflected ray will then meet the opposite surface of the cladding, again at the critical angle θ_c and so is again totally reflected. Therefore, the light wave is propagated along the fibre core by a series of total internal reflections from the core-cladding interface. This is a type of step index fibre, as there are just two values assigned to the refractive indices of the core and cladding.

Light waves originating from a practical optical source such as a LED or a laser have several paths with different angles of incidence at the boundary-layer. It may also contain different colors with different frequencies. Then it is called step index multi mode propagation as shown in the next figure. Any other light wave which is meeting the core-cladding interface at or above the critical angle will also be totally reflected and hence will propagate along the core. However, any light wave that meets the core-cladding interface at an angle less than will pass into and absorbed by the cladding.

Hence, it is obvious that the various light waves travelling along the core will have propagation paths of different lengths. Thus, they will take difference times to reach a given destination. Thus a distortion is produced and is called transit time dispersion. This dispersion sets an upper limit on the rate at which light can be modulated by an analog or digital electrical signal. As a result of this distortion, the variations of successive pulses of light may overlap into each other, and thereby cause distortion of the information being carried. However, this defect can be minimized by making the core diameter of the same order as the wavelength of the light to be propagated. The resultant propagation is a single light wave, as we shall see in the next chapters. However, this type of fibre is called a stepped index monomode fibre. This has very high capacity and large bandwidth.

The propagation of light waves in the graded index fibre is discussed. The individual light waves are gradually refracted in the graded index core instead of being reflected by the cladding. Thus light waves travelling at different incident angles will travel different distances from the horizontal central axis, before being reflected back to recross the central axis.

Light waves with large angle of incidence travel more paths than those with smaller angles. The decrease of refractive index in the cladding allows a higher velocity of propagation. Thus all waves will reach a given point along the fibre at virtually the same time. As a result the transit time dispersion is greatly reduced. This type of wave propagation is referred to as a graded index multimode propagation.

In all three types of optical fibre, the thickness of the cladding material around the fibre core should be at least several wavelengths. This prevents light energy loss due to absorption and scattering.

The optical fibre, as we have mentioned is an optical waveguide. It is a cylindrical waveguide. Hence, we will formulate the laws for wave propagation in an optical fibre using the cylindrical coordinate system.

First generation commercial telephone system was operated at a wavelength of 820 nm, using injection Laser diode, coupled to multimode fibre. The fibre attenuation was in the range of 3 to 8 dB/km. This allowed 90 Mbps of data rate over 8 to 12 km.

Second generation commercial system operated at a wavelength of 1300 nm without repeaters, with single mode fibre. Fibre attenuation was less than 0.5 dB/km. This provided 565 Mbps data rate up to over 45 km. This system was installed by numerous companies, including At & T (now Lucent Technologies), MCI and GTE-Sprint etc.

Third generation system, operating at a wavelength of 1550 nm, is anticipated which will employ single mode fibre with less than 0.3 dB/km. It will provide a transmission rate of 1.3 Gbps over a distance of 45 km.

Fourth generation commercial telephone system, based on single frequency laser at 1550 nm is in laboratory stage. However, transmission rate of 2 GBps over a fibre link of 1330 km has already been successfully demonstrated.

Experimentally additional high bandwidth capabilities have been extended into the gigahertz range, to permit the remote-controlling of satellite microwave signals. These are some of the glimpses of the 5th generation coherent fibre optic system.

An optical fibre link comprises the elements shown in Fig. 1.4. The key sections are a transmitter consisting of a light source and its associated drive circuitry, a cable offering mechanical and environmental protection to the optical fibres contained inside, and a receiver consisting of a photodetector, subsequent amplification of the resulting electrical signal amplification by an amplifier and signal-restoring circuitry. The driver circuit is used to push the requisite current into the optical source for its proper operation. We may note that the transmitter essentially provides an elecrtro optic (E/O) conversion that accepts an electrical signal as the input and furnishes an appropriate optical output by the source. Similarly, the receiver carries out the inverse operation of optoelecrotonic (O/E) conversion. Additional components in the link may include optical connectors, splices, couplers or beam splitters and repeaters. The cabled optical fibre is one of the most important elements in an optical fibre link. In addition to protecting the glass fibres during installation and service, the cable may contain copper wires for powering repeaters which are needed for periodically amplifying and reshaping the signal when the link spans long distances. The performance of an optical receiver for analog communications is the signal to noise ratio (SNR) at the photodetector output while it is probability of bit error (BEP) for a typical digital communication system. Optical receivers are capable of providing a BEP of 10^{-9} for a digital communication system. This figure simply states that out of 1000 mega bits transmitted by the transmitter, only one bit is in error at the receiver.

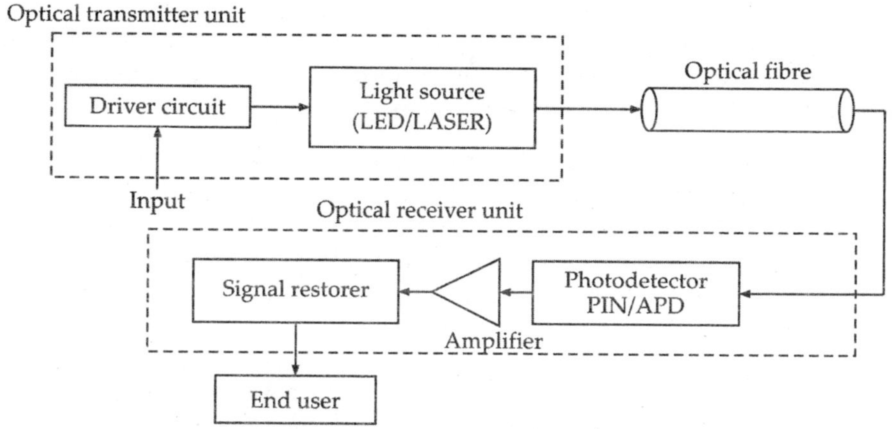

Fig. 1.4: A typical optical fibre based communication system

Similarly, an optical fiber based sensor/instrumentation system is illustrated in Fig. 1.5.

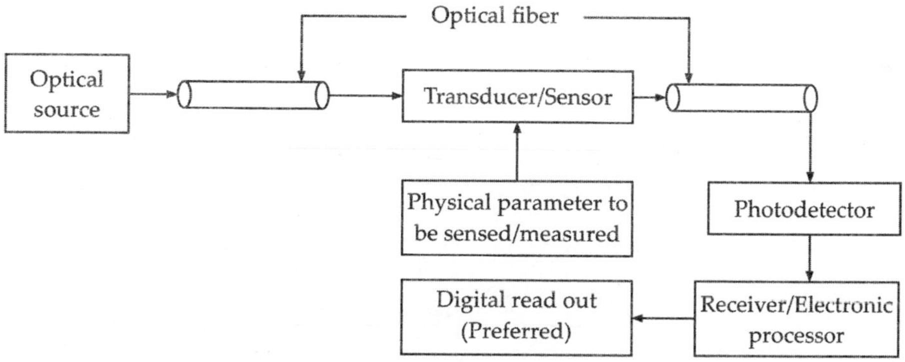

Fig. 1.5: A typical optical fiber based data acquisition/sensor/measurement system

A comparison of Figs 1.4 and 1.5 tells us that an optical source, a connecting optical fiber and an appropriate photodetector are the essential components of both the communication and the instrumentation systems. However, there are important differences between the corresponding transmitters and receivers in both the systems. The communication systems are primarily designed to cater to wideband signals in the form of multiplexed audio/video channels and very high speed data of the order of thousands of Mbps (Gbps) while the sensor based systems are designed to handle low data rates of the order of just a few kbps. The probability of bit error is the principal performance index for communication systems while accuracy, precision, sensitivity and very good resolution are the important performance indices for a typical instrumentation system. In this regard, as we shall see in the next chapters, the properties of light exploited, the associated signal processing to carry out specific tasks are then entirely different for both the systems.

One of the principal characteristics of an optical fibre is its attenuation as a function of wavelength as shown in Fig. 1.6. Early technology made exclusive use of 800- to 900-nm wavelength band, since in this region the fibres made at that time exhibited a local minimum in the attenuation curve, and optical sources and photo detectors operating at these wavelengths were available. This region is often referred to as the first window. By reducing the concentration of hydroxyl ions and metallic ion impurities in the fibre material, fibre manufactures were eventually able to fabricate optical waveguides with very low losses in the 1100- to 1600-nm region. This spectral bandwidth is usually referred to as the long-wavelength region. Windows are defined here, these being the second window centred around 1300 nm and the third window centred around 1550nm.

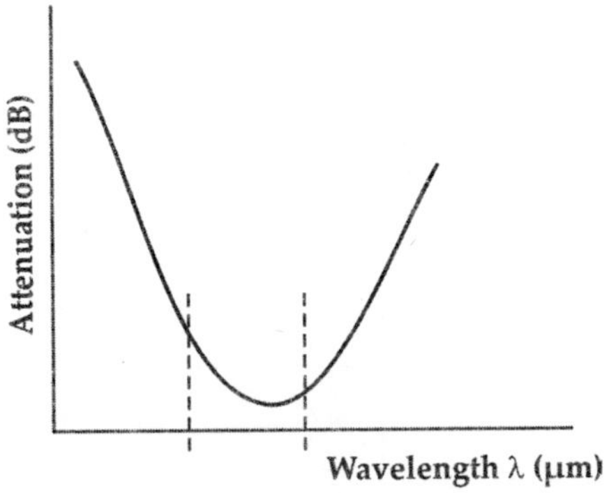

Fig. 1.6: Optical fibre attenuation as a function of wavelength

Researchers have also looked at new types of fibre materials for use in the 3-to 5-μm wavelength band. A fluoride glass with extremely low losses at mid infrared wavelengths was discovered (0.2 to 8μm with the lowest loss being around 2.55 μm). The material that is used to construct such a glass fibre is a heavy-metal fluoride glass which uses ZrF_4 as the major component and glass network former. Although this glass potentially offers intrinsic minimum losses of 0.01 to 0.001 dB/km, fabricating long lengths of these fibres is difficult. First, ultrapure materials must be used to reach this low loss level. Secondly, fluoride glass is prone to devitrification. Fibre-making techniques have to take this into account to avoid the formation of microcrystallines, which have a drastic effect on scattering loss.

Once the cable is installed a light source which is dimensionally compatible with the fibre core is used to launch power into the fibre. Semiconductor light-emitting diodes (LEDs) and other laser diodes are suitable optical sources for this purpose since their light output can be modulated rapidly by simply varying the bias current. The electric input signals to the transmitter can be either analog or digital. The transmitter circuitry converts these electric signals to an optical signal by varying the current flow through the light source. An optical source is a square-law device, which means that a linear variation in drive current results in a corresponding change in the optical output in terms of its power. In the 800- to 900-nm region the light sources are generally alloys of GaAlAs. At the longer wavelengths (1100 to 1600 nm), an InGaAsP alloy is the principal optical source material.

The optical signal will be attenuated and distorted with increasing distance because of scattering, absorption and dispersion mechanisms while propagating along the optical fibre. At the receiver, the attenuated and distorted modulated optical signal emerging from the fibre end will be detected by a photodiode. The photo detector is also a square-law device as it converts the receive optical power directly into an electric current output (photocurrent). Semiconductor pin and avalanche photodiodes (APDs) are the two principal photo detectors used in a fibre optic link. Both types of devices exhibit high efficiency and response speed. For applications where a low-power optical signal is received, an APD is normally used, since it has a greater sensitivity owing to an inherent internal gain mechanism (avalanche effect). Silicon photo detectors are used in the 800- to 900-nm region A variety of optical detectors are potentially available at the longer wavelengths. The prime material candidate in the 1100- to 1600-nm region is an InGaAs alloy.

The design of the receiver is more complex than that of the transmitter, since it has to both amplify and reshape the degraded signal received by the photo detector. The principal figure of merit for a receiver is the minimum optical power necessary at the desired data rate to attain either a given error probability for digital systems or a specified signal-to-noise ratio (SNR) for an analog system. The ability of a receiver to achieve a certain performance level depends on the photo detector type, the effects of noise in the system, and the characteristics of the successive amplification stages of the receiver.

When an optical signal has travelled a certain distance along a fibre, the signal has become attenuated and distorted to such a degree that a repeater is needed in the transmission line to amplify and reshape the signal. An optical repeater consists of a receiver and a transmitter placed back to back. The receiver section detects the optical signal and converts it to an electric signal, which is amplified, reshaped and sent to the electric input of the transmitter. The transmitter section converts this electric signal back to an optical signal and sends it on down the optical fibre waveguide.

The receiver can be a direct detection type or coherent detection type. In the latter, a locally generated optical signal at the receiver is mixed with the incoming information-carrying optical signal. This method enhances the SNR.

The installation of the optical fibre can be either aerial, in ducts, undersea, or buried directly in the ground. As a result of installation and/or manufacturing limitations, individual cable lengths will range from several hundred metres to several kilometres for long-distance applications. Practical considerations such as reel size and cable weight determine the actual length of a single cable section. The shorter lengths tend to be used when the cables are pulled through ducts. Longer cable lengths are used in aerial or direct-burial applications. The complete long-distance transmission line is formed by splicing or connecting together these individual cable sections.

1.5. SIGNAL DISTORTION IN OPTICAL FIBRES

The types of distortion that we encounter in optical fibre communication are:

- ***Intermodal dispersion***: found in multimode step index fibre and graded index fibre. Because many modes can be supported by these fibres, each mode propagates with its own characteristic phase and group velocity. The modes have different propagation constants.
- ***Intramodal dispersion:*** velocity of propagation of any mode varies with frequency, independent of effects due to the materials. This is also known as waveguide dispersion.
- ***Material dispersion:*** arises because of the fact that the refractive index of the fibre becomes a function of wavelength.

As a light wave propagates along an optical fiber, it suffers distortion due to a number of factors. The intramodal dispersion is pulse spreading that occurs within a single mode. This is due to that fact that a the group velocity is a function of the wavelength λ. Since intramodal dispersion depends on the wavelength, its effect on signal distortion increases with the spectral width of the optical source. This spectral width is the band of wavelengths over which the source emits light. It is normally characterized by the root-mean square (RMS) spectral width. For the LEDs, the RMS spectral width is approximately 5% of a central wavelength. For example, if the peak emission wavelength of an LED source is 850 nm, a typical source width would be 40 nm; that is, the source emits most of its optical power in the 830-870-nm wavelength band. Laser diode optical sources have much narrower spectral widths, typical values being 1 to 2 nm.

The two main causes of intramodal dispersion are:

- ***Material dispersion,*** which arises from the variation of the refractive index of the core material as a function of wavelength. (Material dispersion is also called chromatic dispersion or spectral dispersion, since this is the same effect by which a prism spreads out a spectrum.) This causes a wavelength dependence of the group velocity of any given mode; that is, pulse spreading occurs even when different wavelengths follow the same path.
- ***Waveguide dispersion,*** which occurs because a single-mode fiber only confines about 80% of the optical power to the core. Dispersion thus arises, since the 20% of the light propagating in the cladding travels faster than the light confined to the core. The amount of waveguide dispersion depends on the fiber design, since the modal propagation constant β is a function of a/λ (the optical fiber dimension relative to the wavelength λ

The other factor responsible for pulse spreading is intermodal delay, which is a result of each mode having a different value of the group velocity at a single frequency. The waveguide dispersion is usually ignored in multimode fibers. This effect, however, is significant in single mode fibres. The SMFs exhibit a dispersion behavior that limits the long-distance and very high speed transmission. The dispersion of a single mode silica fiber is lowest at 1300 nm whereas its attenuation is a minimum at 1550 nm, where the dispersion is higher. Hence, for achieving a maximum transmission distance of a high-capacity link, the dispersion null should be at the wavelength of minimum attenuation. To achieve this, one can adjust the basic fiber parameters to shift the zero dispersion minimum to longer wavelengths. An alternative to reduce fiber dispersion is by spreading the dispersion minimum out over a broader range. This approach is known as dispersion flattening.

Material dispersion tends to be the dominant pulse-distorting effect in graded index fibers. Intermodal delay distortion can be made very small by carful tailoring of the core refractive-index profile.

A light-pulse will broaden as it travels along a fiber. This pulse broadening eventually causes a pulse to overlap with neighboring pulses. After a certain amount of overlap, has occurred, adjacent pulses can no longer be individually distinguished at the receiver and errors will occur. Thus the dispersive properties determine the limit of the information capacity of the fiber.

Electrical bandwidth is defined as the frequency at which the voltage gain of the circuit falls to 0.707. The optical bandwidth is defined as the frequency at which power gain of the circuit falls to half of the maximum value. In case of optical systems, the power is directly proportional to current and not to the square of the current. Hence, optical bandwidth is more than the electrical bandwidth. In case of an optical fiber, the bandwidth depends upon its length. The fiber bandwidth is normally expressed in terms of the bandwidth times kilometre product. So a 100 MHz.km fiber can usually operate with a 200 MHz bandwidth for a 5 kilometre of optical fiber. A longer fiber results in more pulse spreading and hence leads to a lower bandwidth as a result.

The information-carrying capacity can be determined by examining the deformation of short light pulses propagating along the fiber. A measure of the information capacity of an optical fiber is usually specified by the bandwidth-distance product in MHz.km. For a step index fiber, the various distortion effects tend to limit the bandwidth-distance product to about 20 MHz.km. In graded index fibers, the radial refractive index profile can be carefully selected so that pulse broadening is minimized at a specific operating wavelength. This has led to bandwidth-distance product as high as 2-5 GHz.km. Single-mode fibers can have capacities well in excess of this. The flatness of the attenuation curves for the fibers extend up to the microwave spectrum.

1.6. CHAPTER SUMMARY

An optical fibre is a guided physical medium for carrying information from a transmitter to a receiver through light waves through suitable modulation techniques at the optical transmitter and the corresponding demodulation at the receiver subject to the fulfilment of a certain grade of service. Light waves travel by multiple total internal reflection. As light wave as a carrier occurs in the THz frequency range of the electromagnetic spectrum, large amounts of data can be carried out by an optical fibre as we shall see in the next chapters. Fibre as a communication medium offers many advantages unique to it. At the same time light waves propagating through a fibre are subject to a number of impairments making the theoretical information capacity not realizable. The three essential fibre types offer different data rates and find application in different scenarios.

However, in order to reap the potential benefits offered by an optical fibre fully, advanced signal processing like wavelength division multiplexing, optical fibre based local area network, the use of all optical amplifiers and optical switching must be carried out.

SUGGESTED READING

[1] Keiser, G. "Optical Fiber Communication", McGraw Hill, New York, 1991
[2] Senior, John M. "Optical Fiber Communications Principles and Practice", 3rd Edn., Pearson, 2010
[3] Gowar, J. "Optical Communication Systems", 2nd Edn., PHI, 1995
[4] Ghatak, A. and Thyagrajan, K. "Introduction to Fiber Optics", Cambridge, 1999
[5] Ghatak, A. "Optics", 2nd Edn., Tata McGraw Hill, 2003

❑❑❑

2

Maxell's Equations and Fields

2.1. INTRODUCTION

Electrical circuits make extensive use of voltages and currents for circuit analysis using a multitude of methods. These circuits have fixed or lumped circuit parameters in the form of resistances, capacitances and inductances. However, the concept of voltages and currents become inadequate to explain phenomena occurring in unbounded, homogeneous/heterogeneous media. To understand and analyze certain observed phenomena in those media (for example, free space) we rather resort to fields. These fields, as we have studied in elementary physics, are the electric fields and the magnetic fields. Electric fields are produced by static charges and magnetic fields by dc currents and ac currents. As we may infer, fields are functions in space that requires a certain coordinate axis system to understand and analyze. We have also seen from elementary studies that a magnetic field changing in an abrupt, sudden manner induces an emf whose direction is dictated by Lenz' law. The relationships are governed by Faraday's laws of electromagnetic induction.

In this chapter, our primary concern is to investigate fields in a three dimensional space through the use of appropriate coordinate systems so that the electric and the magnetic fields now become vectors. We invoke a couple of useful vector algebra identities to get meaningful insights into field behavior. From now on, fields would be looked through the window of Maxwell's equations and their solutions in various conditions in unbounded media as well as bounded media subject to the fulfillment of certain boundary conditions sometimes.

2.2. FIELDS IN HOMOGENEOUS UNBOUND MEDIUM

Let us consider a homogeneous, unbound medium without any electric or magnetic source. For this case the medium permittivity ε and permeability μ are scaler quantities which are constants over the entire medium. The source free nature of the medium means that there are no free charges and currents in the medium which means that $\rho = 0$ and $\mathbf{J} = 0$. The Maxwell's equations in the differential form then reduce to the following

$$\nabla \cdot \mathbf{B} = 0 \qquad ...(2.1)$$

$$\nabla \cdot \mathbf{D} = 0 \qquad ...(2.2)$$

$$\nabla \times \mathbf{E} = -\frac{\partial \mathbf{B}}{\partial t} \qquad ...(2.3)$$

$$\nabla \times \mathbf{H} = \frac{\partial \mathbf{D}}{\partial t} \qquad ...(2.4)$$

As we may observe from the foregoing equations, these are differentials in space in the left hand side and on the right hand side we have differentials in the time domain. We can have an alternative form that expresses the above relations in the form of integrals of a given physical quantity. This is presented in Table 2.1 for a concise treatment. For a region having a charge density of ρ C/m^3, the familiar Gauss's law (2.2) become

$$\nabla \cdot \mathbf{D} = \rho \qquad ...(2.5)$$

We have further these relations:

$$\mathbf{B} = \mu\mathbf{H} \quad \text{and} \quad \mathbf{D} = \varepsilon\mathbf{H} \qquad ...(2.6)$$

Maxell's equations are expressed both in the differential as well as integral form. A comprehensive summary is made below.

Table 2.1: Forms of Maxwell's equations

Differential Form	*Integral Form*
$\nabla \cdot \mathbf{D} = \rho$	$\oiint_S \mathbf{D}.d\mathbf{a} = \iiint \rho dv$
$\nabla \cdot \mathbf{B} = 0$	$\oiint_S \mathbf{B}.d\mathbf{a} = 0$
$\nabla \times \mathbf{E} = -\frac{\partial \mathbf{B}}{\partial t}$	$\oint_C \mathbf{E}.d\mathbf{l} = -\iint \frac{\partial \mathbf{B}}{\partial t}.d\mathbf{a}$
$\nabla \times \mathbf{H} = \mathbf{J} + \frac{\partial \mathbf{D}}{\partial t}$	$\oint_C \mathbf{E}.d\mathbf{l} = \iint \left(\mathbf{J} + \frac{\partial \mathbf{D}}{\partial t}\right).d\mathbf{a}$

The differential forms of Maxwell's equations are point relations; they establish the relationship between fields and sources at any point in space. The differential form involves derivatives of the fields with respect to both space as well as time. It may be noted here that, as we assume time harmonic fields of the form of $\exp(j\omega t)$ for subsequent field analysis, the time derivative

$\frac{\partial}{\partial t} \equiv j\omega$ unless otherwise mentioned. Since and $\mathbf{D} = \varepsilon\mathbf{E}$ and $\mathbf{B} = \mu\mathbf{H}$, the derivative form is suitable only in those situations where medium and fields are continuous in space and the space derivatives exist. At media interfaces where the medium properties change abruptly, the differential form of Maxwell's equations cannot, therefore be used. However, the integral form can be used in any situation. Applications of the integral forms of field equations at the media interface give us the relationship between two fields in the two media. These are generally referred to as the boundary conditions. This relationship is valid in any system of coordinates when they are mutually orthogonal to each other. The fibre can be thought of as a planer waveguide having the cross section as a rectangular plane or a cylindrical waveguide having a given radius. Depending upon the type of cross section we are working on, appropriate coordinates axis system is to be invoked and field equations need to be solved in that system of coordinates.

Summary of boundary conditions dictate that:

- Tangential components of the electric field are equal (said to be continuous) at the media interface. If E_{t1} is the tangential component of the electric field in medium 1 and E_{t2} that in medium 2, then we have $E_{t1} = E_{t2}$.
- Normal component of the electric displacement is continuous at the boundary of the two media in the absence of any surface charge. That is we have $D_{n1} = D_{n2}$.
- Normal component of the magnetic flux density is continuous at the boundary of the two media in the absence of any surface charge. That is we have $B_{n1} = B_{n2}$.
- Tangential components of the magnetic field are equal (said to be continuous) at the media interface in the absence of any surface current density across the boundary. If B_{t1} is the tangential component of the magnetic field in medium 1 and B_{t2} that in medium 2, then we have $B_{t1} = B_{t2}$.

These four conditions are extensively used to solve fields at the boundary between two media of different permittivity and permeability.

The medium for which the above equations are satisfied is assumed to be homogeneous and time invariant. Thus, the medium properties ε and μ are constants. Hence, Maxwell's equations become

$$\nabla \cdot \mathbf{B} = \nabla \cdot (\mu\mathbf{H}) = \mu(\nabla \cdot \mathbf{H}) = 0 \Rightarrow \nabla \cdot \mathbf{H} = 0 \qquad \text{...(2.7)}$$

and similarly,

$$\nabla \cdot \mathbf{D} = \nabla \cdot (\varepsilon \mathbf{E}) = \varepsilon (\nabla \cdot \mathbf{E}) = 0 \Rightarrow \nabla \cdot \mathbf{E} = 0 \quad \text{...(2.8)}$$

$$\nabla \times \mathbf{E} = -\frac{\partial \mathbf{B}}{\partial t} = -\frac{\partial (\mu \mathbf{H})}{\partial t} = -\mu \frac{\partial \mathbf{H}}{\partial t} \quad \text{...(2.9)}$$

$$\nabla \times \mathbf{H} = \frac{\partial \mathbf{D}}{\partial t} = -\frac{\partial (\varepsilon \mathbf{E})}{\partial t} = \varepsilon \frac{\partial \mathbf{E}}{\partial t} \quad \text{...(2.10)}$$

We note from (2.7) to (2.10) that the time derivative of the magnetic field is related to the space derivative of the electric field and the time derivative of the electric field is related to the space derivative of the magnetic field. These two equations suggest that a time varying magnetic field cannot exist without the corresponding electric field and *vice versa*. It is therefore, clear that if the fields are time varying, both the electric and the magnetic fields have to co-exist and the existence of just one kind of time varying field is not possible. This is very different from the static electric and magnetic fields which can exist without each other.

We find that the two fields are coupled to each other. In order to decouple those, we differentiate the equations with respect to the space coordinates. Taking the curl of (2.9) and (2.10) on both sides, we have

$$\nabla \times \nabla \times \mathbf{E} = -\mu \times \frac{\partial \mathbf{H}}{\partial t} \quad \text{...(2.11)}$$

$$\nabla \times \nabla \times \mathbf{H} = \varepsilon \times \frac{\partial \mathbf{E}}{\partial t} \quad \text{...(2.12)}$$

As the curl is a linear operator, we can interchange the space and time derivatives and write

$$\nabla \times \nabla \times \mathbf{E} = -\mu \frac{\partial (\nabla \times \mathbf{H})}{\partial t} \quad \text{...(2.13)}$$

and similarly,

$$\nabla \times \nabla \times \mathbf{H} = \varepsilon \frac{\partial (\nabla \times \mathbf{E})}{\partial t} \quad \text{...(2.14)}$$

Substituting the expressions of the curl of the electric and magnetic fields from (2.10) and (2.9) respectively in (2.13) and (2.14) we get,

$$\nabla \times \nabla \times \mathbf{E} = -\mu \frac{\partial \left(\varepsilon \frac{\partial \mathbf{E}}{\partial t} \right)}{\partial t} = -\mu\varepsilon \frac{\partial^2 \mathbf{E}}{\partial t^2} \quad \text{...(2.15)}$$

and similarly,

$$\nabla \times \nabla \times \mathbf{H} = \varepsilon \frac{\partial\left(-\mu \frac{\partial \mathbf{H}}{\partial t}\right)}{\partial t} = -\mu\varepsilon \frac{\partial^2 \mathbf{H}}{\partial t^2} \qquad ...(2.16)$$

Using the vector identity $\nabla \times \nabla \times \mathbf{A} = \nabla(\nabla \cdot \mathbf{A}) - \nabla^2 \mathbf{A}$ where **A** is any arbitrary vector, the LHS of (2.15) and (2.16) are written as

$$\nabla(\nabla \cdot \mathbf{E}) - \nabla^2 \mathbf{E} = -\mu\varepsilon \frac{\partial^2 \mathbf{E}}{\partial t^2} \qquad ...(2.17)$$

$$\nabla(\nabla \cdot \mathbf{H}) - \nabla^2 \mathbf{H} = -\mu\varepsilon \frac{\partial^2 \mathbf{H}}{\partial t^2} \qquad ...(2.18)$$

As the medium is source free, we have $\nabla \cdot \mathbf{H} = 0$ and $\nabla \cdot \mathbf{E} = 0$.

Thus, (2.17) and (2.18) reduce to

$$\nabla^2 \mathbf{E} = \mu\varepsilon \frac{\partial^2 \mathbf{E}}{\partial t^2} \qquad ...(2.19)$$

$$\nabla^2 \mathbf{H} = \mu\varepsilon \frac{\partial^2 \mathbf{H}}{\partial t^2} \qquad ...(2.20)$$

The solution to (2.19) and (2.20) gives us a wavelike function. The physical meaning of the above expressions is that if at all time varying fields have to exist in a homogeneous, unbound medium, they have to exist in the form of a wave. Moreover, both electric and magnetic fields have to exist together. Hence, these coexisting fields in a source free homogeneous medium are called electromagnetic waves. These reduce to the following in a scalar form corresponding to each coordinate of the system under consideration. Let us get some more insight into the nature of the various possible solutions to (2.19) and (2.20). The fields **E** and **H** are vectors as we noted previously and hence these are functions of the space coordinates (x, y, z) as well as time (time harmonic fields) as it is a time varying field only that gives rise to the other. Let us write a generic field vector **F** as

$$\mathbf{F}(x, y, z, t) = \mathbf{F}(x, y, z)e^{j\omega t} \qquad ...(2.21)$$

where ω is the angular frequency of the time varying field. The use of the exponential in (2.21) is simply to convey that this is the simplest form of a time varying field that can also be easily generated. We assume the vector **F** to be a function of a space implying implicit time dependence in the form of $\exp(j\omega t)$.

For time harmonic fields, we have

$$\frac{\partial}{\partial t}\mathbf{F}(x,y,z,t) = \frac{\partial}{\partial t}\{\mathbf{F}(x,y,z)e^{j\omega t}\} = j\omega\mathbf{F}(x,y,z)e^{j\omega t}$$

$$= j\omega\mathbf{F}(x,y,z,t) \quad ...(2.22)$$

Similarly,

$$\int\mathbf{F}(x,y,z,t)dt = \int\mathbf{F}(x,y,z)e^{j\omega t}dt$$

$$= \frac{1}{j\omega}\mathbf{F}(x,y,z)e^{j\omega t} + C = \frac{\mathbf{F}(x,y,z,t)}{j\omega} + C \quad ...(2.23)$$

where C is a constant function of time. As we are interested only in the time varying quantities, we make $C = 0$. The time derivative is equivalent to multiplication by $j\omega$ and time integration is equivalent to division by $j\omega$. We thus have

$$\frac{\partial}{\partial t} \equiv j\omega \quad \text{and} \quad \frac{\partial^2}{\partial t^2} \equiv -j\omega . -j\omega = -\omega^2$$

Therefore, (2.19) and (2.20) simplify to, with this exponential time function,

$$\nabla^2\mathbf{E} = -\omega^2\,\mu\varepsilon\mathbf{E} \quad ...(2.24)$$

$$\nabla^2\mathbf{H} = -\omega^2\,\mu\varepsilon\mathbf{H} \quad ...(2.25)$$

Each of these equations is, in fact, a set of three equations, one for each component of **E** and **H**. In other words, each component of **E** and **H** satisfies the wave equation. Since the ∇^2 operator is a scalar operator, the wave equation for each component is a scalar equation which is written as

$$\nabla^2 \equiv \frac{\partial}{\partial x^2} + \frac{\partial}{\partial y^2} + \frac{\partial}{\partial z^2} \quad ...(2.26)$$

In general, $\nabla^2\Psi = -\omega^2\mu\varepsilon\Psi$ where ψ could be E_x, E_y, E_z or H_x, H_y, H_z.

2.3. POSSIBLE SOLUTIONS OF THE WAVE EQUATION

Here we explore the possibilities of several solutions to (2.24) and (2.25) and try to see if each of the solutions would result in a wavelike function that moves with respect to space. We confine ourselves to the rectangular system of coordinates though any system of mutually orthogonal axes would do. To start with, let us suggest the solution of the second order differential equation a constant and see if it qualifies to become a wave.

- Let us write the electric field as (in the rectangular system of coordinates)

$$\mathbf{E}(x, y, z, t) = \mathbf{K}e^{-j\omega t} \qquad ...(2.27)$$

where **K** is a constant vector having the same orientation and magnitude throughout space. We would like to know whether this would qualify as a wave. To answer this, we write,

$$\nabla^2\mathbf{K} = -\omega^2 \mu\varepsilon\mathbf{K} \qquad ...(2.28)$$

Since **K** is a constant over space, its derivative is zero. Thus,

$$\nabla^2\mathbf{K} = 0 \qquad ...(2.29)$$

As the medium constants μ and ε cannot be zero, either ω or **K** should be zero. When $\omega = 0$, the electric field is a dc field or an electrostatic field. Such a field cannot give rise to a wave. Therefore, the solution to (2.29) can not represent a wave. Moreover, we are dealing with time varying fields for which $\omega \neq 0$ and hence this makes $\mathbf{K} = 0$ that gives rise to $\mathbf{E} = 0$.

The inference from the above is that a time varying uniform electric field in the three-dimensional space cannot exist if it is uniform everywhere in space. Since the magnetic field is also related to the electric field, the magnetic field similarly cannot exist. No time varying field can exist without varying as a function of space.

- Next, let us consider the following:

$$\mathbf{E}(x,y,z,t) = E_0(x)e^{-j\omega t}\hat{\mathbf{x}} \qquad ...(2.30)$$

Substitution of this in (2.3) gives us

$$\nabla \cdot \mathbf{E} = -j\omega\mu\mathbf{H} \qquad ...(2.31)$$

$$= \begin{vmatrix} \hat{\mathbf{x}} & \hat{\mathbf{y}} & \hat{\mathbf{z}} \\ \dfrac{\partial}{\partial x} & \dfrac{\partial}{\partial y} & \dfrac{\partial}{\partial z} \\ E_x & E_y & E_z \end{vmatrix} = -j\omega\mu\mathbf{H}$$

Since **E** is not a function of y and z, we write $\frac{\partial}{\partial y} \equiv 0$ and $\frac{\partial}{\partial z} \equiv 0$. This field is constant in the yz phase and propagates along the x-axts. Further, **E** has only x – component so that $E_y = 0 = E_z$. Hence, (2.31) reduces to

$$\begin{vmatrix} \hat{\mathbf{x}} & \hat{\mathbf{y}} & \hat{\mathbf{z}} \\ \dfrac{\partial}{\partial x} & 0 & 0 \\ E_x & 0 & 0 \end{vmatrix} = -j\omega\mu\mathbf{H} \qquad ...(2.32)$$

The determinant of the above matrix is zero which gives us

$$-j\omega\mu\mathbf{H} = 0 \qquad ...(2.33)$$

Again except when $\omega = 0$, the magnetic field **H** has to be zero. Since no time varying electric field can exist without the corresponding magnetic field, $E_0(x)$ has to be identically zero. We are therefore led to believe that an electric or magnetic field which is time varying and is constant in a plane perpendicular to it's direction also cannot exist. In other words, (2.30) cannot represent a valid wave function in space.

- Let us consider a third category of field aligned along the positive x-axis and constant in the xy – plane and expressed as

$$\mathbf{E}(x,y,z,t) = E_0(z)e^{j\omega t}\hat{\mathbf{x}} \qquad ...(2.34)$$

For such a field, we have $\dfrac{\partial}{\partial x} = \dfrac{\partial}{\partial y} \equiv 0$. Substitution of this in (2.31) gives us

$$\begin{vmatrix} \hat{\mathbf{x}} & \hat{\mathbf{y}} & \hat{\mathbf{z}} \\ 0 & 0 & \dfrac{\partial}{\partial z} \\ E_0(z) & 0 & 0 \end{vmatrix} = -j\omega\mu\mathbf{H} \qquad ...(2.35)$$

Expansion of (2.35) gives us

$$-\frac{\partial E_0(z)}{\partial z}\hat{\mathbf{y}} = -j\omega\mu(H_x\hat{\mathbf{x}} + H_y\hat{\mathbf{y}} + H_z\hat{\mathbf{z}}) \qquad ...(2.36)$$

Equating the respective components on both sides, we therefore get

$$H_x = 0 = \mathrm{H}_z \qquad ...(2.37)$$

$$H_y = \frac{1}{j\omega\mu}\frac{\partial E_0(z)}{\partial z} \qquad ...(2.38)$$

Since $\boldsymbol{E}_0(z)$ is not a constant function of z, $\dfrac{\partial E_0(z)}{\partial z} \neq 0$ and consequently H_y exists and E also exists. Although we have considered an x-oriented field constant in the xy-plane, the arguments are valid for any plane that contains the field vector.

The above discussions are summarized as follow:

(i) A time varying electric or magnetic field which is uniform in the three dimensional space cannot exist.

(ii) A time varying field which is constant in a plane perpendicular to the field direction also cannot exist.

(iii) The simplest form of field that can exist is the field which is constant in a plane containing the field vector, and consequently has spatial variation along the direction perpendicular to the constant field plane.

2.4. UNIFORM PLANE WAVE

Let us consider a case where the electric field is constant in a plane orthogonal to the direction of propagation of the wave. Let us consider an x-directed field which is constant in the plane. The field is expressed as

$$\mathbf{E}(x,y,z,t) = E_x(z)e^{-j\omega t}\hat{\mathbf{x}} \quad ...(2.39)$$

As the field is a constant in the xy-plane, we have $\frac{\partial}{\partial x} = \frac{\partial}{\partial y} \equiv 0$; no variation of the field strength with respect to either the x- or the y-axis.

Substitution of (2.39) in (2.26) gives us

$$\frac{d^2}{dz^2}E_x(z) = -\omega^2\mu\varepsilon E_x(z) \quad ...(2.40)$$

Let us have

$$\psi^2 = -\omega^2\mu\varepsilon \Rightarrow \gamma = j\omega\sqrt{\mu\varepsilon} = j\beta \quad ...(2.41)$$

Therefore,

$$\beta = \omega\sqrt{\mu\varepsilon} \quad ...(2.42)$$

The significance of (2.42) is that it gives us the phase constant of the wave in terms of the medium parameters. This can also be expressed as

$$\beta = 2\pi f\sqrt{\mu\varepsilon} = \frac{2\pi f}{v} = \frac{2\pi}{\lambda} \quad ...(2.43)$$

With the help of (2.41) and (2.42), the solution to (2.40) then may be expressed as

$$E_x = E_x^+ e^{-j\beta z} + E_x^- e^{j\beta z} \qquad \text{...(2.44)}$$

The two terms on RHS of (2.44) represent two travelling waves travelling in opposite directions along the z-axis. However, the second term does not qualify as a wave as it increases monotonically with respect to distance along the z-axis. Farther from the origin, the function exhibits higher amplitude. In a physical system, the wave has to decrease in amplitude farther from the origin where the source is assumed to be present.

Hence, we express the wave as

$$E_x = E_x^+ e^{-j\beta z} \qquad \text{...(2.44)}$$

The magnetic field corresponding to (2.44) is expressed as

$$\begin{vmatrix} \hat{\mathbf{x}} & \hat{\mathbf{y}} & \hat{\mathbf{z}} \\ 0 & 0 & \dfrac{\partial}{\partial z} \\ E_x & 0 & 0 \end{vmatrix} = -j\omega\mu\mathbf{H} \Rightarrow \frac{\partial E_x}{\partial z}\hat{\mathbf{y}}$$

$$= -j\omega\mu\mathbf{H} = -j\omega\mu(H_x\hat{\mathbf{x}} + H_y\hat{\mathbf{y}} + H_z\hat{\mathbf{z}}) \qquad \text{...(2.45)}$$

Equating the respective components on both sides, we get

$$-j\beta E_x^+ e^{-j\beta z}\hat{\mathbf{y}} = -j\omega\mu H_y\hat{\mathbf{y}} \qquad \text{...(2.46)}$$

Thus, the associated magnetic field is

$$H_y = \frac{\beta}{\omega\mu} E_x^+ e^{-j\beta z} \qquad \text{...(2.47)}$$

We observe from (2.44) and (2.47) the existence of a very interesting phenomenon; we have an electric field along the x-axis but not along either the y-axis or the z-axis. Similarly, the corresponding magnetic field exists only along the positive y-axis. This makes the electric field and the associated magnetic field to be in space quadrature with both of them propagating along the axis with the same velocity. The vectors **E, H** and the direction of wave propagation form a mutually orthogonal coordinate system for such a solution as described above that follows the right handed screw rule. Since both the electric and the magnetic fields are perpendicular to the direction of wave propagation, such a wave is known as the "Transverse Electromagnetic" (TEM) wave. These waves known as uniform plane waves which mean that they travel in a plane which is transverse to the direction of wave propagation and any point on this plane has a constant phase. Hence, the waves are said to propagate in constant phase planes. This is illustrated in Fig. 2.1.

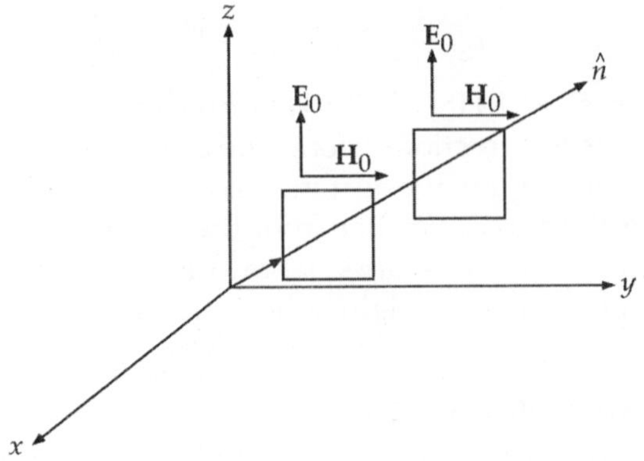

Fig. 2.1: Transverse nature of electromagnetic waves and constant phase planes

It is further interesting to note that, the ratio of the electric field to the magnetic field is a constant for a given medium. This is proved very simply as follows:

$$\frac{E_x}{H_y} = \frac{E_x^+ e^{-j\beta z}}{\left(\frac{\beta}{\omega\mu}\right) E_x^+ e^{-j\beta z}} = \frac{\omega\mu}{\beta} = \frac{2\pi f}{\frac{2\pi}{\lambda}}\mu = f\lambda\mu = c\mu = \frac{\mu}{\sqrt{\mu\varepsilon}} = \sqrt{\frac{\mu}{\varepsilon}} \quad ...(2.48)$$

The wave velocity is expressed as

$$v = \frac{1}{\sqrt{\mu\varepsilon}} \quad ...(2.49)$$

For free space, the wave velocity is computed to be

$$v_0 = \frac{1}{\sqrt{\mu_0\varepsilon_0}} = \frac{1}{\sqrt{4\pi \times 10^{-7} \times \left(\frac{1}{36\pi}\right) \times 10^{-9}}} = 3 \times 10^8 \text{ m/s} \quad ...(2.50)$$

The important consequence of (2.50) is that both the electric and the magnetic fields travel with the velocity of light in free space.

The quantity, the ratio of the electric field to the magnetic field is known as the intrinsic impedance of a medium denoted as

$$\eta = \sqrt{\frac{\mu}{\varepsilon}} \quad ...(2.51)$$

This has the units of ohms as the unit of the electric field is V/m and that of the magnetic field is A/m. The intrinsic impedance is different for different media. For free space, we have

$$\eta_0 = \sqrt{\frac{\mu_0}{\varepsilon_0}} = \sqrt{\frac{4\pi \times 10^{-7}}{\left(\frac{1}{36\pi}\right) \times 10^{-9}}} \approx 120\pi = 377\,\Omega \qquad ...(2.52)$$

For other media, the intrinsic impedance becomes

$$\eta = \sqrt{\frac{\mu}{\varepsilon}} = \sqrt{\frac{\mu_0\mu_r}{\varepsilon_0\varepsilon_r}} = \eta_0\sqrt{\frac{\mu_r}{\varepsilon_r}} \qquad ...(2.53)$$

In an unbounded medium as in free space, the electromagnetic wave is always transverse. As light is an electromagnetic wave and in majority of the situations its wavelength is much smaller compared to the size of the medium, light behaves as a TEM wave in most cases as shown in Fig. 2.2. We may observe from this figure that the electric field is oriented along the y-axis whereas the magnetic field is contained completely along the x-axis the direction of wave propagation being the z-axis. These constitute an orthogonal system giving rise to the TEM nature of light waves in vacuum or free space. Further, the constituent components of the wave, namely the electric and the magnetic fields are contained in the xy-plane and as we have seen earlier, the unit normal to this plane is along the z-axis. This therefore represents the direction of wave propagation.

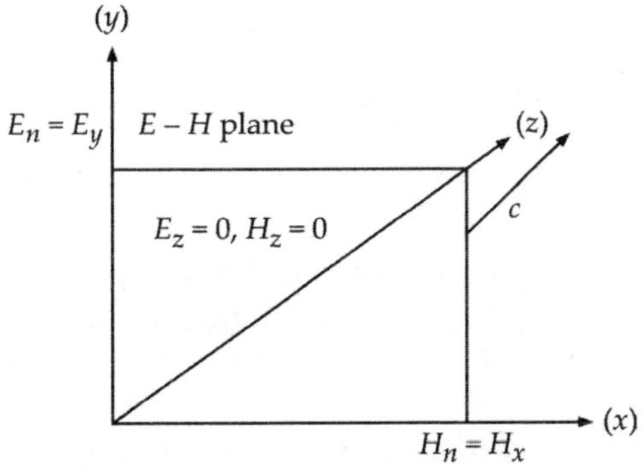

Fig. 2.2: A TEM wave

For a more generalized case, the electric field may be aligned neither along the x-axis nor along either the y-axis. Then it may be decomposed into two mutually orthogonal fields E_x and E_y travelling along the z-axis which will be related to the corresponding magnetic fields as the following.

$$H_y = \frac{E_x}{\eta} \qquad ...(2.54)$$

and
$$H_x = \frac{-E_y}{\eta} \qquad ...(2.55)$$

following the right handed screw rule. Thus, the total magnetic field **H** and the total electric field **E** are expressed as

$$\mathbf{H} = \sqrt{H_x^2 + H_y^2} = \sqrt{\frac{E_x^2}{\eta^2} + \frac{E_y^2}{\eta^2}} = \frac{\mathbf{E}}{\eta} \qquad ..(2.56)$$

This gives us further,

$$\frac{|\mathbf{E}|}{|\mathbf{H}|} = \eta \qquad ...(2.57)$$

Example 2.1: Just inside the surface of a dielectric slab the electric field is 10 V/m and makes an angle of 60° with the surface. If the dielectric constant of the slab is 4, find the electric field and its direction just above the surface.

Solution: We know that, at the boundary of two separate media, the tangential component of the electric field is continuous; $E_{t1} = E_{t2}$

Now, the field just inside the dielectric $E_{t1} = 10 \cos 60° = 5$ V/m

Thus, above the surface of the dielectric $E_{t2} = 5$ V/m

The normal component of electric flux at the interface satisfies

$$D_{n1} = D_{n2} \Rightarrow \varepsilon_1 E_{n1} = \varepsilon_2 E_{n2}$$

Therefore, the normal component of the field just inside the dielectric

$$E_{n1} = 10 \sin 60° = \frac{10\sqrt{3}}{2} \text{ V/m}$$

This gives us

$$E_{n2} = \frac{\varepsilon_1 E_{n1}}{\varepsilon_2} = 4.\frac{10\sqrt{3}}{2} = 20\sqrt{3} \text{ V/m}$$

The total field just above the surface is

$$E = \sqrt{E_{t1}^2 + E_{n1}^2} = \sqrt{25 + 1200} = \sqrt{1225} = 35 \text{ V/m}$$

The direction of this field is $\phi = \left(\frac{20\sqrt{3}}{5}\right) 81.78°$

Example 2.2: A uniform plane wave at a frequency of 300 MHz travels in vacuum along the direction. The electric field of the wave at some instant is given as $\mathbf{E} = 3\hat{\mathbf{x}} + 5\hat{\mathbf{z}}$. Find the phase constant of the wave and also the magnetic field.

Solution: The phase constant is $\frac{2\pi}{\lambda} = \frac{2\pi f}{c} = \frac{2\pi \times 300 \times 10^6}{300 \times 10^6} = 2\pi$ rad/m

The plane wave travels along the $+y$ direction. Thus, the associated magnetic field should lie in the xz plane. Let the magnetic field be given as $\mathbf{H} = H_x\hat{\mathbf{x}} + H_z\hat{\mathbf{z}}$

For a uniform plane wave, we know that the electric field **E** and the magnetic field **H** are orthogonal to each other. Thus, **E.H** = 0

This implies that $3H_x + 5H_y = 0$

We also have $\frac{|\mathbf{E}|}{|\mathbf{H}|} = \eta$ which gives us

$$|\mathbf{H}| = \frac{|\mathbf{E}|}{\eta_0} = \frac{\sqrt{9+25}}{120\pi} = 15.46 \times 10^{-3} \text{ A/m}$$

Hence, we have

$$H_x = \pm\frac{5}{\sqrt{120\pi}} \text{ and } H_y = \mp\frac{5}{\sqrt{120\pi}}$$

Therefore, the magnetic field is

$$\mathbf{H} = \frac{5}{\sqrt{120\pi}}\hat{\mathbf{x}} - \frac{3}{\sqrt{120\pi}}\hat{\mathbf{z}} \text{ A/m}$$

Or, through the use of $\begin{vmatrix} \hat{\mathbf{x}} & \hat{\mathbf{y}} & \hat{\mathbf{z}} \\ \frac{\partial}{\partial x} & \frac{\partial}{\partial y} & \frac{\partial}{\partial z} \\ E_x & E_y & E_z \end{vmatrix} = -j\omega\mu\mathbf{H}$, we write

$$\begin{vmatrix} \hat{\mathbf{x}} & \hat{\mathbf{y}} & \hat{\mathbf{z}} \\ \frac{\partial}{\partial x} & \frac{\partial}{\partial y} & \frac{\partial}{\partial z} \\ E_x & 0 & E_z \end{vmatrix} = -j\omega\mu\mathbf{H} = \frac{\partial E_z}{\partial y}\hat{\mathbf{x}} - \left(\frac{\partial E_z}{\partial x} - \frac{\partial E_x}{\partial z}\right)\hat{\mathbf{y}} - \frac{\partial E_x}{\partial z}\hat{\mathbf{z}}$$

As it is a uniform plane EM wave traveling along the $+y$ direction, obviously both the electric and the magnetic field will have no component along this axis. Hence, the above expression for the magnetic field would consist of only two components; one along the positive x-axis and the other along the negative z-axis. Then also the resultant magnetic field would be orthogonal to the electric field. Further, the variations as the wave travels along the $+y$ direction. We have

$$\frac{\omega\mu}{\beta} = \frac{2\pi f}{\frac{2\pi}{\lambda}}\mu = f\lambda\mu = c\mu = \frac{\mu}{\sqrt{\mu\varepsilon}} = \sqrt{\frac{\mu}{\varepsilon}} = \eta \Rightarrow \omega\mu = \beta\eta$$

For our case,

$$-j\omega\mu\mathbf{H} = \frac{\partial E_z}{\partial y}\hat{\mathbf{x}} - \frac{\partial E_x}{\partial z}\hat{\mathbf{z}} = -j\beta(5\hat{\mathbf{x}} - 3\hat{\mathbf{z}}) = -j\beta\eta\mathbf{H}$$

Thus,

$$\mathbf{H} = \frac{5}{\sqrt{120\pi}}\hat{\mathbf{x}} - \frac{3}{\sqrt{120\pi}}\hat{\mathbf{z}}$$

Example 2.3: A uniform plane wave traveling in the positive z-axis has two components of the electric field $E_x = 5$ V/m and $E_y = 10\angle 30°$ V/m. At some instant say $t = 0$, the E_x component is maximum. Find the magnitude and direction of the field at $t = 0$ and at $t = 0.1$ ns. Frequency of the wave is 2 GHz.

Solution: The two components of the field are expressed as

$$E_x = 5\cos(\omega t) \text{ and } E_y = 10\cos(\omega t + 30°)$$

At $t = 0$, $\quad E_x = 5$ and $E_y = 10\cos 30° = \dfrac{10\sqrt{3}}{2}$

Thus, the resultant electric field is, at

$$E = \sqrt{E_x^2 + E_y^2} = \sqrt{25 + 75} = \sqrt{100} = 10 \text{ V/m}$$

Direction of the electric field is $\phi = \tan^{-1}\left(\dfrac{E_y}{E_x}\right) = \tan^{-1}\left(\dfrac{10\sqrt{3}}{2.5}\right) = 60°$

with the positive x-axis.

At $t = 0.1$ ns, we have $\omega t = 2\pi \times 2 \times 10^9 \times 0.1 \times 10^{-9} = 0.4\pi$ rad

Thus, $\quad E_x = 5\cos(\omega t) = 5\cos(0.4\pi) = 1.545$ V/m and $$E_x = 10\cos\left(0.4\pi + \frac{\pi}{6}\right) = -2.078 \text{ V/m}$$

Therefore, the resultant electric field is, at $t = 0.1$ ns,

$$E = \sqrt{E_x^2 + E_y^2} = \sqrt{1.545^2 + (-2.078)^2} = 2.59 \text{ V/m}$$

Its direction is $\phi = \tan^{-1}\left(\frac{E_y}{E_x}\right) = \tan^{-1}\left(\frac{-2.078}{1.545}\right) = -53.37°$

The same problem could also have been solved by writing the electric field as

$$E_x = \text{Re}\left\{E_0 e^{-j(\beta z - \omega t)}\right\}$$

Example 2.4: Write down the complete expression for a plane wave traveling with its **k** in the *yz*-plane making an angle of 30° with the *z*-axis.

Solution: The uniform plane wave is expressed as

$$\mathbf{E} = E_0 \exp[j(\omega t - \mathbf{k.r} + \theta_0)]\hat{\mathbf{e}}$$

where $\mathbf{k.\hat{e}} = 0$

We understand from the given problem that

$$\mathbf{k.r} = y \sin 30° + z \cos 30° = \frac{1}{2}(y + \sqrt{3}z)$$

and the unit normal is along the *x*-axis. Thus, $\hat{\mathbf{e}} = \hat{\mathbf{x}}$

The plane wave is

$$\mathbf{E} = E_0 e^{j\omega t} \exp\left[-\frac{j}{2}(y + \sqrt{3}z)\right]\hat{\mathbf{x}}$$

Example 2.5: Suppose the electric field of a plane wave is given by

$$\mathbf{E} = E_0(-\hat{x} + \hat{y})e^{-j\beta\left(\frac{x+y}{\sqrt{2}}\right)}$$

Find the direction of wave propagation, the corresponding magnetic field and time average power flow per unit area.

Solution: The wave travels in the *xy*-plane making an angle of 45° with the positive *x*-axis. The electric field has two components; along the negative *x*-axis and along the positive *y*-axis. The component $E_z = 0$. This represents a TM wave and we expect the magnetic field to have only the *z*-component. Further,

$$E_x = -E_0 e^{-\frac{j\beta(x+y)}{\sqrt{2}}}\,\hat{\mathbf{x}} \quad \text{and} \quad E_y = E_0 e^{-\frac{j\beta(x+y)}{\sqrt{2}}}\,\hat{y}$$

Use of $\begin{vmatrix} \hat{\mathbf{x}} & \hat{\mathbf{y}} & \hat{\mathbf{z}} \\ \frac{\partial}{\partial x} & \frac{\partial}{\partial y} & \frac{\partial}{\partial z} \\ E_x & E_y & E_z \end{vmatrix} = -j\omega\mu\mathbf{H}$ gives us

$$\begin{vmatrix} \hat{\mathbf{x}} & \hat{\mathbf{y}} & \hat{\mathbf{z}} \\ \frac{\partial}{\partial x} & \frac{\partial}{\partial y} & \frac{\partial}{\partial z} \\ E_x & E_y & 0 \end{vmatrix} = -j\omega\mu\mathbf{H}$$

$$= -\frac{\partial}{\partial z}E_y\hat{\mathbf{x}} - \left(0 - \frac{\partial}{\partial z}E_x\right)\hat{\mathbf{y}} + \left(\frac{\partial}{\partial x}E_y - \frac{\partial}{\partial y}E_x\right)\hat{\mathbf{z}}$$

We note that, there is no variation along the z-. Thus, $\frac{\partial}{\partial z} = 0$.

We have $\frac{\partial}{\partial x} = -\frac{j\beta}{\sqrt{2}} e^{\frac{-j\beta(x+y)}{\sqrt{2}}}$ and similarly, $\frac{\partial}{\partial y} = -\frac{j\beta}{\sqrt{2}} e^{\frac{-j\beta(x+y)}{\sqrt{2}}}$

$$-j\omega\mu\mathbf{H} = -j\omega\mu(H_x\hat{\mathbf{x}} + H_y\hat{\mathbf{y}} + H_z\hat{\mathbf{z}})$$

Therefore,

$$-j\omega\mu(H_z\hat{\mathbf{z}}) = -\frac{j\beta}{\sqrt{2}} e^{\frac{-j\beta(x+y)}{\sqrt{2}}}(E_y - E_x) = -\frac{j\beta e^{\frac{-j\beta(x+y)}{\sqrt{2}}}[E_0 - (-E_0)]}{\sqrt{2}}\hat{\mathbf{z}}$$

Thus, $$H_z = \frac{\beta}{\sqrt{2}\omega\mu} 2E_0 e^{-\frac{j\beta(x+y)}{\sqrt{2}}}\,\hat{\mathbf{z}}$$

We note that the magnetic field has a z-component only. The power flow is given as

$$P = \frac{1}{2}\text{Re}\{\mathbf{E}\times\mathbf{H}^*\} = \frac{1}{2}\text{Re}\left\{E_0 e^{\frac{-j\beta(x+y)}{\sqrt{2}}}(-\hat{\mathbf{x}} + \hat{\mathbf{y}}) \times \frac{\beta}{\sqrt{2}\omega\mu} 2E_0 e^{\frac{j\beta(x+y)}{\sqrt{2}}}\,\hat{\mathbf{z}}\right\}$$

$$= \frac{E_0^2\beta}{\sqrt{2}\omega\mu}(-\hat{\mathbf{y}} - \hat{\mathbf{x}})$$

We observe that, the power flow occurs in the third quadrant of the xy-plane which also contains the propagation vector. Thus, power contained in the EM wave always flows in a direction containing the propagation vector. The time averaged power flow is simply half of the above value under the assumption of a periodic sinusoidal variation of the electric or the corresponding field.

Example 2.6: In a region the **E** and **H** fields are given as and **E** = 100 $(j\hat{\mathbf{x}}+2\hat{\mathbf{y}}-j\hat{\mathbf{z}})e^{j\omega t}$ and $\mathbf{H}=(-\hat{\mathbf{x}}+j\hat{\mathbf{y}}+\hat{\mathbf{z}})e^{j\omega t}$. Find the average power density and direction of power flow.

Solution: The average Poynting vector

$$P_{av}=\frac{1}{2}\text{Re}\{\mathbf{E}\times\mathbf{H}^*\}=\begin{vmatrix}\hat{\mathbf{x}} & \hat{\mathbf{y}} & \hat{\mathbf{z}}\\ H_x^* & H_x^* & H_z^*\\ E_x & E_y & E_z\end{vmatrix}=\frac{100}{2}\begin{vmatrix}\hat{\mathbf{x}} & \hat{\mathbf{y}} & \hat{\mathbf{z}}\\ -1 & -j & 1\\ j & 2 & -j\end{vmatrix}=150\sqrt{2}\left(\frac{\hat{\mathbf{x}}+\hat{\mathbf{z}}}{\sqrt{2}}\right)$$

The average power density is $150\sqrt{2}$ W/m^2 = 212.132 W/m^2 and the power flows in the direction $\left(\frac{\hat{\mathbf{x}}+\hat{\mathbf{z}}}{\sqrt{2}}\right)$, i.e. in the xz-plane at an angle of 45° with respect to both the and x-axes. There is no power flow along the y-axis. This is due to the fact that, both the electric and magnetic fields have components along the positive y-axis. Hence, curl of the two fields having components along the same axis is zero.

Example 2.7: In a dielectric medium a wave has electric and magnetic fields given as $\mathbf{E}=(j\hat{\mathbf{x}}+2\hat{\mathbf{y}}-j\hat{\mathbf{z}})e^{-j\pi(x+z)}$ V/m and **H** = $\frac{1}{60\pi}$ $(-\hat{\mathbf{x}}+j\hat{\mathbf{y}}+\hat{\mathbf{z}})e^{-j\pi(x+z)}$ A/m. Show that the wave is a uniform plane wave.

Solution: We understand that, for a plane wave as they are orthogonal to each other. Thus,

$$\mathbf{E.H}=(j.-1+2.j-j.1)=0$$

Thus the two fields are orthogonal to each other. The first criterion of a uniform plane wave is satisfied. Next let us consider the average power flow.

$$P_{av}=\frac{1}{2}\text{Re}\{\mathbf{E}\times\mathbf{H}^*\}\begin{vmatrix}\hat{\mathbf{x}} & \hat{\mathbf{y}} & \hat{\mathbf{z}}\\ H_x^* & H_x^* & H_z^*\\ E_x & E_y & E_z\end{vmatrix}=\frac{1}{60\pi}\begin{vmatrix}\hat{\mathbf{x}} & \hat{\mathbf{y}} & \hat{\mathbf{z}}\\ -1 & -j & 1\\ j & 2 & -j\end{vmatrix}=\frac{1}{20\pi}\sqrt{2}\left(\frac{\hat{\mathbf{x}}+\hat{\mathbf{z}}}{\sqrt{2}}\right)$$

The average power density is $\frac{\sqrt{2}}{20\pi}$ W/m^2 and the power flows in the direction $\left(\frac{\hat{\mathbf{x}}+\hat{\mathbf{z}}}{\sqrt{2}}\right)$, *i.e.*, in the *xz*-plane at an angle of 45° with respect to both the *x*- and *z*-axes.

Example 2.8: For a uniform plane wave in a medium, the electric and the magnetic fields are given as

$$\mathbf{E} = (5\hat{\mathbf{x}} + 2\hat{\mathbf{y}} + 6\hat{\mathbf{z}})\ \text{V/m} \quad \text{and} \quad \mathbf{H} = \frac{1}{120\pi}(2\hat{\mathbf{x}} - 2\hat{\mathbf{y}} - \hat{\mathbf{z}}).$$

Find the direction of the wave and the impedance of the medium.

Solution: To find out the direction of the wave, it is necessary to compute the direction of power flow.

$$P_{av} = \frac{1}{2}\text{Re}\{\mathbf{E} \times \mathbf{H}^*\}$$

$$= \begin{vmatrix} \hat{\mathbf{x}} & \hat{\mathbf{y}} & \hat{\mathbf{z}} \\ H_x^* & H_x^* & H_z^* \\ E_x & E_y & E_z \end{vmatrix} = \frac{1}{120\pi}\begin{vmatrix} \hat{\mathbf{x}} & \hat{\mathbf{y}} & \hat{\mathbf{z}} \\ 2 & -2 & -1 \\ 5 & 2 & 6 \end{vmatrix}$$

$$= \frac{1}{120\pi}(-10\hat{\mathbf{x}} - 17\hat{\mathbf{y}} + 14\hat{\mathbf{z}})$$

$$= \frac{1}{120\pi\sqrt{10^2 + 17^2 + 14^2}}(-10\hat{\mathbf{x}} - 17\hat{\mathbf{y}} + 14\hat{\mathbf{z}})$$

$$= \frac{1}{120\pi \times \sqrt{585}}(-10\hat{\mathbf{x}} - 17\hat{\mathbf{y}} + 14\hat{\mathbf{z}})$$

$$= \frac{1}{120\pi(24.1867)}\left(-10\hat{\mathbf{x}} - 17\hat{\mathbf{y}} + 14\hat{\mathbf{z}}\right)$$

Next we compute the direction cosines. To do so, we compute the unit normal along the direction of propagation which is given as

$$\hat{\mathbf{n}} = \frac{-10}{24.1868}\hat{\mathbf{x}} - \frac{17}{24.1868}\hat{\mathbf{y}} + \frac{14}{24.1868}\hat{\mathbf{z}}$$

$$= -0.4134\hat{\mathbf{x}} - 0.7029\hat{\mathbf{y}} + 0.5788\hat{\mathbf{z}}$$

The angles that the wave makes with the axes are:

$$\phi_x = a\cos(-0.4134) = 114.41°$$

$$\phi_y = a\cos(-0.7029) = 134.66°$$

$$\phi_z = a\cos(0.5788) = 54.63°$$

Therefore, the wave travels along a making an angle of 114.41° with the x-axis (negative x-axis), 134.66° with the y-axis (negative y-axis) and 54.63° with the z-axis.

Example 2.9: A uniform plane wave has a power density of 20 W/m^2 and is travelling along the $-y$ direction in free space. If the electric field makes an angle of 30° with the $+x$ axis, find the electric and magnetic field strengths and the direction of the magnetic field.

Solution: For a uniform plane wave **E** and **H** are orthogonal to each other. Therefore, the Poynting vector is

$$\mathbf{P} = \mathbf{E} \times \mathbf{H} = |\mathbf{E}||\mathbf{H}| = \frac{|\mathbf{E}|^2}{\eta_0} \Rightarrow |E| = \sqrt{P\eta_0} = \sqrt{20 \times 120\pi} = 86.83 \text{ V/m}$$

$$|H| = \frac{E}{120\pi} = 0.23 \text{ A/m}$$

Since the electric and magnetic fields are perpendicular to the $-y$ axis, they must lie in the xz-plane. Also **E** × **H** should be in the direction of the wave. The **H** vector will make an angle of 30° with the $+z$ axis.

2.5. WAVE POLARIZATION

We have noted that a uniform plane wave is transverse in nature. The electric and magnetic fields are perpendicular to each other and they lie in a plane transverse to the direction of wave propagation at all times and at all locations in space. However, it is not binding that the two fields should remain fixed at every point in space and at every instant of time. What we require for the plane waves is to maintain the relative orthogonality of the two fields at a given point in space at any instant of time. The ratio of the two fields should also be equal to the intrinsic impedance of the medium in which the wave is traveling. Hence, if the two fields rotate in the transverse plane by the same angle and scale in the same proportion the transverse nature of the wave is not affected. We may then have various possibilities for the two fields as discussed next. In the general case, the electric field may not be oriented just along any one of the orthogonal axis, but in a plane having two components. What we must understand is even though the electric field lies in a plane, the magnetic field still is orthogonal to it in space. This relative orthogonality of the two fields is not violated here. Wave polarization refers to the time domain (temporal) behavior in space.

Let us consider two waves of same frequency propagating along the positive *z*-axis. Let us consider a generalized electric field that is contained in the *xy*-plane; it has an E_x and E_y component. We consider the electric field only as the magnetic field can always be derived from it. Let the amplitudes of the two fields be different with a phase offset of ϕ.

As the two fields are time harmonic, let these be represented as

$$E_x = E_1 = \mathrm{Re}[E_{x0}\exp\{j(\omega t - \beta z)\}] = E_{x0}\cos(\omega t - \beta z) \quad ...(2.58)$$

and $$E_y = E_2 = \mathrm{Re}[E_{y0}\exp\{j(\omega t - \beta z + \phi)\}] = E_{x0}\cos(\omega t - \beta z + \phi) \quad ...(2.59)$$

We note that E_{x0} and E_{y0} represent the amplitudes of the *x*-directed and the *y*-directed fields respectively. We further note that, the *y*-directed field leads the *x*-directed field for a positive value of ϕ and lags behind it for ϕ being negative. We are interested in studying the wave behavior as a function of time at a given point in space. Hence, let us consider the point to be $z = 0$ (the *xy*-plane) without any loss of generality. The two fields then become

$$E_1 = \mathrm{Re}[E_{x0}\exp\{j(\omega t - \beta z)\}] = E_{x0}\cos(\omega t)\big|_{t=0} \quad ...(2.60)$$

and $$E_y = E_2 = \mathrm{Re}[E_{y0}\exp\{j(\omega t - \beta z + \phi)\}] = E_{x0}\cos(\omega t + \phi) \quad ...(2.61)$$

At any time instant, the resultant field is the vector sum of these two fields. Hence, the resultant field is

$$\mathrm{E} = E_{x0}\cos\omega t\,\hat{\mathbf{x}} + E_{y0}\cos(\omega t + \phi)\,\hat{\mathbf{y}}$$

$$= \sqrt{E_{x0}^2\cos^2\omega t + E_{y0}^2\cos^2(\omega t + \phi)}\angle\tan^{-1}\left\{\frac{E_{y0}\cos(\omega t + \phi)}{E_{x0}\cos\omega t}\right\} \quad ...(2.62)$$

It is obvious from (2.62) that the magnitude as well as the direction of the resultant field **E** changes as a function of time. The locus of the tip of the electric field vector can be obtained by eliminating the time parameter from the above two equations. Therefore,

$$\cos\omega t = \frac{E_x}{E_{x0}} \Rightarrow \sin\omega t = \sqrt{1 - \cos^2\omega t} = \sqrt{1 - \left(\frac{E_x}{E_{x0}}\right)^2} \quad ...(2.63)$$

Further,

$$\frac{E_y}{E_{y0}} = \cos\omega t \cos\phi - \sin\omega t \sin\phi$$

$$= \frac{E_x}{E_{x0}}\cos\phi - \sqrt{1-\left(\frac{E_x}{E_{x0}}\right)^2}\sin\phi \qquad ...(2.64)$$

$$\Rightarrow \qquad \left[\frac{E_x}{E_{x0}}\cos\phi - \frac{E_y}{E_{y0}}\right]^2 = \left[1-\left(\frac{E_x}{E_{x0}}\right)^2\right]\sin^2\phi$$

Expansion of (2.64) gives

$$\left(\frac{E_x}{E_{x0}}\right)^2\cos^2\phi + \left(\frac{E_x}{E_{x0}}\right)^2\sin^2\phi - 2\frac{E_x}{E_{x0}}\frac{E_y}{E_{y0}}\cos\phi + \left(\frac{E_y}{E_{y0}}\right)^2 = \sin^2\phi \qquad ...(2.65)$$

$$\Rightarrow \qquad \left(\frac{E_x}{E_{x0}}\right)^2 - 2\frac{E_x}{E_{x0}}\frac{E_y}{E_{y0}}\cos\phi + \left(\frac{E_y}{E_{y0}}\right)^2 = \sin^2\phi$$

(2.65) represents an ellipse. In general, therefore, the tip of the electric field vector **E**, for a time harmonic plane wave draws an ellipse in the *xy*-plane as a function of time which is shown in Fig. 2.3. Since over one time period of the field, ωt changes by 2π, the tip of the field draws the ellipse every one time period. It means that if the ellipse is drawn *f* times a second corresponding to the frequency of the wave which is *f*. This wave is then called 'elliptically polarized wave' or we say that the polarization of the wave is elliptical. The equation of the ellipse and consequently the shape and orientation of the ellipse changes depending upon the values of E_{x0}, E_{y0} and ϕ. We are interested only in the shape of the ellipse and not its absolute size. The shape of the ellipse can be characterized by two parameters, namely the axial ratio (AR), the ratio of the major to the minor axes of the ellipse, and the tilt angle (τ) which is the orientation of the major axis with respect to the *x*-direction. The same ellipse can also be characterized by the ratio of the two amplitudes, E_{y0}/E_{x0}, and the phase difference ϕ between them. Let us investigate the different shapes that an ellipse assumes for different conditions of these two parameters.

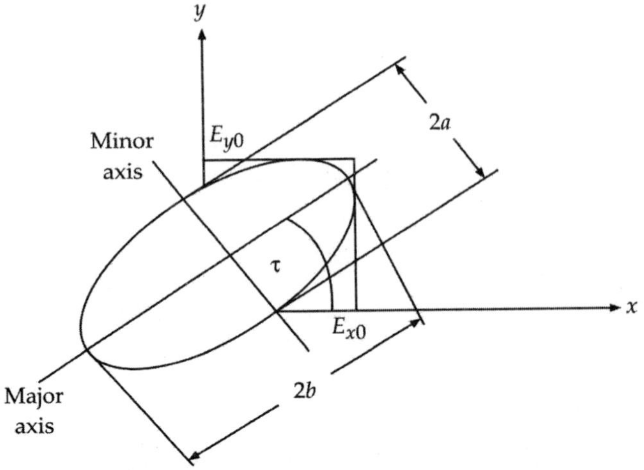

Fig. 2.3: Ellipse drawn out in the plane by the two electric fields

Linear Polarization: ($\phi = 0$) The two components E_x and E_y may or may not have the same amplitude but let us assume that the phase difference ϕ between them is zero. For such a case, the ellipse in (2.65) becomes

$$\left(\frac{E_x}{E_{x0}}\right)^2 - 2\frac{E_x}{E_{x0}}\frac{E_y}{E_{y0}} + \left(\frac{E_y}{E_{y0}}\right)^2 = 0$$

$$\Rightarrow \qquad \left(\frac{E_x}{E_{x0}} - \frac{E_y}{E_{y0}}\right)^2 = 0 \qquad \text{...(2.66)}$$

$$\Rightarrow \qquad \frac{E_x}{E_{x0}} = \frac{E_y}{E_{y0}}$$

$$\Rightarrow \qquad E_y = \left(\frac{E_{y0}}{E_{x0}}\right)E_x$$

As we may observe, this represents the equation of a straight line with a slope of $\left(\frac{E_{y0}}{E_{x0}}\right)$. The tip of the electric field vector, therefore draws a straight line for the case of $\phi = 0$ irrespective of the amplitudes of the two field components. This polarization is hence called 'linear polarization' or the wave is said to be linearly polarized. It is obvious that, if the peak amplitudes of the two fields are equal, then the slope of the straight line in the plane is 45°.

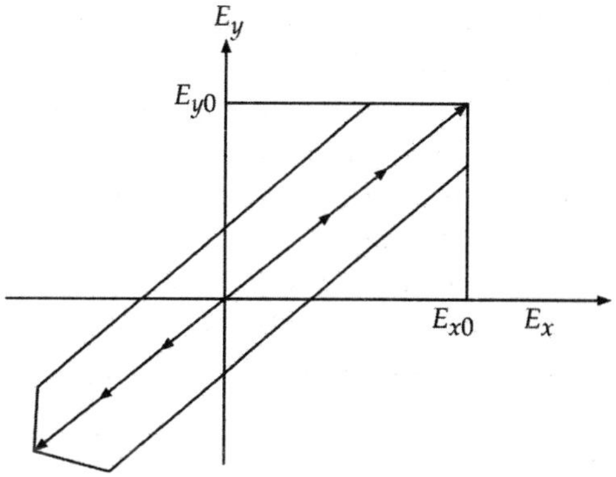

Fig. 2.4: Linear polarization

Let us assume that at some time instant t_0 both components are zero. Then at some later time instant t_1, the two components will be $E_x = E_{x1}$ and $E_y = E_{y1}$ and the resultant field will be E_1 as shown in Fig. 2.4. Similarly, at some other time instant t_2, the field is E_2 and so on. The two fields are shown by the two arrows in the first quadrant. Exactly after one quarter period, $E_x = E_{x0}$ and $E_y = E_{y0}$. The electric field reaches its maximum value E_{max}. In the next quarter period, the field reduces to zero and reverses its direction in the following half period after which it repeats the pattern. We may observe the following from the nature of polarization (the values of E_{x0} and E_{y0}).

(i) The line becomes vertical for $E_{x0} = 0$ and such a wave is known as a vertically polarized wave. In other words, the wave is said to have vertical polarization.

(ii) A horizontally polarized wave is obtained for the case corresponding to $E_{y0} = 0$.

(iii) Further, for the case the amplitudes of the two fields being equal $E_{x0} = E_{y0}$, the wave is said to be linearly polarized with 45° polarization angle.

Circular Polarization: ($\phi = \pm\pi/2$ and $E_{x0} = E_{y0} = E_0$) For this case, (2.65) becomes

$$E_x^2 + E_y^2 = E_0^2 \qquad ...(2.67)$$

This represents the equation of a circle. The tip of the electric field vector draws a circle in the *xy*-plane and the wave is said to be circularly polarized. We are interested in finding out the effect of the sign of phase on polarization. As we may note, it really does not matter whether ϕ is positive or negative as we continue to get a circle. However, the direction of the tracing out of the circle is determined by the sign of this phase. For example, let E_y lag behind E_x by 90°. Then, we note that when $E_x = 0$,

$E_y = -E_0$ and obviously the resultant of these two fields is given by $E_y = -E_0$, point *A* in Fig. 2.5. At some later time instant, E_y becomes less negative and E_x increases to some positive value. The resultant vector assumes a less negative, hence moves from *A* to *B* in a counterclockwise manner. For this case, the circle is drawn by the counterclockwise rotation of the tip of the electric field vector. The sense of polarization is known as "Right Handed Polarization" (RH) as it is equivalent to moving towards our right hand. (The assumption here is that the wave is moving along the +z-axis and hence the wave motion is out of the plane of the paper).

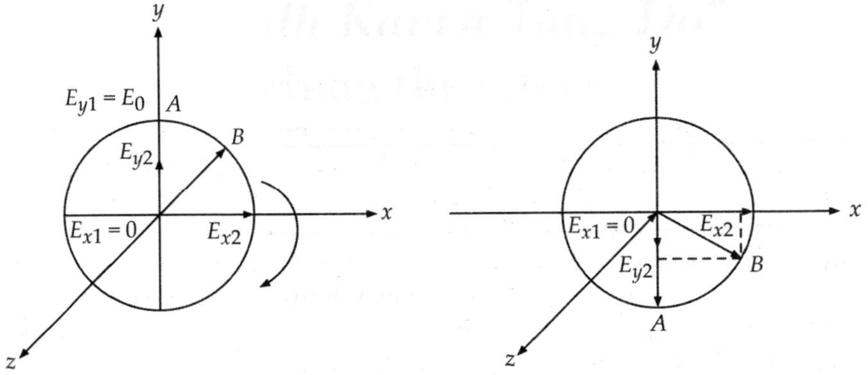

Fig. 2.5: Circular polarization

The situation gets reversed for the case when E_y leads E_x by 90°. Then, we note that when $E_x = 0$, $E_y = E_0$ and obviously the resultant of these two fields is given by $E_y = E_0$, point *A* in Fig. 2.5. At some later time instant, E_y becomes less positive and E_x increases from zero to some positive value. The resultant comes to rest at point *B* which is, therefore a positive value less from the maximum E_0 in magnitude. We note that, the tip of the electric field vector has moved in a clockwise manner from point *A* to point *B* in attaining the resultant values at the two different time instants. We may observe that, this sense of polarization is known as "Left Handed (LH)" polarization. Hence, for a uniform plane wave moving along the positive z-axis, a phase offset of +90° correspond to LH and a phase offset of –90° correspond to RH. Taking into account the sense of rotation, the circular polarization is further classified as Left Circular (LC) and Right Circular (RC) polarization as shown in Fig. 2.6.

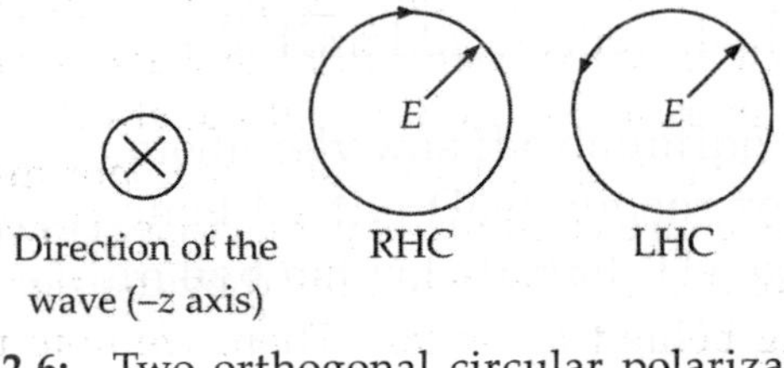

Fig. 2.6: Two orthogonal circular polarizations

Elliptical Polarization: In this case, the amplitudes of the two electric fields are not equal and the phase difference between them is neither zero nor $\pm\pi/2$. As we saw earlier, this corresponds to an elliptically polarized wave. However, the sense of rotation is determined by the sign of the phase angle. We get a left handed polarization corresponding to a positive value of ϕ and a right handed polarization for a negative value of ϕ. Consequently, an elliptically polarized wave is further categorized as left elliptical (LE) and right elliptical (RE) polarizations as shown in Fig. 2.5. We note that the angle ϕ controls the bulge of the ellipse. For $\phi = 0$, the ellipse is compressed into a straight line and for $\phi = \pm\pi/2$, the ellipse is expanded to become a circle. Hence, linear and circular polarizations are two special cases of elliptical polarization.

2.6. CHAPTER SUMMARY

Solution to Maxwell's equation in unbounded, homogeneous media has given rise to time-varying functions that exhibit wave characteristics. These waves are transverse in nature and travel in constant phase planes which are orthogonal to the direction of wave propagation. Hence these are termed as uniform plane waves. Light waves in media having a size greater than the wavelength of light mostly are modeled as uniform plane waves having transverse characteristic. The electric and the magnetic fields are time harmonic fields that are always orthogonal to each other in space. The velocity of propagation is determined by the intrinsic properties of the medium. In free space, both of these fields travel with a velocity equal to light velocity. The average power flow as given by the Poynting vector takes place along the direction of wave propagation.

A perfect conductor is a medium which has infinite conductivity ($\sigma = \infty$). In practice, no material qualifies as a perfect conductor. Metals like silver and copper can be approximated as perfect conductors. The electric field inside such a conductor is zero. This is due to the fact that a material having infinite amount of conductivity would develop an infinite current density $\mathbf{J} = \sigma\mathbf{E}$. For any finite current density, the electric field inside a perfect conductor has to be identically zero. The infinite current density, however, is possible if it is confined to a slab of zero thickness (current sheet), at the surface of a conductor giving finite surface current. The true surface current therefore exists on the surface of a perfect conductor. For a practical conductor, there is no surface current. Since the electric field is zero inside a conductor, there cannot be any charge inside a conductor and a charge has to move to the surface of the conductor. We can have charges only on the surface of a conductor giving rise to surface charge.

SUGGESTED READING

Jordan, E.C. and Balmain, K.G. "Electromagnetic Waves and Radiating Systems", 2nd Edn., PHI Pvt. Ltd., 1989.

❑❑❑

3

Uniform Plane Waves in Different Media

3.1. INTRODUCTION

The waves propagating in free space or in an infinite unguided medium are usually categorized as uniform plane waves from the previous discussion. In an unbound medium the electromagnetic waves are always transverse where the wave propagation takes place in the form of planes having a constant phase and the direction being along a unit normal to the plane. Light is an electromagnetic wave that occurs in the THz region of the spectrum. Therefore, its wavelength being much smaller than the size of the medium, it behaves like a transverse electromagnetic wave (TEM) in most cases. The nature of the waves in different media other than an unbounded homogeneous medium (like free space) will be studied in this chapter. This is because the practical media do not satisfy the condition of unbounded and infinite in size. It is, therefore, of interest to understand the behavior of such waves in practical media. To get more insight, let us first solve a couple of numericals in order to equip ourselves with the requisite tools needed to solve these equations.

Example 3.1: Express the unit vector which points from $z = h$ on the z-axis toward $(\rho, \phi, 0)$ in cylindrical coordinates.

Solution: The position vector $\mathbf{R}$ is expressed as

$$\mathbf{R} = (\rho - 0)\hat{\rho} + (0 - h)\hat{\mathbf{z}} = \rho\hat{\rho} - h\hat{\mathbf{z}}$$

The unit normal $$\mathbf{a}_R = \frac{\mathbf{R}}{|\mathbf{R}|} = \frac{\rho\hat{\rho} - h\hat{\mathbf{z}}}{\sqrt{\rho^2 + h^2}}$$

Example 3.2: Express the unit vector which is directed toward the origin from an arbitrary point on the plane $z = -5$.

Solution: The position vector **R** is expressed as

$$\mathbf{R} = (0-x)\hat{\mathbf{x}} + (0-y)\hat{\mathbf{y}} + [0-(-5)]\hat{\mathbf{z}} = -x\hat{\mathbf{x}} - y\hat{\mathbf{y}} + 5\hat{\mathbf{z}}$$

$$\mathbf{a}_R = \frac{\mathbf{R}}{|\mathbf{R}|} = \frac{-x\hat{\mathbf{x}} - y\hat{\mathbf{y}} + 5\hat{\mathbf{z}}}{\sqrt{x^2+y^2+5^2}} = \frac{-x\hat{\mathbf{x}} - y\hat{\mathbf{y}} + 5\hat{\mathbf{z}}}{\sqrt{x^2+y^2+25}}$$

Example 3.3: A current sheet, $\mathbf{K} = \left(\frac{8.0}{\mu_0}\right)\hat{\mathbf{y}}$ at $x = 0$, separates region 1, $x < 0$ and $\mu_{r1} = 3$, from region 2, $x > 0$ and $\mu_{r1} = 1$. Given $\mathbf{H}_1 = \left(\frac{10.0}{\mu_0}\right)(\hat{\mathbf{y}} + \hat{\mathbf{z}})$ A/m, find $\mathbf{H}_2$.

Solution: We know that $(\mathbf{H}_1 - \mathbf{H}_2) \times a_{n12} = \mathbf{K}$ $a_{n12} = \hat{\mathbf{x}}$ in this problem

Let $$\mathbf{H}_2 = H_x\hat{\mathbf{x}} + H_y\hat{\mathbf{y}} + H_z\hat{\mathbf{z}}$$

$$\left[\left(\frac{10.0}{\mu_0}\right)(\hat{\mathbf{y}} + \hat{\mathbf{z}}) - \mathbf{H}_2\right] \times \hat{\mathbf{x}} = \left(\frac{8.0}{\mu_0}\right)\hat{\mathbf{y}}$$

$$\Rightarrow \quad \left(\frac{10.0}{\mu_0}\right)(\hat{\mathbf{y}} - \hat{\mathbf{z}}) - \left[H_x\hat{\mathbf{x}} + H_y\hat{\mathbf{y}} + H_z\hat{\mathbf{z}}\right] \times \hat{\mathbf{x}} = \left(\frac{8.0}{\mu_0}\right)\hat{\mathbf{y}}$$

$$= \left(\frac{10.0}{\mu_0}\right)(\hat{\mathbf{y}} - \hat{\mathbf{z}}) - (-H_y\hat{\mathbf{z}} + H_z\hat{\mathbf{y}}) = \left(\frac{8.0}{\mu_0}\right)\hat{\mathbf{y}}$$

$$= \left(\frac{10.0}{\mu_0}\right)(\hat{\mathbf{y}} - \hat{\mathbf{z}}) - \left(\frac{8.0}{\mu_0}\right)\hat{\mathbf{y}} = (-H_y\hat{\mathbf{z}} + H_z\hat{\mathbf{y}})$$

$$= \frac{1}{\mu_0}(2.0\hat{\mathbf{y}} - 10.0\hat{\mathbf{z}}) = (H_z\hat{\mathbf{y}} - H_y\hat{\mathbf{z}})$$

Therefore,

$$H_y = 10.0 \text{ A/m} \quad \text{and} \quad H_z = 2.0 \text{ A/m}$$

$$\mathbf{H}_2 = \frac{1}{\mu_0}(10.0\hat{\mathbf{y}} + 2.0\hat{\mathbf{z}})$$

Example 3.4: The $x = 0$ plane contains a current sheet of density A current sheet $\mathbf{K}$ which separates region 1, $x < 0$ and $\mu_{r1} = 3$, from region 2, $x > 0$ and $\mu_{r2} = 7$. Given $\mathbf{B}_1 = 6.0\hat{\mathbf{x}} + 4.0\hat{\mathbf{y}} + 10.0\hat{\mathbf{z}}$ (T), $\mathbf{B}_2 = 6.0\hat{\mathbf{x}} - 50.96\hat{\mathbf{y}} + 8.96\hat{\mathbf{z}}$ (T). Find $\mathbf{K}$.

Solution: We know that at the boundary, the tangential component of $\mathbf{H}$ is continuous. Hence,

$$\frac{\mathbf{B}_1}{\mu_0\mu_{r1}} = \frac{\mathbf{B}_2}{\mu_0\mu_{r2}}$$

As the $x = 0$ plane is being considered, the tangential components are the $\hat{\mathbf{y}}$ and the $\hat{\mathbf{z}}$ components and the $\hat{\mathbf{x}}$ is discontinuous.

Thus,

$$\mathbf{H}_1 = \frac{1}{\mu_0}(2.0\hat{\mathbf{y}} + 5.0\hat{\mathbf{z}})$$

and similarly,

$$\mathbf{H}_2 = \frac{1}{7\mu_0}(-50.96\hat{\mathbf{y}} + 8.96\hat{\mathbf{z}}) = \frac{1}{\mu_0}(-7.28\hat{\mathbf{y}} + 1.28\hat{\mathbf{z}})$$

Therefore, $(\mathbf{H}_1 - \mathbf{H}_2) \times a_{n12} = \mathbf{K}$

$$= \frac{1}{\mu_0}[2.0\hat{\mathbf{y}} + 5.0\hat{\mathbf{z}} - (-7.28\hat{\mathbf{y}} + 1.28\hat{\mathbf{z}})] \times \hat{\mathbf{x}}$$

$$= \frac{1}{\mu_0}(9.28\hat{\mathbf{y}} + 3.72\hat{\mathbf{z}}) \times \hat{\mathbf{x}}$$

$$= \frac{1}{\mu_0}(3.72\hat{\mathbf{y}} - 9.28\hat{\mathbf{z}}) \text{ A/m}$$

Example 3.5: Inside a right circular cylinder, $\mu_{r1} = 1000$. The exterior is free space. If $\mathbf{B}_1 = 2.5\hat{\boldsymbol{\varphi}}$ (T) inside the cylinder, determine $\mathbf{B}_2$ just outside.

Solution: We know that the tangential component of $\mathbf{H}$ is continuous across the boundary. Thus,

$$\frac{\mathbf{B}_1}{\mu_0\mu_{r1}} = \frac{\mathbf{B}_2}{\mu_0\mu_{r2}} \text{ and hence, } \mathbf{B}_2 = \frac{2.5}{1000}\hat{\boldsymbol{\varphi}} \text{ mT}$$

Example 3.6: In free space, $\mathbf{B} = B_m e^{j(\omega t+\beta z)}\hat{\mathbf{y}}$. Find the corresponding electric field.

Solution: We know that

$$\nabla \times \mathbf{H} = \varepsilon_0 \frac{\partial \mathbf{E}}{\partial t} = \nabla \times \frac{\mathbf{B}}{\mu_0} = \varepsilon_0 \frac{\partial \mathbf{E}}{\partial t} \Rightarrow \nabla \times \mathbf{B} = \mu_0 \varepsilon_0 \frac{\partial \mathbf{E}}{\partial t} = j\omega \mu_0 \varepsilon_0 \mathbf{E}$$

If we expand the above expression we get

$$\begin{bmatrix} \hat{\mathbf{x}} & \hat{\mathbf{y}} & \hat{\mathbf{z}} \\ \frac{\partial}{\partial x} & \frac{\partial}{\partial y} & \frac{\partial}{\partial z} \\ B_x & B_y & B_z \end{bmatrix}$$

$$= \left(\frac{\partial B_z}{\partial y} - \frac{\partial B_y}{\partial z}\right)\hat{\mathbf{x}} - \left(\frac{\partial B_z}{\partial x} - \frac{\partial B_x}{\partial z}\right)\hat{\mathbf{y}} + \left(\frac{\partial B_y}{\partial x} - \frac{\partial B_x}{\partial y}\right)\hat{\mathbf{z}} = j\omega\, \mu_0\, \varepsilon_0\, \mathbf{E}$$

In the above expression we understand that the magnetic field has an y component only and further $\frac{\partial}{\partial x} = 0 = \frac{\partial}{\partial y}, \frac{\partial}{\partial z} = j\beta$. Therefore,

$$\frac{\partial B_y}{\partial z} = j\beta B_m e^{j(\omega t+\beta z)}$$

and
$$\mathbf{E} = -\frac{j\beta B_m}{j\omega \mu_0 \varepsilon_0} e^{j(\omega t+\beta z)}\hat{\mathbf{x}} = -\frac{\beta B_m}{\omega \mu_0 \varepsilon_0} e^{j(\omega t+\beta z)}\hat{\mathbf{x}}$$

As
$$\frac{1}{\mu_0 \varepsilon_0} = \frac{\omega^2}{\beta^2} \Rightarrow \mu_0 \varepsilon_0 = \frac{\beta^2}{\omega^2}$$

Substitution of the above in the electric field expression we get

$$\mathbf{E} = -\frac{\beta B_m}{\omega \mu_0 \varepsilon_0} e^{j(\omega t+\beta z)}\hat{\mathbf{x}} = -\frac{\beta B_m}{\omega\left(\frac{\beta^2}{\omega^2}\right)} e^{j(\omega t+\beta z)}\hat{\mathbf{x}} = -\frac{\omega B_m}{\beta} e^{j(\omega t+\beta z)}\hat{\mathbf{x}}$$

This expression shows that the electric field **E** and the magnetic field **H** are in space quadrature but in time phase. Both of the fields propagate with the same velocity in the positive z-direction.

Example 3.7: Region 1, where $\mu_{r1} = 5$, is on the side of the plane $6x + 4y + 3z = 12$ that includes the origin. In region 2, $\mu_{r2} = 3$. Given

$\mathbf{H}_1 = \frac{1}{\mu_0}(3.0\hat{\mathbf{x}} - 0.5\hat{\mathbf{y}})$ A/m, find $\mathbf{B}_2$ and θ_2.

Solution: $\mathbf{B}_1 = \mu_0\mu_{r1}\mathbf{H}_1 = 5\mu_0.\frac{1}{\mu_0}(3.0\hat{\mathbf{x}} - 0.5\hat{\mathbf{y}}) = 15.0\hat{\mathbf{x}} - 2.5\hat{\mathbf{y}}$

The unit normal to the plane is

$$a_{n1} = \frac{6x + 4y + 3z}{\sqrt{6^2 + 4^2 + 3^2}} = \frac{6\hat{\mathbf{x}} + 4\hat{\mathbf{y}} + 3\hat{\mathbf{z}}}{\sqrt{61}}$$

The magnitude of the normal component of the magnetic field is

$$B_{n1} = \mathbf{B}_1.a_{n1} = 15.0\hat{\mathbf{x}} - 2.5\hat{\mathbf{y}}.\frac{6\hat{\mathbf{x}} + 4\hat{\mathbf{y}} + 3\hat{\mathbf{z}}}{\sqrt{61}} = \frac{90 - 10}{\sqrt{61}} = \frac{80}{\sqrt{61}}$$

Therefore, the normal component of the magnetic field is

$$\mathbf{B}_{n1} = B_{n1}.a_{n1} = B_{n1}.\mathbf{a}_{\mathbf{n1}} = \frac{80}{\sqrt{61}}.\frac{6\hat{\mathbf{x}} + 4\hat{\mathbf{y}} + 3\hat{\mathbf{z}}}{\sqrt{61}} = \frac{480\hat{\mathbf{x}} + 320\hat{\mathbf{y}} + 240\hat{\mathbf{z}}}{61}$$

$$= 7.868\hat{\mathbf{x}} + 5.245\hat{\mathbf{y}} + 3.934\hat{\mathbf{z}} = \mathbf{B}_{n2}$$

The tangential component of the magnetic field in region 1 is given as

$$\mathbf{B}_{t1} = \mathbf{B}_1 - \mathbf{B}_{n1} = 15.0\hat{\mathbf{x}} - 2.5\hat{\mathbf{y}} - (7.868\hat{\mathbf{x}} + 5.245\hat{\mathbf{y}} + 3.934\hat{\mathbf{z}})$$

$$= 7.132\hat{\mathbf{x}} - 7.745\hat{\mathbf{y}} - 3.934\hat{\mathbf{z}}$$

Hence,

$$\mathbf{H}_{t1} = \frac{\mathbf{B}_{t1}}{\mu_0\mu_{r1}} = \frac{7.132\hat{\mathbf{x}} - 7.745\hat{\mathbf{y}} - 3.934\hat{\mathbf{z}}}{5\mu_0}$$

$$= \frac{1}{\mu_0}(1.426\hat{\mathbf{x}} - 1.549\hat{\mathbf{y}} - 0.7886\hat{\mathbf{z}}) = \mathbf{H}_{t2}$$

$$\mathbf{B}_{t2} = \mu_0\mu_{r2}\mathbf{H}_{t2} = 3\mu_0.\frac{1}{\mu_0}(1.426\hat{\mathbf{x}} - 1.549\hat{\mathbf{y}} - 0.7886\hat{\mathbf{z}})$$

$$= 4.279\hat{\mathbf{x}} - 4.647\hat{\mathbf{y}} - 2.3658\hat{\mathbf{z}}$$

Therefore,

The total magnetic field in region 2 is

$$\mathbf{B}_2 = \mathbf{B}_{n2} + \mathbf{B}_{t2} = 7.868\hat{\mathbf{x}} + 5.245\hat{\mathbf{y}} + 3.934\hat{\mathbf{z}} + 4.279\hat{\mathbf{x}} - 4.647\hat{\mathbf{y}} - 2.3658\hat{\mathbf{z}}$$

$$= 12.147\hat{\mathbf{x}} + 0.598\hat{\mathbf{y}} + 1.56\hat{\mathbf{z}} = 12.15\hat{\mathbf{x}} + 0.6\hat{\mathbf{y}} + 1.56\hat{\mathbf{z}}$$

The angle between the unit normal a_{n1} and the field $\mathbf{B}_1$ is computed as

$$\cos\alpha_1 = \frac{\mathbf{B}_1 . a_{n1}}{|\mathbf{B}_1|} = \frac{15.0\hat{\mathbf{x}} - 2.5\hat{\mathbf{y}}}{\sqrt{15^2 + (-2.5)^2}} . \frac{6\hat{\mathbf{x}} + 4\hat{\mathbf{y}} + 3\hat{\mathbf{z}}}{\sqrt{61}} = \frac{90 - 10}{\sqrt{231.25}\sqrt{61}}$$

$$= \frac{80}{15.206 \times 7.81} = 0.67359$$

So, $\alpha_1 = \cos^{-1}(0.67359) = 47.65°$ and that gives us the angle between the unit normal and the surface of separation as

$$\theta_1 = 90° - \alpha_1 = 90° - 47.65° = 42.35°$$

Further, $$\frac{\mu_{r1}}{\mu_{r2}} = \frac{\tan\theta_2}{\tan\theta_1}$$

$$\Rightarrow \tan\theta_2 = \frac{\mu_{r1_1}\tan\theta_1}{\mu_{r2}} = \frac{5\tan(42.35)}{3} = 1.66 \times 0.91152 = 1.5131$$

And therefore, $\theta_2 = \tan^{-1}(1.5131) = 56.5°$

Example 3.8: The interface between two different regions is normal to one of the three Cartesian axes. If $\mathbf{B}_1 = \mu_0(43.5\hat{\mathbf{x}} + 24.0\hat{\mathbf{z}})$ and $\mathbf{B}_2 = \mu_0(22.0\hat{\mathbf{x}} + 24.0\hat{\mathbf{z}})$, what is the ratio $\tan\theta_1/\tan\theta_2$?

Solution: We observe that in both the fields are contained in the *xz*-plane. In these two fields, the *z*-component remains same in both the regions. Hence this is the tangential component. The *x*-component becomes the normal component. Hence,

$$\tan\theta_1/\tan\theta_2 = \frac{B_{n2}}{B_{n1}} = \frac{22.0}{43.5} = 0.5057 \approx 0.506$$

Example 3.9: In spherical coordinates region 1 is $r < a$, region 2 is $a < r < b$ and region 3 is $r > b$. Regions 1 and 3 are free space, while $\mu_{r2} = 500$. Given $\mathbf{B}_1 = 0.20\hat{\mathbf{r}}$, find $\mathbf{H}$ in each region.

Solution: We know that, the normal components are related as

$$\mu_0\mu_{r1}\mathbf{H}_1 = \mu_0\mu_{r2}\mathbf{H}_2 = \mu_0\mu_{r3}\mathbf{H}_3$$

Hence,

$$\mathbf{H}_1 = \frac{\mathbf{B}_1}{\mu_0\mu_{r1}} = \frac{0.2}{\mu_0} = \mu_0\mu_{r2}\mathbf{H}_2$$

$$\Rightarrow \qquad \mathbf{H}_2 = \frac{0.2}{500\mu_0} = 0.04 \times 10^{-2} = \frac{4 \times 10^{-4}}{\mu_0} \text{ A/m}$$

and the corresponding magnetic field in region 3 is

$$\mu_0\,\mu_{r1}\,\mathbf{H}_1 = \mu_0\mu_{r3}\mathbf{H}_3 \Rightarrow \mathbf{H}_1 = \mathbf{H}_3 = \frac{0.2}{\mu_0}$$

3.2. PLANE WAVES IN LOSSY DIELECTRICS

All dielectric materials have some amount of conductivity and while it may be neglected in many cases, even then it is necessary to establish the criteria for doing so. Under the assumption of a sinusoidal time variation, the Maxwell's curl equations are

$$\nabla \times \mathbf{H} = \mathbf{J} + j\omega\varepsilon\mathbf{E} = (\sigma + j\omega\varepsilon)\mathbf{E} \qquad ...(3.1)$$

and

$$\nabla \times \mathbf{E} = -j\omega\mu\mathbf{H} \qquad ...(3.2)$$

The only effect of including the conductivity σ is that the factor $j\omega\varepsilon$ is changed to $(\sigma + j\omega\varepsilon)$. Therefore, the new value of the propagation constant becomes

$$\gamma^2 = (\sigma + j\omega\varepsilon)j\omega\mu \qquad ...(3.3)$$

or

$$\gamma = \pm\sqrt{j\omega\mu(\sigma + j\omega\varepsilon)} = \sqrt{(j\omega)^2\mu\varepsilon + j\omega\mu\sigma}$$

$$= \sqrt{(j\omega)^2\mu\varepsilon\left(1 + \frac{\sigma}{j\omega\varepsilon}\right)}$$

$$= j\omega\sqrt{\mu\varepsilon}\sqrt{1 - j\frac{\sigma}{\omega\varepsilon}} \qquad ...(3.4)$$

where we have retained only the positive sign of the root. As we observe from (3.4), this expression differs from the lossless case in the presence of the second radical factor, which becomes unity as σ vanishes. In the general case, values for σ, μ, ε and ω may be inserted into (3.4). The real and the imaginary parts of the propagation constant are

$$\gamma = \alpha + j\beta \qquad ...(3.5)$$

The x-component of the electric field propagating along the positive z-axis is expressed as

$$E_x = E_o e^{-\gamma z} = E_o e^{-\alpha z} e^{-j\beta z} \qquad ...(3.6)$$

The use of the plus sign for the radical in (4) leads to positive numerical values for α and β, and hence corresponds to propagation along the positive z-axis. Once we have the electric field, the corresponding magnetic field becomes

$$H_y = \frac{E_x}{\eta} e^{-\alpha z} e^{-j\beta z} \qquad ...(3.7)$$

where the intrinsic impedance is now a complex quantity given as

$$\eta = \sqrt{\frac{j\omega\mu}{\sigma + j\omega\varepsilon}} \qquad ...(3.8a)$$

Dividing the numerator and the denominator of (3.8) by $j\omega$, we obtain

$$\eta = \sqrt{\frac{\mu}{\varepsilon + \dfrac{\sigma}{j\omega}}} = \sqrt{\frac{\mu}{\varepsilon\left(1 - j\dfrac{\sigma}{\omega\varepsilon}\right)}} = \sqrt{\frac{\mu}{\varepsilon}} \frac{1}{\sqrt{1 - j(\sigma/\omega\varepsilon)}} \qquad ...(3.8b)$$

This relation clearly establishes that the electric and magnetic fields are no longer in time phase in a lossy dielectric as was the case in an unbounded infinite source free medium. This is the first difference between the wave propagation in the two cases we just considered. For a wave propagating along the positive z-axis, the factor $e^{-\gamma z}$ results in an exponential decrease in amplitude with increasing values of displacement from the origin; z. The attenuation constant is measured in nepers per metre in order that the exponent of *e* be measured in the dimensionless units of nepers. Therefore, for a value of $\alpha = 0.01$ Np/m, the peak amplitude of the wave at $z = 50$ m becomes $e^{-(0.01\times 50)} = e^{-0.5} = 0.607$. In traveling a distance of $1/\alpha$, the wave amplitude gets reduced by a factor of 37%.

3.3. PROPAGATION OF LIGHT WAVE THROUGH A QUARTER-WAVE PLATE

A quarter wave plate (QWP) is a device that is used in many experiments involving filter and integrated optics and is basically used to change the state of polarization (SOP) of a propagating optical wave. The term quarter here refers to a relative phase difference of $\pi/2$ between the two electric field components. For example, using this device we may transform a circularly polarized wave to a linearly polarized wave and *vice versa*. A QWP is made of an anisotropic medium like calcite or quartz. In such type of crystals there is a preferred direction known as the optic axis. A QWP the crystal is cut in such a way that the optic axis is at right angles to the thickness of the plate. In Fig. 3.1, we have shown a QWP with its optic axis along the x-direction. The modes of the QWP are x- and y-polarized; the x-polarized mode, which is polarized along the optic axis, is usually referred to as the extraordinary wave, and the y-polarized mode is referred to as the ordinary wave. By modes we imply that if the incident wave is x-polarized, then it will propagate as an x-polarized wave; similarly, if the incident wave is y-polarized then it will propagate as a y-polarized wave without any change in the SOP but with a velocity slightly different from the velocity of the x-polarized wave. On the other hand, if a circularly polarized wave is incident, we must express this as a linear combination of x- and y-polarized waves and then consider the independent propagation of the two waves.

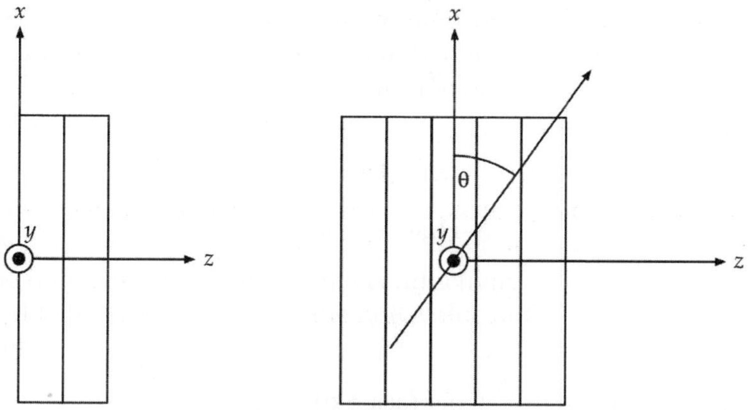

Fig. 3.1: A plane polarized wave is incident normally on a quarter wave plate with the plane of polarization making an angle θ with the optic axis which is the x-axis here. The left diagram is the side view while the right diagram represents its front view.

As an example, let us consider the incidence of a wave polarized at an angle of 45° to the x-axis. At $z = 0$, it is expressed as

$$E_x = E_0 \cos \theta \cos \omega t \qquad ...(3.9)$$

and

$$E_y = E_0 \sin \theta \cos \omega t \qquad ...(3.10)$$

where $\theta = 45°$. The refractive indices seen by the x- and y-polarized waves are different and are usually denoted by n_e and n_0 respectively. These are known as the extraordinary and ordinary refractive indices and are known constants of the crystal. Thus, the x- and y-polarized waves propagate with velocities c/n_e and c/n_0 respectively, and therefore inside the crystal we have

$$E_x = \frac{E_0}{\sqrt{2}}\cos(\omega t - \beta_e z) \qquad ...(3.11)$$

and
$$E_y = \frac{E_0}{\sqrt{2}}\cos(\omega t - \beta_o z) \qquad ...(3.12)$$

where
$$\beta_e = \frac{\omega}{c} n_e \qquad ...(3.13)$$

and
$$\beta_0 = \frac{\omega}{c} n_o \qquad ...(3.14)$$

Thus as the wave propagates, a phase difference is set up between the two components. We assume that $n_0 > n_e$ (which is true for negative crystals like calcite). Let the thickness d of the crystal be such that

$$(\beta_0 - \beta_e)d = \pi/2 \qquad ...(3.15)$$

$$d = \frac{\lambda_0}{4(n_0 - n_e)} \qquad ...(3.16)$$

where $\lambda_0 = \omega/2\pi c$ represents the free space wavelength. Then at $z = d$, the x- and y-components of polarization are given by

$$E_x = \frac{E_0}{\sqrt{2}}\cos(\omega t - \phi) \qquad ...(3.17)$$

and
$$E_y = \frac{E_0}{\sqrt{2}}\cos\left[\omega t - \left(\phi + \frac{\pi}{2}\right)\right] = \frac{E_0}{\sqrt{2}}\sin(\omega t - \phi) \qquad ...(3.18)$$

where $\phi = \beta_e d$, (3.17) and (3.18) represent an RCP wave.

Similarly, a half plate device is a device that introduces a phase difference of π between the two electric field components; thus, its thickness is twice the thickness of the QWP. A linearly polarized light wave remains linearly polarized after passing through a HWP; however, the plane of orientation of polarization can be changed by a desired amount using an HWP.

3.4. PLANE WAVE REFLECTION AND REFRACTION FROM MEDIA INTERFACE

When an EM wave propagating in one medium comes across a different medium, a part of it gets reflected at the boundary of the two media while another part gets either transmitted or absorbed by the second medium. In this section we consider the case of normal incidence of the plane waves. By normal incidence, we understand that the wave propagation is taking place along the normal to the boundary of the two media.

3.4.1. Lossy Conducting Medium

We assume plane waves propagating along the positive z-axis with the electric field vector along the y-axis without any loss of generality. We further assume the plane $z < 0$ to be region 1 so that the naturally the plane $z > 0$ is the second region. Normal incidence here refers to the fact that the wave propagation is along the positive z-axis which is the normal to the xy-plane containing the fields. The first medium is assumed to be a lossy conducting medium. Hence, the incident electric and magnetic fields become

$$\mathbf{E}_i = E_0 e^{-\gamma_1 z}\hat{\mathbf{y}} \qquad ...(3.19)$$

and

$$\mathbf{H}_i = -\frac{1}{\eta_1} E_0 e^{-\gamma_1 z}\hat{\mathbf{x}} \qquad ...(3.20)$$

where η_i is the intrinsic impedance of medium 1 and E_0 is the peak η_1 amplitude of the electric field. We can see from Fig. 3.2 that the incident magnetic field is aligned along the positive x-axis which is for the present case out of the plane of the paper.

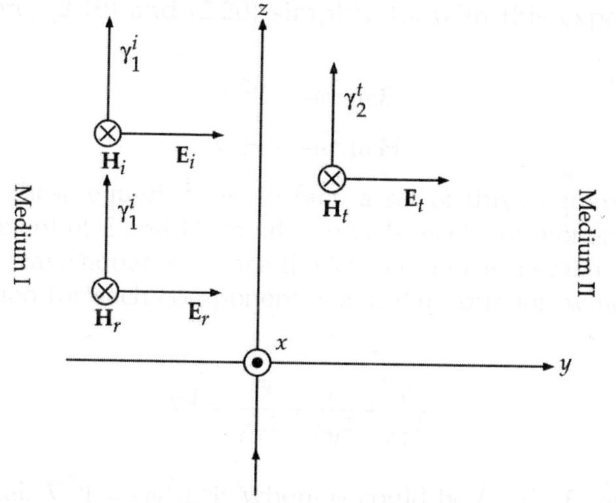

Fig. 3.2: Normal incidence of a plane wave at the boundary of two media (the z-axis)

The incident wave from medium 1 is partially reflected and transmitted at the boundary of separation of the two media. The reflected wave is contained in the same plane as incidence; $z < 0$, but the wave direction becomes $-z$. Hence, the reflected electric field is given as

$$\mathbf{E}_r = \Gamma E_0 e^{\gamma_1 z}\hat{\mathbf{y}} \qquad ...(3.21)$$

where Γ is the reflection coefficient defined as

$$\Gamma = \frac{|E_r|}{|E_i|} \quad \text{...(3.22)}$$

The reflected magnetic field is obtained from the reflected electric field by taking its curl first. These are the fields in medium I.

$$\nabla \times \mathbf{E}_r = j\omega\,\mu_1\,\mathrm{H}_r \quad \text{...(3.23)}$$

Hence,

$$\mathbf{H}_r = -\frac{\nabla \times \mathbf{E}_r}{j\omega\mu_1} = j\frac{\nabla \times \mathbf{E}_r}{\omega\mu_1} = \frac{j}{\omega\mu_1}\begin{vmatrix} \hat{\mathbf{x}} & \hat{\mathbf{y}} & \hat{\mathbf{z}} \\ \dfrac{\partial}{\partial x} & \dfrac{\partial}{\partial y} & \dfrac{\partial}{\partial z} \\ E_x & E_y & E_z \end{vmatrix}$$

$$= \frac{j}{\omega\mu_1}\begin{vmatrix} \hat{\mathbf{x}} & \hat{\mathbf{y}} & \hat{\mathbf{z}} \\ \dfrac{\partial}{\partial x} & \dfrac{\partial}{\partial y} & \dfrac{\partial}{\partial z} \\ 0 & E_y & 0 \end{vmatrix} \quad \text{...(3.24)}$$

This is because the electric field neither has a x-component nor a z-component. Therefore,

$$\mathbf{H}_r = \frac{j}{\omega\mu_1}\begin{vmatrix} \hat{\mathbf{x}} & \hat{\mathbf{y}} & \hat{\mathbf{z}} \\ \dfrac{\partial}{\partial x} & \dfrac{\partial}{\partial y} & \dfrac{\partial}{\partial z} \\ 0 & E_y & 0 \end{vmatrix} = \frac{j}{\omega\mu_1}\left[-\left(\frac{\partial}{\partial z}E_y\right)\right]\hat{\mathbf{x}}$$

$$= \left[-\frac{j}{\omega\mu_1}\frac{\partial}{\partial z}(\Gamma E_0 e^{\gamma_1 z})\right]\hat{\mathbf{x}} \quad \text{...(3.25)}$$

$$= -\frac{j\Gamma}{\omega\mu_1}.\gamma_1 E_0 e^{\gamma_1 z}\hat{\mathbf{x}} = \frac{\gamma_1}{j\omega\mu_1}\Gamma E_0 e^{\gamma_1 z}\hat{\mathbf{x}} = \frac{\Gamma}{\eta_1}E_0 e^{\gamma_1 z}\hat{\mathbf{x}}$$

The Poynting vectors for the incident and the reflected waves in region I become

$$P_{I,i} = \mathbf{E}_i \times \mathbf{H}_i^* = E_0 e^{\gamma_1 z}\hat{\mathbf{y}} \times -\frac{1}{\eta_1}E_0^* e^{\gamma_1 z}\hat{\mathbf{x}} = \frac{1}{\eta_1}|E_0|^2 e^{2\gamma_1 z}\hat{\mathbf{z}} \quad \text{...(3.26)}$$

$$P_{I,r} = \mathbf{E}_r \times \mathbf{H}_r^* = \Gamma E_0 e^{\gamma_1 z}\hat{\mathbf{y}} \times \frac{\Gamma^*}{\eta_1}E_0^* e^{\gamma_1 z}\hat{\mathbf{x}} = -\frac{1}{\eta_1}|\Gamma|^2|E_0|^2 e^{2\gamma_1 z}\hat{\mathbf{z}} \quad \text{...(3.27)}$$

We have adopted here the right handed cork screw coordinate system. The expressions in (3.26) and (3.27) show that the incident power is oriented along the positive *z*-axis which is also the direction of wave propagation. Intuitively, this is also correct. This is because for a wave traveling along the positive *z*-axis, we expect the power to be along this direction only. Similarly, the reflected power in (3.27) shows that it is traveling along the negative *z*-axis. This condition must be satisfied for the other fields as well. For the transmitted waves, the region of travel is in region II corresponding to z > 0. The fields in the second lossy medium become

$$\mathbf{E}_t = \tau E_0 e^{-\gamma_2 z}\hat{\mathbf{y}} \qquad \text{...(3.28)}$$

and

$$\mathbf{H}_t = -\frac{\tau}{\eta_2} E_0 e^{-\gamma_2 z}\hat{\mathbf{x}} \qquad \text{...(3.29)}$$

where η_2 is the intrinsic impedance of the second region. The transmission coefficient is defined as

$$\tau = \frac{|E_t|}{|E_i|} \qquad \text{...(3.30)}$$

We know that η_2 is given as

$$\eta_2 = \sqrt{\frac{j\omega\mu_2}{\sigma_2 + j\omega\varepsilon_2}} \qquad \text{...(3.31)}$$

and the corresponding propagation constant is

$$\gamma_2 = \alpha_2 + j\beta_2 = j\omega\sqrt{\mu_2\varepsilon_2}\sqrt{1 - j\frac{\sigma_2}{\omega\varepsilon_2}} \qquad \text{...(3.32)}$$

A comprehensive summary of the incident, reflected and the transmitted or the refracted fields at the boundary of two lossy media is presented in Table 3.1.

Table 3.1: Summary of fields at the interface of lossy conducting media for normal incidence

Region I (lossy medium 1)	*Region II (lossy medium 2)*
$\mathbf{E}_i = E_0 e^{-\gamma_1 z}\hat{\mathbf{y}}$, $\mathbf{H}_i = -\frac{1}{\eta_1} E_0 e^{-\gamma_1 z}\hat{\mathbf{x}}$ $\mathbf{E}_r = \Gamma E_0 e^{\gamma_1 z}\hat{\mathbf{y}}$, $\mathbf{H}_r = \frac{\Gamma}{\eta_1} E_0 e^{\gamma_1 z}\hat{\mathbf{x}}$	$\mathbf{E}_t = \tau E_0 e^{-\gamma_2 z}\hat{\mathbf{y}}$, $\mathbf{H}_t = -\frac{\tau}{\eta_2} E_0 e^{-\gamma_2 z}\hat{\mathbf{x}}$

A close look at Table 3.1 tells us that all the fields of interest at the boundary namely, the incident, the reflected field and the transmitted fields are contained in the *xy*-plane. However, the direction of the reflected magnetic undergoes a 180° phase reversal at the boundary while that of the incident and the refracted magnetic fields remain unaltered. This is to make sure that power flow for the incident and the refracted fields still take place along the + *z*-axis while it is along the negative *z*-axis for the reflected field. The reader is advised to verify this for all the three fields.

The reflection coefficient and the transmission coefficients are computed in terms of the medium parameters. To do so, we note that the tangential component of the electric field **E** as well as the magnetic field **H** is continuous across the boundary. As we are considering the separation of the two regions as the interface $z = 0$; the *xy*-plane, the tangential components are E_x, E_y and H_x, H_y. However, we have only the *y*-component for the electric field. The tangential component E_x in medium I consists of two parts; E_i and E_r whereas this is E_t in medium II. Therefore,

$$E_i + E_r = E_t|_{z=0} \qquad \text{...(3.33)}$$

$$\Rightarrow \quad (1 + \Gamma)E_0 = \tau E_0$$

$$\Rightarrow \quad 1 + \Gamma = \tau$$

Similarly, the tangential component across the boundary has the H_x only. As there is no surface current,

$$H_i + H_r = H_t|_{z=0}$$

$$\Rightarrow \quad \left(\frac{-1+\Gamma}{\eta_1}\right)E_0 = -\frac{\tau}{\eta_2}E_0$$

$$\Rightarrow \quad \left(\frac{1-\Gamma}{\eta_1}\right)E_0 = \frac{\tau}{\eta_2}E_0 \qquad \text{...(3.34)}$$

$$\Rightarrow \quad \frac{1-\Gamma}{\eta_1} = \frac{\tau}{\eta_2}$$

Substituting the expression of from (3.33), we get

$$\frac{1-\Gamma}{\eta_1} = \frac{1+\Gamma}{\eta_2}$$

$$\Rightarrow \quad (1 - \Gamma)\eta_2 = (1 + \Gamma)\eta_1 \qquad \text{...(3.35)}$$

$$\Rightarrow \quad \Gamma = \frac{\eta_2 - \eta_1}{\eta_2 + \eta_1}$$

And the reflection coefficient is expressed as

$$\tau = 1+\Gamma = 1+\frac{\eta_2-\eta_1}{\eta_2+\eta_1} = \frac{2\eta_2}{\eta_2+\eta_1} \qquad ...(3.36)$$

3.4.2. Lossless Medium

If the two regions consist of lossless dielectrics, then $\sigma = 0$ and μ, ε are real quantities. The propagation constant for this case is purely imaginary and is expressed as

$$\gamma = j\beta = j\omega(\sqrt{\mu\varepsilon}) \qquad ...(3.37)$$

The intrinsic impedance of the dielectric is

$$\eta = \frac{j\omega\mu}{\gamma} = \frac{j\omega\mu}{j\beta} = \frac{\omega\mu}{\omega\sqrt{\mu\varepsilon}} = \sqrt{\frac{\mu}{\varepsilon}} = \eta_0\sqrt{\frac{\mu_r}{\varepsilon_r}} \qquad ...(3.38)$$

A brief summary of the three fields of incidence, reflection and transmission is presented in Table 3.2 for the case of normal incidence at the boundary of two lossless media.

Table 3.2: A comprehensive summary of fields in two lossless dielectric regions corresponding to normal incidence

Region I (lossless medium 1)	***Region II (lossless medium 2)***
$\mathbf{E}_i = E_0 e^{-j\beta_1 z}\hat{\mathbf{y}}$, $\mathbf{H}_i = -\frac{1}{\eta_1}E_0 e^{-j\beta_1 z}\hat{\mathbf{x}}$ $\mathbf{E}_r = \Gamma E_0 e^{j\beta_1 z}\hat{\mathbf{y}}$, $\mathbf{H}_r = \frac{\Gamma}{\eta_1}E_0 e^{j\beta_{11} z}\hat{\mathbf{x}}$	$\mathbf{E}_t = \tau E_0 e^{-j\beta_2 z}\hat{\mathbf{y}}$, $\mathbf{H}_t = -\frac{\tau}{\eta_2}E_0 e^{-j\beta_2 z}\hat{\mathbf{x}}$

A comparison of the fields presented in Table 3.1 and 3.2 shows us that they look very similar except for the propagation vector. For a lossless medium, η is real; so both Γ and τ are real. The electric and magnetic fields are in phase with each other. The wavelength in the dielectric is

$$\lambda = \frac{2\pi}{\beta} = \frac{2\pi}{\omega\sqrt{\mu\varepsilon}} \qquad ...(3.39)$$

and the corresponding phase velocity is

$$v_p = \frac{\omega}{\beta} = \frac{1}{\sqrt{\mu\varepsilon}} = \frac{c}{\sqrt{\mu_r\varepsilon_r}} \qquad ...(3.40)$$

Applying the continuity of the tangential components of fields across the boundary, the reflection coefficient and the transmission coefficient are obtained for this case as follows

$$\Gamma = \frac{\eta_2 - \eta_1}{\eta_2 + \eta_1} \qquad ...(3.41)$$

and

$$\tau = \frac{2\eta_2}{\eta_2 + \eta_1} \qquad ...(3.42)$$

The expression (3.35) giving the reflection coefficients corresponding to the lossy conducting medium and (3.41) for the lossless medium are identical in form. However, the values of the intrinsic impedances are calculated very differently for these two cases. The intrinsic impedances are real numbers given as

$$\eta = \eta_0 \sqrt{\frac{\mu_r}{\varepsilon_r}}$$

for the computation of these two coefficients as given by the above expressions.

Power conservation for both the cases: The Poynting vector for the incident wave in region I is

$$P_{I,i} = \mathbf{E}_i \times \mathbf{H}_i^* = E_0 e^{-\gamma_1 z}\hat{\mathbf{y}} \times -\frac{1}{\eta_1} E_0^* e^{(-\gamma_1 z)^*}\hat{\mathbf{x}}$$

$$= \frac{1}{\eta_1}|E_0|^2 e^{-2\alpha_1 z}\hat{\mathbf{z}} \qquad ...(3.43)$$

Similarly, The Poynting vector for the reflected wave in region I is

$$P_{I,r} = \mathbf{E}_r \times \mathbf{H}_r^* = \Gamma E_0 e^{\gamma_1 z}\hat{\mathbf{y}} \times \frac{\Gamma^*}{\eta_1} E_0^* e^{(\gamma_1 z)^*}\hat{\mathbf{x}}$$

$$= -\frac{1}{\eta_1}|\Gamma|^2 |E_0|^2 e^{2\alpha_1 z}\hat{\mathbf{z}} \qquad ...(3.44)$$

The total power of the wave in region I is therefore,

$$P_{II,i} = \frac{1}{\eta_1}|E_0|^2 e^{-2\gamma_1 z}\hat{\mathbf{z}} - \frac{1}{\eta_1}|\Gamma|^2 |E_0|^2 e^{2\gamma_1 z}\hat{\mathbf{z}}$$

$$= \left(\frac{1-\Gamma^2}{\eta_1}\right)|E_0|^2 (e^{-2\alpha_1 z} - e^{2\alpha_1 z})\hat{\mathbf{z}} \qquad ...(3.45)$$

The total power of the wave in region II is,

$$P_{II,t} = \mathbf{E}_t \times \mathbf{H}_t^* = \tau E_0 e^{-\gamma_2 z}\hat{\mathbf{y}} \times -\frac{\tau}{\eta_2} E_0^* e^{(-\gamma_2 z)^*}\hat{\mathbf{x}}$$

$$= \frac{1}{\eta_2}|\tau|^2 |E_0|^2 e^{-2\alpha_2 z}\hat{\mathbf{z}} \qquad ...(3.46)$$

Now, $$1-\Gamma^2 = 1-\left(\frac{\eta_2-\eta_1}{\eta_2+\eta_1}\right)^2 = \frac{4\eta_1\eta_2}{(\eta_2+\eta_1)^2} \qquad ...(3.47)$$

$$\Rightarrow \qquad \frac{1-\Gamma^2}{\eta_1} = \frac{4\eta_2}{(\eta_2+\eta_1)^2}$$

We have used (3.35) in (3.47).

Similarly,

$$\tau^2 = \left[\frac{2\eta_2}{(\eta_1+\eta_2)}\right]^2 = \frac{4\eta_2^2}{(\eta_1+\eta_2)^2} \qquad ...(3.48)$$

$$\Rightarrow \qquad \frac{\tau^2}{\eta_2} = \frac{4\eta_2}{(\eta_1+\eta_2)^2} = \frac{1-\Gamma^2}{\eta_1}$$

(3.36) has been used in (3.48).

3.4.3. Reflection by a Perfect Conductor-Normal Incidence

When an electromagnetic wave travelling in one medium impinges upon a second medium having a different dielectric constant, permeability, or conductivity, the wave in general will be partially transmitted and partially reflected. In the case of a plane wave in air incident normally upon the surface of a perfect conductor, the wave is entirely reflected. For fields that vary with time neither **E** nor **H** can exist within a perfect conductor so that none of the energy of the incident wave can be transmitted. Since there can be no loss within a perfect conductor, none of the energy is absorbed. As a result the amplitude of **E** and **H** in the reflected wave are the same as in the incident wave, and the only difference is in the direction of power flow. Now for a wave travelling in the positive axis, we write the incident wave as

$$E_i e^{-j\beta z} \qquad ...(3.49)$$

and the surface of the perfect conductor is taken to be the $z = 0$ plane. The expression for the reflected wave is

$$E_r e^{j\beta z} \qquad ...(3.50)$$

where E_r must be determined from the boundary conditions. In as much as the tangential component of **E** must be continuous across the boundary and **E** is zero within the conductor, the tangential component of **E** just outside the conductor must also be zero. This requires that the sum of the electric field strengths in the incident and reflected waves add to give zero resultant field strength in the plane $z = 0$. Therefore,

$$E_r = -E_i \qquad ...(3.51)$$

The amplitude of the reflected electric field strength is equal to that of the incident electric field strength, but its phase has been reversed on reflection. The resultant electric field at any point a distance $-z$ from the $z = 0$ plane is the sum of the field strengths of the incident and the reflected waves at that point and is given by

$$E_T(z) = E_i e^{-j\beta z} + E_r e^{j\beta z} = E_i(e^{-j\beta z} - e^{j\beta z}) = -2jE_i \sin \beta z \qquad ...(3.52)$$

Therefore,

$$E_T(z, t) = \text{Re}\{-2jE_i \sin\beta z e^{j\omega t}\} \qquad ...(3.53)$$

If E_i is chosen to be real,

$$E_T(z, t) = 2E_i \sin \beta z \sin \omega t \qquad ...(3.54)$$

This equation (3.54) indicates that the incident and reflected waves combine to produce a standing wave, which does not progress. The magnitude of the electric field varies sinusoidally with distance from the reflecting plane. It is zero at the surface and at multiples of half wavelength from the surface. It has a maximum value of twice the electric field strength of the incident wave at distances from the surface that are odd multiples of a quarter wavelength.

The magnetic field strength must be reflected at the surface without reversal of phase, however. If both magnetic and electric field strengths were reversed, there would be no reversal of direction of energy propagation which is required in this case. Therefore, the phase of the reflected magnetic field strength H_r is the same as that of the incident magnetic field H_i at the surface of reflection $z = 0$. The expression for the resultant magnetic field is

$$H_T(z) = H_i e^{-j\beta z} + H_r e^{j\beta z} = H_i(e^{-j\beta z} + e^{j\beta z}) = -2H_i \cos \beta z \qquad ...(3.55)$$

H_i is real since it is in phase with E_i.

$$H_T(z, t) = \text{Re}\{H_T(z)e^{j\omega t}\} = 2H_i \cos\beta z \cos\omega t \qquad ...(3.56)$$

The resultant magnetic field also has a standing wave distribution. In this case, however, it has maximum value at the surface of the conductor and at multiples of half wavelength from the surface, whereas the zero points occur at odd multiples of a quarter wavelength from the surface. From the boundary conditions for **H** it follows that there must be a surface current of J_s A/m such that $J_s = H_T$ at $(z = 0)$.

Since E_i and H_i were in time phase in the incident plane wave, a comparison of (3.54) and (3.56) shows that E_T and H_T are 90° out of time phase because of the factor j in (3.53). This is as it should be, for it indicates no average flow of power. This is the case when the energy transmitted in the forward direction is equaled by that reflected back.

That E_T and H_T are 90° apart in time phase is seen more clearly by rewriting (3.53) and (3.55). Replacing $-j$ by its equivalent $e^{j\pi/2}$ and combining this with $e^{j\omega t}$ term to give $e^{j(\omega t-\pi/2)}$, (3.53) becomes

$$E_T(z, t) = \text{Re}\{2E_i \sin \beta z e^{-j\pi/2}\, e^{j\omega t}\} = 2E_i \sin \beta z \cos(\omega t - \pi/2) \quad ...(3.57)$$

Similarly, (3.55) can be written as

$$H_T(z, t) = \text{Re}\{2H_i \cos\beta z e^{-j\pi/2} e^{j\omega t}\} = 2H_i \cos\beta z \cos\omega t \quad ...(3.58)$$

We observe from (3.57) and (3.58) that E_T and H_T differ in time phase by 90°.

3.5. PLANE WAVE REFRACTION FROM MEDIA INTERFACE AT OBLIQUE INCIDENCE

Here we consider the problem of a plane wave incident obliquely at the boundary of two different media. Two particular cases are considered;

(a) The electric field is in the *xz*-plane (parallel polarization) and
(b) The electric field is normal to the *xz*-plane (perpendicular polarization)

Any arbitrary incident plane wave can be expressed as a linear combination of these two principal polarizations. The reflection and transmission coefficients for oblique incidence are also known as Fresnel coefficients. The plane of incidence is the plane containing the normal vector to the plane and the direction of propagation of the incident wave. For parallel polarization, the electric field has a longitudinal component; a component along the direction of propagation. Such types of waves are known as transverse magnetic (TM) waves, as the magnetic field is orthogonal to the direction of propagation. The electric field is entirely transverse in case of perpendicular polarization. This is called a TE wave.

3.5.1. Perpendicular Polarization (TE Wave)

Let us assume that the plane wave propagates in the first quadrant of the *xz*-plane without any loss of generality. As the name suggests the electric field vector is perpendicular to the plane which means it has a component along the y-axis.

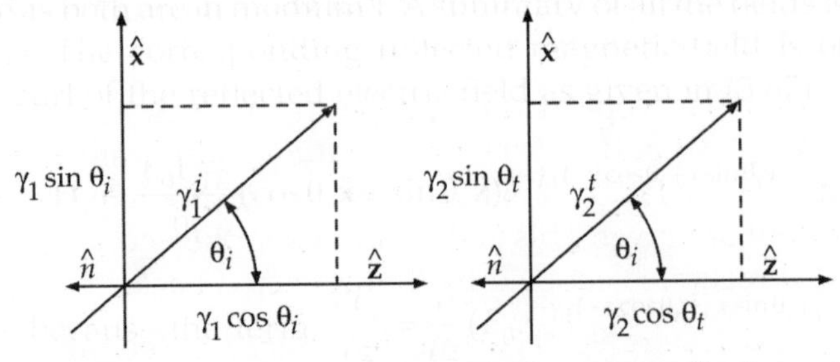

Fig. 3.5: Orientation of the electric fields for perpendicular polarization

Let the propagation vector in medium I γ_1^i make an angle of θ_i with the normal as shown in Fig. 3.5. We, therefore, write

$$\gamma_1^i \cdot \mathbf{z}' = (\gamma_1 \cos\theta_i \hat{\mathbf{z}} + \gamma_1 \sin\theta_i \hat{\mathbf{x}})(z\hat{\mathbf{z}} + x\hat{\mathbf{x}}) = z\gamma_1 \cos\theta_i$$
$$+x\gamma_1 \sin\theta_i = \gamma_1(z\cos\theta_i + x\sin\theta_i) \qquad ...(3.59)$$

As the electric field is entirely transverse to the plane of propagation, let it be expressed as

$$E_i = E_0 e^{-\gamma_1(z\cos\theta_i + x\sin\theta_i)}\hat{\mathbf{y}} \qquad ...(3.60)$$

The electric field for such a wave is transverse to the plane containing the propagation vector. No tangential component of the electric field exists in this propagation plane. Such a wave is known as "Transverse Electric" (TE) wave.

The corresponding magnetic field becomes

$$\nabla \times \mathbf{E}_i = -j\omega\mu_1 \mathbf{H}_i \Rightarrow \mathbf{H}_i = \frac{\nabla \times \mathbf{E}_i}{-j\omega\mu_1} \qquad ...(3.61)$$

As the electric field has a y-component only, we expect to have both the x- and z- components for the magnetic field. Let us verify our intuition.

$$\mathbf{H}_i = \frac{1}{-j\omega\mu_1}\begin{vmatrix} \hat{\mathbf{x}} & \hat{\mathbf{y}} & \hat{\mathbf{z}} \\ \frac{\partial}{\partial x} & \frac{\partial}{\partial y} & \frac{\partial}{\partial z} \\ 0 & E_i & 0 \end{vmatrix} \qquad ...(3.62)$$

$$= -\frac{1}{j\omega\mu_1}\left\{\left[-\left(-\frac{\partial}{\partial z}E_i\right)\right]\hat{\mathbf{x}} + \left(\frac{\partial}{\partial x}E_i\right)\hat{\mathbf{z}}\right\}$$

Now,

$$\frac{\partial E_i}{\partial z} = \frac{\partial}{\partial z}(E_0 e^{-\gamma_1(z\cos\theta_i + x\sin\theta_i)})$$

$$= -\gamma_1 \cos\theta_i E_0 e^{-\gamma_1(z\cos\theta_i + x\sin\theta_i)} \quad ...(3.63)$$

and

$$\frac{\partial E_i}{\partial x} = \frac{\partial}{\partial x}(E_0 e^{-\gamma_1(z\cos\theta_i + x\sin\theta_i)})$$

$$= -\gamma_1 \sin\theta_i E_0 e^{-\gamma_1(z\cos\theta_i + x\sin\theta_i)} \quad ...(3.64)$$

Hence substitution of (3.63) and (3.64) in (3.61) gives us

$$\mathbf{H}_i = -\frac{1}{j\omega\mu_1}\left\{\begin{matrix}\left[-(-\gamma_1 \cos\theta_i E_0 e^{-\gamma_1(z\cos\theta_i + x\sin\theta_i)})\right]\hat{\mathbf{x}} \\ +(-\gamma_1 \sin\theta_i E_0 e^{-\gamma_1(z\cos\theta_i + x\sin\theta_i)})\hat{\mathbf{z}}\end{matrix}\right\}$$

$$= \frac{\gamma_1}{j\omega\mu_1} E_0(-\cos\theta_i\hat{\mathbf{x}} + \sin\theta_i\hat{\mathbf{z}})e^{-\gamma_1(z\cos\theta_i + x\sin\theta_i)}$$

$$= \frac{E_0}{\eta_1}(-\cos\theta_i\hat{\mathbf{x}} + \sin\theta_i\hat{\mathbf{z}})e^{-\gamma_1(z\cos\theta_i + x\sin\theta_i)} \quad ...(3.65)$$

Let us assume that the reflected wave is contained in the second quadrant of the *xz*-plane as shown and the reflected propagation vector $\gamma_1{}^r$ makes an angle of θ_r with the normal as shown in Fig. 3.6. Hence,

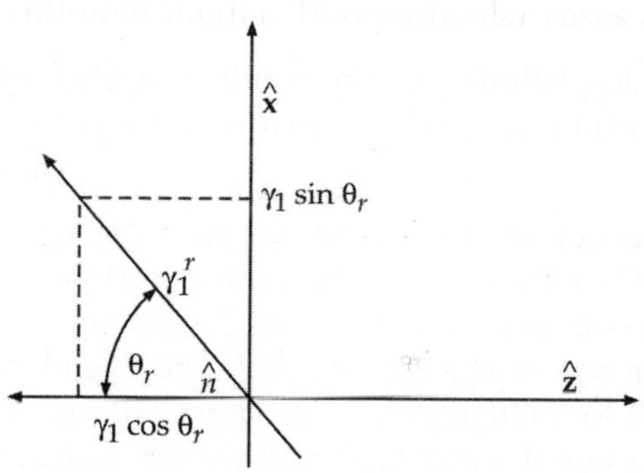

Fig. 3.6: The field components corresponding to the reflected wave of Fig. 3.5

$$\gamma_1^r \cdot \mathbf{z}' = (-\gamma_1 \cos\theta_r\hat{\mathbf{z}} + \gamma_1 \sin\theta_r\hat{\mathbf{x}})(z\hat{\mathbf{z}} + x\hat{\mathbf{x}})$$

$$= -z\gamma_1 \cos\theta_r + x\gamma_1 \sin\theta_r = \gamma_1(-z\cos\theta_r + x\sin\theta_r) \quad ...(3.66)$$

Thus, the reflected electric field is given by

$$\mathbf{E}_r = \Gamma_{TE} E_0 e^{-\gamma_1(-z\cos\theta_r + x\sin\theta_r)}\hat{\mathbf{y}} \quad ...(3.67)$$

The use of Γ_{TE} in (3.67) is simply so state that we are looking at the reflection coefficient of a *TE* wave. We note that $\gamma_1{}^i$ and $\gamma_1{}^r$ have the same magnitude as both are in medium I. A summary of all the fields is presented in Table 3.3. The corresponding reflected magnetic field is obtained by taking the curl of the reflected electric field as given in (3.67).

$$\mathbf{H}_r = \frac{E_0\Gamma_{TE}}{\eta_1}(\cos\theta_r\hat{\mathbf{x}} + \sin\theta_r\hat{\mathbf{z}})e^{-\gamma_1(-z\cos\theta_i + x\sin\theta_i)} \qquad ...(3.68)$$

This is because the term $\frac{\partial E_r}{\partial z} = \frac{\partial}{\partial z}(E_0 e^{-\gamma_1(-z\cos\theta_r +_r x\sin\theta_r)}) = \gamma_1 \cos\theta_r$ $E_0 e^{-\gamma_1(-z\cos\theta_r +_r x\sin\theta_{ri})}$ and the term $\frac{\partial E_r}{\partial x}$ remains the same as the incident field. Hence, we get a sign change in the first term corresponding to the $\hat{\mathbf{x}}$ term only.

Next we derive the expressions for the transmitted field as experienced by the second medium. As we might guess, the amplitude of the electric field would be multiplied by the transmission coefficient τ_{TE}. Hence, we write

$$\mathbf{E}_t = \tau_{TE}E_0 e^{-\gamma_2(z\cos\theta_t + x\sin\theta_t)}\hat{\mathbf{y}} \qquad ...(3.69)$$

$$\Rightarrow \mathbf{H}_t = -\frac{\nabla\times\mathbf{E}_t}{j\omega\mu_2} = \frac{\tau_{TE}E_0}{\eta_2}e^{-\gamma_2(z\cos\theta_t + x\sin\theta_t)}(-\cos\theta_t\hat{\mathbf{x}} + \sin\theta_t\hat{\mathbf{z}})$$

Equating the tangential components of the electric field and the magnetic field at the boundary of $z = 0$, we get some interesting results. We note that, at this interface, the tangential component of the electric field has the y-component only whereas the magnetic field has the x-component. Hence, continuity of the tangential components at the media boundary requires that,

$$e^{-\gamma_1 x\sin\theta_i} + \Gamma_{TE}e^{-\gamma_1 x\sin\theta_r} = \tau_{TE}e^{-\gamma_{12} x\sin\theta_t} \text{ (for the electric field)} \qquad ...(3.70)$$

and

The components of the magnetic fields at the boundary require that

$$-\frac{1}{\eta_1}\cos\theta_i e^{-\gamma_1 x\sin\theta_i} + \frac{\Gamma_{TE}}{\eta_1}\cos\theta_r e^{-\gamma_1 x\sin\theta_r} = -\frac{\tau_{TE}}{\eta_2}\cos\theta_t e^{-\gamma_2 x\sin\theta_t} \qquad ...(3.71)$$

As E_x and E_y are continuous at the interface $z = 0$ for all x, this x variation must remain the same on both sides of the boundary. Hence,

$$\gamma_1 \sin\theta_i = \gamma_1 \sin\theta_r = \gamma_2 \sin\theta_t \qquad ...(3.72)$$

$$\Rightarrow \qquad \theta_i = \theta_r$$

Also, $$\gamma_1 \sin\theta_i = \gamma_2 \sin\theta_t \qquad ...(3.73)$$

The first result given in (3.72) is the familiar law of reflection for plane mirrors which states that the angle of incidence is equal to the angle of reflection. Moreover, the incident ray (wave here) and the reflected wave must lie in the same plane. This plane is the xz-plane in the present case.

The second law in (3.73) represents the much familiar Snell's law of refraction (refraction or transmission of the wave to be synonymously used here). Refraction is the change in the direction of the wave due to the change in velocity of the wave in moving from one medium to other. We further note that, the refractive index n of the second medium relative to the first medium is defined as

$$n = \frac{c}{v_p} = \sqrt{\frac{1}{\mu_0\varepsilon_0} \div \frac{1}{\mu_r\varepsilon_r\mu_0\varepsilon_0}} = \sqrt{\mu_r\varepsilon_r} \qquad ...(3.74)$$

For a lossless dielectric media interface, from (3.73),

$$\frac{\sin\theta_i}{\sin\theta_t} = \frac{\gamma_2}{\gamma_1} = \frac{j\beta_2}{j\beta_1} = \sqrt{\frac{\mu_2\varepsilon_2}{\mu_1\varepsilon_1}} = \frac{v_1}{v_2} = \sqrt{\frac{\varepsilon_2}{\varepsilon_1}} = \frac{n_2}{n_1} \qquad ...(3.75)$$

$$\because \qquad \mu_2 = \mu_1$$

Table 3.3: A comprehensive summary of fields in two lossless dielectric regions corresponding to oblique incidence (TE waves)

Region 1 (Lossy Medium 1)	*Region 2 (Lossy Medium 2)*
$\mathbf{E}_i = E_0 e^{-\gamma_1(z\cos\theta_i + x\sin\theta_i)}\hat{\mathbf{y}}$	$\mathbf{E}_t = \tau_{TE}E_0 e^{-\gamma_2(z\cos\theta_t + x\sin\theta_t)}\hat{\mathbf{y}}$ $\mathbf{H}_t = \frac{\tau_{TE}E_0}{\eta_2} e^{-\gamma_2(z\cos\theta_t + x\sin\theta_t)}$ $(-\cos\theta_t\hat{\mathbf{x}} + \sin\theta_t\hat{\mathbf{z}})$
$\mathbf{H}_i = \frac{E_0}{\eta_1}(-\cos\theta_i\hat{\mathbf{x}} + \sin\theta_i\hat{\mathbf{z}})e^{-\gamma_1(z\cos\theta_i + x\sin\theta_i)}$ $\mathbf{E}_r = \Gamma_{TE}E_0 e^{-\gamma_1(-z\cos\theta_r + x\sin\theta_r)}\hat{\mathbf{y}}$ $\mathbf{H}_r = \frac{E_0\Gamma_{TE}}{\eta_1}(\cos\theta_r\hat{\mathbf{x}} + \sin\theta_r\hat{\mathbf{z}})$ $e^{-\gamma_1(-z\cos\theta_i + x\sin\theta_i)}$	

It is obvious from the above that, if the permittivity of the second medium is higher than that of the first medium, then,

$$\sin\theta_i > \sin\theta_t \Rightarrow \theta_i > \theta_t \Rightarrow \theta_t < \theta_i \qquad ...(3.76)$$

The angle of refraction is less than the angle of incidence. This further means that the refracted ray bends more toward the normal (drawn at the interface) when a wave moves from a rarer medium to a denser medium. The wave velocity decreases in the denser medium. For the reverse case, we note that $\sin\theta_i > \sin\theta_t \Rightarrow \theta_i > \theta_t \Rightarrow \theta_t < \theta_i$ making the wave bend away from the normal. The wave velocity increases in the rarer medium in moving from the denser medium. This is the verification of the much familiar Snell's law.

3.5.2. Reflection and Transmission Coefficient for Perpendicular polarization

We next proceed to derive appropriate expressions for the reflection coefficient and the transmission coefficient for the case of perpendicular polarization. Use of Snell's law gives us

$$1 + \Gamma_{TE} = \tau_{TE} \qquad \text{...(3.77)}$$

from (3.70)

$$-\frac{\cos\theta_i}{\eta_1} + \Gamma_{TE}\frac{\cos\theta_r}{\eta_1} = -\frac{\tau_{TE}}{\eta_2}\cos\theta_t \qquad \text{...(3.78)}$$

from (3.71)

The above two equations (3.77) and (3.78) has two unknowns Γ_{TE} and τ_{TE}. Hence to solve for the two unknowns, we write

$$1+\Gamma_{TE} = \tau_{TE} \text{ and } \tau_{TE} = \frac{\eta_2}{\cos\theta_t}\left(\frac{\cos\theta_i}{\eta_1} - \Gamma_{TE}\frac{\cos\theta_r}{\eta_1}\right)$$

$$\Rightarrow \quad 1+\Gamma_{TE} = \frac{\eta_2\cos\theta_i}{\eta_1\cos\theta_t} - \Gamma_{TE}\frac{\eta_2\cos\theta_r}{\eta_1\cos\theta_t}$$

$$\Rightarrow \quad \Gamma_{TE}\left(1+\frac{\eta_2\cos\theta_r}{\eta_1\cos\theta_t}\right) = \frac{\eta_2\cos\theta_i}{\eta_1\cos\theta_t} - 1 \qquad \text{...(3.79)}$$

$$\Rightarrow \quad \Gamma_{TE}\left(\frac{\eta_1\cos\theta_t + \eta_2\cos\theta_r}{\eta_1\cos\theta_t}\right) = \frac{\eta_2\cos\theta_i - \eta_1\cos\theta_t}{\eta_1\cos\theta_t}$$

$$\Rightarrow \quad \Gamma_{TE} = \frac{\eta_2\cos\theta_i - \eta_1\cos\theta_t}{\eta_1\cos\theta_t + \eta_2\cos\theta_r}$$

The transmission coefficient is derived by using the following

$$1 + \Gamma_{TE} = \tau_{TE}$$

$$= 1 + \frac{\eta_2 \cos\theta_i - \eta_1 \cos\theta_t}{\eta_1 \cos\theta_t + \eta_2 \cos\theta_r} \qquad \text{...(3.80)}$$

$$= \frac{\eta_1 \cos\theta_t + \eta_2 \cos\theta_r + \eta_2 \cos\theta_i - \eta_1 \cos\theta_t}{\eta_1 \cos\theta_t + \eta_2 \cos\theta_r}$$

$$= \frac{2\eta_2 \cos\theta_i}{\eta_1 \cos\theta_t + \eta_2 \cos\theta_i} \qquad \because \theta_i = \theta_r$$

Thus, we obtain a set of elegant results for the reflection coefficient and the transmission coefficient corresponding to perpendicular polarization (*TE* waves) as the following

$$\Gamma_{TE} = \frac{\eta_2 \cos\theta_i - \eta_1 \cos\theta_t}{\eta_1 \cos\theta_t + \eta_2 \cos\theta_r} \qquad \text{...(3.81)}$$

$$\tau_{TE} = \frac{2\eta_2 \cos\theta_i}{\eta_1 \cos\theta_t + \eta_2 \cos\theta_i} \qquad \text{...(3.82)}$$

These coefficients are also known as Fresnel's coefficients.

Example 3.4: Let us consider a waveguide whose bounding surfaces are made of perfect conductors so that E_y may be assumed to vanish at $x = \pm\frac{d}{2}$ as shown in Fig. 3.7. Consider the *TE* modes and show that

$$\beta_m^2 = k_0^2 n_1^2 - \left(\frac{m\pi}{d}\right)^2$$

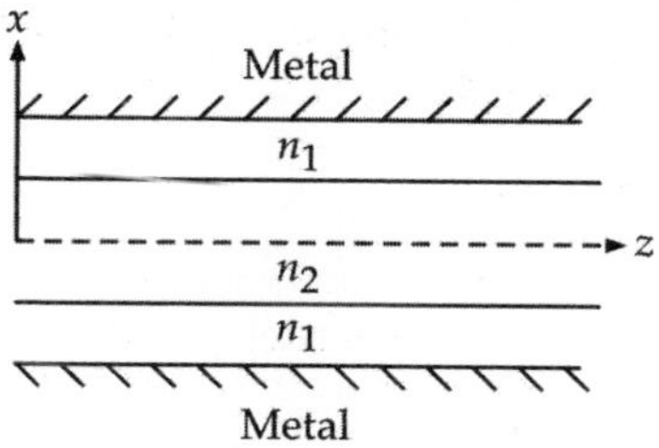

Fig. 3.7: Diagram for Example 3.4.

Solution: Continuity of the boundary conditions require that the tangential components of the electric field must be continuous across the boundary. Hence,

$$E_y(x) = \begin{cases} A\cos\kappa x\,; |x| < -\dfrac{d}{2} \\ C\exp\left[-\gamma|x|\right]\,; |x| > \dfrac{d}{2} \end{cases}$$

At the boundary we have

$$E_y(x) = A\cos\kappa\left(\frac{d}{2}\right) = 0$$

$$\Rightarrow \quad \kappa\left(\frac{d}{2}\right) = (2m+1)\frac{\pi}{2}$$

$$\Rightarrow \quad \kappa d = (2m+1)\pi$$

$$\Rightarrow \quad \sqrt{k_0^2 n_1^2 - \beta_m^2}\,d = (2m+1)\pi$$

$$\Rightarrow \quad k_0^2 n_1^2 - \beta_m^2 = \left[\frac{(2m+1)\pi}{d}\right]^2$$

$$\Rightarrow \quad \beta_m^2 = k_0^2 n_1^2 - \left[\frac{(2m+1)\pi}{d}\right]^2$$

for the symmetric modes.

For the antisymmetric modes

$$E_y(x) = \begin{cases} B\sin\kappa x\,; |x| < -\dfrac{d}{2} \\ \dfrac{x}{|x|} D\exp\left[-\gamma|x|\right]\,; |x| > \dfrac{d}{2} \end{cases}$$

At the boundary we have

$$E_y(x) = B\sin\kappa\left(\frac{d}{2}\right) = 0$$

$$\Rightarrow \quad \kappa\left(\frac{d}{2}\right) = m\frac{\pi}{2}; \quad m = 2n \text{ (an even integer)}$$

$$\Rightarrow \quad \kappa d = m\pi$$

$$\Rightarrow \quad \sqrt{k_0^2 n_1^2 - \beta_m^2}\,d = m\pi$$

$$\Rightarrow \qquad k_0^2 n_1^2 - \beta_m^2 = \left(\frac{m\pi}{d}\right)^2$$

$$\Rightarrow \qquad \beta_m^2 = k_0^2 n_1^2 - \left(\frac{m\pi}{d}\right)^2$$

3.5.3. Parallel Polarization (TM Waves)

We next consider the case of parallel polarization. To make a study of this, we assume the incident wave is propagating in the first quadrant of the *xz*-plane without the loss of generality and $\gamma_1{}^i$ makes an angle of θ_i with the normal. As the name suggests, the electric field vector is parallel to the *xz*-plane; it lies in it. However, the magnetic field is transverse to the plane of incidence; it has a *y*-directed component. As we may observe such a wave is termed transverse magnetic (TM) wave.

Hence,

$$\begin{aligned} \gamma_1^i \cdot \mathbf{z}' &= (\gamma_1 \cos\theta_i \hat{\mathbf{z}} + \gamma_1 \sin\theta_i \hat{\mathbf{x}}).(z\hat{\mathbf{z}} + x\hat{\mathbf{x}}) \\ &= z\gamma_1 \cos\theta_i + x\gamma_1 \sin\theta_i = \gamma_1(z\cos\theta_i + x\sin\theta_i) \end{aligned} \qquad ...(3.83)$$

To analyze the wave, we start by writing the expression for the magnetic field vector which is oriented along the positive *y*-direction. Thus,

$$\mathbf{H}_i = \frac{E_0}{\eta_1} e^{-\gamma_1(z\cos\theta_i + x\sin\theta_i)} \hat{\mathbf{y}} \qquad ...(3.84)$$

$$\nabla \times \mathbf{H}_i = j\omega\varepsilon_1 \mathbf{E}_i \Rightarrow \mathbf{E}_i = \frac{\nabla \times \mathbf{H}_i}{j\omega\varepsilon_1} \qquad ...(3.85)$$

Expansion of (3.85) gives

$$\nabla \times \mathbf{H}_i = \begin{vmatrix} \hat{\mathbf{x}} & \hat{\mathbf{y}} & \hat{\mathbf{z}} \\ \dfrac{\partial}{\partial x} & \dfrac{\partial}{\partial y} & \dfrac{\partial}{\partial z} \\ H_x & H_y & H_z \end{vmatrix} = j\omega\varepsilon_1 \mathbf{E}_i \qquad ...(3.86)$$

$$\mathbf{E}_i = \frac{1}{j\omega\varepsilon_1} \begin{vmatrix} \hat{\mathbf{x}} & \hat{\mathbf{y}} & \hat{\mathbf{z}} \\ \dfrac{\partial}{\partial x} & \dfrac{\partial}{\partial y} & \dfrac{\partial}{\partial z} \\ 0 & \dfrac{E_0}{\eta_1} e^{-\gamma_1(z\cos\theta_i + x\sin\theta_i)} & 0 \end{vmatrix}$$

$$= \frac{E_0}{j\omega\varepsilon_1\eta_1}\left[\begin{array}{l}\left\{-\frac{\partial}{\partial z}\left(\frac{E_0}{\eta_1}e^{-\gamma_1(z\cos\theta_i + x\sin\theta_i)}\right)\right\}\hat{\mathbf{x}} \\ +\left\{\frac{\partial}{\partial x}\left(\frac{E_0}{\eta_1}e^{-\gamma_1(z\cos\theta_i + x\sin\theta_i)}\right)\right\}\hat{\mathbf{z}}\end{array}\right]$$

$$= \frac{E_0}{j\omega\varepsilon_1\eta_1}\left[-(-\gamma_1\cos\theta_i)\hat{\mathbf{x}} + (-\gamma_1\sin\theta_i)\hat{\mathbf{z}}\right]e^{-\gamma_1(z\cos\theta_i + x\sin\theta_i)}$$

$$= \frac{E_0 e^{-\gamma_1(z\cos\theta_i + x\sin\theta_i)}\gamma_1}{j\omega\varepsilon_1\eta_1}(\cos\theta_i\hat{\mathbf{x}} - \sin\theta_i\hat{\mathbf{z}}) \quad \text{...(3.87)}$$

Further, we have

$$\frac{\gamma_1}{j\omega\varepsilon_1\eta_1} = \frac{j\omega\mu_1/\eta_1}{j\omega\varepsilon_1\eta_1} = \frac{\mu_1}{\varepsilon_1\eta_1^2} = \frac{\eta_1^2}{\eta_1^2} = 1$$

Thus, the incident electric field has a form that is given by

$$\mathbf{E}_i = \frac{E_0 e^{-\gamma_1(z\cos\theta_i + x\sin\theta_i)}\gamma_1}{j\omega\varepsilon_1\eta_1}(\cos\theta_i\hat{\mathbf{x}} - \sin\theta_i\hat{\mathbf{z}})$$

$$= E_0 e^{-\gamma_1(z\cos\theta_i + x\sin\theta_i)}(\cos\theta_i\hat{\mathbf{x}} - \sin\theta_i\hat{\mathbf{z}}) \quad \text{...(3.88)}$$

The electric field in (3.88) gives the following observations:

The peak amplitude of the field is E_0.

- This field lies in the second quadrant of the xz-plane, We expected the electric field to have longitudinal components. This is confirmed mathematically in (3.88).
- As the magnetic field is completely oriented along the positive y-axis, the curl of this field is expected to give us components in the positive x-axis and the negative z-axis. This is verified mathematically which has given us the electric field along two directions.
- The electric field has tangential components E_x and E_z.
- The three incident fields associated with a TM mode are, E_x, E_z and H_y that are mutually orthogonal to each other.

The reflected magnetic field is obtained by writing

$$\mathbf{H}_r = -\frac{\Gamma E_0}{\eta_1}e^{-\gamma_1(z\cos\theta_r + x\sin\theta_r)}\hat{\mathbf{y}} \quad \text{...(3.89)}$$

where θ_r is the angle of reflection. The negative sign in (3.89) for the reflected field justifies the fact that power flow for the reflected wave takes place along the direction of the propagation vector of the reflected wave. The reflected electric field is obtained from the reflected magnetic field by carrying out the procedure as outlined above for the incident field.

$$\mathbf{E}_r = \frac{1}{j\omega\varepsilon_1}\begin{vmatrix} \hat{\mathbf{x}} & \hat{\mathbf{y}} & \hat{\mathbf{z}} \\ \dfrac{\partial}{\partial x} & \dfrac{\partial}{\partial y} & \dfrac{\partial}{\partial z} \\ 0 & -\dfrac{\Gamma_{TM}E_0}{\eta_1}e^{-\gamma_1(-z\cos\theta_i + x\sin\theta_i)} & 0 \end{vmatrix}$$

$$= \frac{E_0}{j\omega\varepsilon_1\eta_1}\left[\begin{matrix} \left\{-\dfrac{\partial}{\partial z}\left(\dfrac{-\Gamma_{TM}E_0}{\eta_1}e^{-\gamma_1(-z\cos\theta_i + x\sin\theta_i)}\right)\right\}\hat{\mathbf{x}} \\ +\left\{\dfrac{\partial}{\partial x}\left(\dfrac{-\Gamma_{TM}E_0}{\eta_1}e^{-\gamma_1(-z\cos\theta_i + x\sin\theta_i)}\right)\right\}\hat{\mathbf{z}} \end{matrix}\right]$$

$$= -\frac{\Gamma_{TM}E_0}{j\omega\varepsilon_1\eta_1}[-(\gamma_1\cos\theta_r)\hat{\mathbf{x}} + (-\gamma_1\sin\theta_r)\hat{\mathbf{z}}]e^{-\gamma_1(-z\cos\theta_i + x\sin\theta_i)}$$

$$= \frac{\Gamma_{TM}E_0e^{-\gamma_1(-z\cos\theta_i + x\sin\theta_i)}\gamma_1}{j\omega\varepsilon_1\eta_1}(\cos\theta_r\hat{\mathbf{x}} + \sin\theta_r\hat{\mathbf{z}}) \qquad \text{...(3.90)}$$

The reflected electric field is given as

$$\mathbf{E}_r = \frac{\Gamma_{TM}E_0e^{-\gamma_1(-z\cos\theta_r + x\sin\theta_r)}\gamma_1}{j\omega\varepsilon_1\eta_1}(\cos\theta_r\hat{\mathbf{x}} + \sin\theta_r\hat{\mathbf{z}})$$

$$= \Gamma_{TM}E_0e^{-\gamma_1(-z\cos\theta_r + x\sin\theta_r)}(\cos\theta_r\hat{\mathbf{x}} + \sin\theta_r\hat{\mathbf{z}}) \qquad \text{...(3.91)}$$

The transmitted fields have similar expressions as the incident fields except for the replacement of the incident angle θ_i by θ_t (the angle that the transmitted propagation vector makes with the normal), γ_1 by γ_2 and multiplication of the amplitude of the fields by a factor τ_{TM}. Hence the transmitted or the refracted fields are expressed as

$$\mathbf{E}_t = \tau_{TM}E_0e^{-\gamma_2(-z\cos\theta_t + x\sin\theta_t)}(\cos\theta_t\hat{\mathbf{x}} - \sin\theta_t\hat{\mathbf{z}}) \qquad \text{...(3.92a)}$$

$$\mathbf{H}_t = \frac{\tau_{TM}E_0}{\eta_2}e^{-\gamma_2(z\cos\theta_t + x\sin\theta_t)}\hat{\mathbf{y}} \qquad \text{...(3.92b)}$$

3.5.3 Reflection and Transmission on Coefficients for TM waves

To establish the relationships between the various angles we start by noting that the tangential component of the electric field is continuous across the boundary. The tangential components of the electric field are E_x and E_z as noted previously. Hence, at the boundary, $z = 0$,

$$\mathbf{E}_i + \mathbf{E}_r = \mathbf{E}_t \quad \text{...(3.93)}$$

Substitution of the field expressions from (3.88), (3.90) and (3.92) into (3.93) gives us

$$e^{-\gamma_1 x \sin\theta_i} \cos\theta_i + \Gamma_{TM} e^{-\gamma_1 x \sin\theta_r} \cos\theta_r = \tau_{TM} e^{-\gamma_2 x \sin\theta_t} \cos\theta_t \quad \text{...(3.94)}$$

for the x – fields at $z = 0$

Similarly, equating the tangential components of the magnetic field at the interface,

$$\frac{e^{-\gamma_1 x \sin\theta_i}}{\eta_1} - \frac{\Gamma_{TM} e^{-\gamma_1 x \sin\theta_{ri}}}{\eta_1} = \frac{\tau_{TM}}{\eta_2} e^{-\gamma_2 x \sin\theta_t} \quad \text{...(3.95)}$$

Hence, continuity of the tangential components at the boundary requires that

$$-\gamma_1 x \sin\theta_i = -\gamma_1 x \sin\theta_r = -\gamma_2 x \sin\theta_t \quad \text{...(3.96)}$$

$$\Rightarrow \qquad \gamma_1 \sin\theta_i = \gamma_1 \sin\theta_r = \gamma_2 \sin\theta_t$$

Snell's law of refraction is represented by (3.96). Further,

$$\gamma_1 \sin\theta_i = \gamma_1 \sin\theta_r = \gamma_2 \sin\theta_t$$

$$\Rightarrow \qquad \gamma_1 \sin\theta_i = \gamma_2 \sin\theta_t \quad \text{...(3.97)}$$

$$\Rightarrow \qquad \frac{\sin\theta_i}{\sin\theta_t} = \frac{\gamma_2}{\gamma_1} = \frac{j\omega\mu_2}{\eta_2} \div \frac{j\omega\mu_1}{\eta_1} = \frac{\eta_1}{\eta_2}\frac{\mu_2}{\mu_1}$$

Moreover, from (3.94), we have

$$\cos\theta_i + \Gamma_{TM} \cos\theta_r = \tau_{TM} \cos\theta_t \quad \text{...(3.98)}$$

Also, (3.95) gives us

$$\frac{(1 - \Gamma_{TM})}{\eta_1} = \frac{\tau_{TM}}{\eta_2} \quad \text{...(3.99)}$$

We have two unknowns Γ_{TM} and τ_{TM} and two equations. Hence, from (3.98), we get

$$\tau_{TM} = \frac{\cos\theta_i + \Gamma_{TM} \cos\theta_r}{\cos\theta_t} \quad \text{...(3.100)}$$

From (3.99), we get

$$\tau_{TM} = \frac{\eta_2}{\eta_1}(1-\Gamma_{TM}) \qquad \text{...(3.101)}$$

Manipulation of (3.100) and (3.101) give us

$$\tau_{TM} = \frac{\cos\theta_i + \Gamma_{TM}\cos\theta_r}{\cos\theta_t} = \frac{\eta_2}{\eta_1}(1-\Gamma_{TM})$$

$$\Rightarrow \quad \frac{\eta_2}{\eta_1}(1-\Gamma_{TM}) = \frac{\cos\theta_i}{\cos\theta_t} + \frac{\Gamma_{TM}\cos\theta_r}{\cos\theta_t}$$

$$\Rightarrow \quad \Gamma_{TM}\left(-\frac{\eta_2}{\eta_1} - \frac{\cos\theta_r}{\cos\theta_t}\right) = \left(\frac{\cos\theta_i}{\cos\theta_t} - \frac{\eta_2}{\eta_1}\right) \qquad \text{...(3.102)}$$

$$\Rightarrow \quad \Gamma_{TM}\left(\frac{-\eta_2\cos\theta_t - \eta_1\cos\theta_r}{\eta_1\cos\theta_t}\right) = \left(\frac{\eta_1\cos\theta_i - \eta_2\cos\theta_t}{\eta_1\cos\theta_t}\right)$$

$$\Rightarrow \quad \Gamma_{TM} = \frac{\eta_2\cos\theta_t - \eta_1\cos\theta_i}{\eta_2\cos\theta_t + \eta_1\cos\theta_r}$$

The reflection coefficient corresponding to the TM mode is expressed as

$$\tau_{TM} = \frac{\eta_2}{\eta_1}(1-\Gamma_{TM}) = \frac{\eta_2}{\eta_1}\left(1 - \frac{\eta_2\cos\theta_t - \eta_1\cos\theta_i}{\eta_2\cos\theta_t + \eta_1\cos\theta_r}\right)$$

$$= \frac{\eta_2}{\eta_1}\left(\frac{\eta_2\cos\theta_t + \eta_1\cos\theta_r - \eta_2\cos\theta_t + \eta_1\cos\theta_i}{\eta_2\cos\theta_t + \eta_1\cos\theta_r}\right) \quad (\theta_i = \theta_r) \qquad \text{...(3.103)}$$

$$= \frac{\eta_2}{\eta_1} \cdot \frac{2\eta_1\cos\theta_i}{\eta_2\cos\theta_t + \eta_1\cos\theta_i}$$

$$= \frac{2\eta_2}{\eta_2\cos\theta_t + \eta_1\cos\theta_i}\cos\theta_i$$

We note that, for the case of normal incidence (when the wave strikes the interface along the normal to the plane of propagation)

$$\theta_i = \theta_r = 0 = \theta_t$$

Hence, the reflection coefficient as given in (3.103) becomes

$$\Gamma_{TM} = \frac{\eta_2 - \eta_1}{\eta_1 + \eta_2} \qquad ...(3.104)$$

and the transmission coefficient, similarly becomes

$$\tau_{TM} = \frac{2\eta_2}{\eta_1 + \eta_2} \qquad ...(3.105)$$

3.5.4. Brewster Angle

The Brewster angle is defined as the angle of incidence for which there occurs no reflection of the plane wave from the medium in which the wave is travelling. All of the wave is either transmitted into the other medium or absorbed by it. This implies that the reflection coefficient Γ is zero. We consider the cases of parallel polarization and perpendicular polarization separately.

Parallel Polarization (TM waves)

For parallel polarization, use of (3.102) gives us

$$\Gamma_{TM} = \frac{\eta_2 \cos\theta_t - \eta_1 \cos\theta_i}{\eta_2 \cos\theta_t + \eta_1 \cos\theta_i} = 0 \qquad ...(3.106)$$

$$\theta_i = \theta_B^{TM}$$

$$\Rightarrow \quad \eta_2 \cos\theta_t = \eta_1 \cos\theta_i$$

$$\Rightarrow \quad \eta_2 \cos\theta_t = \eta_1 \cos\theta_B^{TM} \qquad ...(3.107)$$

$$\Rightarrow \quad \cos\theta_t = 1 - \sin^2\theta_t$$

$$= \frac{\eta_1}{\eta_2} \cos\theta_B^{TM} = \sqrt{1 - \left(\frac{\gamma_1}{\gamma_2} \sin\theta_B^{TM}\right)^2}$$

The last expression follows from the continuity of the tangential components at the surface which gives us

$$\gamma_1 \sin\theta_B^{TM} = \gamma_2 \sin\theta_t \qquad ...(3.108)$$

$$\Rightarrow \quad \sin\theta_t = \frac{\gamma_1 \sin\theta_B^{TM}}{\gamma_2}$$

The last expression in (3.107) gives

$$\eta_1 \cos\theta_B^{TM} = \eta_2\sqrt{1-\left(\frac{\gamma_1}{\gamma_2}\sin\theta_B^{TM}\right)^2} = \eta_1\sqrt{1-\sin^2\theta_B^{TM}}$$

$$\Rightarrow \quad \eta_2^2\left(1-\frac{\gamma_1^2}{\gamma_2^2}\sin^2\theta_B^{TM}\right) = \eta_1^2\left(1-\sin^2\theta_B^{TM}\right)$$

$$\Rightarrow \quad \frac{1-\dfrac{\gamma_1^2}{\gamma_2^2}\sin^2\theta_B^{TM}}{1-\sin^2\theta_B^{TM}} = \frac{\eta_1^2}{\eta_2^2}$$

$$\Rightarrow \quad 1-\frac{\gamma_1^2}{\gamma_2^2}\sin^2\theta_B^{TM} = \frac{\eta_1^2}{\eta_2^2}(1-\sin^2\theta_B^{TM})$$

$$\Rightarrow \quad \left(\frac{\eta_1^2}{\eta_2^2}-\frac{\gamma_1^2}{\gamma_2^2}\right)\sin^2\theta_B^{TM} = \frac{\eta_1^2}{\eta_2^2}-1$$

$$\sin^2\theta_B^{TM} = \left[\frac{\dfrac{\eta_1^2}{\eta_2^2}-1}{\left(\dfrac{\eta_1^2}{\eta_2^2}-\dfrac{\gamma_1^2}{\gamma_2^2}\right)}\right] \qquad ...(3.109)$$

For lossless media, we get

$$\frac{\eta_1^2}{\eta_2^2} = \frac{\mu_1^2\gamma_2^2}{\mu_2^2\gamma_1^2} = \frac{\mu_1^2\mu_2\varepsilon_2}{\mu_2^2\mu_1\varepsilon_1} = \frac{\mu_1}{\mu_2}\frac{\varepsilon_2}{\varepsilon_1} \qquad ...(3.110)$$

Substitution of (3.11) in (3.109) leads to

$$\sin^2\theta_B^{TM} = \left[\frac{\dfrac{\mu_1\varepsilon_2}{\mu_2\varepsilon_1}-1}{\left(\dfrac{\mu_1\varepsilon_2}{\mu_2\varepsilon_1}-\dfrac{\mu_1\varepsilon_1}{\mu_2\varepsilon_2}\right)}\right] \qquad ...(3.111)$$

For a wave incident at the boundary between two magnetic media; $\varepsilon_1 = \varepsilon_2 = \varepsilon$.

Therefore (3.111) becomes

$$\sin^2\theta_B^{TM} = \left[\frac{\frac{\mu_1}{\mu_2}-1}{\left(\frac{\mu_1}{\mu_2}-\frac{\mu_1}{\mu_2}\right)}\right] = \frac{1}{0} \qquad ...(3.112)$$

The significance of (3.112) is that Brewster angle does not exist for magnetic media. However, at the boundary of two media consisting of dielectrics of different permittivities on both sides, we get the following

$$\sin^2\theta_B^{TM} = \left[\frac{\frac{\mu_1\varepsilon_2}{\mu_2\varepsilon_1}-1}{\frac{\mu_1\varepsilon_2}{\mu_2\varepsilon_1}-\frac{\mu_1\varepsilon_1}{\mu_2\varepsilon_2}}\right] = \left[\frac{\frac{\varepsilon_2}{\varepsilon_1}-1}{\frac{\varepsilon_2}{\varepsilon_1}-\frac{\varepsilon_1}{\varepsilon_2}}\right]$$

$$\Rightarrow \quad \frac{\varepsilon_2(\varepsilon_2-\varepsilon_1)}{\varepsilon_2^2-\varepsilon_1^2} = \frac{\varepsilon_2}{\varepsilon_1+\varepsilon_2} \qquad ...(3.113)$$

$$\Rightarrow \quad \sin\theta_B^{TM} = \sqrt{\frac{\varepsilon_2}{\varepsilon_1+\varepsilon_2}}$$

$$\Rightarrow \quad \tan\theta_B^{TM} = \sqrt{\frac{\varepsilon_2}{\varepsilon_1}} = \frac{n_2}{n_1}$$

Perpendicular Polarization (TE waves)

For perpendicular polarization that is responsible for giving the *TE* waves, we get from (3.81), However, for dielectric media,

$$\Gamma_{TE} = \frac{\eta_2\cos\theta_i - \eta_1\cos\theta_t}{\eta_1\cos\theta_t + \eta_2\cos\theta_r} = 0 \qquad ...(3.114)$$

Now, (3.114) become

$$\eta_2\cos\theta_B^{TE} - \eta_1\cos\theta_t = 0$$

$$\Rightarrow \quad (\eta_2\cos\theta_B^{TE})^2 = (\eta_1\cos\theta_t)^2 \qquad ...(3.115)$$

$$\Rightarrow \quad \left(\frac{\eta_2}{\eta_1}\right)^2\cos^2\theta_B^{TE} = \cos^2\theta_t$$

From (3.73), we have

$$\gamma_1 \sin\theta_B^{TE} = \gamma_2 \sin\theta_t$$

$$\Rightarrow \qquad (\gamma_1 \sin\theta_B^{TE})^2 = (\gamma_2 \sin\theta_t)^2 \qquad \text{...(3.116)}$$

$$\Rightarrow \qquad \left(\frac{\gamma_1 \sin\theta_B^{TE}}{\gamma_2}\right)^2 = \sin^2\theta_t$$

Combining (3.115) and (3.116) gives

$$\left(\frac{\eta_2}{\eta_1}\right)^2 \cos^2\theta_B^{TE} + \left(\frac{\gamma_1 \sin\theta_B^{TE}}{\gamma_2}\right)^2 = \cos^2\theta_t + \sin^2\theta_t = 1$$

$$\left(\frac{\eta_2}{\eta_1}\right)^2 (1 - \sin^2\theta_B^{TE}) + \left(\frac{\gamma_1 \sin\theta_B^{TE}}{\gamma_2}\right)^2 = 1$$

$$\Rightarrow \qquad \sin^2\theta_B^{TE}\left[\frac{\gamma_1^2}{\gamma_2^2} - \left(\frac{\eta_2}{\eta_1}\right)^2\right] = 1 - \left(\frac{\eta_2}{\eta_1}\right)^2 \qquad \text{...(3.117)}$$

$$\Rightarrow \qquad \sin^2\theta_B^{TE}\left(\frac{\gamma_1^2\eta_1^2 - \gamma_2^2\eta_2^2}{\gamma_2^2\eta_1^2}\right) = \frac{\eta_1^2 - \eta_2^2}{\eta_1^2}$$

$$\Rightarrow \qquad \sin^2\theta_B^{TE} = \frac{(\eta_1^2 - \eta_2^2)\gamma_2^2}{\gamma_1^2\eta_1^2 - \gamma_2^2\eta_2^2}$$

As $\eta = \frac{j\omega\mu}{\gamma} \Rightarrow \gamma = \frac{j\omega\mu}{\eta}$, substituting this in (3.117), we have

$$\sin^2\theta_B^{TE} = \frac{(\eta_1^2 - \eta_2^2)\gamma_2^2}{\gamma_1^2\eta_1^2 - \gamma_2^2\eta_2^2}$$

$$= \frac{\gamma_2^2}{\left(\gamma_1 . \frac{j\omega\mu_1}{\gamma_1}\right)^2 - \left(\gamma_2 . \frac{j\omega\mu_2}{\gamma_2}\right)^2}\left[\left(\frac{j\omega\mu_1}{\gamma_1}\right)^2 - \left(\frac{j\omega\mu_2}{\gamma_2}\right)^2\right]$$

$$= \frac{\gamma_2^2 (j\omega)^2}{(j\omega)^2(\mu_1^2 - \mu_2^2)}\left[\left(\frac{\mu_1}{\gamma_1}\right)^2 - \left(\frac{\mu_2}{\gamma_2}\right)^2\right]$$

$$= \frac{\gamma_2^2}{(\mu_1^2 - \mu_2^2)}\left[\frac{(\mu_1\gamma_2)^2 - (\mu_2\gamma_1)^2}{(\gamma_1\gamma_2)^2}\right] = \frac{(\gamma_2/\gamma_1)^2(\mu_1)^2 - \mu_2^2}{(\mu_1^2 - \mu_2^2)}$$

$$= \frac{(\mu_2\varepsilon_2/\mu_1\varepsilon_1)(\mu_1)^2 - \mu_2^2}{(\mu_1^2 - \mu_2^2)} = \frac{\dfrac{\mu_1\mu_2\varepsilon_2}{\varepsilon_1} - \mu_2^2}{(\mu_1^2 - \mu_2^2)} \qquad ...(3.118)$$

For the boundary between two dielectric media, we have $\mu_1 = \mu_2 = \mu$. Therefore, the denominator of (3.118) vanishes. In other words, we do not get a value for Brewster angle for two dielectric media corresponding to perpendicular polarization or the TE waves. However, if the boundary separates two different magnetic media, then we have the following result, with the substitution of $\varepsilon_1 = \varepsilon_2 = \varepsilon$

$$\sin^2\theta_B^{TE} = \frac{\mu_1\mu_2 - \mu_2^2}{\mu_1^2 - \mu_2^2} = \frac{\mu_2}{\mu_1 + \mu_2} \qquad ...(3.119)$$

$$\Rightarrow \qquad \sin\theta_B^{TE} = \sqrt{\frac{\mu_2}{\mu_1 + \mu_2}}$$

We also have

$$\tan\theta_B^{TE} = \sqrt{\frac{\mu_2}{\mu_1}} = \frac{n_2}{n_1}$$

3.6. TOTAL INTERNAL REFLECTION

We know that when a ray of light strikes the boundary of two different media (air-water for example) at an angle greater than some critical angle, all of the light is reflected back to the first medium in which it was travelling and none of light gets refracted or transmitted into the second medium. The critical angle is obtained from Snell's law through the use of the following:

$$\frac{\sin\theta_i}{\sin\theta_t} = \frac{\gamma_2}{\gamma_1} = \frac{j\omega\mu_2}{\eta_2} \div \frac{j\omega\mu_1}{\eta_1} = \frac{\eta_1}{\eta_2}\frac{\mu_2}{\mu_1} = \frac{\sqrt{\mu_1\varepsilon_2}}{\sqrt{\mu_2\varepsilon_1}}\frac{\mu_2}{\mu_1} \qquad ...(3.120)$$

$$= \frac{\sqrt{\varepsilon_2\mu_2}}{\sqrt{\varepsilon_1\mu_1}} = \frac{v_{p1}}{v_{p2}} = \frac{n_2}{n_1}$$

For total internal reflection to take place, the minimum value of the angle of refraction is 90°. Hence, we write

$$n_1 \sin\theta_i = n_2 \sin\theta_t$$

$$\Rightarrow \quad n_1 \sin\theta_c = n_2 \sin 90$$

$$\Rightarrow \quad n_1 \sin\theta_c = n_2$$

$$\Rightarrow \quad \sin\theta_c = \frac{n_2}{n_1}$$

$$\Rightarrow \quad \theta_c = \sin^{-1}\left(\frac{n_2}{n_1}\right) \qquad ...(3.121)$$

We have

$$\gamma_1 \sin\theta_i = \gamma_2 \sin\theta_t$$

$$\Rightarrow \quad \sin\theta_i = \frac{\gamma_2}{\gamma_1}\sin\theta_t$$

$$\Rightarrow \quad \sin\theta_c = \frac{\gamma_2}{\gamma_1} \quad \because \theta_t = 90°$$

$$\Rightarrow \quad \theta_c = \sin^{-1}\left(\frac{\gamma_2}{\gamma_1}\right) = \sin^{-1}\left(\sqrt{\frac{\mu_0\mu_{r2}\varepsilon_0\varepsilon_{r2}}{\mu_0\mu_{r1}\varepsilon_0\varepsilon_{r1}}}\right)$$

$$= \sin^{-1}\left(\sqrt{\frac{\mu_{r2}\varepsilon_{r2}}{\mu_{r1}\varepsilon_{r1}}}\right) = \sin^{-1}\left(\frac{n_2}{n_1}\right) \qquad ...(3.122)$$

It is thus obvious that the refractive index of a medium is expressed as

$$n = \sqrt{\mu_0\,\mu_r\,\varepsilon_0\,\varepsilon_r} \qquad ...(3.123)$$

For nonmagnetic materials we have $\mu_1 = \mu_2 = \mu_0$, in that case

$$\sin\theta_i = \sqrt{\frac{\mu_1\varepsilon_2}{\mu_2\varepsilon_1}\frac{\mu_2}{\mu_1}}\sin\theta_t = \sqrt{\frac{\varepsilon_2}{\varepsilon_1}}\sin\theta_t \qquad ...(3.124)$$

For an incident angle equal to the critical angle, we understand that the angle of refraction called the grazing angle is $\theta_i = \theta_c$ and $\theta_t = 90°$. Therefore, for such a case,

$$\sin\theta_c = \sqrt{\frac{\varepsilon_2}{\varepsilon_1}}\sin 90 = \sqrt{\frac{\varepsilon_2}{\varepsilon_1}}$$

$$\Rightarrow \qquad \theta_c = \sin^{-1}\left(\sqrt{\frac{\varepsilon_2}{\varepsilon_1}}\right) \qquad ...(3.125)$$

It is also clear from the above discussion that for a dielectric media interface, the value of the dielectric constant of the first medium should be higher than that of the second medium so that the critical angle as given by $\sin^{-1}\left(\sqrt{\frac{\varepsilon_2}{\varepsilon_1}}\right)$ is a real number. We are interested in the case corresponding to the incidence angle being higher than the critical angle. For the case when waves move from the denser to the rarer medium we get $\theta_i > \theta_c$.

$$\Rightarrow \qquad \sin\theta_t = \frac{\gamma_1}{\gamma_2}\sin\theta_i = \sqrt{\frac{\mu_1\varepsilon_1}{\mu_2\varepsilon_2}}\sin\theta_i > 1$$

$$\Rightarrow \qquad \cos\theta_t = \pm\sqrt{1-\sin^2\theta_t} \qquad ...(3.126)$$

As $\theta_t > 1$, the two values of cosine are imaginary.

Hence, the transmitted or refracted field in the second medium (assumed lossless) is expressed as, corresponding to the case of perpendicular polarization

$$\begin{aligned}\mathbf{E}_t &= \tau E_0 \exp[-\gamma_2(z\cos\theta_t + x\sin\theta_t)]\hat{\mathbf{y}} \\ &= \tau E_0 \exp[-j\beta_2(z\cos\theta_t + x\sin\theta_t)]\hat{\mathbf{y}} \\ &= \tau E_0 \exp(-j\beta_2 z\cos\theta_t)\exp(-j\beta_2 x\sin\theta_t)\hat{\mathbf{y}} \qquad ...(3.127)\end{aligned}$$

As the value of the cosine is imaginary in this case, the first exponential $(-j\beta_2 z \cos\theta_t)$ term is a real number whereas the second exponential term $(-j\beta_2 x \sin\theta_t)$ is an imaginary number. The cosine has both positive and negative roots. We choose the negative sign so that $(-j\beta_2 z \cos\theta_t)$ is negative since waves growing with distance are physically unrealizable. Hence, the first exponential term represents a wave that attenuates very fast with respect to distance z (evanescent) and the second exponential term represents a wave that propagates along the x-axis. Such waves are known as surface waves that are bound to the surface. In order to understand the power flow occurring in the second medium, we write the associated magnetic field as

$$\begin{aligned}\mathbf{H}_t &= \frac{\tau E_0}{\eta_2}\exp[-j\beta_2(z\cos\theta_t + x\sin\theta_t)](-\cos\theta_t\hat{\mathbf{x}} + \sin\theta_t\hat{\mathbf{z}}) \\ &= -\cos\theta_t \frac{\tau E_0}{\eta_2}\exp[-j\beta_2(z\cos\theta_t + x\sin\theta_t)]\hat{\mathbf{x}}\end{aligned}$$

$$+\ \sin\theta_t \frac{\tau E_0}{\eta_2} \exp[-j\beta_2(z\cos\theta_t + x\sin\theta_t)]\hat{\mathbf{z}} \qquad \text{...(3.128)}$$

Therefore,

$$\mathbf{H}_t^* = -(\cos\theta_t)^* \left(\frac{\tau E_0}{\eta_2}\right)^* \exp(-j\beta_2 z\cos\theta_t)\exp(j\beta_2 x\sin\theta_t)\hat{\mathbf{x}}$$

$$+\ (\sin\theta_t)^* \left(\frac{\tau E_0}{\eta_2}\right)^* \exp(-j\beta_2 z\cos\theta_t)\exp(j\beta_2 x\sin\theta_t)\hat{\mathbf{z}}$$

$$= -\frac{1}{\eta_2}(\tau E_0)^* \exp(-j\beta_2 z\cos\theta_t)\exp(j\beta_2 x\sin\theta_t)\cos^*(\theta_t)\hat{\mathbf{x}}$$

$$+\ \frac{1}{\eta_2}(\tau E_0)^* \exp(-j\beta_2 z\cos\theta_t)\exp(j\beta_2 x\sin\theta_t)\sin^*(\theta_t)\hat{\mathbf{z}} \qquad \text{...(3.129)}$$

The power flow in the second medium as given by Poynting vector becomes

$$\mathbf{P} = \mathbf{E}_t \times \mathbf{H}_t^* = \begin{bmatrix} \hat{\mathbf{x}} & \hat{\mathbf{y}} & \hat{\mathbf{z}} \\ E_x & E_y & E_z \\ H_x^* & H_y^* & H_z^* \end{bmatrix} = \begin{bmatrix} \hat{\mathbf{x}} & \hat{\mathbf{y}} & \hat{\mathbf{z}} \\ 0 & E_y & 0 \\ H_x^* & 0 & H_z^* \end{bmatrix}$$

$$= E_y H_z^* \hat{\mathbf{x}} - E_y H_x^* \hat{\mathbf{z}} = E_y H_z^* \hat{\mathbf{x}} + E_y H_x^* \hat{\mathbf{z}} \text{ as } H_x \text{ is negative} \qquad \text{...(3.130)}$$

The first term in the RHS of (3.130) becomes

$$E_y H_z^* = \tau E_0 \exp(-j\beta_2 z\cos\theta_t)\exp(-j\beta_2 x\sin\theta_t) \times \frac{1}{\eta_2}(\tau E_0)^*$$

$$\exp(-j\beta_2 z\cos\theta_t)\exp(j\beta_2 x\sin\theta_t)\sin^*(\theta_t)$$

$$= \frac{1}{\eta_2}|\tau E_0|^2 \exp(-j2\beta_2 z\cos\theta_t)\sin^*(\theta_t) \qquad \text{...(3.131)}$$

Similarly, the second term in RHS of (3.130) becomes

$$E_y H_x^* = \tau E_0 \exp(-j\beta_2 z\cos\theta_t)\exp(-j\beta_2 x\sin\theta_t)$$

$$\times \frac{1}{\eta_2}(\tau E_0)^* \exp(-j\beta_2 z\cos\theta_t)\exp(j\beta_2 x\sin\theta_t)\cos^*(\theta_t)$$

$$= \frac{1}{\eta_2}|\tau E_0|^2 \exp(-j2\beta_2 z\cos\theta_t)\cos^*(\theta_t) \qquad \text{...(3.132)}$$

Substitution of (3.131) and (3.132) in (3.130) gives us

$$\mathbf{P} = E_y H_z^* \hat{\mathbf{x}} + E_y H_x^* \hat{\mathbf{z}}$$

$$= \frac{1}{\eta_2} |\tau E_0|^2 \exp(-j2\beta_2 z \cos\theta_t)[\sin^*(\theta_t)\hat{\mathbf{x}} + \cos^*(\theta_t)\hat{\mathbf{z}}] \quad ...(3.133)$$

The power flow in the second medium is contained in the *xz*-plane as observed from (3.133); same as the plane containing the propagation vector. The power changes at the rate of $-2z$. For $z > 0$, we have noted that $\cos\theta_t$ is an imaginary number which means that $\cos^*(\theta_i)$ is also an imaginary number. Hence, the second term is imaginary representing no real power flow along the *z*-axis. The first term represents a real number and hence real power flow along the positive *x*-axis that decays exponentially with respect to. At the boundary of the two media, $z = 0$ and therefore,

$$\mathbf{P} = E_y H_z \hat{\mathbf{x}} + E_y H_x \hat{\mathbf{z}}$$

$$= \frac{1}{\eta_2} |\tau E_0|^2 [\sin^*(\theta_t)\hat{\mathbf{x}} + \cos^*(\theta_t)\hat{\mathbf{z}}] \quad ...(3.134)$$

The phase velocity of the plane waves is reduced by a factor of n, the refractive index of the medium. Hence, we write

$$v_p = c/n \quad ...(3.135)$$

It is customary to talk of optics in terms of wavelength λ that is obviously defined as $\lambda = \frac{c}{f}$. This is also known as the free space wavelength. Within a refractive medium, the wavelength is reduced locally to λ_m such that

$$\lambda_m = \lambda/n \quad ...(3.136)$$

and therefore,

$$v_p = c/n = \lambda_f/n = \lambda_m f \quad ...(3.137)$$

The phase velocity of a plane wave is

$$v_p = f\lambda = (\omega/2\pi)\lambda = \left(\frac{\omega}{2\pi/\lambda}\right) = \omega/\beta \quad ...(3.138)$$

Equating the two expressions for the phase velocity in (3.137) and (3.138), we have

$$v_p = \frac{c}{n} = \frac{\omega}{\beta}$$

$$\Rightarrow \qquad n = c\frac{\beta}{\omega} = \left(\frac{c}{\omega}\right)\beta \qquad \text{...(3.139)}$$

Wave propagation through optical fibers requires wave solutions that are subject to the radial symmetric variations of the relative permittivity in the fiber. The propagating waves inside a fiber are not pure TEM waves because they have an axial component of the electric field or the magnetic field. The boundary condition imposed by the fiber is the prescribed variation of the optical frequency permittivity.

3.7. NUMERICAL APERTURE OF AN OPTICAL FIBER

An offshoot of the phenomenon of total internal reflection is the wave propagation inside an optical fiber. We may note that, an optical fiber acts like a guiding medium for light waves. Common sense suggests that all of the light getting coupled to the fiber from a given source must be propagate through the core region only as light travelling in the cladding would serve no useful purpose. Light waves, hence become guided waves once they are confined to travel inside the core of the fiber and therefore exhibit behavior similar to TE, TM or hybrid waves. The process that ensures all light waves to be guided by the core and none by the cladding is this phenomenon of total internal reflection. This is achieved by making the refractive index of the core higher than that of the cladding; in other words the optical fiber core is optically denser than its cladding. This is illustrated in Fig. 3.8.

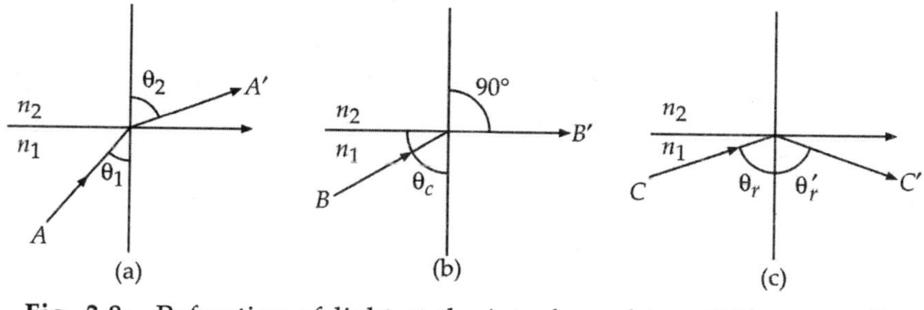

Fig. 3.8: Refraction of light at the interface of two different media with refractive indices n_1 and n_2 (a) normal refraction wherein the incident light beam bends away from the normal to the interface as $n_2 < n_1$, (b) onset of grazing ($\theta_r = 90°$) and therefore all the light beams graze the interface, (c) total internal reflection wherein all the light beams come back to the first medium as $n_2 < n_1$

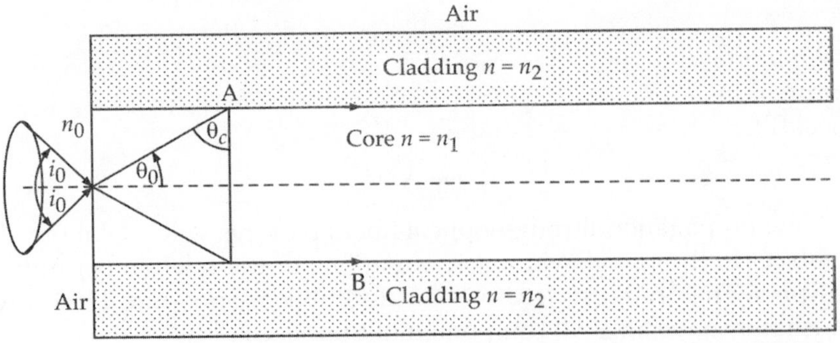

Fig. 3.9: Calculating the numerical aperture of an optical fiber

Applying Snell's law to the core-cladding interface at *A* in Fig. 3.9 we write

$$n_1 \sin \theta_c = n_2 \qquad \text{...(3.140)}$$

$$\Rightarrow \qquad \sin \theta_c = \frac{n_2}{n_1}$$

At the air-core interface, we have

$$\sin i_0 = n_1 \sin\theta_0 = n_1 \sin(90 - \theta_c) = n_1 \cos\theta_c \qquad \text{...(3.141)}$$

However,

$$\cos \theta_c = \sqrt{1 - \sin^2 \theta_c} = \sqrt{1 - \left(\frac{n_2}{n_1}\right)^2} = \sqrt{\frac{n_1^2 - n_2^2}{n_1^2}} = \frac{\sqrt{n_1^2 - n_2^2}}{n_1} \qquad \text{...(3.142)}$$

Substituting the expression of $\cos \theta_c$ from (3.142) in (3.141), we get

$$\sin i_0 = n_1 \cos\theta_c = n_1 \frac{\sqrt{n_1^2 - n_2^2}}{n_1} = \sqrt{n_1^2 - n_2^2} \qquad \text{...(3.143)}$$

This gives us an expression for the acceptance angle defined as $2i_0$ at the air-core interface as

$$i_0 = \sin^{-1} \sqrt{n_1^2 - n_2^2} \qquad \text{...(3.144)}$$

$$\Rightarrow \qquad 2i_0 = 2\sin^{-1} \sqrt{n_1^2 - n_2^2}$$

Light that propagates within a cone defined by the acceptance angle is trapped by the core of the optical fiber and therefore guided by its. This cone is referred to as the acceptance cone. This gives us the light gathering capability of an optical fiber.

An important parameter that is associated with the fibers is numerical aperture. This sometimes has also been called as the figure of merit for an optical fiber. This is defined as the following

$$N_A = \sin i_0(\max) = \frac{\sqrt{n_1^2 - n_2^2}}{n_0} = \sqrt{n_1^2 - n_2^2} \qquad ...(3.145)$$

if light is traveling in air and falls on the air-core interface.

It is observed that, n_1 and n_2 do not differ very much. Therefore, (3.145) may be approximated as

$$NA = \sin i_0(\max) = \sqrt{n_1^2 - n_2^2} = \sqrt{(n_1 + n_2)(n_1 - n_2)}$$

$$= \sqrt{2n_1(n_1 - n_2)} = \sqrt{2n_1^2\left(\frac{n_1 - n_2}{n_1}\right)} = n_1\sqrt{2\Delta} \qquad ...(3.146)$$

We have approximated in (3.146) and the quantity is defined as

$$\Delta = \frac{n_1 - n_2}{n_1} \qquad ...(3.147)$$

As we may observe, Δ is known as the fractional difference between the core and the cladding refractive indices.

It has been found that the numerical apertures of fibers used for short distance communication lie in the range of 0.4 to 0.5 whereas for long distances the range gets lowered to 0.1 to 0.3. It may be mentioned here that the numerical aperture derived in (3.147) is with reference to a stepped index fiber wherein the refractive indices of the core and the cladding vary in a step like manner. This will be derived for a graded index fiber later on.

3.8. CHAPTER SUMMARY

Light waves are transverse electromagnetic waves in unbounded homogeneous media such as free space. They travel with a velocity of '$c = 3 \times 10^8$' m/s. The orientation of the electric field in the light wave gives its polarization that changes when it travels through either a quarter wave plate or a half wave plate. The polarization has been categorized as linear, circular or elliptical. Light waves incident at the interface between two different media have a part reflected from the boundary the amount of which is decided by the reflectivity of the surface. The rest of it gets transmitted into the other medium thereby constituting the refracted waves that obey Snell's law. The two laws of refraction are derived by considering the continuity of the electric and magnetic fields across the boundary. The incident angle responsible for the condition of no reflection at the boundary is termed as Brewster angle. This angle does not exist at the boundary of two different dielectric media corresponding to TE waves. Similarly, we do not have any Brewster angle existing at the boundary of two different magnetic media corresponding to TM waves. The phenomenon of total internal reflection occurs at the interface of optically denser medium to an optically rarer medium in which we have obtained the surface waves.

SUGGESTED READING

[1] Jordan, E.C. and Balmain, K.G. "Electromagnetic Waves and Radiating Systems", 2nd Edn., PHI Pvt. Ltd., 1989.

❑❑❑

4

Optical Sources and Characteristics

4.1. INTRODUCTION

The optical communication instrumentation system is characterized by a source, the optical fibre serving as the physical channel and a suitable photodetector as the receiver. We need to generate light and further modulate it by some baseband signal in such a system in order to carry information. The optical source broadly is categorized into two types; the light emitting diodes (LED) and the lasers. These two sources along with their different attributes are described next.

4.2. OPTICAL SOURCES

The sources available for generating optical beams are the LED and the LASER (Light Amplification by Stimulated Emission of Radiation). They work on the principle of electro-luminance. It is a phenomenon of emission of optical radiation by converting electrical energy into light. These devices as optical sources are preferred for reasons stated below.

(a) Their small size which is compatible with the small diameter of the fibers.

(b) A simple solid state construction

(c) Low power requirements which are compatible with the state-of-the art solid state electronics.

For optical systems operating at 100 to 200 Mbps data rate constructed with multimode optical fiber-coupled optical power in the range of tens of microwatts, the LEDs are usually the best light source. They require less complex drive circuitry than the laser diodes as no thermal or optical stabilization circuits are needed.

In choosing an optical source compatible with the optical waveguide, various characteristics of the fiber such as its geometry, its attenuation as a function of wavelength, its group delay distortion (bandwidth), and its modal characteristics must be taken into account. The interplay of these factors with the optical source power, bandwidth, spectral width, radiation pattern, and modulation capability needs to be considered. The spatially directed coherent optical output from a laser diode can be coupled into either single-mode or multimode fibers. In general, LEDs are used with multimode fibers, since normally the incoherent optical power from an LED can only be coupled into a multimode fiber in sufficient quantities to be useful. However, some applications have used specially fabricated LEDs with single-mode fibers for data transmission at bit rates up to 1.2 Gbps over several kilometres.

A major difference between LEDs and laser diodes is that the optical output from an LED is incoherent whereas that from a laser diode is coherent. In a coherent source the optical energy is produced in an optical resonant cavity. The optical energy released from this cavity has spatial and temporal coherence, which means it is highly monochromatic and the output beam is very directional. In an incoherent LED source no optical cavity exists for wavelength selectivity. The output radiation has a broad spectral width, since the emitted photon energies range over the energy distribution of the recombining electrons and holes, which usually lie between 1 and $2K_BT$. In addition, the incoherent optical energy is emitted into a hemisphere according to a cosine power distribution and thus has a large beam divergence. We next discuss the LED as an optical source from a number of parameters.

4.3. LIGHT EMITTING DIODES

This is a pn junction semiconductor diode. When forward biased, it emits light. The fundamental steps involved in light emission are electron-hole pair generation due to the applied excitation, process of recombination in which the excited charge carriers give up their energy either through a radiative or a through non radiative process and finally the process of extraction of emitted photons from the active region of the semiconductor to the observer.

When carriers are injected across a forward biased junction, the current is usually accounted for by recombination in the transition region and in the neutral regions near the junction. In a semiconductor with an indirect band gap, the recombination releases heat to the crystal lattice. On the other hand, indirect bandgap semiconducting materials give off light under a forward bias. This effect is known as injection electroluminescence which provides an important application of diodes as generators of light. The method which is adopted for excitation of the LEDs is the injection of minority carriers through a forward biased junction. The avalanche excitation and tunneling are also used to excite the LEDs. The overall efficiency of the LEDs depends on the efficiencies of each of these three processes and is observed to be the product of the three.

To be useful as an optical source, an LED must have a high radiance output, a fast emission response time and a high quantum efficiency. The radiance or brightness is a measure in watts of optical power radiated into a unit solid angle per unit area of the emitting surface. High radiations are desirable to couple sufficiently high optical power levels into a fiber. The emission response time is the time delay between the application of a current pulse and the onset of optical radiation. This time delay is the factor limiting the bandwidth with which the source can be modulated directly by varying the injected current. The quantum efficiency is related to the fraction of injected electron-hole pairs that recombine radiatively.

The LED structure must provide a means of confining the charge carriers and the stimulated optical radiation to the active region of the PN junction where radiative recombination takes place in. This is necessary in order to achieve a high radiance and high quantum efficiency. Carrier confinement is used to achieve a high level of radiative recombination in the active region of the device, which yields a high quantum efficiency. Optical confinement is of importance for preventing absorption of the emitted radiation by the material surrounding the pn junction.

To achieve carrier and optical confinement, the homojunctions and single and double heterojunctions are widely used LED configurations. The PN junctions formed by introducing relatively small concentrations of impurity atoms into what is basically the same semiconductor material are known as homojunctions. The other category of semiconductor materials grown on a single crystal has a change in the band-gap energy at the boundary between the different materials. Such junctions in a single crystal between semiconductors whose lattices match but which are otherwise different, are known a heterojunctions. A heterojunction may result due to four possible combinations of the doped n-type or p-type material. These may be n-N, p-P, n-P and p-N type of semiconductors. In these, we have chosen the bold letter to indicate a doped semiconductor having a wider band gap whereas the narrower band gap is indicated by the lower case; n-or p. Each type of material shows a different behavior.

The spectral output of an LED may have a relatively wide wavelength bandwidth of between 30 and 40 nm. However, this emission is narrow enough so that a particular color will be observed, provided the output is in the visible range.

4.4. RECOMBINATION PROCESS

The recombination of electrons and holes in a semiconductor may be brought about by many independent processes. It is sometimes convenient to distinguish between direct band-to-band transitions on the one hand, and transitions which involve immediate steps on the other. More important is the different distinction that can be made between the radiative and the non-radiative recombination processes. In a non radiative transition most of the energy of recombination is released into the crystal in the form of heat. In a radiative process most of the energy of recombination is released as a single quantum of radiation. When recombination takes place in several stages it is possible for more than one quantum of longer wavelength radiation to be emitted. A couple of different recombination process are illustrated in Fig. 4.1. Of these, the one which concerns us mainly is the direct, radiative, band-to-band transition shown in (a) in the diagram. The main competing non-radiative transitions are those involving trapping levels lying deep in the band gap, shown as (c) and (d). Such trapping levels may be caused by impurity atoms such as gold in silicon, or by dislocations or other crystal lattice defects, and they occur in large numbers at the semiconductor interface.

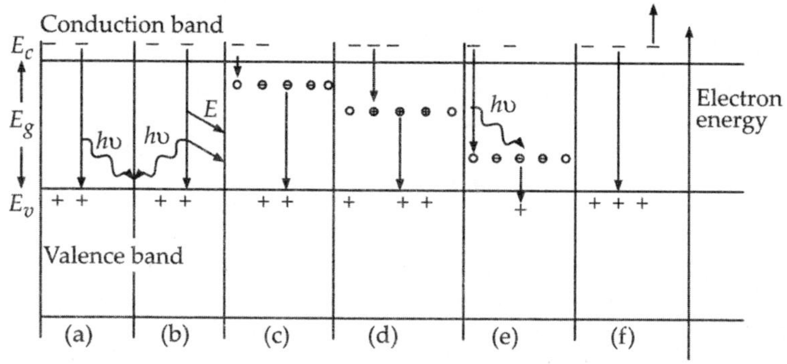

Fig. 4.1: Illustration of possible electron-hole recombination methods in a semiconductor with a bandgap energy of E_g: (a) direct, ban-to-band E, radiative transition, (b) radiative, band-to-band transition involving one or more photons of energy , (c) non radiative transitions involving a deep lying acceptor trap, (d) non radiative transitions involving a deep lying donor level, (e) both radiative and non radiative transitions involving a shallow acceptor level and (f) non radiative 'Augur' recombination transition.

The perturbation of the band structure at a free surface causes a much higher proportion of the trapping levels there to act as effective recombination centers. The proximity of the surface, or of a macroscopic flaw in the material, or of an interface where there is a discontinuity in the crystal structure, thus has a considerable effect on the recombination processes. The Augur process illustrated in (f), has the energy of the recombination taken up by another free carrier. As an interaction involving three carriers, its probability is proportional to the cube of the carrier concentration, with the result that it becomes significant at high carrier concentrations.

4.5. SPECTRUM OF RECOMBINATION RADIATION

When the recombination energy is released into a single photon of energy E_{ph}, the frequency υ and the free space wavelength of the radiation are related as

$$E_{ph} = h\upsilon = \frac{hc}{\lambda} \qquad ...(4.1)$$

The peak emission wavelength λ in micrometres is expressed as a function of the band-gap energy E_g in eV by

$$\lambda = \frac{1.240}{E_g(\text{eV})} \qquad ...(4.2)$$

In order to determine the distribution with wavelength or photon energy of the radiated recombination energy, we proceed as follows. The probability that an electron of energy E_2 recombines with a hole at energy E_1 is proportional to the concentration of electrons at E_2, $n(E_2)$, and to the concentration of holes at E_1, $p(E_1)$. The probability of a photon of energy E_{ph}, being radiated may be obtained by integrating the product $n(E_2)\, p(E_1)$ over all values of E_1 and E_2 subject to the constraint that $E_2 - E_1 = E_{ph}$. We need to have the knowledge of the carrier concentration distribution in the valence and conduction bands, $p(E_1)$ and $n(E_2)$.

An indication of these is illustrated in Fig. 4.2 for the regions close to a forward biased PN junction. According to the simple model used in this diagram, the distributions rise very sharply at the band edge; they then change relatively slightly over an energy range of about K_BT, and finally they decrease exponentially as $\exp(-E/K_BT)$. The concentration distribution of electrons at energy E_2 in the conduction band is the product of two terms. The first is the probability distribution of allowed energy states per unit volume per unit range of energy at energy E_2 denoted by $\rho_c(E_2)$. The density of states $\rho(E)$ in a semiconductor refers to the number of allowed electron states per unit volume per unit energy interval. Thus for example, if $\rho_c(E)$ represents the density of states in the conduction band, then $\rho_c(E)dE$ gives the number of states per unit volume between energy values E and $E + dE$. The second term is the probability that the energy states are occupied which is denoted by $f(E_2)$. Hence,

$$n(E_2) = \rho_c(E_2)f(E_2) \qquad \text{...(4.3)}$$

Similarly, the concentration distribution of holes at an energy, E_1, in the valence band is expressed as

$$p(E_1) = \rho_v(E_2)[1 - f(E_1)] \qquad \text{...(4.4)}$$

where $\rho_v(E_1)$ is the distribution of allowed energy states per unit volume per unit range of energies at E_1.

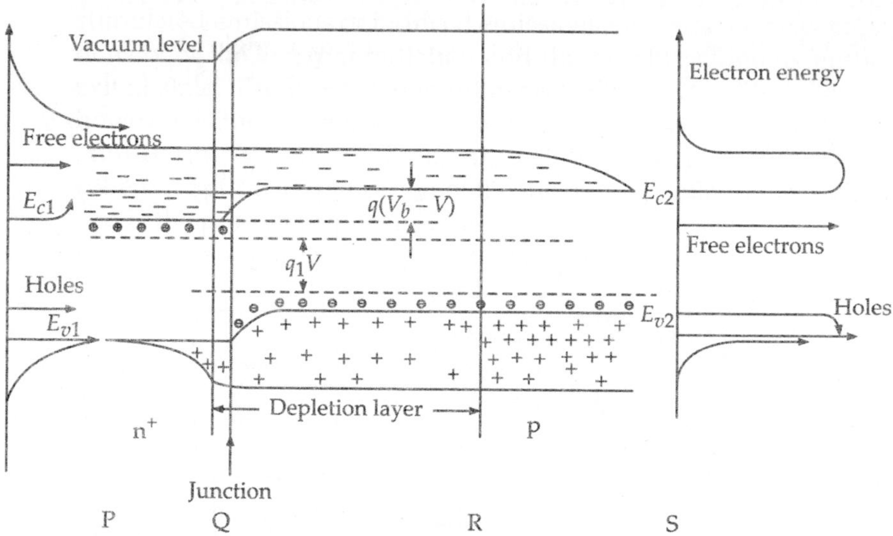

Fig. 4.2: Electron energy levels across a forward biased n^+p junction. The concentration of free electrons and holes in any given range of energies changes with position due to carrier diffusion that takes place outside the depletion layer. For electrons in the conduction band, the distribution remains in quasi-equilibrium from *P* to *R*. Diffusion between *R* and *S* causes the electron concentration to fall until the equilibrium concentration distribution characteristic of the bulk material is reached

The density of states is usually given by

$$\rho_c(E_2) = C_1(E_2 - E_c)^{1/2} \qquad \text{...(4.5)}$$

in the conduction band and

$$\rho_v(E_1) = C_2(E_v - E_1)^{1/2} \qquad \text{...(4.6)}$$

in the valence band. In these expressions, C_1 and C_2 are constants that depend on the effective mass of the charge carriers. High concentration of impurities distort the density of states functions $\rho_c(E_2)$ and $\rho_v(E_1)$, at the band-edges. Typically band tails are formed as shown in Fig. 4.2 with the result that the band-gap is effectively made narrower. An unbiased *pn*-junction is shown in Fig. 4.3(a) along with its energy band diagram. As we may observe from this, the Fermi level E_f lies in the middle of the two energy bands corresponding to the *p*- as well as the *n*-side. However, an external bias (dc voltage) applied to the diode has an effect of altering the location of the Fermi level. The effect of forward bias and reverse bias on the width of the depletion layer is illustrated in Figs. 4.3(b) and (c) along with a shift of the location of the Fermi level.

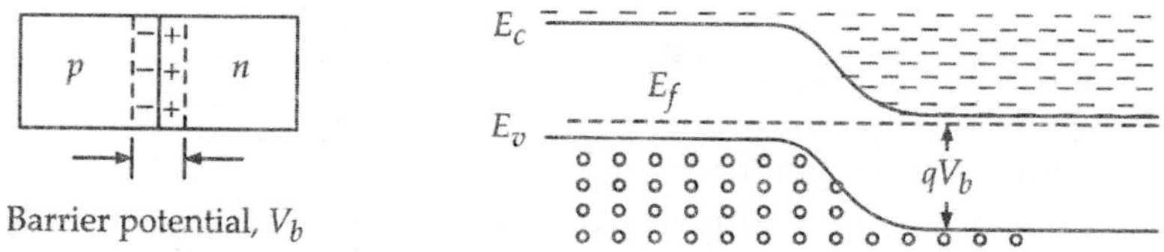

Fig. 4.3: (a) Unbiased *pn* junction and the associated energy band diagram

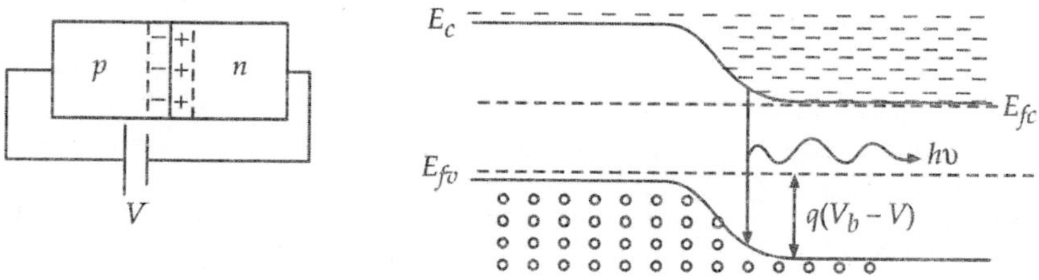

Fig. 4.3: (b) Forward biased *pn* junction and the associated energy band diagram showing a lowered potential barrier

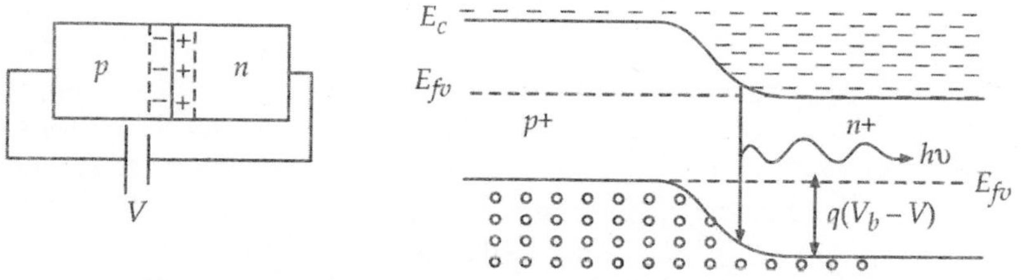

Fig. 4.3: (c) Forward biased heavily doped $p^+ n^+$ junction and the associated energy band diagram showing a lowered potential barrier

We assume the concentration of free electrons to decrease exponentially with increasing energy from the conduction band edge at E_c. Similarly, the concentration of holes is assumed to decrease exponentially with the energy difference from the top of the valence band E_v.

Thus, we write

$$n(E_2) \approx A\exp[-(E_2 - E_c)/K_BT] \qquad ...(4.7)$$

$$p(E_1) \approx B\exp[-(E_v - E_1)/K_BT] \qquad ...(4.8)$$

where A and B are constants related to the total concentration of free electrons and holes. These distributions are shown in Fig. 4.4.

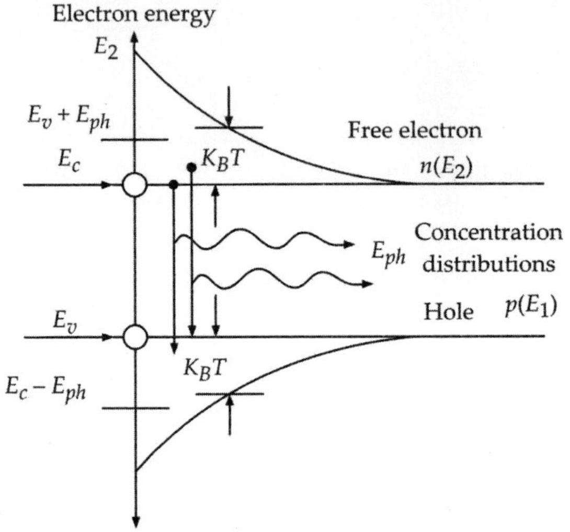

Fig. 4.4: Simplified exponential carrier concentration given by (4.7) and (4.8)

The spectral distribution of the radiated power density as a function of photon energy is given by

$$P(E_{ph}) \propto \int n(E_2)p(E_1)dE_2 \qquad ...(4.9)$$

subject to the constraint

$$E_2 - E_1 = E_{ph} \qquad ...(4.10)$$

so that

$$P(E_{ph}) = K \int_{E_2=E_c}^{E_v+E_{ph}} \exp[-(E_2 - E_c)/K_BT]\exp[-(E_v - E_1)/K_BT]dE_2 \qquad ...(4.11)$$

where K is a constant

The integration has been performed over the allowed range of energies in the conduction band. If E_2 were to exceed $(E_v + E_{ph})$, then E_1 would be in the band-gap, so $p(E_1)$ would be zero. The constants A and B have been assumed in a new constant K that includes the transition probability, which takes into account the radiative lifetime τ_{rr} of the electrons. Thus, we have

$$P(E_{ph}) = K \int_{E_2=E_c}^{E_v+E_{ph}} \exp[-(E_2 - E_c)/K_B T]\exp[-(E_v - E_1)/K_B T] dE_2$$

$$= K\exp(E_g/K_B T)\exp(-E_{ph}/K_B T) \int_{E_c}^{E_v+E_{ph}} dE_2 \qquad \text{...(4.12)}$$

$$= K(E_{ph} - E_g)\exp[-(E_{ph} - E_g)/K_B T]$$

On the basis of this analysis the spectral distribution of the emitted recombination radiation is shown in Fig. 4.5. The emission intensity is expected to be peaked at $E_g + K_B T$, and the spectral width at half-peak intensity at 2.4 $K_B T$.

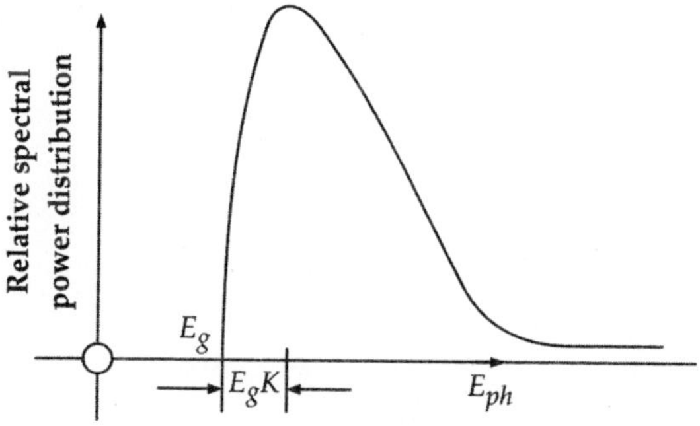

Fig. 4.5: Spectral distribution as given by (4.12)

The observed spectrum, however, is much more symmetrical than this due to (a) the use of high donor and acceptor impurity concentrations in normal LEDs and lasers causing band-edge distortion, (b) transitions involving an interaction with the crystal lattice as well as the emission of a photon. In that case a small fraction of the recombination energy may be carried away as the quantized energy associated with the lattice vibration that is excited at the same time, (c) the transition involving an intermediate step, one of the impurity levels near to the band-edge. As a result rather less than the band-gap energy is released in the main radiative transition. The recombination probability has been assumed to be a constant which may be expected to be a function of the electron and hole energies, (d) the analysis as outlined above gives the internal spectrum of the recombination radiation. The externally observed radiation is emitted after suffering selective absorption in the surface layers of the semiconductor.

As a result of these effects, the peak intensity is often shifted to wavelengths rather longer than those corresponding to $E_g + K_B T$. The spectral line-width is usually found to lie between 1.5 $K_B T$ and 3.5 $K_B T$

increasing with temperature and with doping concentration in the active region. The corresponding spread in terms of the wavelength of the emitted radiation may be expressed as the following:

$$\lambda = \frac{hc}{E_{ph}} \qquad \text{...(4.13)}$$

$$\Rightarrow \qquad \Delta\lambda = -\frac{hc}{E_{ph}^2}\Delta E_{ph}$$

The fractional spectral linewidth is

$$\gamma = \left|\frac{\Delta\lambda}{\lambda}\right| = \frac{\Delta E_{ph}}{E_{ph}} \approx \frac{2.4K_BT\lambda}{1.24[\mu m.eV]}$$

taking 2.4 K_BT as the spread in photon energies. At room temperature this is 0.062 eV and leads to the following values for γ and $\Delta\lambda$:

At 0.85 μm, $\gamma = 0.043$ and $\Delta\lambda = 36$ nm

At 1.30 μm, $\gamma = 0.065$ and $\Delta\lambda = 85$ nm

At 1.55 μm, $\gamma = 0.078$ and $\Delta\lambda = 120$ nm

These values are an approximate guide to what is observed in practice from LEDs emitting at these wavelengths. However, it may be pointed out that they are influenced by many factors, and there is considerable variation among different types of device.

4.6. LED MATERIALS WITH DIRECT AND INDIRECT BAND-GAP SEMICONDUCTORS

Good LED materials should have the following characteristics:

(a) Materials should have the ability to fabricate a PN junction.

(b) They should have good injection and luminescence efficiencies

(c) They should be able to extract a reasonable fraction of the generated light

With these considerations, we now explore the possibility of different kinds of semiconductor materials for use as LEDs. Semiconductors such as silicon, germanium and gallium phosphide have what is known as an indirect band-gap. In simple terms this means that an electron in an energy state near to the bottom of the conduction band has a momentum in the crystal that is quite different from that of an electron in an energy state near to the top of the valence band. It shows that a direct band-to-band transition can take place only if some means can be found to absorb the surplus momentum. One way would be for a high energy lattice vibration to be excited during recombination. While this might carry away the excess momentum, it would also take up a significant proportion of the recombination energy, up to 0.05 eV. The need to initiate two events simultaneously (producing a phonon a well as a photon) greatly reduces the probability of the type of recombination transition. As a result, non-radiative processes, in particular those that make use of trapping levels near to the centre of the band-gap, tend to dominate recombination in indirect band-gap semiconductors and the internal quantum efficiency is very low.

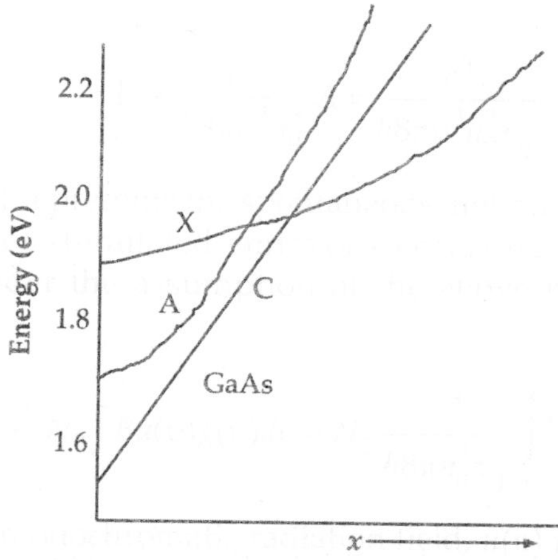

Fig. 4.6: Conduction band energies as a function of alloy composition for the ternary compound of $GaAs_{1-x}P_x$

The bandgaps of a couple of possible candidate semiconductor compound materials is shown in Fig. 4.6. There is a wide variation in band gaps and therefore, in available photon energies extending from the ultraviolet (GaN, 3.4 eV) into the infrared (InSb, 0.18 eV). In fact, by utilizing ternary and quaternary compounds the number of available energies can be increased significantly. A good example of the variation in photon energy obtainable from the compound semiconductors is the ternary alloy GaAsP, illustrated in Fig. 4.6. When the percentage of As is reduced and P is increased in this material, the resulting band gap varies from the direct 1.43 eV gap of GaAs (infrared) to the indirect 2.26 eV of GaP (green). The band gap of $GaAs_{1-x}P_x$ varies almost linearly with x until the 0.45 composition is reached, and electron-hole recombination is direct over this range. The most common alloy composition used in LED displays is $x \approx 0.4$. For this composition the band gap is direct. This results in efficient radiative recombination, and the emitted photons (~1.9 eV) are in the red portion of the spectrum.

For $GaAs_{1-x}P_x$ with phosphorous concentrations above 45%, the band gap is due to the indirect X minimum. Radiative recombination in such indirect materials is not very likely, because electrons in the conduction band have different momentum from holes in the valence band as illustrated in Fig. 4.7(b). Indirect $GaAs_{1-x}P_x$ doped with nitrogen can be used in LEDs with light output in the yellow to green portions of the spectrum. This is possible due to the tight binding of the impurity nitrogen atom with an electron. This helps in a spreading out of momentum space that takes care of the inhibition preventing radiative recombination in such type of materials.

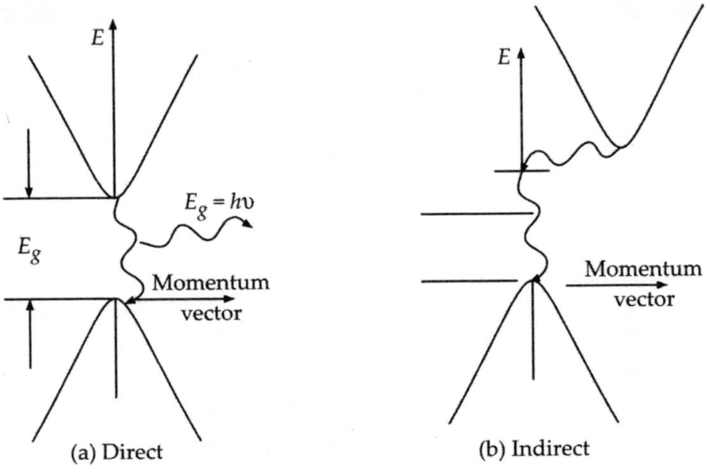

Fig. 4.7: Band gap in direct and indirect semiconductor materials

Other semiconductor materials may have a direct band-gap of the type illustrated in Fig. 4.7(a). In this case an electron with the lowest allowed energy in the conduction band does have a crystal momentum very similar to that of the electrons at the highest allowed valence-band energies. There is, then, a high probability of direct, band-to-band transitions, and the internal quantum efficiency for photon generation is also relatively high. The nature of the band gap of the III-V binary compounds is shown in Table 4.1.

Some of the ternary and quaternary compounds may change from direct to indirect band-gap materials as their composition is varied. One such instance is shown in Fig. 4.8.

There is a general tendency for the narrower band gaps to be direct and the wider ones to be indirect. This is good for the development of near-infra-red sources for optical fibre communications, but presents a serious problem in the production of efficient, visible LEDs. For the latter,

Table 4.1: Some direct and indirect bandgap semiconductor materials with their bandgap energy

Semiconductor material	*Energy bandgap (eV)*	*Cutoff Wavelength μm*
Ge	Indirect: 0.67	1.88
Si	Indirect: 1.12	1.15
GaP	Indirect: 2.26	0.55
InSb	Direct: 0.18	7.3
InAs	Direct: 0.35	3.5
GaSb	Direct: 0.73	1.70
InP	Direct: 1.35	0.92
GaAs	Direct: 1.43	0.87
AlSb	Indirect: 1.58	0.75
AlAs	Indirect: 2.16	0.57
AlP	Indirect: 2.45	0.52

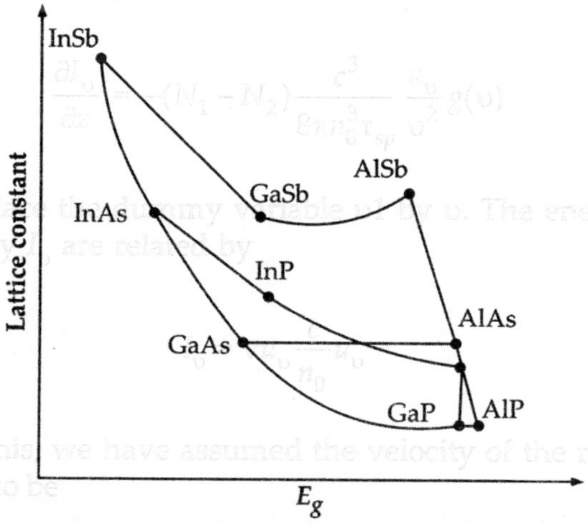

Fig. 4.8: Variation of lattice parameter with band gap energy for a class of group III-V ternary alloys

the indirect band-gap material, gallium phosphide, is often used, heavily doped with impurities such as N and ZnO. These give rise to trapping levels near to the band edges which promote radiative transitions like Fig. 4.1. The photon energy of the emitted light is not much less than E_g. But the optical conversion efficiency is still too low to make a good source for optical communication. Hence we restrict our discussion to direct band-gap type of semiconductors.

Gallium arsenide is a direct band gap material and it has been used for many years as the basis of a number of different types of semiconductor device: transferred electron devices, field effect transistors, LEDs and lasers.

Of all the III-V semiconductors, it is the one of which there is the greatest technological experience and expertise. It is thus natural that it should have been the first choice for sources for optical communications. We expect the wavelength of the peak emission to be in the region of

$$\frac{\lambda}{[\mu m]} \approx \frac{1.24}{E_g(\text{eV})} \approx \frac{1.24}{1.42} = 0.87 \qquad ...(4.14)$$

Because of the effect of high doping concentrations on the band-edge and the recombination processes, the peak wavelength is often slightly longer than this and may be as high as 0.92 μm.

The ternary system $Ga_{1-x}Al_xAs$ has a direct band gap over the composition range, $0 < x < 0.37$, giving a range of values for E_g of 1.42-1.92 eV. The corresponding wavelength range is 0.87-0.65 μm. One of the great advantages of this material derives from the very small difference between the lattice parameters of GaAs and AlAs, 0.5653 and 0.5661 nm respectively, which enables single-crystal layers of GaAlAs to be grown on a substrate of GaAs. In consequence, GaAlAs sources dominated the early phases of optical fibre transmission.

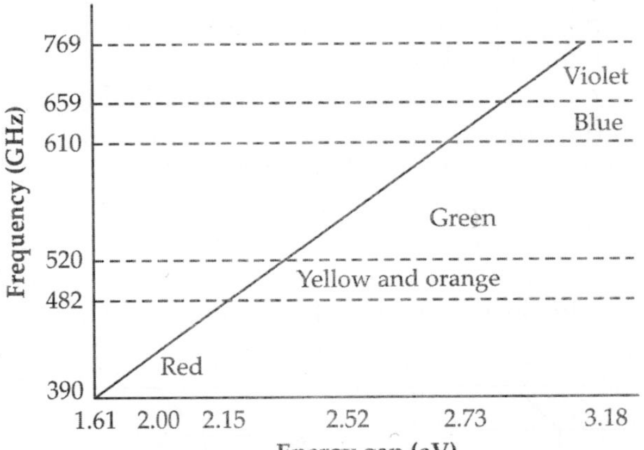

Fig. 4.9: Different energy gaps are needed to produce LEDs emitting different colors of light

The compound $GaAs_{0.6}P_{0.4}$ is typically used for red LEDs. The InAlGaP system is useful for yellow and green LEDs. Blue LEDS have been hardest to develop. GaN with its energy gap of about 3.4 eV emits very efficiently blue light, but it has not been possible to develop monocrystalline GaN wafers. The recently developed blue LEDs use GaN deposited on either SiC or sapphire wafers. When excited by electrons, GaN emits blue and ultraviolet light. Growing usable GaN was made possible by placing an aluminium nitride layer on a sapphire substrate and then growing crystals on that. The second method is to dope GaN crystals with magnesium atoms. Blue LEDS became commercially available in 1993. The development of blue LEDs led to the development of blue semiconductor lasers which proved useful for high capacity devices such as Blu-ray discs. Different energy gaps required to produce LEDs of different colors is shown in Fig. 4.9.

Sources for radiation of longer wavelengths, in particular, 1.3 and 1.55 μm require the use of semiconductors with narrower band-gaps, around 0.95 and 0.8 eV respectively. Possible III-V compound semiconductors may be identified in Fig. 4.8. Most interest has been garnered in the quaternary system. GaAs may be used as a substrate for a range of lattice-matched compounds in the InGaAsP system having energy gaps between 1.42 eV(GaAs) and 2.0 eV($In_{0.}5\ Ga_{0.5}\ P$). Devices made from these materials might be expected to cover the range of wavelengths from 0.87 to 0.62 μm. Much more important is the range of lattice-matched compounds in this system that can be grown on a substrate of InP. Their band-gaps vary from 1.35 eV down to 0.74 eV and indicate a possible wavelength range of 0.92 to 1.67 μm.

The semiconductor material used for the active layer of an optical source must have a direct band gap. In such a material, the electrons and holes can recombine directly across the band gap without needing a third particle to conserve momentum. The radiative recombination is sufficiently high to produce an adequate level of optical emission. Many binary compounds are direct band gap type of semiconductor materials. For operation in the 800- to 900-nm spectrum the principal material used is the ternary alloy $Ga_{1-x}Al_xAs$. The ratio x of aluminium arsenide to gallium arsenide determines the band gap of the alloy and, correspondingly, the wavelength of the peak emitted radiation.

At longer wavelengths, the quaternary alloy $In_{1-x}Ga_xAsyP_{1-y}$ is one of the primary material. By varying the mole fractions x and y in the active area, LEDs with peak output powers at any wavelength between 1.0 and 1.7 μm can be constructed. For the sake of simplicity, notations such as ALGaAs and GaInPAs are used. These alloys are considered to make semiconductor light sources as it is possible to match the lattice parameters of the heterostructure interfaces by using a proper combination of binary, ternary and quaternary materials.

4.7. DIFFERENT LED STRUCTURES

The LED structures can be classified into five types. These are (i) planer (ii) dome (iii) double heterostructure surface emitter (iv) double heterostructure edge emitter and (v) super luminescence type. Let us briefly discuss each type.

(i) Planer LED

This type of LED has the simplest structure. It is fabricated by either liquid or vapor phase epitaxial process and the junction is formed by the diffusion technique. The LED is so designed that most of the radiative recombination takes place in the side of the junction nearest to the surface and therefore, the probability of re absorption is reduced to a great extent. This is accomplished by arranging most of the current flowing through the device to be carried by those carriers injected to the surface layer. Figs. 4.10(a) and (b) show the LED structure made by epitaxial growth and the one fabricated by the diffusion technique respectively. When a forward bias is applied to the diode, the device emits light from all surfaces and only a small amount of light can escape to total internal reflection. The radiance of this type of LED is usually low.

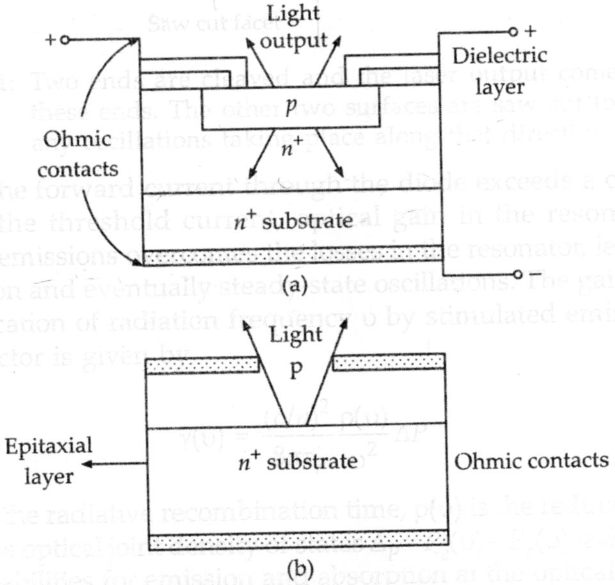

Fig. 4.10: (a) A planer LED structure fabricated by the epitaxial growth technique (b) A planer surface emitting LED structure fabricated by diffusion technique

(ii) Dome LED

In this type of LED the semiconductor surface is shaped in the form of a hemisphere. This is quite expensive and impractical for most applications. A modified dome shaped LED which comparatively cheaper is illustrated in Fig. 4.11(a) and (b). A dielectric encapsulation shaped as a dome is used with a transparent material of high refractive index in fabricating this type of LED. Higher output may be obtained from this LED than that obtainable from a planer LED. These dome shaped LEDS are useful for high volume low priced applications.

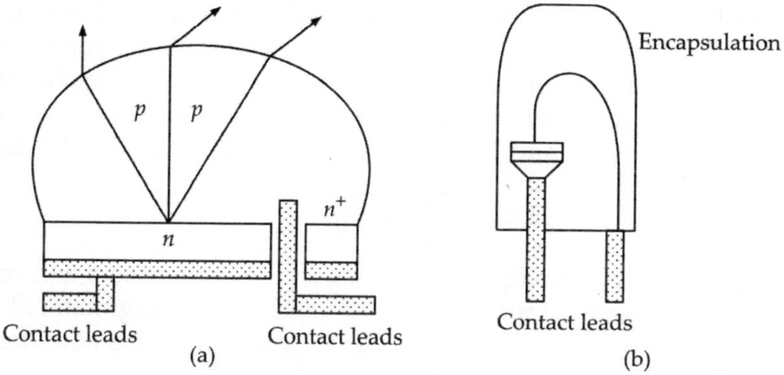

Fig. 4.11: (a) A dome LED with semiconductor surface shaped in the form of a hemisphere (b) A low cost dome LED

Planer and dome LEDs are used as cheap alarm indicators, display, in calculators, TV channel changer and industrial counting.

(iii) Double Heterostructure Surface Emitter LED (SLED, Burrous Type)

The two basic LED configurations being used are surface emitters and edge emitters. In the surface emitter the plane of the active light-emitting region is oriented perpendicularly to the axis of the fiber as shown in Fig. 4.12.

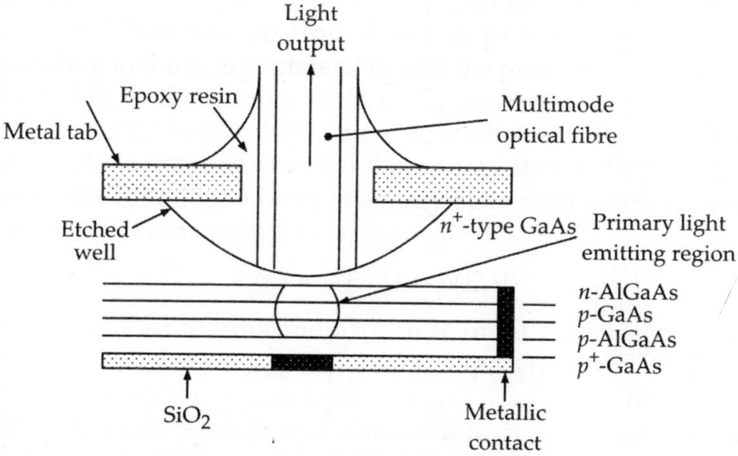

Fig. 4.12: Surface emitting LED, Burrus type

A double heterostructure device as shown in Fig. 4.12 has two different alloy layers on each side of the active region. By means of this sandwich structure composed alloy layers, both the carriers and the optical field are confined in the central active layer. The bottom p^+ GaAs and the top n^+ GaAs are included for getting low resistance ohmic contacts. However, in order to avoid the reabsorption of the emitted radiation, a deep well is etched in the top n^+ GaAs layer so as to reach the top top n^- AlGaAs layer. This well is used to support the fiber which is butt coupled to the LED and the fiber is held in proper place with an epoxy resin of appropriate refractive index to increase the external power efficiency of the device. Photons are generated in the thin *p*-GaAs region under the condition of forward bias and the emission from the top surface is ensured by the heterostructure by sandwiching *p*-GaAs layer between *n*-AlGaAs and *p*-AlGaAs layer and the reflection from the back crystal face. This results in a higher radiance of the device in the forward direction. The top n^+ GaAs contact layer ensures low electrical contact resistance and thermal resistance and therefore the current density are high. The fiber is properly aligned to optimize the coupling of the emitted radiation with the fiber. However, there is some loss of the emitted radiation due to its Lambertian distribution nature.

The circular active area in practical surface emitters is nominally 50 μm in diameter and up to 2.5 μm thick. This LED usually operates in the 600 nm to 900 nm wavelength range. The power coupled to the multimode step index fiber may be estimated from the following relationship

$$P_C = \pi(1 - R)AR_d(\text{NA})^2 \qquad \text{...(4.15)}$$

In the above expression, R is the Fresnel reflection coefficient at the surface and R_d is the source radiance, A is the fiber cross section or emission area of the device (whichever is less), NA is the numerical aperture of the fiber. The radiance is defined as the measure of the optical power radiated per unit area of the surface into a unit solid angle. The Fresnel coefficient R is evaluated as

$$R = \left(\frac{n_1 - n_2}{n_1 + n_2}\right)^2 \qquad \text{...(4.16)}$$

where n_1 and n_2 are the refractive indices the device material and the medium in which the device is placed respectively. The emission pattern is isotropic with a 120° half-power beam width. The thin SiO_2 layer at the back isolates the contact layer from the gold heat sink.

However, the power coupled to the fiber depends on the distance and the alignment between the emission area and the fiber. For example, when epoxy resin is added to the etched well, it has the tendency to reduce the refractive index mismatch and increase the external efficiency of the LED.

The band-gap differences of adjacent layers confine the charge carriers while the different refractive indices of the adjoining layers confine the optical field to the central active layer. This dual confinement leads to both high efficiency and high radiance. Other parameters affecting the device performance are the optical absorption in the active region (self absorption), carrier recombination at the heterostructure interfaces, doping concentration of the active layer, injection carrier density and active layer thickness.

(iv) Stripe Geometry AlGaAs Edge-Emitter LED

The edge emitter shown in Fig. 4.13 consists of an active junction region, which is the source of incoherent light, and two guiding layers. The active region is lightly doped or undoped, and consists of thin *n*-AlGaAs layer which is sandwiched between a *p*-AlGaAs and an *n*-AlGaAs which act as carrier confinement layers. Above and below these carrier confinement layers there are two transparent p^+AlgaAS and n^-AlgaAS layers working as the optical guiding layers. The guiding layers both have a refractive index which is lower than that of the active region but higher than the index of the surrounding material. This structure forms a waveguide channel that directs the optical radiation toward the fiber core.

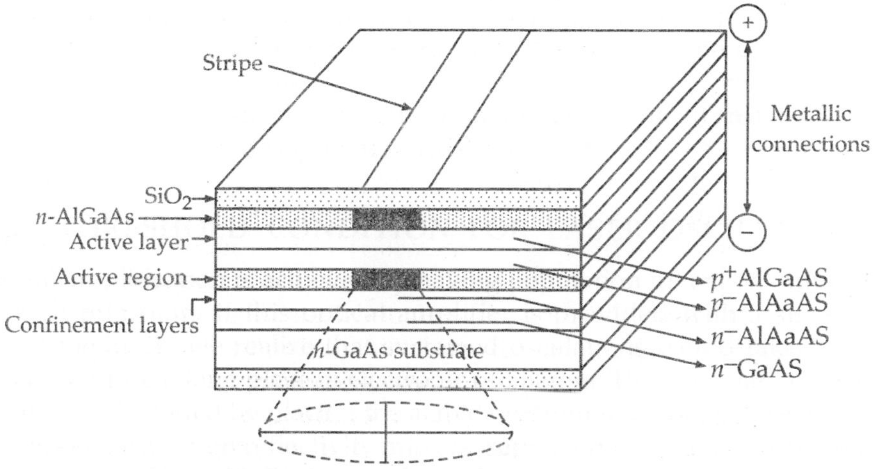

Fig. 4.13: Stripe Geometry AlGaAS type edge emitter LED

The stripe geometry is introduced in the structure to provide optical confinement in the horizontal plane. The active region is within the stripe. Generally the stripe is formed by the creation of high resistance areas on either side by proton bombardment technique or oxide isolation technique. To match the typical fiber core diameters (50 μm to 100 μm), the contact stripes for the edge emitters are 50 to 70 μm wide. The device takes advantage of the transparent guiding layer with a very thin active layer to aid light produced in the active layer to spread into the transparent guiding layers reducing self-absorption in the active layer. Lengths of the active regions usually range from 100 to 150 μm. The top of the device should be connected to the positive terminal and the bottom face to the negative terminal of the external voltage source.

The emission pattern of this LED configuration is more directional than that of the surface emitter. In the plane parallel to the junction where there is no waveguide effect, the emitted beam varies as cos? with a half-power width of $\theta = 120°$. In the plane perpendicular to the junction the half-power beam width has been made as small as 25 to 35° by a proper choice of the waveguide thickness. The enhanced waveguiding of the edge emitter enables to couple more power to a low (NA) fiber than a comparable surface emitter fiber. The stripe geometry of the edge emitter allow a very high carrier injection densities for a given drive current. Thus, it is possible to couple mW of optical power into a low NA (~0.14) multimode step index fiber with edge emitting LEDS operating with a high drive current of 500 mA.

These types of LEDs have also been investigated to have a substantially better bandwidth of the order of several MHz.

4.7.1. LED Parameters (Internal Quantum Efficiency)

A parameter that needs to be made as large as possible in an optical source is the internal quantum efficiency. This is defined as the ratio of the number of photons generated to the number of carriers crossing the junction. Clearly, its value depends on the relative probabilities of the radiative and non-radiative recombination processes. These in turn depend on the structure of the junction, the level of impurities in the semiconductor, and most of all on the type of semiconductor. This internal quantum efficiency of an LED is the fraction of diode current that produces luminescence. It is a function of the injection efficiency and a function of the percentage of radiative combination events compared with the total number of recombination events.

The three current components in a forward biased diode are the minority carrier electron diffusion current, the minority carrier hole diffusion current, and the space charge recombination current. These current densities are, respectively written as

$$J_n = \frac{qD_n n_{po}}{L_n}\left[\exp\left(\frac{qV}{K_B T}\right) - 1\right] \qquad ...(4.17)$$

$$J_p = \frac{qD_p p_{no}}{L_p}\left[\exp\left(\frac{qV}{K_B T}\right) - 1\right] \qquad ...(4.18)$$

$$J_R = \frac{qn_i W}{2\tau_0}\left[\exp\left(\frac{qV}{2K_B T}\right) - 1\right] \qquad ...(4.19)$$

D_n, D_p, n_{po}, p_{no}, L_n, L_p, n_i and W are the diode constants.

The recombination of electron and holes within the space charge region, is, in general, through traps near midgap and is a nonradiative process. Since luminescence is due primarily to the recombination of minority carrier electrons in GaAs, we define an injection efficiency as the fraction of electron current to the total current. Therefore,

$$\gamma = \frac{J_n}{J_n + J_p + J_R} \qquad ...(4.20)$$

Here γ is the injection efficiency. We can make γ approach unity by using an n^+p diode so that J_p is a small fraction of diode current and by forward biasing the diode sufficiently so that the recombination current component J_R is a small fraction of the total diode current.

Once the electrons are injected into the p-region, not all electrons recombine radiatively. The radiative and the nonradiative recombination rates are defined as

$$R_r = \frac{\delta n}{\tau_r} \qquad ...(4.21)$$

and

$$R_{nr} = \frac{\delta n}{\tau_{nr}} \qquad ...(4.22)$$

where τ_r and τ_{nr} are the radiative and nonradiative recombination lifetimes, respectively and δn is the excess carrier concentration. The total recombination rate is

$$R = R_r + R_{nr} = \frac{\delta n}{\tau} = \frac{\delta n}{\tau_r} + \frac{\delta n}{\tau_{nr}} \qquad ...(4.23)$$

where τ is the net excess carrier lifetime.

The radiative efficiency is defined as the fraction of recombination that are radiative. Hence we write

$$\eta = \frac{R_r}{R_r + R_{nr}} = \frac{1/\tau_r}{\frac{1}{\tau_r} + \frac{1}{\tau_{nr}}} = \frac{\tau}{\tau_r} \qquad ...(4.24)$$

where η is the radiative efficiency. The nonradiative recombination rate is proportional to N_t which is the density of nonradiative trapping sites within the forbidden band gap. Obviously, the radiative efficiency increases as N_t is reduced. The internal quantum efficiency is written as

$$\eta_{int} = \gamma\eta \qquad ...(4.25)$$

The radiative recombination rate is proportional to the *p*-type doping. As the *p*-type doping increases, the radiative recombination rate increases. However, the injection efficiency decreases as the *p*-type doping increases; therefore there is an optimum doping that maximizes the internal quantum efficiency.

For a heterojunction material like a *p-n* junction, it may be shown that the electron current density is given by

$$J_n = \frac{qD_n}{L_p}\frac{n_{i1}^2}{N_A}\left[\exp\left(\frac{qV}{K_BT}\right)-1\right] \qquad ...(4.26)$$

where n_{i1}^2 refers to the intrinsic carrier concentration in region1 formed by the *p*-type material having a narrow band gap with an acceptor concentration N_A, and equilibrium carrier concentrations being given as $p_{po} \approx N_A$ and thus, $n_{po} \approx n_{i1}^2/N_A$. Similarly, region 2 is the *N*-type material with a wide band gap having a donor concentration of $n_{No} \approx N_D$ and $p_{No} \approx n_{i2}^2/N_D$. The hole current density is expressed as

$$J_p = \frac{qD_p}{L_N}\frac{n_{i2}^2}{N_D}\left[\exp\left(\frac{qV}{K_BT}\right)-1\right] \qquad ...(4.27)$$

Hence, the ratio of the electron-to-hole current density is

$$\frac{J_p}{J_n} = \frac{J_p}{J_n} = C\frac{D_n}{D_p}\frac{L_N}{L_p}\frac{N_D}{N_A}\exp[(E_{g2}-E_{g1})/K_BT] \qquad ...(4.28)$$

where C is a constant. The corresponding injection efficiency for such a heterojunction, hence is

$$\eta_{int} = \frac{1}{(1+J_p/J_n)} \qquad ...(4.29)$$

For example, let us consider two materials having an energy-gap difference of 0.37 eV. This corresponds to the system GaAs-$Ga_{0.7}Al_{0.3}As$. At room temperature, the value of the exponential factor in (4.28) is 1.5×10^6 and easily dominates all the other terms in the expression for J_n/J_p. It is thus possible to maintain a high injection efficiency even though the ratios N_D/N_A and L_N/L_p may be small for other reasons.

4.7.2. Light Intensity

The light intensity is the number of photons emitted per unit tme. It is directly related to the optical power, $P_{opt}A_j$ where P_{opt} is the optical power density in W/m^2 and A_j is the junction area. By observing that the optical power is equal to the energy of photons emitted per unit time and by knowing that the energy of each photon is $h\upsilon$, we find that the number of photons emitted per unit time – the light intensity-is equal to $P_{opt}A_j/h\upsilon$. If every recombined minority carrier was emitting a photon, then $P_{opt}A_j/h\upsilon$ would be equal to the number of injected (and consequently recombined) minority carriers per unit time, I_d/q. However, only a fraction of the recombination events result in light emission so that

$$P_{opt}A_j/h\upsilon = \frac{\eta}{q}I_D \qquad ...(4.30)$$

The parameter η gives the efficiency of an LED and is referred to as the radiative recombination efficiency.

The radiative recombination efficiency is not equal to unity due to some of the recombination events releasing energy in the form of heat (phonons). The ratio of energy released as photons (light) and phonons (heat) depends to a large extent on whether the semiconductor is direct or indirect.

Given a certain radiative recombination efficiency, the light intensity is directly proportional to the concentration of electron-hole pairs recombined per unit time, which is the recombination rate. The recombination rate is directly proportional to the concentration of available electrons and the concentration of available holes. In the *p*-type region, there are plenty of available holes, so the recombination rate is basically limited by the excess concentration of holes (again the minority carriers).

The excess concentrations of the minority carriers make the stored charge. In a steady-state situation (a constant diode current I_D), the stored charge density does not change in time. This is because the recombined minority carriers are replaced by the new minority carriers that are injected over the junction barrier as the forward-bias current I_D. Therefore, the light intensity is directly proportional to the diode current I_D. The diode current becomes proportional to the applied input voltage once the diode is turned on.

A simple LED driving circuit has been shown in Fig. 4.14(a) whereas its operating point is illustrated in (b). The operating point in Fig. 4.14(b) is defined in terms of a dc voltage V_{DO} and a current I_{DO}. The voltage is chosen to be the independent variable and current becomes the dependent variable which can be verified by looking at (4.26). Therefore, voltage is plotted along the x-axis. For $v_{in} >> V_{DO}$, the light intensity becomes proportional to the input voltage v_{in}.

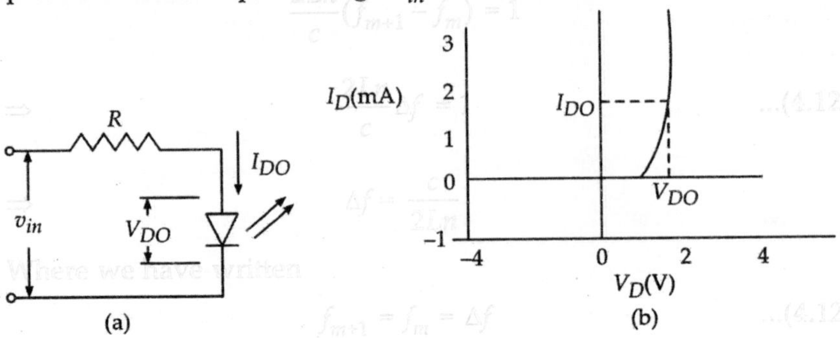

Fig. 4.14: (a) LED driving circuit (b) Operating point

$$\text{Light intensity} \propto I_{DO} = \frac{v_{in} - V_{DO}}{R} \approx \frac{v_{in}}{R} \qquad ...(4.31)$$

4.7.3. Modulation Capability

The LED, as a source of light needs to be examined for carrying information by subjecting it to a sinusoidal variation; we need to examine the modulation capability of the LED. We assume that a small sinusoidal modulation is superimposed on the normal forward bias. We consider a forward biased ntp LED where electron injection dominates the carrier flow and the excess electron concentration falls to zero far from the junction. In practice, the modulation behaviour of an LED is likely to be influenced by the proximity of the emitting surface. Our objective here is to find out an expression for the optical output power to the electrical input as a function of frequency. The concentration of carriers within an elementary volume $\partial A \partial x$ under dynamic conditions is given by

$$D_n \frac{\partial^2 n(x,t)}{\partial x^2} = \frac{[n(x,t) - n_{p0}]}{\tau_p} + \frac{\partial n(x,t)}{\partial t} \qquad ...(4.32)$$

D_n is the diffusion constant for elections, τ_p is the mean lifetime of minority roles.

We assume the electron concentration at the edge of the depletion layer in the p-type region ($x = 0$) to be controlled by the concentration of majority electrons in the n-type region and the potential change across the depletion layer. Let the potential be $V_0 + \tilde{V} \cos \omega t$. As we may observe, the forward bias is V_0 whereas the sinusoidal variation superimposed on the dc bias is $V_1 \cos \omega t$ and the modulation applied to the LED has a frequency of ω. Then we have, the electron concentration at the edge as

$$n(0, t) = n_{p0} \exp[q(V_0 + \tilde{V} \cos \omega t)/K_B T] \qquad ...(4.33)$$

$$= n_{p0} \exp(qV_0/K_BT) \exp[(q\tilde{V} \cos\omega t)/K_BT]$$

The time-varying part of the voltage is taken to be the real part of $\tilde{V} \exp(j\omega t)$. Hence, under the assumption of $qV_1 << K_BT$, we have

$$n(0, t) \approx n_{p0} \exp(qV_0/K_BT)\left[1 + \frac{q\tilde{V}}{K_BT} \exp(j\omega t)\right] \quad ...(4.34)$$

As we may note, we have used the following approximation

$$\exp(x) \approx 1 + x \text{ for } x << 1 \quad ...(4.35)$$

In the above expression for carrier concentration, we have the steady-state part as

$$n(0) = n_{p0} \exp(qV_0/K_BT) \quad ...(4.36)$$

and the time-varying part as

$$n_1(0, t) = n(0)\frac{q\tilde{V}}{K_BT} \exp(j\omega t) \quad ...(4.37)$$

where

$$n_1(0) = n_{p0} \frac{q\tilde{V}}{K_BT} \exp\left(\frac{qV_0}{K_BT}\right) \quad ...(4.38)$$

The solution to the carrier concentration, under the application of a sinusoidal modulating voltage, hence becomes

$$n(x, t) = n(x) = n_1(x, t) \quad ...(4.39)$$

It is more convenient to work in terms of excess carrier concentration, $\Delta n(x, t)$

$$\Delta n(x, t) = n(x, t) - n_{p0} = n(x) + n_1(x, t) - n_{p0} = n(x) - n_{p0} + n_1(x, t)$$

Therefore,

$$D_n \frac{\partial^2 \Delta n(x, t)}{\partial x^2} = \frac{\Delta n(x, t)}{\tau_p} + \frac{\partial \Delta n(x, t)}{\partial t} \quad ...(4.40)$$

Both parts of (4.28) must satisfy (4.40). Thus the steady-state part remains unchanged:

$$D_n \frac{d^2 \Delta n(x)}{dx^2} = \frac{\Delta n(x)}{\tau_p} \quad ...(4.41)$$

giving

$$\Delta n(x) = \Delta n(0)\exp(-x\sqrt{D_n\tau_p}) \qquad ...(4.42)$$

But the time-varying part is

$$D_n\frac{\partial^2 n_1(x,t)}{\partial x^2} = \frac{n_1(x,t)}{\tau_p} + \frac{\partial n_1(x,t)}{\partial t} \qquad ...(4.43)$$

Because of the boundary condition at $x = 0$, the solution to (4.43) takes the form of

$$n_1(x, t) = n_1(x) \exp(j\omega t) \qquad ...(4.44)$$

Substituting (4.44) into (4.43) gives us

$$D_n\frac{d^2 n_1(x)}{dx^2} = \frac{n_1(x)}{\tau_p} + j\omega n_1(x) \qquad ...(4.45)$$

Therefore,

$$\frac{d^2 n_1(x)}{dx^2} = \frac{n_1(x)}{L_p^{*2}} \qquad ...(4.46)$$

where

$$L_p^{*2} = \frac{D_n\tau_p}{(1+j\omega\tau_p)} \qquad ...(4.47)$$

Thus,

$$n_1(x) = n_1(0) \exp(-x/L_p^*) \qquad ...(4.48)$$

This expression shows that we have an effective, time-varying diffusion length that is a complex function of frequency. Using this, let us calculate the time-varying components of the current density J_1, and the optical output power density P_1 that result from the sinusoidal component of the driving voltage V_1.

The current density is derived from the carrier concentration gradient at $x = 0$:

$$\tilde{J} = -qD_n\left(\frac{dn_1}{dx}\right)_{x=0} = \frac{qD_n n_1(0)}{L_p^*}$$

$$= \frac{qD_n n_{p0}}{L_p^*}\frac{q\tilde{V}}{K_B T}\exp(qV_0/K_B T) \qquad ...(4.49)$$

$$\approx J_0\sqrt{1+j\omega\tau_p}\,\frac{q\tilde{V}}{K_BT}$$

Here we have used (4.47). In the above expression, we have

$$J_0 = \frac{qD_n}{L_p}\Delta n(0) = \frac{qD_n n_{p0}}{L_p}[\exp(qV_0/K_BT)-1]$$

$$\approx \frac{qD_n n_{p0}}{L_p}\exp(qV_0/K_BT) \qquad \text{...(4.50)}$$

$$\Rightarrow \qquad \Delta n(0) = \frac{J_0L_p}{qD_n}$$

for the steady state current density. The admittance of the forward-biased diode is given by

$$Y = \frac{1}{r_D} + j\omega C_D = \frac{\tilde{J}A}{\tilde{V}} = \frac{\tilde{I}}{\tilde{V}} \qquad \text{...(4.51)}$$

where A_1 is the junction area and I_1 is the ac component of the current. Therefore,

$$Y = \frac{qJ_0A}{K_BT}\frac{1}{r_D} + j\omega C_D = \frac{\tilde{J}A}{\tilde{V}} = \frac{\tilde{I}}{\tilde{V}} \qquad \text{...(4.52)}$$

$$= \frac{qJ_0A}{K_BT}\sqrt{1+j\omega\tau_p}$$

Let us substitute

$$I_0 = J_0A \qquad \text{...(4.53)}$$

Squaring both the sides, we get

$$Y_2 = \frac{1}{r_D^2} + j\frac{2\omega C_D}{r_D} - \omega^2C_D^2 \qquad \text{...(4.54)}$$

$$= \left(\frac{qI_0}{K_BT}\right)^2(1+j\omega\tau_p)$$

Equating the real parts, we get

$$\frac{1}{r_D^2} - \omega^2C_D^2 = \left(\frac{qI_0}{K_BT}\right)^2 \qquad \text{...(4.55)}$$

so that for $\omega r_D C_D << 1$,

$$r_D \approx \frac{K_B T}{q I_0} \qquad ...(4.56)$$

Equating the imaginary parts, we get

$$\frac{2C_D}{r_D} = \tau_p \left(\frac{q I_0}{K_B T} \right)^2 \qquad ...(4.57)$$

so that

$$C_D \approx \frac{\tau_p}{2} \left(\frac{q I_0}{K_B T} \right) \approx \frac{\tau_p}{2 r_D} \qquad ...(4.58)$$

In deriving the above expressions for the LED resistance r_D and capacitance C_D, we have assumed that $\omega r_D C_D \approx \frac{\omega \tau_p}{2}$. Let us consider high frequencies when $\omega \tau_p >> 1$.

We note that, for $\omega \tau_p >> 1$,

$$\frac{2C_D}{r_D} = \tau_p \left(\frac{1}{r_D^2} - \omega^2 C_D^2 \right)$$

$$\Rightarrow \qquad 1 - \omega^2 C_D^2 r_D^2 = \frac{2 C_D r_D}{\tau_p} \qquad ...(4.59)$$

$$\Rightarrow \qquad \omega^2 C_D^2 r_D^2 + \frac{2 C_D r_D}{\tau_p} - 1 = 0$$

This is a quadratic expression in the diode capacitance. Hence, the solutions are obtained as

$$C_D = \frac{-(2r_D/\tau_p) \pm \sqrt{(2r_D/\tau_p)^2 - 4\omega^2 r_D^2(-1)}}{2\omega^2 r_D^2}$$

$$= \frac{-(2r_D/\tau_p) \pm \left(\frac{4r_D^2}{\tau_p^2} + 4\omega^2 r_D^2 \right)^{1/2}}{2\omega^2 r_D^2}$$

$$= \frac{-(2r_D/\tau_p) \pm (2r_D/\tau_p)(1 + \omega^2 \tau_p^2)^{1/2}}{2\omega^2 r_D^2} \qquad ...(4.60)$$

$$= \frac{-(2r_D/\tau_p) \pm \omega\tau_p(2r_D/\tau_p)}{2\omega^2 r_D^2}$$

$$= \frac{2r_D}{\tau_p}\left(\frac{-1 \pm \omega\tau_p}{2\omega^2 r_D^2}\right)$$

As we have assumed very high frequency operation, we further simplify the above LED capacitance as

$$C_D = \frac{r_D}{\tau_p}\left(\frac{-1 \pm \omega\tau_p}{\omega^2 r_D^2}\right)$$

$$= \frac{r_D}{\tau_p} \cdot \frac{\omega\tau_p}{\omega^2 r_D^2} \qquad \text{...(4.61)}$$

$$= \frac{1}{\omega r_D}$$

$$\Rightarrow \qquad \omega_{CD} = \frac{1}{r_D}$$

Thus, use of (4.57) gives us

$$2C_D \cdot \omega C_D = \tau_p\left(\frac{qI_0}{K_B T}\right)^2$$

$$\Rightarrow \qquad \omega C_D^2 = \frac{\tau_p}{2}\left(\frac{qI_0}{K_B T}\right)^2$$

$$\Rightarrow \qquad \sqrt{\omega} C_D = \sqrt{\frac{\tau_p}{2}}\left(\frac{qI_0}{K_B T}\right) \qquad \text{...(4.62)}$$

$$\Rightarrow \qquad C_D = \sqrt{\frac{\tau_p}{2\omega}}\left(\frac{qI_0}{K_B T}\right)$$

$$\Rightarrow \qquad \omega C_D = \omega . \sqrt{\frac{\tau_p}{2\omega}}\left(\frac{qI_0}{K_B T}\right)$$

$$= \left(\frac{qI_0}{K_B T}\right)\sqrt{\frac{\omega\tau_p}{2}}$$

The LED resistance is calculated as

$$r_D = \frac{1}{\omega C_D} \qquad ...(4.63)$$

$$= \left(\frac{K_B T}{q I_0}\right)\sqrt{\frac{2}{\omega \tau_p}}$$

From the above simplified expressions for the LED capacitance we observe that, the diode capacitance $C_D \propto \frac{1}{\sqrt{\omega}}$ and at high frequencies, this causes a drop in the capacitance value. Hence the junction behavior is dominated by the depletion layer capacitance C_j.

The total optical power density is evaluated as the total number of radiative recombination per second per unit area multiplied by the mean photon energy. Thus,

$$P = \int_0^\infty -\bar{E}_{ph}\left(\frac{dn}{dt}\right)_{rr} dx \qquad ...(4.64)$$

Further, the power may be separated into a steady-state component P_0 and a time-varying component $\tilde{P}$. Thus, the rate of radiative recombination become

$$\left(\frac{dn}{dt}\right)_{rr} = \eta_{\text{int}}\left(\frac{dn}{dt}\right) = -\eta_{\text{int}}\frac{\Delta n(x,t)}{\tau_p} \qquad ...(4.65)$$

This makes

$$P = P_0 + \tilde{P} = \bar{E}_{ph}\eta_{\text{int}}\int_0^\infty \frac{\Delta n(x,t)}{\tau_p} dx \qquad ...(4.66)$$

The steady-state power is

$$P_0 = \frac{\bar{E}_{ph}\eta_{\text{int}}}{\tau_p}\int_0^\infty \Delta n(x) dx \qquad ...(4.67)$$

and

$$\tilde{P} = \frac{\bar{E}_{ph}\eta_{\text{int}}}{\tau_p}\int_0^\infty n_1(x,t) dx \qquad ...(4.68)$$

We make use of the expression for the excess carrier concentration to evaluate the integrals

$$\Delta n(x) = \Delta n(0)\exp\left[-\frac{x}{\sqrt{D_n \tau_p}}\right] = \Delta n(0)\exp\left(-\frac{x}{L_p}\right) \qquad ...(4.69)$$

Hence,

$$P_0 = \frac{\bar{E}_{ph}\eta_{\text{int}}}{\tau_p}\int_0^\infty \Delta n(0)\exp\left(-\frac{x}{L_p}\right)dx \qquad ...(4.70)$$

$$= \frac{\bar{E}_{ph}\eta_{\text{int}}}{\tau_p}\Delta n(0)L_p$$

We further make use of the steady-state current density J_0 to simplify P_0.

$$P_0 = \frac{\bar{E}_{ph}\eta_{\text{int}}}{\tau_p}\Delta n(0)L_p \qquad ...(4.71)$$

$$= \frac{\bar{E}_{ph}\eta_{\text{int}}}{\tau_p}L_p.\frac{J_0 L_p}{qD_n}$$

$$= J_0\frac{\bar{E}_{ph}\eta_{\text{int}}}{\tau_p}\frac{L_p^2}{qD_n}$$

$$= J_0\frac{\bar{E}_{ph}\eta_{\text{int}}}{q} \quad \because L_p^2 = D_n\tau_p$$

The time-varying component is similarly obtained as

$$\tilde{P} = \frac{\bar{E}_{ph}\eta_{\text{int}}}{\tau_p}\int_0^\infty n_1(x,t)dx \qquad ...(4.72)$$

$$= \frac{\bar{E}_{ph}\eta_{\text{int}}}{\tau_p}\int_0^\infty n_1(0)\exp(-x/L_p^*)dx$$

$$= \frac{\bar{E}_{ph}\eta_{\text{int}}}{\tau_p}n_1(0)L_p^*$$

Elimination of $n_1(0)$ gives us

$$\tilde{P} = \frac{\bar{E}_{ph}\eta_{int}}{\tau_p} n_1(0)L_p^* \qquad ...(4.73)$$

$$= \frac{\bar{E}_{ph}\eta_{int}}{\tau_p} \frac{\tilde{J}L_p^*}{qD_n} L_p^*$$

$$= \tilde{J}\frac{\bar{E}_{ph}\eta_{int}}{qD_n\tau_p} L_p^{*2}$$

$$= \frac{\bar{E}_{ph}\eta_{int}}{q} \frac{\tilde{J}}{(1+j\omega\tau_p)} \qquad \because L_p^{*2} = \frac{D_n\tau_p}{(1+j\omega\tau_p)}$$

Hence, the time-varying a.c. frequency response of the LED may be expressed as

$$\left|\frac{\tilde{P}}{\tilde{J}}\right| = \frac{P_0}{J_0} \frac{1}{\sqrt{(1+\omega^2\tau_p^2)}} \qquad ...(4.74)(A2)$$

We may note in passing that there is always a linear relationship between $\tilde{I}$ and $n_1(0, t)$ and $\tilde{P}$ that does not depend on the small-signal assumption that produced (4.73). The only part of this analysis that does depend on the assumption of small signals is the relationship between $\tilde{I}$ and $\tilde{V}$. If the assumption of $q\tilde{V} << K_BT$ does not hold good, but we assume that $n(0, t) = n(0) + n_1(0) \cos \omega t$, then

$$\tilde{V}(t) = \frac{K_BT}{q} \ln\left[1 + \frac{n_1(0)}{n(0)} \cos \omega t\right] \qquad ...(4.75)$$

However, $\tilde{I}$ and $\tilde{P}$ are still related by (4.74). As the boundary between the depletion layer and the diffusion region is very difficult to model correctly, the assumptions made in the previous analysis have to be treated carefully when applied to large-signal, transient conditions.

The implications of (4.74) for the modulated power that can be produced in an optical source at high modulation frequencies are important. At frequencies above $f_m = 1/(2\pi\tau_p)$, the source efficiency falls as shown in Fig. 4.15.

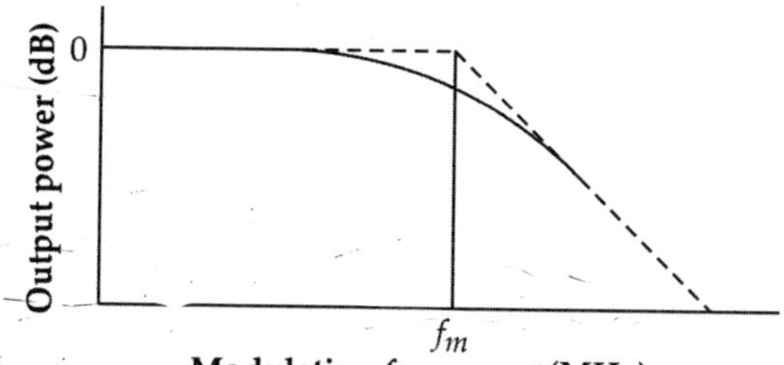

Fig. 4.15: Variation of LED output power as a function of modulating frequency

We may observe from Fig. 4.15 that the output power decreases with increase in modulation frequency. This characteristic is typical of a low pass filter and we therefore may think that the LED is not a suitable source choice for operating frequencies in excess of thousands of MHz. Decreasing τ_p increases this cutoff frequency, but decreasing τ_p by decreasing the lifetime for non-radiative τ_{nr} recombination does not increase the optical power obtainable at high frequencies; it merely decreases the power at low frequencies by reducing the internal quantum efficiency. This follows from (4.24), since

$$f_m = \frac{1}{2\pi\tau_p} = \frac{1}{2\pi\eta_{int}\tau_{rr}} \qquad ...(4.76)$$

It is thus essential to minimize the radiative lifetime. This maximizes both the quantum efficiency at low modulation frequencies and the high frequency cutoff. A high doping concentration and a high injection level minimize τ_{rr} and thus maximize the $\eta_{int} f_m$ product. A limit is reached when the doping level becomes so large that the non-radiative lifetime is decreased as well and so the quantum efficiency falls. In GaAs this occurs at impurity levels in the region of 10^{24} per cubic metre at which $\tau_{rr} \approx 10$ ns. Thus, modulation frequencies much above about 20 MHz can be obtained only at the expense of quantum efficiency. Modulation bandwidths about four times higher than this can be obtained without sacrificing efficiency in longer wavelength InGaAsP-InP LEDs.

4.7.4. Output Power Characteristics

This characteristic curve represents the variation of total optical power with the drive or bias current. There is the possibility of reduction of the optical power output by as much as 3 dB over the device life time. The life time is normally 10^5 hours. The E-LEDs are operated at lower drive current that the S-LEDs. If the drive current is 100 mA, the optical output power is found to lie between 0.5 μW to 10 μW. The variation of the drive current can produce the modulation of the optical power level, as shown in Fig. 4.16.

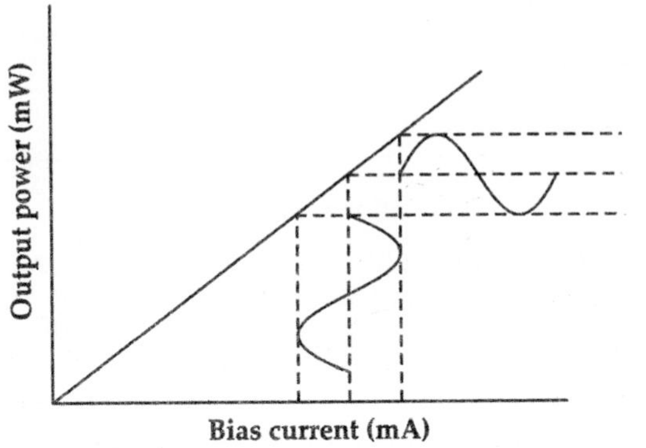

Fig. 4.16: Optical output power as a function of bias current in an LED

We may observe from Fig. 4.16 that a sinusoidal current superimposed on the bias current of the diode results in a change or modulation of the power delivered by the device. As we have assumed a linear operation of the device, a sinusoidal drive results a similar change in the voltage and a sinusoidally varying output power from the device. Needless to say, the bias current is needed to make sure that the device is operating in the linear portion of its volt-ampere characteristics as illustrated in Fig. 4.14(b).

4.7.5. Spectral Emission

LED spectral emission curve is represented by Fig. 4.17. It is a plot of relative power with wavelength. The optical power is measured either in terms of the RMS width or the full width at half maximum abbreviated as FWHM expressed as

$$\text{RMS width} = \frac{\text{FWHM}}{2.35} \qquad \text{...(4.77)}$$

At a wavelength of 1.3 μm the spectral width of the S-LEDs can be as great as 1.30 nm (FWHM). This limits the fibre bandwidth. But E-LEDs have greater material absorption and reemission in the optical channel. Due to all these, the high frequency side (short wavelength) of the spectrum is removed. As a result, the spectrum is narrowed to less than 50 nm at the operating wavelength of 850 nm and less than 70 nm at the wavelength of 1300 nm. However, S-LED with a preferential optical gain at the centre wavelength, can show a spectral width of 10 nm (approx.).

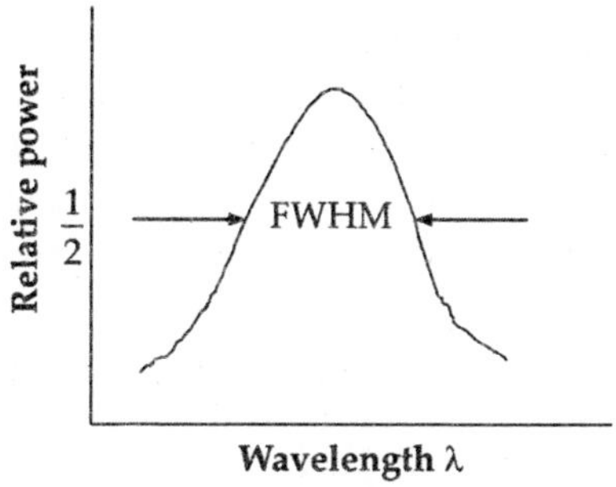

Fig. 4.17: Emission spectral characteristics of LED

4.8. LASER DIODES

LASER is an acronym for light amplification by stimulated emission of radiation. As we may note, this comprises of two terms; stimulated emission and light amplification. Amplification of light takes place under some suitable conditions due to this stimulated emission. Laser may be defined as a highly monochromatic coherent source of optical radiation. In this sense, it may be thought of as being an analogue to an electronic oscillator that acts as a source of periodic sine waves (usually) in the lower frequency range of the electromagnetic spectrum. A laser consists of an active medium that is capable of providing optical amplification and an optical resonator that provides the necessary optical feedback. The active medium may be a collection of atoms, molecules or ions in the solid, liquid or gaseous form and hence this may be addressed as an atomic system. The optical resonator in its simplest form consists of two plane or spherical mirrors aligned suitably to confine optical energy as light propagates back and forth between mirrors. The basic emission and absorption in the atomic system processes are reviewed first and conditions for light amplification and laser oscillation are discussed first. They come in many forms with dimensions ranging from the size of a grain of salt to one that occupies an entire room. The lasing medium can be a gas, a liquid, an insulating crystal (solid state) or a semiconductor. For optical fiber systems the laser sources used almost exclusively are semiconductor laser diodes. They are similar to other lasers, such as the conventional solid-state and gas lasers, in that the emitted radiation has spatial and temporal coherence; that is the output radiation is highly monochromatic and the light beam is very directional.

In fact, Townes, Basov and Prochorov were awarded the 1964 Nobel prize in physics for the invention of laser and maser. The phenomenon of stimulated emission was studied by Towens in 1954 while constructing the maser (similar to laser, however used at microwave frequencies).

For optical fiber communication systems requiring bandwidths greater than approximately 200 MHz, the semiconductor injection laser diode is preferred over LED. Laser diodes typically have response times less than 1 ns, have optical bandwidths of 2 nm or less, and, in general, are capable of coupling several mW of useful luminescent power into optical fibres with small cores and small numerical apertures. Virtually all laser diodes in use and being researched upon at present are multilayered heterojunction devices. The construction of a laser diode is more complicated from that of a LED due to the additional requirement of current confinement in a small lasing cavity.

Lasers come in many forms with dimensions ranging from the size of a grain of a salt to one that will occupy an entire room. The lasing medium can be gas, liquid, an insulating crystal or a semiconductor. For optical fiber systems the laser sources used almost exclusively are semiconductor laser diodes. They are similar to other lasers, such as the conventional solid-state and gas lasers, in that the emitted radiation has spatial and temporal coherence; that is the output radiation is highly monochromatic and the light beam is very directional. For optical fiber communication systems requiring bandwidths greater than approximately 200MHz, the semiconductor injection laser diode is preferred over the LED. Laser diodes typically have response times less than 1ns, have optical bandwidths of 2nm or less, and, in general, are capable of coupling several milliwatts of useful luminescent power into optical fibers with small cores and small numerical apertures. Usually, all lasers are multilayered heterojunction devices. The LEDS have an edge over the laser diodes in terms of inherently simpler construction, smaller temperature dependence of the emitted optical power and the absence of catastrophic degradation. The construction of the laser diodes is more complicated mainly because of the additional requirement of current confinement in a small lasing cavity. The radiation in a laser diode is generated within a Fabry-Perot resonator cavity. This cavity has dimensions of the order of 250 to 500 μm in length, 50 to 15 μm in width and 0.1 to 0.2 μm in thickness. These dimensions are commonly referred to as the longitudinal, lateral and transverse dimensions respectively.

For high speed long distance communications one needs single-mode lasers which must contain only a single longitudinal mode and a single transverse mode. Consequently the spectral width of the optical emission is very narrow. One way of restricting a laser to have only one longitudinal mode is to reduce the length L of the lasing cavity to the point where the frequency separation of the adjacent modes is larger than the laser transition line width, that is, only a single longitudinal mode falls within the gain bandwidth of the device.

Laser action is the result of three key processes. These are photon absorption, spontaneous emission, and stimulated emission. These three processes are represented by the simple two-energy-level diagrams as shown in Fig. 4.18 where E_1 is the ground-state energy and E_2 is the excited-state energy. According to Plank's law, a transition between these two states involves the absorption or emission of a photon of energy $h\upsilon_{12} = E_2 - E_1$. Normally the system is in the ground state. When a photon of energy $h\upsilon_{12}$ impinges on the system, the electron in state E_1 can absorb the photon energy and be excited to the state E_2 as shown in Fig. 4.18(a). Since this is an unstable state, the electron will shortly return to the ground state, thereby emitting a photon of energy $h\upsilon_{12}$. This occurs without any external stimulation and is called spontaneous emission. These emissions are isotropic and of random phase, and thus appear as a narrowband Gaussian output.

The electron can also be induced to make a downward transition from the excited level to the ground state level by an external stimulation. If a photon of energy $h\upsilon_{12}$ impinges on the system while the electron is still in its excited state, the electron is immediately stimulated to drop to the ground state and give off a photon of energy $h\upsilon_{12}$. This emitted photon is in phase with the incident photon, and the resultant emission is known as stimulated emission.

Despite their differences the basic principle of operation is same for each type of laser. Laser action is the result of three key processes. These are photon absorption, spontaneous emission and stimulated emission. These are explained next.

4.8.1. Absorption and Emission of Radiation

We know from elementary physics that an atomic system is characterized by discrete energy levels and the atoms/molecules can exist in one of the allowed levels or energy states. In the energy domain, atoms/molecules can make upward or downward transitions between any two allowed states by absorbing or releasing, respectively, an amount of energy equal to the difference between two energy levels. Thus, if we consider just two energy levels of an atomic system that participate in an interaction with optical radiation of photon energy $h\upsilon = E_2 - E_1$, then there would be three different types of interactions.

(a) *Absorption:* An atom in level 1 can absorb photons of frequency $(E_2 - E_1)/h$ and make an upward transition to the higher energy level. The rate of absorption depends on the number of atoms present in the lower level 1 and the energy density of radiation present in the system.

(b) *Spontaneous Emission:* An atom in an excited state can make a down ward transition spontaneously (on its own) by emitting a photon corresponding to the energy difference between the two levels. The rate of such transition depends only on the number of atoms present in the excited level.

(c) *Stimulated Emission:* An atom in an excited state can also make a downward transition in the presence of external radiation of frequency $(E_2 - E_1)/h$. As in the case of absorption, the rate of such stimulated emissions or induced emissions depends on the energy density of the external radiation and the number of atoms in the exited state. The most important aspect of this type of transition, however, is the fact that the emitted radiation is coherent with the stimulating radiation. Thus, stimulated emission leads to a coherent optical amplification of the incident stimulating radiation and is the process responsible for amplification of optical radiation in a laser.

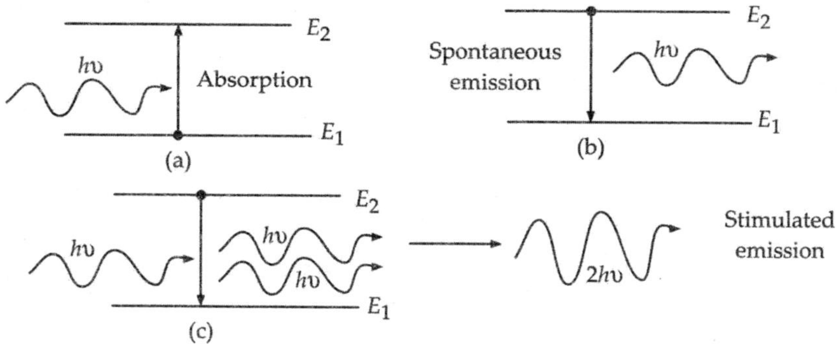

Fig. 4.18: Various ways in which radiation interacts with an atomic system existing at just two energy levels E_1 and E_2

Transitions involving emission or absorption of photons are known as radiative transitions. Atoms can also make an upward or downward transition without involving a photon. For example, an excited atom in the gaseous state can come down to a lower level through an inelastic collision with the walls of the container. Similarly, an excited atom in a solid can make a downward transition by giving the excess energy to the lattice. These are known as nonradiative transitions.

For an atomic system in thermal equilibrium, the occupation probabilities of various atomic energy levels at any temperature are given by the Maxwell-Blotzmann statistic.

$$P(E_i) \propto g_i \exp[-E_i/K_B T]$$

where E_i is the energy associated with the i-th level, K_B is the Boltzmann constant, and g_i is the degeneracy of the level. For nondenegerate levels,

$g_i = 1$ and its value is equal to the number of sublevels (levels having the same energy value) for a degenerate level. Hence, for an atomic system with nondegenerate energy levels, the population density in the various levels under thermal equilibrium is given by $N_1 = C \exp(-E_1/K_BT)$ and $N_2 = C \exp(-E_2/K_BT)$ and and so on where the constant depends on the total number of atoms in the system. Using the above equations, we get

$$\frac{N_2}{N_1} = \exp[-(E_2 - E_1)/K_BT]$$

We observe that, for $E_2 > E_1$, the population density of N_2 energy level E_2 is less than N_1. In other words, an atomic system in thermal equilibrium has its lower energy level more densely populated than a state with higher energy, and the population has an exponential dependence on energy.

In thermal equilibrium, the density of the excited electrons is very small. Most photons incident on the system will therefore be absorbed, so that stimulated emission is essentially negligible. Stimulated emission will exceed absorption only if the population of the excited states is greater than that of the ground state. This condition is known as population inversion. Since this is not an equilibrium condition, population inversion is achieved by various pumping techniques. In a semiconductor laser, population inversion is accomplished by injecting electrons into the material at the device contacts to fill the lower energy states of the conduction band.

In the laser diode Fabry-Perot resonator a pair of flat, partially reflecting mirrors are directed toward each other to enclose the cavity. The mirror facets are constructed by making two parallel cleaves along natural cleavage planes of the semiconductor material. The purpose of these mirrors is to provide strong optical feedback in the longitudinal direction, thereby converting the device into an oscillator with a gain mechanism that compensates for optical losses in the cavity. The laser cavity can have many resonant frequencies. The device will oscillate at those resonant frequencies for which the gain is sufficient to overcome the losses. The sides of the cavity are simply formed by roughing the edges of the device to reduce unwanted emissions in these directions.

In another laser diode type, commonly referred to as the distributed-feedback type, the cleaved facets are not required for optical feedback. The fabrication of this device is similar to the Fabry-Perot type, except that the lasing action is obtained from Bragg reflectors (gratings) or periodic variations of the refractive index which are incorporated into the multilayer structure along the length of the diode.

The optical radiation pattern within the resonance cavity of a laser diode sets up a pattern of electric and magnetic field lines called modes of the cavity. These can conveniently separated into two independent sets of TE and TM modes (to be discussed in later chapters). Each set of modes can be described in terms of the longitudinal, lateral and transverse half-sinusoidal variations of the electromagnetic fields along the major axes of the cavity. The longitudinal modes are related to the length L of the cavity and determine the principal structure of the frequency spectrum of the emitted optical radiation. Since L is much larger than the lasing wavelength of approximately 1 μm, many longitudinal modes can exist. Lateral modes lie in the plane of the pn junction. These modes depend on the side wall preparation and the width of the cavity, and determine the shape of the lateral profile of the laser beam. Transverse modes are associated with the electromagnetic field and beam profile in the direction perpendicular to the plane of the pn junction. These modes are of great importance, since they largely determine laser characteristics such as the radiation pattern and the threshold current density.

Lasing is the condition at which light amplification becomes possible in the laser diode. The requirement for this is to satisfy population inversion. Optical amplification of selected modes is provided by the feedback mechanism of the optical cavity. In the repeated passes between the two partially reflecting mirrors, a portion of the radiation associated with those modes having the highest optical gain coefficient is retained and further amplified during each trip through the cavity. Lasing occurs when the gain of one or several guided modes is sufficient to exceed the optical loss during one roundtrip through the cavity, that is $z = 2L$. The radiation intensity at a photon energy $h\upsilon$ varies exponentially with the distance z that it traverses along the lasing cavity according to the relationship

$$I(z) = I(0)\exp\{[\Gamma g(h\upsilon) - \bar{\alpha}(h\upsilon)]z\} \qquad \text{...(4.78)}$$

where $\bar{\alpha}$ is the effective absorption coefficient of the material in the optical path and Γ is the optical confinement factor (the fraction of optical power in the active layer).

4.8.2. Laser Types

Stimulated emission in semiconductor lasers arises from optical transitions between distributions of energy states in the valence and conduction bands. This differs from gas and solid-state lasers, in which radiative transitions occur between discrete isolated atomic or molecular levels. The radiation in the laser diode is generated within what is known as a 'Fabry-Perot' cavity shown in Fig. 4.19 as in most other types of lasers. However, this cavity is much smaller, being approximately 250 to 500 μm long, 5 to 15 μm wide, and 0.1 to 0.2 μm thick. These dimensions are referred to as the longitudinal, lateral and transverse dimensions of the cavity respectively.

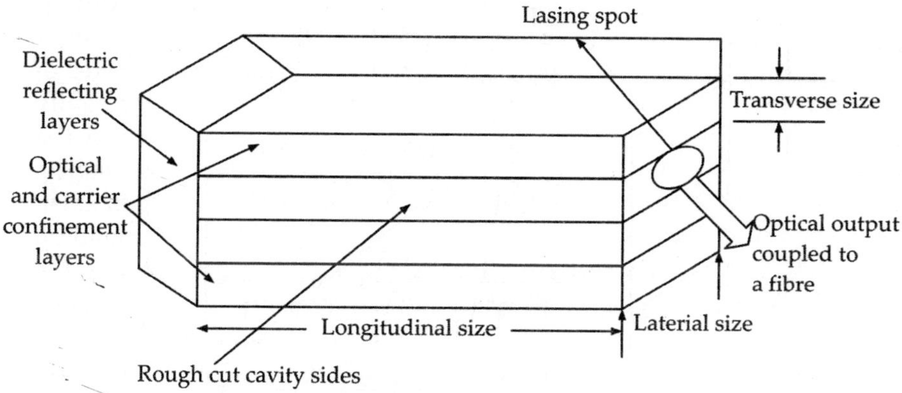

Fig. 4.19: Fabry Perot resonator cavity

4.8.3. Conditions for Amplification by Stimulated Emission

In a nongenerate atomic system, the probability of absorption per atom is the same as the probability of stimulated emission per atom in the presence of an external radiation. Therefore, if the number of atoms in the upper level is more than the number of atoms in the lower level, the rate of stimulated emission would exceed the rate of absorption. In other words, there will be a net emission of radiation, leading to light amplification.

If $u(\upsilon)d\upsilon$ represents the radiation energy per unit volume between υ and $\upsilon + d\upsilon$, then

Rate of absorption = $Bu(\upsilon)/N_1$ and rate of emission = $AN_2 + Bu(\upsilon)N_2$

where A are B Einstein coefficients and represent proportionality constants associated with the spontaneous and stimulated emissions, respectively. At steady state, the rate of absorption is equal to the rate of emission; that is

$$Bu(\upsilon)N_1 = AN_2 + Bu(\upsilon)N_2 \qquad ...(4.79)$$

If we ignore the spontaneous emission for the moment, since we are interested in amplification process due to stimulated emission, then it is clear that for the rate of emission to exceed the rate of absorption of photons (for net emission or light amplification), the number of atoms per unit volume (N_2) in the upper level must be greater than the number of atoms per unit volume (N_1) in the lower level. Since this does not correspond to the normal population distribution in the atomic system in thermal equilibrium, the condition in which a higher level is more populated than a lower level is known as population inversion. Hence we infer that population inversion is the condition for light amplification. The number of upward transitions is equal to the number of downward transitions at thermal equilibrium. Hence, under the assumption of thermal equilibrium conditions to prevail we have the steady state conditions given by (4.79). A simple manipulation of this gives us

$$u(\upsilon) = \frac{AN_2}{B(N_1 - N_2)} = \frac{A}{B\left(\frac{N_1}{N_2} - 1\right)} \quad ...(4.80)$$

The next step would be to understand how to achieve population inversion in an atomic system. The population of various levels is given by the Boltzmann distribution at thermal equilibrium. However, if we irradiate an atomic system by an external radiation of appropriate frequency, this population distribution changes, and in some atomic systems, under certain conditions, it is possible to achieve population inversion between a pair of levels.

Boltzman's law gives us the populations of these two energy levels as

$$\frac{N_2}{N_1} = \exp[-(E_2 - E_1)/K_B T] = \exp\left[-\frac{h\upsilon}{K_B T}\right] \quad ...(4.81)$$

Substitution of (4.81) in the energy density (4.80) gives us further,

$$u(\upsilon) = \frac{A}{B\left(\frac{N_1}{N_2} - 1\right)} = \frac{A}{B\left[\exp\left(\frac{h\upsilon}{K_B T}\right) - 1\right]} \quad ...(4.82)$$

From Plank's law, the energy density of radiation is given as

$$u(\upsilon) = \frac{h(2\pi\upsilon)^3 n_0^3}{\pi^2 c^3} \frac{1}{\exp\left(\frac{h\upsilon}{K_B T}\right) - 1} \quad ...(4.83)$$

We may immediately observe that,

$$\frac{A}{B} = \frac{4\pi\upsilon^3}{c^3} n_0^3 \quad ...(4.84)$$

We have assumed that an atom is capable of interacting with radiation of a particular frequency υ. However, in general, an atom can interact with radiation over a range of frequencies and the strength of the interaction is a function of the frequency. This function is known as the line shape function. If $g(\upsilon)$ represents the normalized line shape function corresponding to the transition between levels 1 and 2, then $N_1 g(\upsilon)d\upsilon$ and $N_2 g(\upsilon)d\upsilon$ would represent the number of atoms per unit volume in levels 1 and 2 respectively that are capable of interacting with radiation of frequency between υ and $\upsilon + d\upsilon$. Since the atomic system is likely to interact with radiation at one frequency or the other, we must have the normalized line shape function is hence, expressed as $\int_0^\infty g(\upsilon)d\upsilon = 1$ so that the total number of atoms in levels 1 and 2 is N_1 and N_2 respectively. We have, further

$$B = \frac{c^3}{h8\pi\upsilon^3 n_0^3} A = \frac{c^3}{h8\pi\upsilon^3 n_0^3 \tau_{sp}} \qquad \text{...(4.85)}$$

where $\tau_{sp} = 1/A$ represents the spontaneous lifetime of the upper level. The total number of stimulated emissions occurring in a unit time in unit volume is, under the assumption of the above normalized line shape function

$$N_{21} = N_2 \int_0^\infty Bu(\upsilon)g(\upsilon)d\upsilon = N_2 \frac{c^3}{h8\pi n_0^3 \tau_{sp}} \int_0^\infty \frac{u(\upsilon)}{\upsilon^3} g(\upsilon)d\upsilon \qquad \text{...(4.86)}$$

For a near monochromatic radiation field, $u(\upsilon)$ is very sharply peaked at a particular value of υ(say some $\upsilon1$). Hence, in carrying out the above integration, we may think that only the term corresponding to $u(\upsilon1)$ would contribute significantly to the integrand taking smaller contributions from other frequency components. In other words, the overall integrand is dominated by this peak frequency contribution only so that we may approximate the above expression as

$$N_{21} \approx N_2 \frac{c^3}{h8\pi n_0^3 \tau_{sp}} \frac{u_{\upsilon1}}{\upsilon1^3} g(\upsilon1) \qquad \text{...(4.87)}$$

where $u_{\upsilon1}$ represents the energy density associated with the radiation field and $g(\upsilon1)$ represents the value of the line shape function evaluated at the radiation frequency $\upsilon1$. We may note here that the units of $u_{\upsilon1}$ and $u(\upsilon1)$ are different. The first represents the total energy in a unit volume whereas the second energy represents the energy contained in a unit volume at unit frequency. In a similar manner, the number of stimulated emissions in unit time corresponding to a unit volume of the active medium is expressed as

$$N_{12} = N_1 \int_0^\infty Bu(\upsilon)g(\upsilon)d\upsilon = N_1 \frac{c^3}{h8\pi n_0^3 \tau_{sp}} \frac{u_{\upsilon 1}}{\upsilon 1^3} g(\upsilon 1) \qquad ...(4.88)$$

Let us consider a collection of atoms and let a monochromatic light of frequency υ be propagating through it along the z-direction. We wish to obtain en expression for the rate of change of intensity of the beam as it propagates. To do so, we consider two planes of area A transverse to the propagation axis at locations of z and $z + dz$ respectively. The volume of the medium between planes P_1 and P_2 is Adz and hence the number of stimulated absorptions in a unit time is $N_{12}Adz$. Since each photon has an energy $h\upsilon$, the energy absorbed in unit time in the volume element of Adz becomes, therefore,

$$N_{12}\, h\upsilon Adz$$

In a similar manner, the energy radiated back (energy gain due to stimulated radiations) becomes

$$N_{21}\, h\upsilon Adz$$

In the energy expression for stimulated emissions, the energy content due to the spontaneous emissions has been neglected. This is due to the reason that the spontaneous emissions take place in random directions which results in a very small amount of overall or net energy available from such emissions. The net amount of energy absorbed in unit time in the volume element of Adz is, therefore,

$$(N_{12} - N_{21})h\upsilon Adz \qquad ...(4.89)$$

Let the beam intensity in the plane P_1 be $I_\upsilon(z)$ so that the total energy entering the volume element Adz in unit time is

$$I_\upsilon(z)A$$

Similarly, the total energy leaving this volume element in one unit time is, corresponding to the beam intensity of $I_\upsilon(z + dz)$ in plane P_2 is,

$$I_\upsilon(z + dz)A = I_\upsilon(z)\mathrm{A} + \frac{\partial I_\upsilon}{\partial z} dz\mathrm{A}$$

From all of this, we have the combined result that the net amount of energy leaving the volume element in one unit time is

$$\frac{\partial I_\upsilon}{\partial z} dz A \qquad ...(4.90)$$

This must be equal to the negative of net energy absorbed by the medium present between z and $z + dz$. Thus, we write

$$\frac{\partial I_\upsilon}{\partial z} dz A = -(N_{12} - N_{21}) h\upsilon A dz$$

$$\Rightarrow \qquad \frac{\partial I_\upsilon}{\partial z} = -(N_{12} - N_{21}) h\upsilon$$

$$= -\left[N_1 \frac{c^3}{h 8\pi n_0^3 \tau_{sp}} \frac{u_{\upsilon 1}}{\upsilon 1^3} g(\upsilon 1) - N_2 \frac{c^3}{h 8\pi n_0^3 \tau_{sp}} \frac{u_{\upsilon 1}}{\upsilon 1^3} g(\upsilon 1) \right] h\upsilon \qquad \text{...(4.91)}$$

$$= -(N_1 - N_2) \frac{c^3}{8\pi n_0^3 \tau_{sp}} \frac{u_{\upsilon 1}}{\upsilon 1^2} g(\upsilon 1)$$

This may be written as

$$\frac{\partial I_\upsilon}{\partial z} = -(N_1 - N_2) \frac{c^3}{8\pi n_0^3 \tau_{sp}} \frac{u_\upsilon}{\upsilon^2} g(\upsilon) \qquad \text{...(4.92)}$$

We can replace the dummy variable $\upsilon 1$ by υ. The energy density u_υ and the intensity I_υ are related by

$$I_\upsilon = \text{v} u_\upsilon \frac{c}{n_0} u_\upsilon \qquad \text{...(4.93)}$$

In writing this, we have assumed the velocity of the radiation in the active medium to be

$$c/n_0 = \text{v} \qquad \text{...(4.94)}$$

The differential equation simplifies to

$$\frac{\partial I_\upsilon}{\partial z} = -(N_1 - N_2) \frac{c^3}{8\pi n_0^3 \tau_{sp}} \frac{I_\upsilon}{\upsilon^2 \text{v}} g(\upsilon)$$

$$= -(N_1 - N_2) \frac{c^2}{8\pi n_0^2 \tau_{sp}} \frac{I_\upsilon}{\upsilon^2} g(\upsilon) \qquad \text{...(4.95)}$$

$$= -\gamma(\upsilon)I_\upsilon$$

In the above expression, we have substituted

$$\gamma(\upsilon) = (N_1 - N_2)\frac{c^2}{8\pi n_0^2 \tau_{sp}}\frac{g(\upsilon)}{\upsilon^2} \quad \text{...(4.96)}$$

As the line shape function is very sharply peaked, the function $\gamma(\upsilon)$ is also a sharply peaked function. The integration of (4.95) gives us

$$I_\upsilon(z) = I_\upsilon(0)\exp[-\gamma(\upsilon)z] \quad \text{...(4.97)}$$

Let us consider three levels-the ground state and two excited states-of an atomic system shown in Fig. 4.20(a). If we irradiate the system with a radiation of frequency $\upsilon = (E_3 - E_1)/h$, atoms will get excited from level 1 to level 3. Excitation of an atomic by an external means is often referred to as pumping. The excited atoms in level 3 can make a downward transition to level 1 by spontaneous and stimulated emissions. Also, the transitions either may be one-step direct transition from level 3 to level 1 or may involve two steps: an initial transition from level 3 to level 2, followed by a second transition from level 2 to level 1. In the latter case, at a given pumping rate, if the rate of transition from level 3 to level 2 exceeds that between level 2 and level $1(T_{21})$, then atoms would start accumulating in level 2, and the population of level 2 would increase with the pumping rate. When the pumping rate exceeds a certain threshold, the steady-state population in level 1 becomes higher than that in level 1. In this state of population inversion, radiation at frequency $\upsilon_1 = (E_2 - E_1)/h$ will get amplified by the phenomenon of stimulated emission.

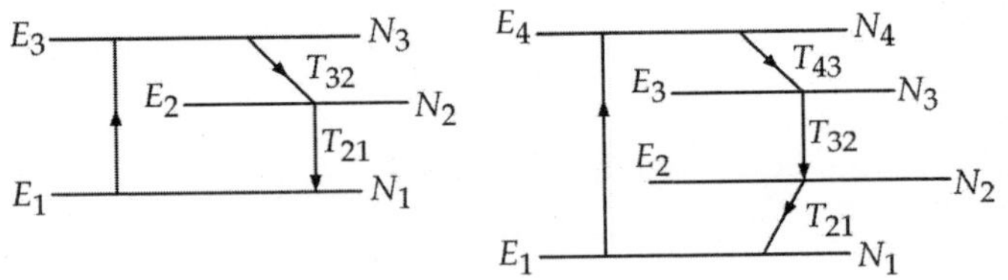

Fig. 4.20: (a) A three level laser (b) A four level laser

The above scheme of transitions to realize laser action is known as a three level pumping scheme, and the atomic system is identified as a three-level system. Because population inversion is to be achieved between an excited state (level 2) and the ground state (level 1), considering the fact that the population of the ground level at thermal equilibrium is very much larger than that in an excited level, the required threshold pumping power for population inversion is very large (~ kW). A more efficient scheme employed in most lasers is the four-level scheme illustrated in Fig. 4.20(b). Here, the population inversion and the lasing transitions occur between levels 3 and 2. If the transition rates T_{43} and T_{21} are large compared with the transition rate T_{32}, then population inversion can be achieved between levels 3 and 2 even with low to moderate pumping powers. The small signal gain coefficient of a laser amplifier at frequency υ is given by, we have derived above as

$$\gamma(\upsilon) = \frac{(c/n)^2}{8\pi\tau_{sp}} \frac{g(\upsilon)}{\upsilon^2} \Delta N \qquad ...(4.98)$$

Where τ_{sp} is the spontaneous emission lifetime (the average time spent by an atom in the excited state before making a downward transition); ΔN is the population inversion density between the two levels taking part in laser transition-that is $\Delta N = N_2 - N_1$ in a three level system and $\Delta N = N_3 - N_2$ in a four-level system. The function $g(\upsilon)$ is known as the normalized line shape function and is defined so that $g(\upsilon)d\upsilon$ represents the probability that an atom interacts with radiation between frequencies υ and $\upsilon + d\upsilon$.

We may see from the small signal gain expression that a positive value implies that $N_1 > N_2$ and the beam intensity decreases exponentially so that it becomes $1/e$ times its value $z = 0$ at after traversing a distance of $1/\gamma(\upsilon)$. Under thermal equilibrium conditions, the number of atoms in the lower level is greater than that in the higher level and this explains the exponential decay of the beam as it travels down the medium. However, the onset of population inversion creates more number of atoms in the higher level than in the lower level which makes this gain a negative quantity. In other words, optical amplification is accorded to the light beam as it travels through the active medium.

The function $g(\upsilon)$ also gives the absorption or emission spectrum of the atomic system. In general, this function is sharply peaked in the form of a Lorentzian or Gaussian around a central resonant frequency. The gain coefficient at any frequency and the bandwidth of the laser amplifier are essentially determined by $g(\upsilon)$.

4.9. SEMICONDUCTOR LASER

A semiconductor laser consists of a forward biased PN junction formed in a direct bandgap semiconductor. The recombination of the injected carriers-namely electrons and holes in the junction region result in the emission of photons. The cleaved ends of the laser structure act like mirrors, forming a Fabry-Perot resonator along the P-N junction. The other two ends in the perpendicular direction are saw-cur to reduce reflections from these ends and prevent lasing along the perpendicular direction. This is illustrated in Fig. 4.21.

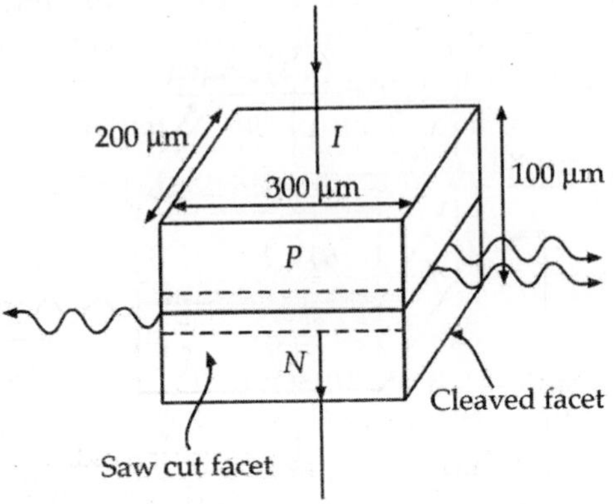

Fig. 4.21: Two ends are cleaved and the laser output comes out of these ends. The other two surfaces are saw cut to prevent any oscillations taking place along that direction.

When the forward current through the diode exceeds a critical value known as the threshold current, optical gain in the resonator due to stimulated emissions overcomes the losses in the resonator, leading to net amplification and eventually steady-state oscillations. The gain coefficient for amplification of radiation frequency υ by stimulated emission of in a semiconductor is given by

$$\gamma(\upsilon) = \frac{(c/n)^2}{8\pi\tau_r} \frac{\rho(\upsilon)}{\upsilon^2} \Delta P \qquad \text{...(4.99)}$$

where τ_r is the radiative recombination time, $\rho(\upsilon)$ is the reduced density of states, or the optical joint density of states $\Delta_P = P_e(\upsilon) - P_a(\upsilon)$ is the difference of the probabilities for emission and absorption at the optical frequency υ and is known as the Fermi inversion factor. For amplification, $\gamma(\upsilon) > 0$ which in turn requires $\Delta P > 0 \Rightarrow P_e(\upsilon) > P_a(\upsilon)$. This means that the probability of emission must be higher than the probability of absorption. This is the necessary condition for light amplification.

The density of states in a semiconductor refers to the number of allowed electron states per unit volume per unit energy interval. Thus, for example, if $\rho_c(E)$ represents the density of states of electrons in the conduction band, then $\rho_c(E)dE$ is the number of states having energy levels between E and dE. A similar condition prevails for holes occupying the valence band and we have $\rho_v(E)$ as the density of energy states corresponding to the holes. When a photon interacts with an electron and a hole in a semiconductor, the number of states available for the interaction is limited by the requirement of energy conservation. For example, if the photon energy is just equal to the bandgap energy E_g, then the electron states well above the conduction band edge and well below the valence band edge cannot participate in the interaction. The optical joint density of states takes into account the number of states available in both the conduction band and the valence bands with which a photon of energy $h\upsilon$ can interact and is given by

$$\rho(\upsilon) = K.\sqrt{(h\upsilon - E_g)};h\upsilon \geq E_g \qquad ...(4.100)$$

where K is a constant. We observe that only photons with energy higher than the bandgap energy of the semiconductor material can take part in the emission and the absorption process.

This gain expression for a semiconductor laser is very similar to (4.98) for the gain coefficient of general lasers with the Fermi inversion factor ΔP taking up the place of population inversion ΔN. We are henceforth interested in the conditions leading to a positive value for ΔP.

Let E_1 and E_2 be the energies of two allowed electron states in the valence band and the conduction band respectively. An electron should be present in the energy state E_1 in the valence band and the energy state E_2 in the conduction band should be empty in order for an incident photon of energy $h\upsilon$ to be absorbed. The probability of absorption is given by

$$P_a(\upsilon) = p(E_1)[1 - p(E_2)] \qquad ...(4.101)$$

where $p(E)$ represents the probability of energy state E occupancy and it is obvious that $[1 - p(E)]$ is the probability that the states with energy E is unoccupied. Thus we do find that the probability of occupancy corresponding to the energy state of E_1 and the probability of unoccupancy corresponding to the energy state of E_2 are two independent events. Similarly, for emission of a photon because of a hole-electron recombination, we require a hole in the valence band at energy state with value E_1 and an electron in the conduction band at energy state with value E_2. Therefore, the probability of emission is given by

$$P_e(\upsilon) = p(E_2)[1 - p(E_1)] \qquad ...(4.102)$$

We find that for a semiconductor in thermal equilibrium at practical temperatures, the values of $p(E_1)$ and $[1 - p(E_1)]$ are much larger than $p(E_2)$ and $[1 - p(E_1)]$. Thus, $P_e(\upsilon) < P_a(\upsilon)$ which means that optical amplification is not possible in a semiconductor at thermal equilibrium.

We next consider a semiconductor in quasi-equilibrium. For such a case the probabilities of absorption and emission are given by

$$P_e(\upsilon) = p_c(E_2)[1 - p_v(E_1)] \qquad ...(4.103)$$

and $$P_a(\upsilon) = p_v(E_1)[1 - p_c(E_2)] \quad ...(4.104)$$

where the probabilities p_v and p_c denote the quasi-Fermi distributions in the valence band and the conduction band respectively.

For a net emission to occur, we know that the probability of emission should be more than that of absorption. This means that

$$p_c(E_2)[1 - p_v(E_1)] > p_v(E_1)[1 - p_c(E_2)] \quad ...(4.105)$$

Further, the probability distribution of the carriers in the valence band and the conduction band is given as

$$p_v(E) = \frac{1}{1+\exp\left[-\frac{(E-E_{fv})}{K_BT}\right]} \quad ...(4.106)$$

for the valence band and the corresponding probability of the energy state E being empty is

$$1 - p_v(E_1) = 1 - \frac{1}{1+\exp\left[-\frac{(E_1-E_{fv})}{K_BT}\right]} = \frac{e^{-\frac{(E_1-E_{fv})}{K_BT}}}{1+e^{-\frac{(E_1-E_{fv})}{K_BT}}} \quad ...(4.107)$$

and $$p_c(E) = \frac{1}{1+\exp\left[-\frac{(E-E_{fc})}{K_BT}\right]} \quad ...(4.108)$$

for the conduction band and the corresponding probability of having an empty energy state becomes

$$1 - p_c(E_2) = 1 - \frac{1}{1+e^{-\frac{(E_2-E_{fv})}{K_BT}}} = \frac{e^{-\frac{(E_2-E_{fv})}{K_BT}}}{1+e^{-\frac{(E_2-E_{fv})}{K_BT}}} \quad ...(4.109)$$

The probability of absorption becomes

$$P_a(\upsilon) = p_v(E_1)[1 - p_c(E_2)] \quad ...(4.110)$$

$$= \frac{1}{1+e^{-\frac{(E_1-E_{fv})}{K_BT}}} \cdot \frac{e^{-\frac{(E_1-E_{fv})}{K_BT}}}{1+e^{-\frac{(E_1-E_{fv})}{K_BT}}} = \frac{e^{-\frac{(E_1-E_{fv})}{K_BT}}}{\left[1+e^{-\frac{(E_1-E_{fv})}{K_BT}}\right]^2}$$

The probability of emission can be written in a similar manner.

In order for a net emission to take place, we should ensure

$$\frac{e^{\frac{(E_1-E_{fv})}{K_BT}}}{\left[1+e^{\frac{(E_1-E_{fv})}{K_BT}}\right]^2} \geq \frac{e^{\frac{(E_2-E_{fcv})}{K_BT}}}{\left[1+e^{\frac{(E_2-E_{fc})}{K_BT}}\right]^2} \qquad ...(4.111)$$

This in turn gives us

$$E_{fc} - E_{fv} > h\upsilon \qquad ...(4.112)$$

Since

$$E_{fc} - E_{fv} > E_g \qquad ...(4.113)$$

Physically it means that when the separation between the quasi-Fermi levels in a semiconductor exceeds the bandgap, then for all frequencies that satisfy (4.113) it is possible to have light amplification.

The above condition is equivalent to the requirement of population inversion in an atomic system. The quasi-Fermi levels should lie close to or into the respective bands in order to satisfy the above equation. This implies that there is a large concentration of electrons in the conduction band and holes in the valence band which is unlike the normal carrier distributions in a semiconductor at thermal equilibrium. In this sense, it is indeed equivalent to population inversion.

4.10. THRESHOLD CONDITION FOR LASER OSCILLATION

Optical amplification is achieved by having population inversion induced through pumping. If this optical amplifier is provided with a suitable optical feedback, we realize that sustained oscillations would take place similar to those taking place in electronic oscillators. The necessary optical feedback is provided by placing the active medium in an optical resonator composed of two high reflectivity mirrors, separated by a suitable distance d as shown in Fig. 4.22. Plane or spherical mirrors may be used and may take the form of discrete components located outside the gain medium or be attached to the ends of the gain medium. Usually, one of the mirrors (say M_1) is almost 100% reflecting ($R_1 \approx 1.0$) and the other mirror M_2 is partially reflecting ($R_2 < 1.0$) so that a small fraction of the optical energy comes out of the resonator. Obviously, from the resonator's point of view, this output is a loss but forms the useful output of the laser for the users.

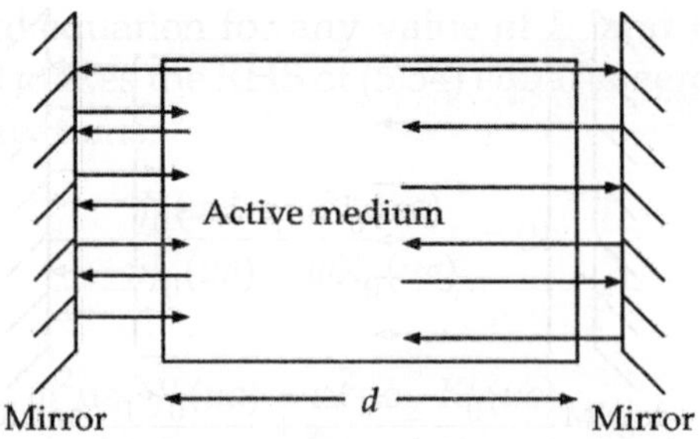

Fig. 4.22: A plane parallel resonator consisting of a pair of plane mirrors facing each other. The active medium is placed inside the cavity. One of the mirrors is made partially reflecting to obtain the laser output

Let us consider the optical resonator shown in Fig. 4.22. In the absence of the gain medium, there is no gain in the resonator, and the resonator is referred to as a passive resonator; it is a lossy device because of the finite reflectivities $(R_1, R_2) < 1$ of the mirrors at the ends of the resonator and the propagation loss associated with the light bouncing back and forth between the mirrors. The propagation loss consists of loss mechanisms such as scattering loss in the medium and diffraction loss due to the finite size of the mirrors. In the presence of pumping, there would be both amplification and attenuation of radiation at the lasing transition within the resonator. Let I_0 be the irradiance of the light beam leaving the mirror M_1 and propagating toward M_2. Let the average loss coefficient per unit length of the resonator due to all mechanisms other than the finite reflectivities of the mirrors be a. The irradiance of the beam after one complete round trip is given by

$$I_1 = I_0 e^{-ad} e^{\gamma d} R_2 e^{-ad} e^{\gamma d} R_1 \qquad ...(4.114)$$

The factors multiplying I_0 on the RHS of (4.114) indicate the various processes occurring in one complete round trip of the light beam. Thus, the above equation may be written as

$$I_1 = I_0 e^{2(\gamma - a)d} R_1 R_2 \qquad ...(4.115)$$

For sustained oscillations to be built up in the resonator, we must have

$$R_1 R_2 e^{2(\gamma - a)d} \geq 1 \qquad ...(4.116)$$

The equality sign in (4.116) corresponds to the situation when the roundtrip loss is exactly compensated by the amplification provided by the gain medium, so that $I_1 = I_0$. This condition is known as the threshold condition for laser oscillation and the corresponding value of the gain coefficient is known as the threshold gain coefficient γ_{Th}. Hence, at threshold we should have

$$R_1R_2e^{2(\gamma_{Th}-a)d} = 1$$

$$\Rightarrow \qquad e^{2(\gamma_{Th}-a)d} = \frac{1}{R_1R_2}$$

$$2(\gamma_{Th} - a)d = \ln\left(\frac{1}{R_1R_2}\right) \qquad \text{...(4.117)}$$

$$\Rightarrow \qquad \gamma_{Th} = \frac{1}{2d}\left[2ad + \ln\left(\frac{1}{R_1R_2}\right)\right] = a - \frac{1}{2d}\ln(R_1R_2)$$

We may also observe from the foregoing that the absence of an active medium the energy stored in the resonator would have an exponential decay profile due to finite reflectivities of the mirrors, losses due to scattering, diffraction. This loss profile may be modelled as $E(t) = E(0)$ exp $(-t/\tau_c)$ where τ_c is known as the passive cavity lifetime. The time taken for one round trip in the active medium is $t = 2d/(c/n_0) = 2n_0d/c$, we may write

$$R_1R_2\exp(-2ad) = \exp(-2n_0d/c\tau_c)$$

$$\Rightarrow \qquad 2n_0d/c\tau_c = 2ad - \ln R_1R_2 \qquad \text{...(4.118)}$$

$$\Rightarrow \qquad a = \frac{n_0}{c\tau_c} + \frac{\ln R_1R_2}{2d}$$

Substitution of this loss coefficient in the threshold gain expression gives us

$$\gamma_{Th} = a - \frac{1}{2d}\ln(R_1R_2) \qquad \text{...(4.119)}$$

$$\Rightarrow \qquad \frac{n_0}{c\tau_c} + \frac{\ln R_1R_2}{2d} - \frac{1}{2d}\ln(R_1R_2) = \frac{n_0}{c\tau_c}$$

Further manipulation gives us

$$\frac{n_0}{ct_c} = \frac{(c/n_0)^2}{8\pi\tau_{sp}}\frac{g(\upsilon)}{\upsilon^2}\Delta N_{Th}$$

$$\Rightarrow \qquad \Delta N_{th} = \frac{8\pi\tau_{sp}}{(c/n_0)^2}\frac{\upsilon^2}{g(\upsilon)}\frac{n_0}{c\tau_c} \qquad \text{...(4.120)}$$

$$= \frac{8\pi\upsilon^2}{c^3}\frac{n_0^3}{g(\upsilon)}\frac{\tau_{sp}}{\tau_c}$$

The above equation gives the threshold population inversion required for the oscillations to be initiated in the laser. Clearly, the minimum threshold value corresponds to the centre of the line shape function where it is a maximum. As the laser medium is pumped harder and harder, the population inversion between the two levels goes on increasing. The mode that lies nearest to the resonance frequency of the atomic system reaches threshold first and begins to oscillate. As the pumping is further increased, the nearby modes may also reach the threshold and start oscillating.

When the gain coefficient becomes larger than the threshold value, the "greater than" sign in (4.116) applies and the radiation builds up after every round trip. The mode that satisfies (4.117) reaches threshold first. Theoretically, at the onset of this condition, all additional energy introduced into the laser should augment the growth of this particular mode. In practice, various phenomena lead to the excitation of more than one mode. Studies on the conditions of longitudinal single-mode operation have shown that important factors are thin active regions and a high degree of temperature stability.

In the laser diode Fabry-Perot resonator a pair of flat, partially reflecting mirrors face each other to enclose a cavity. The purpose of these mirrors is to provide a strong optical feedback in the longitudinal direction thereby converting the device into an oscillator with a gain mechanism that compensates for optical losses in the cavity. The laser cavity has many resonant frequencies. The device oscillates (emits light) at those resonant frequencies for which the gain is sufficient to overcome the losses. The sides of the cavity are formed by roughening the edges of the device to reduce unwanted emissions in these directions.

The optical radiation within the resonance cavity of a laser diode sets up a pattern of electric and magnetic field lines called the 'modes of the cavity'. These can conveniently be separated into two independent sets of TE and TM modes. Each of these modes further can be described in terms of the longitudinal, lateral and transverse half-sinusoidal variations of the electromagnetic fields along the major axes of the cavity. The longitudinal modes are related to the length L of the cavity. Since L is much larger than the lasing wavelength of approximately 1 μm, many longitudinal modes can exist. Lateral modes lie in the plane of the P-N junction. These modes depend on the side wall preparation and width of the cavity, and determine the shape of the lateral profile of the laser beam. Transverse modes are associated with the electromagnetic field and beam profile in the direction perpendicular to the plane of the PN junction.

These modes are of great importance, since they largely determine laser characteristics such as the radiation pattern and the threshold current density.

4.10.1. Resonant Frequencies

Let us examine the resonant frequencies for the laser. We have

$$2\beta L = 2\pi m \quad m \text{ being an integer} \qquad \text{...(4.121)}$$

Use of $\beta = \frac{2\pi n}{\lambda}$ for the propagation constant we get

$$m = \frac{L}{\lambda/2n} = \frac{2Ln}{c} f \qquad \text{...(4.122)}$$

where $c = f\lambda$. This states that the cavity resonates when an integer number of half-wavelengths spans the region between the mirrors.

In all lasers the gain is a function of frequency or conversely the wavelength, there would be a range of frequencies or wavelengths for which (4.122) holds. Each of these frequencies corresponds to a mode of oscillation of the laser. Depending on the laser structure, any number of frequencies can satisfy (4.121) and (4.122). Thus some lasers are single mode and some are multimode. The relationship between gain and frequency is assumed to be of a Gaussian form given as

$$\gamma(\lambda) = g(0)\exp\left[-\frac{(\lambda-\lambda_0)^2}{2\sigma^2}\right] \qquad \text{...(4.123)}$$

Where λ_0 is the wavelength at the centre of the spectrum as shown in Fig. 4.23, σ is the spectral width of the gain $g(0)$ and the maximum gain is proportional to population inversion.

Let us now look at the spacing between the modes of a multimode laser. We consider only the longitudinal modes here. For each longitudinal mode, there may be several transverse modes that arise from one or more reflections of the propagating wave at the sides of the resonator cavity. To find the frequency spacing, we consider two successive modes of frequencies f_m and f_{m+1} represented by integers m and $m + 1$. Thus we write

$$m = \frac{2Ln}{c} f_m \qquad \text{...(4.124)}$$

and

$$m + 1 = \frac{2Ln}{c} f_{m+1} \qquad \text{...(4.125)}$$

Subtracting these two equations we get

$$\frac{2Ln}{c}(f_{m+1} - f_m) = 1$$

$$\Rightarrow \qquad \frac{2Ln}{c}\Delta f = 1 \qquad \text{...(4.126)}$$

$$\Rightarrow \qquad \Delta f = \frac{c}{2Ln}$$

Where we have written

$$f_{m+1} = f_m = \Delta f \qquad \text{...(4.127)}$$

We can relate this to the wavelength spacing $\Delta\lambda$ through the relationship

$$\frac{\Delta f}{f} = \frac{\Delta\lambda}{\lambda}$$

$$\Rightarrow \qquad \lambda = \frac{\Delta\lambda}{\Delta f} f = \frac{\Delta\lambda}{(c/2Ln)}\frac{c}{\lambda} \qquad \text{...(4.128)}$$

$$\Rightarrow \qquad \Delta\lambda = \frac{\lambda^2}{2Ln}$$

Thus given (4.123) and (4.128), the output spectrum of a multimode laser has a gain versus frequency characteristic as depicted in Fig. 4.23 where the exact number of modes, their heights and their spacing depend on the laser construction.

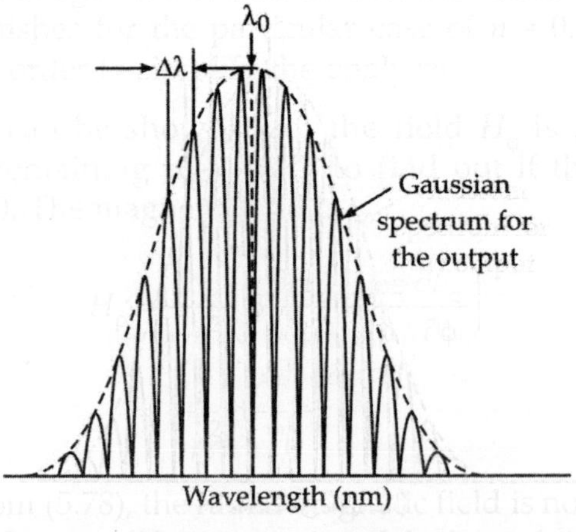

Fig. 4.23: Gaussian profile for the output spectrum of a typical laser

The relationship between optical output power and diode drive current is presented in Fig. 4.24.

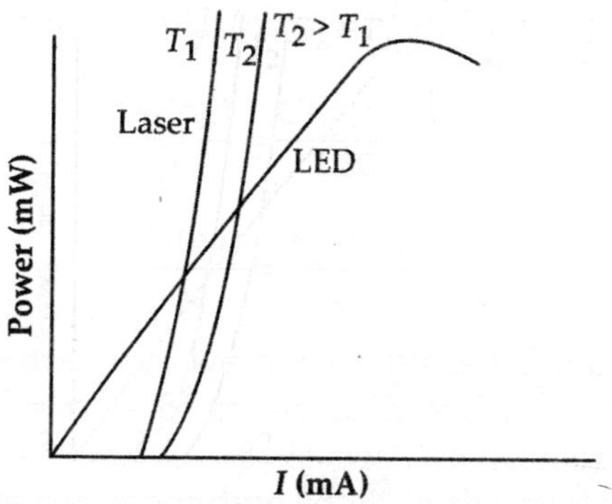

Fig. 4.24: Variation of output power with input current for a laser and an LED

We observe from Fig. 4.24 that at low diode currents only spontaneous emission is emitted. Both the spectral range and the lateral beam width of this emission are broad like that of an LED. A dramatic and sharply defined increase in the power output occurs at the lasing threshold. As this transition point is approached, the spectral range and the beam width both narrow with increasing drive current. The final spectral width of approximately 1 nm and the fully narrowed lateral beam width of nominally 5 to 10° are reached just pass the threshold point. The threshold current I_{Th} is conventionally defined by extrapolation of the lasing region of the power-*versus*-current curve as shown in Fig. 4.25. At high power outputs the slope of the curve decreases because of junction heating.

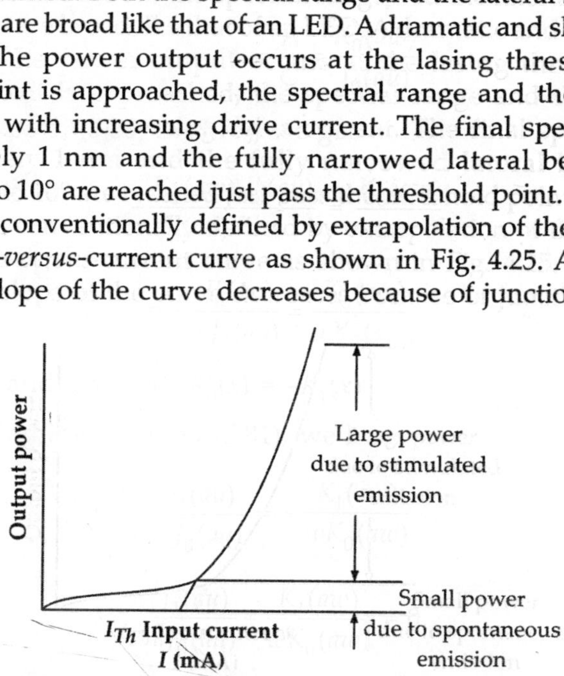

Fig. 4.25: Extrapolation of Fig. 4.24 to show the existence of a threshold drive current in a typical laser that demarcates the spontaneous emission and the stimulated emission

The external differential quantum efficiency η_{ext} is defined as the number of photons emitted per radiative electron-hole pair recombination above threshold. Under the assumption that above threshold the gain coefficient remains fixed at γ_{Th}, η_{ext} is given by

$$\eta_{ext} = \frac{\eta_i (g_{Th} - \bar{\alpha})}{g_{Th}} \qquad ...(4.129)$$

where η_i is the internal quantum efficiency. This is not a well defined quantity in laser diodes, but most measurements show that $\eta_i \approx 0.6$ to 0.7 at room temperature. Experimentally, η_{ext} is calculated from the straight-line portion of the curve for the emitted optical P power with respect to the drive current I, which gives

$$\eta_{ext} = \frac{q}{E_g}\frac{dP}{dI} = 0.8065\lambda(\mu m)\frac{dP(mW)}{dI(mA)} \qquad ...(4.130)$$

where E_g is the bandgap energy in eV, dP is the incremental change in the emitted optical power in mW for an incremental change dI in the drive current in mA and λ is the emission wavelength in µm. For standard semiconductor lasers, external differential quantum efficiency of 15 to 20% per facet is typical. High-quality devices have differential quantum efficiencies of 30 to 40%.

4.11. OPTICAL TRANSMITTERS

The unique properties and characteristics of the injection laser and the LEDs that make them attractive for use as optical sources are the following. Although both devices exhibit a number of similarities in terms of their general performance and compatibility with optical fibres, striking differences exist between them in relation to both system application and transmitter design. It is useful to consider these differences as well as the limitations of the two types of sources, prior to discussion of transmitter circuits for various applications. Threshold current as applied to a laser is that current that must be supplied to it so that the spontaneous emission from the laser is changed to the coherent or stimulated emission. As we may note the system must be designed to deliver to the laser this threshold current so that the laser gives off stimulated emissions. This is carried out by a driver circuit. The block schematic in Fig. 4.26 shows the essential components of a driver circuit. A driver circuit limits the amount of current being fed to the laser in order for it to function properly. Therefore, a laser diode driver circuit is a linear, noiseless constant current source in the ideal form. The primary requirement of this circuit is to deliver a given amount of current to the laser depending upon the application. As we may observe, the laser acts as the load for such a circuit.

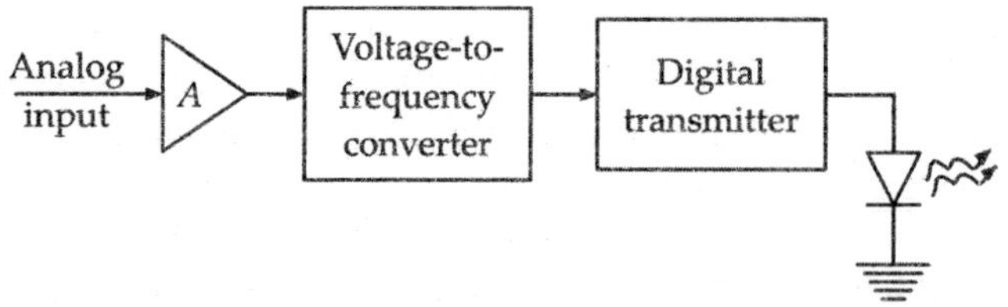

Fig. 4.26: Simple drive circuit for digital transmission

4.11.1. Source Limitations

(a) Power: The electrical power required to operate both injection lasers and LEDS is generally similar with typical current levels of between 20 and 300 mA with exceptions wherein certain type of lasers may require substantially higher currents in terms of 1 to 2A and voltage drops across the terminals of 1.5 to 2.5V. However, the optical power output against current characteristics for the two devices varies considerably. The injection laser is a threshold device that must be operated in the region of stimulated emission (above threshold) where continuous optical power levels are typically in the range of 1 to 10 mW.

Much of this light output may be coupled into an optical fibre because the isotropic distribution of the narrow linewidth, coherent radiation is relatively directional. In addition, the spatial coherence of the laser emission allows it to be readily focused by appropriate lenses within the numerical aperture of the fibre. Coupling efficiencies near 30% may be obtained by placing a fibre close to a laser mirror and these can approach 90% with a suitable lens and optical coupling arrangement. Therefore, injection lasers are capable of launching between 0.5 and several milliwatts of optical power into a fibre.

LEDs are capable of similar optical output power levels to injection lasers depending on their structure and quantum efficiency. However, the spontaneous emission of radiation over a wide linewidth from the LED generally exhibits a Lambertian intensity distribution which gives poor coupling into optical fibres. Consequently, only between 1% and up to 30% of the emiited optical power from an LED may be launched into a multimode fibre, even with appropriate lens coupling. These considerations translate into optical power levels from a few to several hundred microwatts lauched into individual multimode fibres. Thus the optical power launched into a fibre from an LED can be 10 to 20 dB below that obtained with a typical injection laser. The power advantage gained with the injection laser is a major factor in the choice of source, especially when considering a long-haul optical fibre link.

(b) Linearity: Linearity of the optical output power against current characteristic is an important consideration with both injection laser and the LED. It is especially pertinent to the design of analog optical fibre communication systems where source nonlinearities may cause severe distortion of the transmitted signal. At first sight the LED may appear to be ideally suited to analog transmission as its output is approximately proportional to the drive current. However, most LEDs display some degree of nonlinearity in their optical output against current characteristic because of junction heating effects which may prohibit their use or necessitate the incorporation of a linearizing circuit within the optical transmitter. Certain LEDs do display good linearity, with distortion products between 35 and 45 dB below the signal level.

An alternative approach to obtaining a linear source characteristic is to operate an injection laser in the light-generating region above its threshold. This may prove more suitable for analog transmission than would the use of certain LEDS. However, gross nonlinearities due to mode instabilities may occur in this region. These are known as kinks in the laser output characteristic. Therefore, many of the multimode injection lasers have a limited use for transmission without additional linearizing circuits within the transmitter, although some of the single mode structures have demonstrated linearity suitable for most analog applications. Alternatively, digital transmission, especially that utilizing a binary format is far less sensitive to source nonlinearities and is therefore often preferred when using both injection lasers and LEDs.

4.11.2. LED Drive Circuits

The driver circuits can be configured using BJTs or differential amplifiers. Here we discuss a number of circuits implemented with BJTs first. A low impedance drive circuit with a shunt configuration is illustrated in Fig. 4.27.

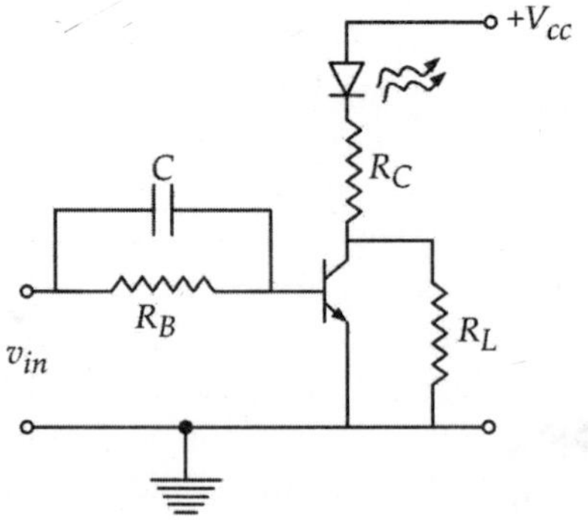

Fig. 4.27: Low impedance drive circuit

A simple drive circuit for binary data transmission using a BJT switch in the CC configuration is shown in Fig. 4.28.

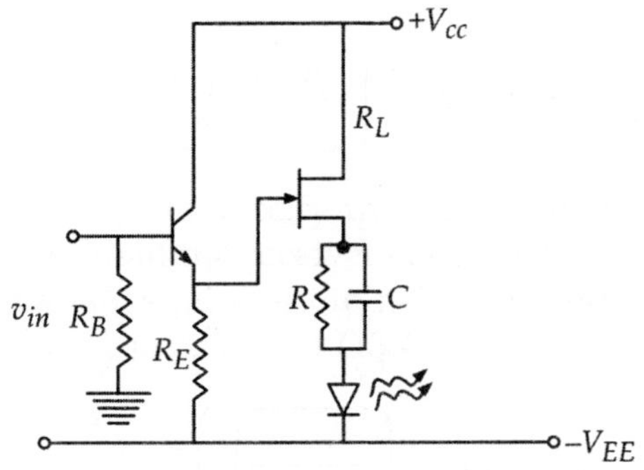

Fig. 4.28: A drive circuit configured with a BJT in the CC configuration

Next a transimpedance drive circuit in common emitter configuration suitable for analog signal transmission is illustrated in Fig. 4.29.

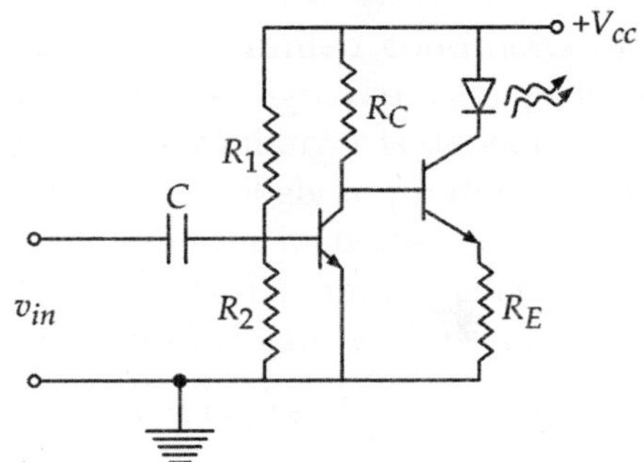

Fig. 4.29: Transimpedance drive circuit

Differential Amplifier Based Drive Circuit

The differential amplifier based drive circuit as shown in Fig. 4.30 consists of a differential amplifier operated over its linear region which modulates the LED directly. The LED operating point is controlled by a reference voltage while the current generator provided by the transistor feeding the differential stage comprising of the BJTs and limits the maximum current through the device. The transimpedance of the driver is reduced through current series feedback provided by the two resistors and which are normally assigned equal values. Furthermore, variation between these feedback resistors can be used to compensate for the transfer function of both the drive circuit and the LED.

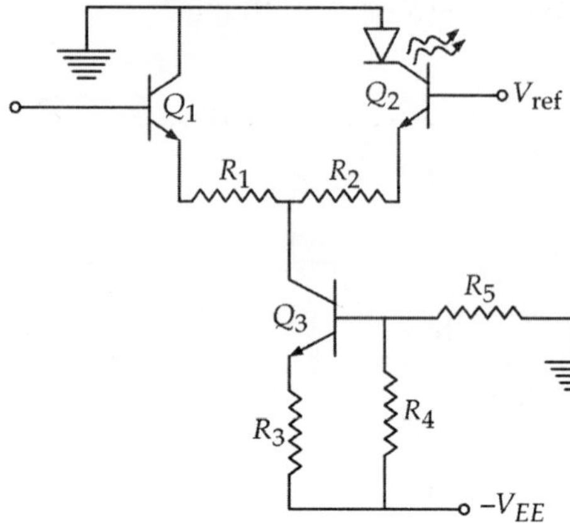

Fig. 4.30: Differential amplifier based drive circuit

A low impedance drive circuit in a shunt configuration is discussed in Fig. 4.31.

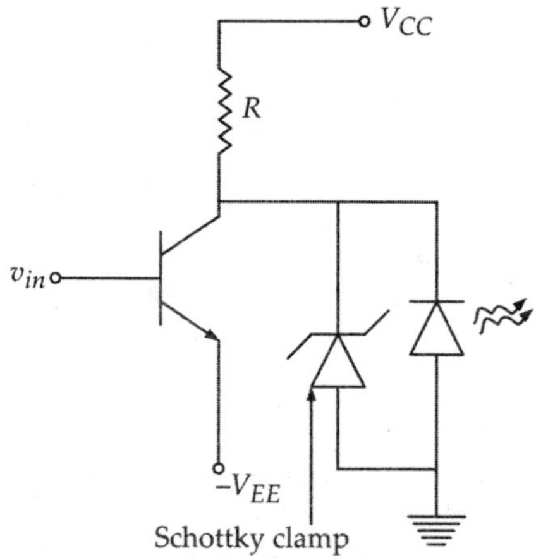

Fig. 4.31: Shunt configured driver circuit

The name shunt used in Fig. 4.31 refers to the fact that, the switching transistor is placed in parallel with the LED thereby providing a low impedance path for switching off the LED by shunting current around it. The switching-on time of this kind of circuit is decided primarily by the resistance R and the LED capacitance. The stored space charge may be removed by applying a small reverse bias to the LED when the device is switched off. This is achieved by applying a reverse bias to the emitter that is less than the ground potential. A Schottky clamp may be placed in order to limit the amplitude of the reverse bias without introducing any minority carrier stored charge in the circuit.

Driver with Emitter Follower

The driver shown in Fig. 4.32 is suitable for working at high switching frequencies. The parallel combination of the capacitor and the resistor at the emitter leg of the second transistor makes possible fast modulation of LEDs with comparatively lower driver power.

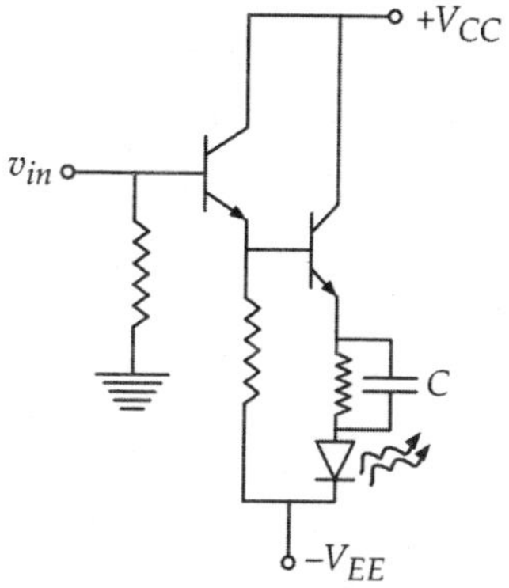

Fig. 4.32: Emitter follower based driver

4.11.3. Laser Drive Circuit

A number of configurations suitable for use as driver circuits for LEDS for both analog and digital data transmission may also be adapted for injection laser diode applications with minor modifications in the circuit. The laser requires a certain threshold current for its operation and hence a different circuit may be designed for establishing the appropriate drive current that exceeds the threshold required for lasing. When digital transmission is used, the laser usually is given a substantial applied bias in the off state. Reasons for biasing the laser near but below threshold in the off state are motivated by the following factors:

(a) It reduces the switch-on delay and minimizes any relaxation oscillations.

(b) It allows easy compensation for changes in the ambient temperature and device aging.

(c) It reduces the junction heating caused by the digital drive current since the on and off currents are not widely different for most lasers.

Biasing near threshold causes spontaneous emission of light in the off state. However this usually is not a problem for digital transmission because of the large number of stimulated emissions made from the device. A simple drive circuit for digital transmission is illustrated in Fig. 4.33. This is a shunt driver made of a field effect transistor (FET) to provide high speed laser operation. Sufficient voltage is maintained with the laser using the resistor R_2 and the compensating capacitor C such that the FET is biased into its active or pinch-off region. For a particular input voltage v_{in} a specific amount of the total current flowing through R_1 is diverted around the laser leaving the balance of the current to flow through R_2 and provide the off state for the device. This type of circuit implemented with GaAS MESFET is capable of modulating lasers at a rate in excess of 1 Gbps.

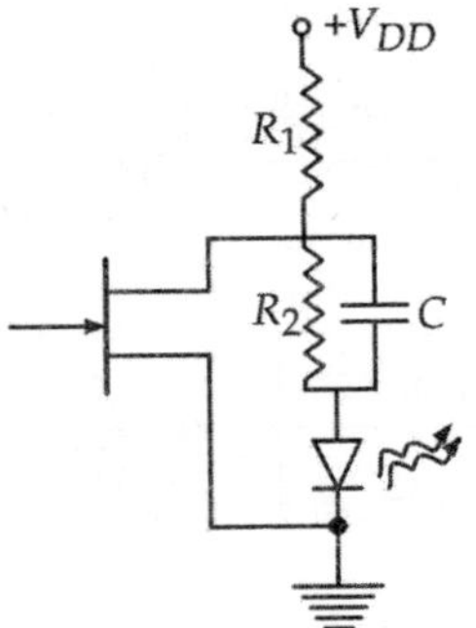

Fig. 4.33: A shunt drive circuit for use with an injection laser

4.11.4. Analog Transmission Drive Circuit

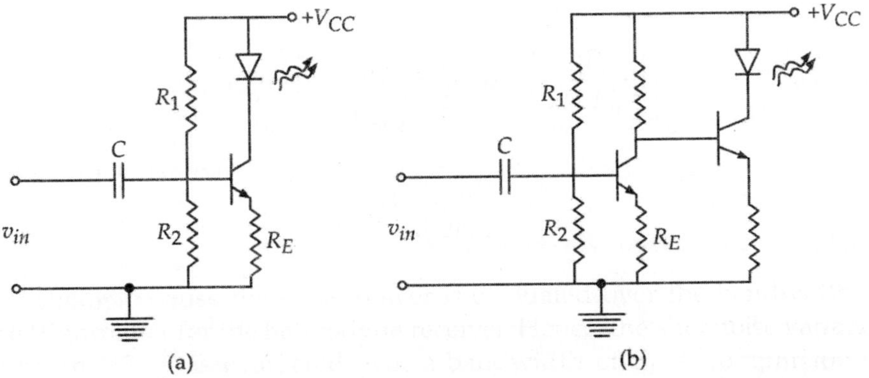

Fig. 4.34: (a) Low impedance driver in shunt configuration with a single transistor, (b) Driver using a Darlington transistor pair

Two high speed drive circuit configurations suitable for analog signal transmission are illustrated in Fig. 4.34(a) and (b). A transconductance amplifier that converts the input base voltage to a collector current is shown in Fig. 4.34(a). As we may observe, this circuit takes a voltage at its input or the base terminal and converts this to the collector current that is designed to drive the LED. This circuit is biased for a class A operation. The quiescent collector current is fixed at about half of the peak value. Another driver realized with a Darlington pair is shown in Fig. 4.34(b) which presents a higher value of input impedance and a larger current gain suitable for driving an LED.

The circuit shown in Fig. 4.35 converts the input analog signal to its corresponding digital equivalent by means of a suitable analog-to-digital converter. Analog signals from similar sources are subjected to digitization so that they can be transmitted as a common stream through the digital multiplexer for which an appropriate selection signal is to be designed and generated. Digital transmission has a number of advantages over the conventional analog transmission in the sense that these are enjoy more noise immunity, simple detector circuits at the receiver, provision of error detection and correction, availability of a large number of chip choices for carrying out the same, faster speed, programmability besides having low cost and light weight. Hence, digital multiplexers have been used in place of their analog counterparts. This circuit is suitable for digital laser transmission. A number of signals are combined onto one serial output stream which helps to keep the transmitter at a very low cost compared to its analog counterpart.

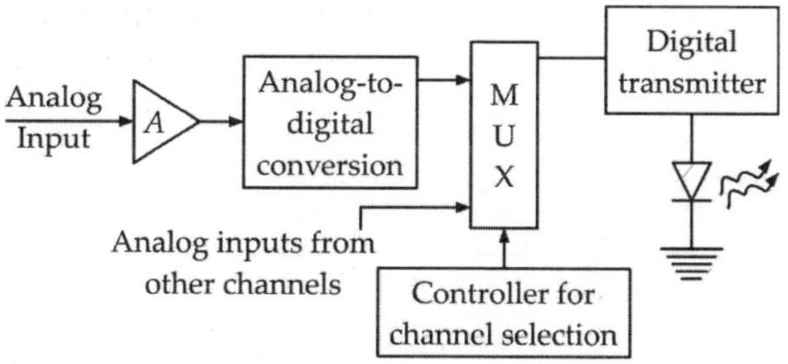

Fig. 4.35: Digital laser transmitter with ADC and multiplexing

4.11.5. Linearizing Circuits

The drive circuits described so far using either shunt configurations of discrete BJT, FET or differential amplifier circuits are more suitable for the transmission of a single analog channel. Here amplitude and phase distortions observed in the output signal as produced by the fibre optic channel may be tolerated to a certain extent. However, when it comes to a frequency division multiplexed system involving the transmission and reception of multiple channels, this kind of amplitude and phase distortions are not tolerable. A high degree of linearity is required to minimize interference between individual channels because of the generation of the intermodulation products. Also, baseband video transmission transmission of TV signals requires setting up extremely low levels of both amplitude and phase distortions. The drive circuits discussed so far need to be modified by incorporating linearizing circuits in order to make them suitable for such applications. The linearizing circuits should be designed to take care of both LED and drive circuit nonlinearities. A complementary distortion method has been shown in Fig. 4.36. where additional nonlinear devices are included in the system. It may take the form of predistortion (before the source driver) compensation or postdistortion compensation (after the receiver).

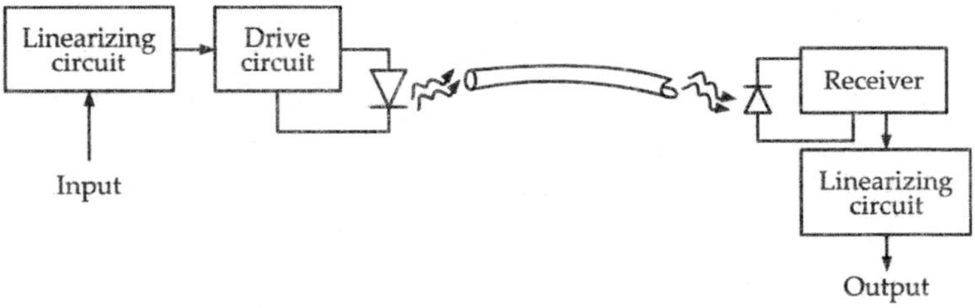

Fig. 4.36: Complementary distortion linearizing circuit

The LED is included in the linearization scheme for schemes that employ negative feedback such as one shown in Fig. 4.37. The optical output is detected and compared with the input waveform in such a scheme. The amount of compensation depends on the gain of the feedback loop. This method has large bandwidth requirements that may be problematic at high frequencies.

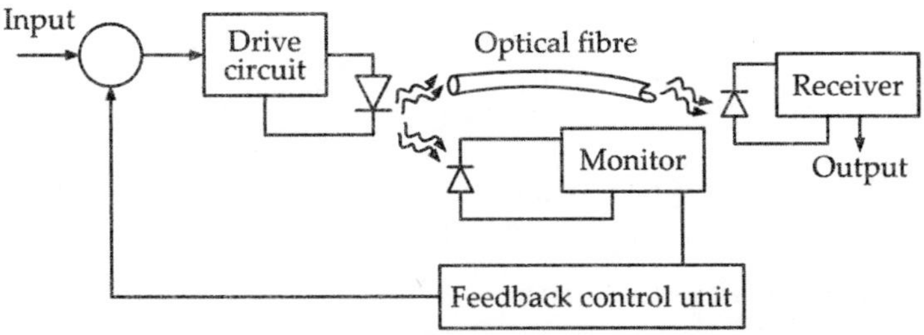

Fig. 4.37: Negative feedback compensation technique

Another method of linearization is the phase shift modulation for selective harmonic compensation using a pair of LEDS with similar characteristics as shown in Fig. 4.38. Here the input signal is divided into equal parts which are subjected to phase shifts of 90° and 60° (for example). This results in a relatively higher amount of distortion cancellation.

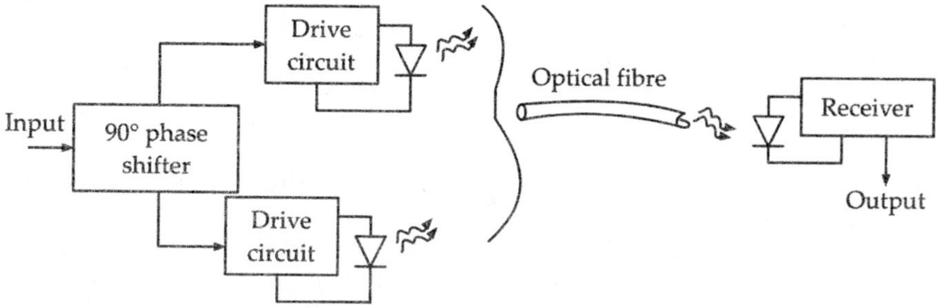

Fig. 4.38: Selective harmonic compensation technique

4.12. CHAPTER SUMMARY

Semiconductor junction devices are used exclusively for the sources and detectors of optical fibre communication systems. A key semiconductor parameter is the band-gap energy, Band gap energy and lattice spacing vary continuously with the composition of ternary and quaternary III-V compounds. Using quaternary and some ternary materials, it is possible to grow, as a single crystal, semiconductors having different bandgap energies. GaAlAs-GaAs sources, used with silicon detectors are well established for wavelengths around 0.85 μm. For longer wavelengths InGaAsP-InP sources and germanium or InGaAs-InP detectors are widely used over the range of 1.0-1.7 μm. Efficient injection luminescence depends on direct, band-to-band radiative recombination dominating the competing recombination mechanisms. This requires a direct band-gap material. The LEDs emit light by spontaneous emission which is incoherent in nature whereas the lasers produce coherent light through stimulated emission which requires the onset of a phenomenon of population inversion. This is achieved by pumping external power at a given frequency.

A number of optical transmitter circuits that transmit analog or digital data have been discussed along with the corresponding driver and linearizing circuits.

SUGGESTED READING

[1] Streetman, B.G. and Banerjee, S. "Solid State Electronic Devices", Pearson LPE, 5th Edn., 2006.

[2] Dimitrijev, S. "Principles of Semiconductor Devices", Oxford University Press, 2010.

[3] Keiser, G. "Optical Fiber Communication", 3rd Edn., McGraw Hill, 1999.
[4] Ghatak, A. "Optics", 2nd Edn, Tata McGraw Hill, 1995.
[5] Bergh, A.A. and Dean, P.J. "Light-Emitting Diodes", Oxford University Press, 1976.
[6] Burrus, C.A. and Miller, B.I. "Small-area double heterostructure aluminium-gallium arsenide electroluminescent diode sources for optical fibre transmission lines", Opt. Comm., 4, pp. 307-369, 1971.
[7] Gloge, D. "LED design for fibre systems", Electron Lett., vol. 4, no. 13, pp. 399-400,1977.
[8] Strauss, J. "Linearized transmitters for analog fiber links", *Laser Focus,* vol. 10, no. 14, pp. 54-61, 1978.
[9] Prochazka, A., Lancaster, P. and Neumann, R. "Amplifier linearization by complementary pre or post distortion", *IEEE Trans. Cable Television,* vol. 1, no. 1, pp. 31-39, 1976.
[10] Asatani, K. and Kimura, T. "Nonlinear distortions and their compensations in light emitting diodes", *Proc. of Int. Conf. on Integrated Optics and Optical Fiber Communications"*, p. 105, 1977.
[11] Strauss, J. Springthorpe, A.J. and Szentesi, O.I. "Phase shift modulation technique for the linearization of analog optical transmitters", *Electron. Letters,* vol. 5, no. 13, pp. 149-151, 1977.
[12] Asatani, K. and Kimura, T. "Linearization of LED nonlinearity by predistortions", *IEEE Jou. Solid State Circuits"*, vol. 1, no. 13, pp. 133-138, 1978.

❑❑❑

5

Guided Waves in Step Index Fibers

5.1. INTRODUCTION

The optical fiber, as we have mentioned is an optical waveguide. It acts as a guiding medium for light waves. It is modeled as a cylindrical waveguide. Propagation of light through such guiding structures requires an approach completely different from the unguided case. Hence, we will formulate appropriate expressions for wave propagation in an optical fiber by solving the Helmholtz equations subject to various boundary conditions using the cylindrical coordinate system.

The wave propagation in a waveguide is different from unguided waves (as in free space) in the sense that in the guided case, we have waves propagate in the form of what are known as modes; the transverse electric and magnetic (TEM) waves in unbounded homogeneous media. Incidence of such waves at the boundary of two different media gives rise to either the transverse electric (TE) and the transverse magnetic (TM) waves depending upon whether the electric field is normal to the plane of incidence or parallel to it. For the TE waves, the electric field component along the direction of propagation is zero but has a longitudinal component of the magnetic field. The electric field is entirely transverse to the direction of propagation. For the TM case, the magnetic field component along the direction of propagation is zero whereas the electric field has a longitudinal component. The magnetic field is entirely transverse to the direction of propagation. We develop appropriate expressions for propagation of light waves in the subsequent sections.

5.2. GUIDED WAVES

Our starting point here is the use of Maxwell's equations derived earlier. However, unlike the uniform plane waves characterized by typical TEM mode as in free space, here we have to look at certain boundary conditions in order to be able to derive the wave expressions. We are interested to determine if the waves have a component along the direction of propagation (longitudinal waves).

$$\nabla \cdot \mathbf{B} = 0 \quad ...(5.1)$$

$$\nabla \cdot \mathbf{D} = 0 \quad ...(5.2)$$

$$\nabla \times \mathbf{E} = -\frac{\partial \mathbf{B}}{\partial t} \quad ...(5.3)$$

$$\nabla \times \mathbf{H} = \frac{\partial \mathbf{D}}{\partial t} \quad ...(5.4)$$

5.2.1. Transverse Wave Components

We assume the light wave to be propagating along the z direction. The cross section of the core is assumed to be circular. Hence, wave equations and field components are expressed in the cylindrical system of coordinates. As the core is a dielectric, we write the generalized expression for the electric field along the z direction as

$$\mathbf{E} = E_0 \exp(-j\beta z) \quad ...(5.5)$$

where $\beta = \frac{2\pi}{\lambda}$ is the propagation constant along the z-axis. To include the effects of time variation and assuming the electric field to be functions of the ρ and ϕ directions, we write

$$\mathbf{E} = E_0(\rho, \phi) e^{-j\beta z} \cdot e^{j\omega t} \quad ...(5.6)$$

For a sinusoidal time variation we have seen previously that

$$\nabla \times \mathbf{E} = -j\omega\mu\mathbf{H} \quad ...(5.7)$$

Substitution of this field in Maxwell's curl equation (5.7) gives us

$$\frac{1}{\rho}\left(\frac{\partial E_z}{\partial \varphi} + j\rho\beta E_\phi\right) = -j\omega\mu H_\rho \quad ...(5.8)$$

$$j\beta E_\rho + \frac{\partial E_z}{\partial \rho} = j\omega\mu H_\phi \quad ...(5.9)$$

and

$$\frac{1}{\rho}\left[\frac{\partial}{\partial \rho}(\rho E_\phi) - \frac{\partial E_\rho}{\partial \phi}\right] = j\omega\mu H_z \quad ...(5.10)$$

Using $\nabla \times \mathbf{H} = \frac{\partial \mathbf{D}}{\partial t}$, we obtain

$$\frac{1}{\rho}\left(\frac{\partial H_z}{\partial \varphi} + j\beta\rho H_\phi\right) = j\omega\varepsilon E_\rho \qquad ...(5.11)$$

$$j\beta H_\rho + \frac{\partial H_z}{\partial \rho} = -j\omega\varepsilon E_\phi \qquad ...(5.12)$$

and

$$\frac{1}{\rho}\left[\frac{\partial}{\partial \rho}(\rho H_\phi) - \frac{\partial H_\rho}{\partial \varphi}\right] = j\omega\varepsilon E_z \qquad ...(5.13)$$

From (5.11), we obtain

$$E_\rho = \frac{1}{j\rho\omega\varepsilon}\left(\frac{\partial H_z}{\partial \phi} + j\beta\rho H_\phi\right) \qquad ...(5.14)$$

H_ϕ from (5.9) gives us

$$H_\phi = \frac{1}{j\omega\mu}\left(j\beta E_\rho + \frac{\partial E_z}{\partial \rho}\right) = \frac{\beta}{\omega\mu}E_\rho + \frac{1}{j\omega\mu}\frac{\partial E_z}{\partial \rho} \qquad ...(5.15)$$

and thus, using (5.14)

$$E_\rho = \frac{1}{j\rho\omega\varepsilon}\left[\frac{\partial H_z}{\partial \varphi} + j\beta\rho\left(\frac{\beta}{\omega\mu}E_\rho + \frac{1}{j\omega\mu}\frac{\partial E_z}{\partial \rho}\right)\right] \qquad ...(5.16)$$

$$= \frac{1}{j\rho\omega\varepsilon}\frac{\partial H_z}{\partial \varphi} + \frac{\beta^2}{\omega^2\mu\varepsilon}E_\rho + \frac{\beta}{j\omega^2\mu\varepsilon}\frac{\partial E_z}{\partial \rho}$$

Hence,

$$E_\rho - \frac{\beta^2}{\omega^2\mu\varepsilon}E_\rho = \frac{1}{j\rho\omega\varepsilon}\frac{\partial H_z}{\partial \phi} + \frac{\beta}{j\omega^2\mu\varepsilon}\frac{\partial E_z}{\partial \rho} = \left(\frac{\omega^2\mu\varepsilon - \beta^2}{\omega^2\mu\varepsilon}\right)E_\rho \qquad ...(5.17)$$

$$p^2 E_\rho = \frac{1}{j}\left(\beta\frac{\partial E_z}{\partial \rho} + \frac{\mu\omega}{\rho}\frac{\partial H_z}{\partial \varphi}\right)$$

$$= -j\beta\frac{\partial E_z}{\partial \rho} - j\frac{\mu\omega}{\rho}\frac{\partial H_z}{\partial \phi}$$

where $p^2 = \omega^2\mu\varepsilon - \beta^2$

Thus, the radial component of the electric field is

$$E_\rho = \frac{1}{jp^2}\left(\beta\frac{\partial E_z}{\partial \rho} + \frac{\mu\omega}{\rho}\frac{\partial H_z}{\partial \phi}\right) \qquad ...(5.18)$$

$$= \frac{-j}{p^2}\left(\beta\frac{\partial E_z}{\partial \rho} + \frac{\mu\omega}{\rho}\frac{\partial H_z}{\partial \phi}\right)$$

Similarly from (5.8), we obtain

$$H_\rho = \frac{1}{-j\rho\omega\mu}\left(\frac{\partial E_z}{\partial \phi} + j\rho\beta E_\phi\right) \quad ...(5.19)$$

(5.12) gives us

$$E_\phi = \frac{1}{-j\omega\varepsilon}\left(j\beta H_\rho + \frac{\partial H_z}{\partial \rho}\right) = -\frac{\beta}{\omega\varepsilon}H_\rho - \frac{1}{j\omega\varepsilon}\frac{\partial H_z}{\partial \rho} \quad ...(5.20a)$$

Substitution of (5.20a) into (5.19) gives

$$H_\rho = \frac{1}{-j\rho\omega\mu}\left[\frac{\partial E_z}{\partial \phi} + j\rho\beta\left(-\frac{\beta}{\omega\varepsilon}H_\rho - \frac{1}{j\omega\varepsilon}\frac{\partial H_z}{\partial \rho}\right)\right] \quad ...(5.20b)$$

$$= -\frac{1}{j\rho\omega\mu}\frac{\partial E_z}{\partial \phi} + \frac{\beta^2}{\omega^2\mu\varepsilon}H_\rho + \frac{\beta}{j\omega^2\mu\varepsilon}\frac{\partial H_z}{\partial \rho}$$

Collecting all H_ρ terms together at LHS of (5.20) (a) and (b), we have

$$H_\rho\left(1 - \frac{\beta^2}{\omega^2\mu\varepsilon}\right) = \left(\frac{\beta}{j\omega^2\mu\varepsilon}\frac{\partial H_z}{\partial \rho} - \frac{1}{j\rho\omega\mu}\frac{\partial E_z}{\partial \phi}\right) \quad ...(5.20c)$$

$$\Rightarrow \qquad H_\rho\left(\frac{\omega^2\mu\varepsilon - \beta^2}{\omega^2\mu\varepsilon}\right) = \left(\frac{\beta}{j\omega^2\mu\varepsilon}\frac{\partial H_z}{\partial \rho} - \frac{1}{j\rho\omega\mu}\frac{\partial E_z}{\partial \phi}\right)$$

$$H_\rho = \frac{1}{jp^2}\left(\beta\frac{\partial H_z}{\partial \rho} - \frac{\omega\varepsilon}{\rho}\frac{\partial E_z}{\partial \phi}\right)$$

$$= -\frac{j}{p^2}\left(\beta\frac{\partial H_z}{\partial \rho} - \frac{\omega\varepsilon}{\rho}\frac{\partial E_z}{\partial \phi}\right) \quad ...(5.21)$$

Evaluation of the ϕ – component of the fields is carried out as follows. Substitution of H_ρ from (5.21) in (5.20a) gives us

$$E_\phi = -\frac{\beta}{\omega\varepsilon}\left[-\frac{j}{p^2}\left(\beta\frac{\partial H_z}{\partial \rho} - \frac{\omega\varepsilon}{\rho}\frac{\partial E_z}{\partial \phi}\right)\right] - \frac{1}{j\omega\varepsilon}\frac{\partial H_z}{\partial \rho}$$

$$= j\frac{\beta^2}{p^2\omega\varepsilon}\frac{\partial H_z}{\partial \rho} - j\frac{\beta}{p^2\rho}\frac{\partial E_z}{\partial \phi} - \frac{1}{j\omega\varepsilon}\frac{\partial H_z}{\partial \rho}$$

$$= \frac{j}{\omega\varepsilon}\left(\frac{\beta^2}{p^2}+1\right)\frac{\partial H_z}{\partial \rho} - j\frac{\beta}{p^2\rho}\frac{\partial E_z}{\partial \phi} \qquad ...(5.22)$$

$$= \frac{j}{\omega\varepsilon}\left(\frac{\beta^2+p^2}{p^2}\right)\frac{\partial H_z}{\partial \rho} - j\frac{\beta}{p^2\rho}\frac{\partial E_z}{\partial \phi}$$

$$= \frac{j}{p^2\omega\varepsilon}.\omega^2\mu\varepsilon\frac{\partial H_z}{\partial \rho} - j\frac{\beta}{p^2\rho}\frac{\partial E_z}{\partial \phi}$$

$$= -\frac{j}{p^2}\left(\frac{\beta}{\rho}\frac{\partial E_z}{\partial \phi} - \mu\omega\frac{\partial H_z}{\partial \rho}\right)$$

The magnetic field is obtained by substituting the expression for E_ϕ as obtained from (5.22) in (5.15)

$$E_\phi = \frac{1}{j\omega\mu}\left(j\beta E_\rho + \frac{\partial E_z}{\partial \rho}\right)$$

$$= \frac{\beta}{\omega\mu}E_\rho + \frac{1}{j\omega\mu}\frac{\partial E_z}{\partial \rho}$$

$$= \frac{\beta}{\omega\mu}\left[\frac{-j}{p^2}\left(\beta\frac{\partial E_z}{\partial \rho} + \frac{\mu\omega}{\rho}\frac{\partial H_z}{\partial \phi}\right)\right] + \frac{1}{j\omega\mu}\frac{\partial E_z}{\partial \rho}$$

$$= -\frac{j\beta^2}{p^2\omega\mu}\frac{\partial E_z}{\partial \rho} - j\frac{\beta}{p^2\rho}\frac{\partial H_z}{\partial \phi} + \frac{1}{j\omega\mu}\frac{\partial E_z}{\partial \rho}$$

$$= -\frac{j}{\omega\mu}\left(\frac{\beta^2}{p^2}+1\right)\frac{\partial E_z}{\partial \rho} - j\frac{\beta}{p^2\rho}\frac{\partial H_z}{\partial \phi}$$

$$= -\frac{j}{\omega\mu}\left(\frac{\beta^2+p^2}{p^2}\right)\frac{\partial E_z}{\partial \rho} - j\frac{\beta}{p^2\rho}\frac{\partial H_z}{\partial \phi}$$

$$= -\frac{j}{p^2\omega\mu}.\omega^2\mu\varepsilon\frac{\partial E_z}{\partial \rho} - j\frac{\beta}{p^2\rho}\frac{\partial H_z}{\partial \phi}$$

$$= -\frac{j}{p^2}\omega\varepsilon\frac{\partial E_z}{\partial \rho} - j\frac{\beta}{p^2\rho}\frac{\partial H_z}{\partial \phi}$$

$$= -\frac{j}{p^2}\left(\omega\varepsilon\frac{\partial E_z}{\partial \rho} + \frac{\beta}{\rho}\frac{\partial H_z}{\partial \phi}\right)$$

Thus, $$H_\phi = -\frac{j}{p^2}\left(\omega\varepsilon\frac{\partial E_z}{\partial \rho} + \frac{\beta}{\rho}\frac{\partial H_z}{\partial \phi}\right) \qquad ...(5.23)$$

The fields so obtained are summarized in Table 5.1. The entries are valid for wave propagation along the positive z-axis.

Table 5.1: The four fields (transverse components) in an optical fiber

Field Type	*Expression*
E_ρ	$\frac{-j}{p^2}\left(\beta\frac{\partial E_z}{\partial \rho} + \frac{\mu\omega}{\rho}\frac{\partial H_z}{\partial \phi}\right)$
H_ρ	$-\frac{j}{p^2}\left(\beta\frac{\partial H_z}{\partial \rho} - \frac{\omega\varepsilon}{\rho}\frac{\partial E_z}{\partial \phi}\right)$
E_ϕ	$-\frac{j}{p^2}\left(\frac{\beta}{\rho}\frac{\partial E_z}{\partial \phi} - \mu\omega\frac{\partial H_z}{\partial \rho}\right)$
H_ϕ	$-\frac{j}{p^2}\left(\omega\varepsilon\frac{\partial E_z}{\partial \rho} + \frac{\beta}{\rho}\frac{\partial H_z}{\partial \phi}\right)$

We observe from Table 5.1 that ρ and the ϕ components of the electric and the magnetic fields have been considered. These are transverse to the direction of wave propagation the positive z-axis.

5.2.2. Longitudinal Wave Components

The objective is to evaluate the fields propagating along the direction of propagation. To do so, we need to develop the necessary wave equation first. This is done by the substitution of (5.21) and (5.23) in (5.13) which gives us the following wave equation:

$$\frac{1}{\rho}\left[\frac{\partial}{\partial \rho}(\rho H_\phi) - \frac{\partial H_\rho}{\partial \varphi}\right] = j\omega\varepsilon E_z \qquad ...(5.24)$$

$$\Rightarrow \qquad E_z = \frac{1}{j\rho\omega\varepsilon}\left[\frac{\partial}{\partial \rho}(\rho H_\phi) - \frac{\partial H_\rho}{\partial \phi}\right]$$

$$\Rightarrow \qquad E_z = \frac{1}{j\omega\varepsilon}\left[\begin{array}{l}\frac{1}{\rho}\frac{\partial}{\partial\rho}\left\{-\frac{j}{p^2}\left(\omega\varepsilon\rho\frac{\partial E_z}{\partial\rho}+\rho.\frac{\beta}{\rho}\frac{\partial H_z}{\partial\phi}\right)\right\} \\ -\frac{\partial}{\rho\partial\phi}\left\{-\frac{j}{p^2}\left(\beta\frac{\partial H_z}{\partial\rho}-\frac{\omega\varepsilon}{\rho}\frac{\partial E_z}{\partial\phi}\right)\right\}\end{array}\right]$$

$$\Rightarrow \qquad E_z = \frac{1}{j\omega\varepsilon}\left[-\frac{j\omega\varepsilon}{p^2}\left\{\left(\frac{1}{\rho}\right)\frac{\partial}{\partial\rho}\left(\frac{\rho\partial E_z}{\partial\rho}\right)\right\}-\frac{j\beta}{p^2}\frac{\partial}{\rho\partial\rho}\left(\frac{\partial H_z}{\partial\phi}\right)\right]$$

$$-\frac{1}{j\omega\varepsilon}\left[-\frac{j\beta}{p^2}\frac{\partial}{\rho\partial\phi}\left(\frac{\partial H_z}{\partial\rho}\right)-\frac{j\omega\varepsilon}{p^2}\frac{\partial}{\rho\partial\phi}\left(\frac{\partial E_z}{\partial\phi}\right)\right]$$

the 2nd and 3rd terms cancel each other

$$E_z = -\frac{\partial^2 E_z}{p^2\partial\rho^2}-\frac{1}{p^2\rho}\frac{\partial E_z}{\partial\rho}-\frac{1}{p^2}\frac{\partial^2 E_z}{\rho^2\partial\phi^2}$$

Substitution of (5.24) into the Helmholtz equation gives us

$$\frac{\partial^2 E_z}{\partial\rho^2}+\frac{1}{\rho^2}\frac{\partial^2 E_z}{\partial\varphi^2}+\frac{1}{\rho}\frac{\partial E_z}{\partial\rho} = -p^2 E_z \qquad ...(5.25)$$

$$\Rightarrow \qquad \frac{\partial^2 E_z}{\partial\rho^2}+\frac{1}{\rho}\frac{\partial E_z}{\partial\rho}+\frac{1}{\rho^2}\frac{\partial^2 E_z}{\partial\varphi^2}+p^2 E_z = 0$$

Let us use the method of separation of variables. Then the wave equation is expressed as

$$E_z = P(\rho)Q(\phi)e^{-j\beta z} \qquad ...(5.26)$$

where $P(\rho)$ is a function of ρ alone and $P(\phi)$ is a function of ϕ alone. Substituting (5.26) in (5.25) gives us

$$Q\frac{d^2P}{d\rho^2}+\frac{Q}{\rho}\frac{dP}{d\rho}+\frac{P}{\rho^2}\frac{d^2Q}{d\phi^2}-PQ\beta^2+\omega^2\mu\varepsilon PQ = 0 \qquad ...(5.27)$$

Dividing (5.27) by PQ, we obtain

$$\frac{1}{P}\frac{d^2P}{d\rho^2}+\frac{1}{\rho P}\frac{dP}{d\rho}+\frac{1}{Q\rho^2}\frac{d^2Q}{d\varphi^2}+p^2 = 0 \qquad ...(5.28)$$

In (5.28) we have $\omega^2\mu\varepsilon - \beta^2 = p^2$

We break up (5.28) into two parts, one part being an exclusive function of ρ and the other being a function of ϕ only.

$$\frac{1}{Q\rho^2}\frac{d^2Q}{d\phi^2} + p^2Q = 0$$

$$\Rightarrow \qquad \frac{d^2Q}{d\phi^2} = -p^2\rho^2 Q \qquad \text{...(5.29)}$$

$$\frac{d^2Q}{d\phi^2} = -n^2 Q$$

This (5.29) is a function of ϕ only. Also, we note that $n = p\rho$ should be an integer. The solution to (5.29) is

$$Q = (A_n \cos n\phi + B_n \sin n\phi) \qquad \text{...(5.30)}$$

Similarly,

$$\frac{d^2P}{d\rho^2} + \frac{1}{\rho}\frac{dP}{d\rho} + \left(p^2 - \frac{n^2}{\rho^2}\right)P = 0 \qquad \text{...(5.31a)}$$

$$\frac{d^2P}{d(\rho p)^2} + \frac{1}{\rho p}\frac{dP}{d(\rho p)} + \left[1 - \frac{n^2}{(\rho p)^2}\right]P = 0 \qquad \text{...(5.31b)}$$

which is a function of ρ only. This is a standard form of Bessel's function in terms of (ρp). As we know, a step index fiber consists of a homogeneous core of refractive index n_1 and radius a surrounded by an infinite cladding of refractive index n_2. The reason for assuming an infinitely thick cladding is that the guided modes in the core have exponentially decaying fields outside the core which must have insignificant values at the outer boundary of the cladding. In practice, optical fibers are designed with claddings that are sufficiently thick so that the guided-mode field does not reach the outer boundary of the cladding. The wave propagation as a single axial ray corresponding to a step index monomode fibre is shown in Fig. 5.1. The term 'monomode' here refers to the existence of just one mode for the wave which happens to be the dominant mode while eliminating the existence of all other modes. We shall learn more about the modes later. The electric field distributions for several lower order guided modes in a symmetrical slab waveguide are shown subsequently. The fields vary harmonically in the guiding region of refractive index n_1 and decay exponentially outside of this region.

The solution to (5.31) is

$$P = AJ_n(\rho p) + Bk_n(\rho p) \qquad \text{...(5.32)}$$

The wave equation (5.28) is solved for the regions inside and outside the core. A solution of the form of $AJ_n(p\rho)$ tends to infinity as ρ tends to zero. This implies that the field strength is infinite on the axis of the fiber core. For the inside region the solutions for the guided modes must remain finite as, $\rho \to 0$ whereas on the outside the solutions must decay to zero as $\rho \to \infty$. Thus for $\rho \to a$, the solutions are Bessel's function of the first kind of order n. For the optical fiber, the constant $u = \sqrt{\omega^2\mu\varepsilon_1 - \beta^2} = \sqrt{p_1^2 - \beta^2}$ is used to express the fields inside the core. The equivalent of $\omega\sqrt{\mu\varepsilon}$ is $\omega n_1 = \frac{2\pi}{\lambda}n_1$ corresponding to the core of refractive index n_1. For these functions we use the common designation $J_n(\rho u)$. Here with $u^2 = p_1^2 - \beta^2$, we have $p_1 = 2\pi n_1/\lambda$.

The wave propagation constant in free space is

$$\beta = \frac{2\pi}{\lambda} = \frac{2\pi}{(c/f)} = \frac{2\pi f}{c} = \frac{\omega}{c}$$

In any other medium, the wave propagation constant is

$$p_1 = \frac{\omega}{v_1} = \frac{2\pi f}{(c/n_1)} = 2\pi n_1/\lambda$$

where λ is the free space wavelength.

The expression for the longitudinal fields E_z and H_z are (making use of 5.30 and 5.32)

$$E_z(\rho < a) = J_n(\rho u)(A_n \cos n\phi + B_n \sin n\phi)e^{-j\beta z} \quad \text{...(5.33)}$$

The solution for will have exactly the same form as for and can therefore be written as the magnetic fields (5.33)

$$H_z(\rho < a) = J_n(\rho u)(C_n \cos n\phi + D_n \sin n\phi)e^{-j\beta z} \quad \text{...(5.34)}$$

Outside the core, the solutions of the wave equations are given by modified Bessel functions of the second kind $K_n(w\rho)$. The associated electric and magnetic fields are given as

$$E_z(\rho > a) = AK_n(w\rho)(A_n \cos n\phi + B_n \sin n\phi)e^{j(\omega t - \beta z)} \quad \text{...(5.35)}$$

$$H_z(\rho > a) = BK_n(w\rho)(A_n \cos n\phi + B_n \sin n\phi)e^{j(\omega t - \beta z)} \quad \text{...(5.36)}$$

where

$$w^2 = \beta^2 - p_2^2 = \beta^2 - \omega^2\mu\varepsilon_2 \text{ and } p_2 = \frac{2\pi}{\lambda}n_2 \quad \text{...(5.37)}$$

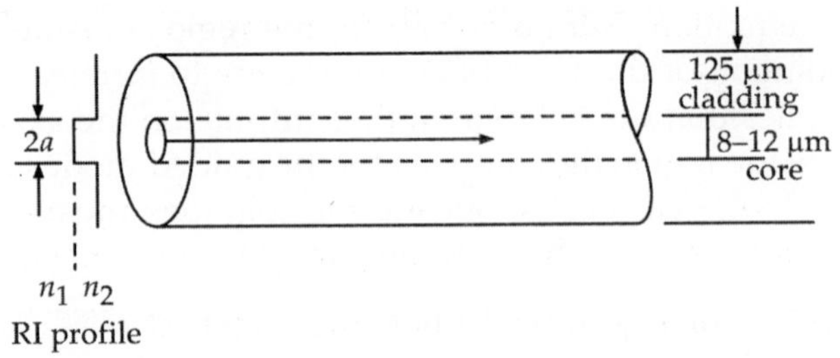

Fig. 5.1: Fiber cross section and typical ray path in a monomode step index fiber

The definition of $J_n(\rho u)$ and $K_n(w\rho)$ and various recursion relations are given in App. V. From the definition of the modified Bessel function, it is observed that $K_n(w\rho) \to e^{-w\rho}$ as $w\rho \to \infty$. Since $K_n(w\rho)$ must go to zero as $\rho \to \infty$, it follows that $w > 0$. This in turn implies that $\beta \geq p_2$. The cutoff condition is the point at which a mode is no longer bound to the core region. A second condition on β can be inferred from the behavior of $J_n(\rho u)$. Inside the core the parameter must be a real number which implies that $p_1 > \beta$. The permissible range of β for bound solutions is $n_1\beta = p_1 \geq \beta \geq p_2 = n_2\beta$ where $\beta = 2\pi/\lambda$ is the free space propagation constant. The waves propagating along the axis are spiraling waves, but their transverse non-uniformity gradually dies out with radial distance away from the core cladding interface. The modified Bessel's function $K_n(w\rho)$ is related to the original Bessel function $J_n(u\rho)$ in the same way as hyperbolic functions are linked with trigonometric functions. $J_n(u\rho)$ is an oscillating function whereas $K_n(w\rho)$ behaves like a decaying exponential. The modified Bessel function tends to $e^{-w\rho}$ when $w\rho$ tends to infinity. Moreover, $K_n(w\rho)$ must be zero for very high (infinite) value of ρ, so $w > 0$. This indicates that $\beta \geq p_2$ and it gives a cutoff condition.

5.3. MODAL EQUATION

A solution of the form of $AJ_n(p\rho)$ tends to infinity as ρ tends to zero. This implies that the field strength is infinite on the axis of the fiber core. Hybrid modes are present in optical waveguides.

The boundary conditions require that the tangential components of E_ϕ and E_z of the electric field are continuous across the dielectric boundary defined by $\rho = a$. This is also true for the magnetic field components. Continuity of the tangential components at the boundary dictates that the z component, at the inner core-cladding boundary $E_z = E_{z1}$ and at the outside of the boundary $E_z = E_{z2}$. Thus, we obtain,

$$E_{z1} - E_{z2} = AJ_n(ua) - CK_n(wa) = 0 \qquad ...(5.38)$$

To find the φ component, we note that, inside the core,

$$u^2 = p_1^2 - \beta^2 \text{ with } p_1 = 2\pi n_1/\lambda \quad \text{...(5.39a)}$$

and outside the core

$$w^2 = \beta^2 - p_2^2 = \beta^2 - \left(\frac{2\pi}{\lambda}n_2\right)^2 \quad \text{...(5.39b)}$$

with $p_2 = 2\pi n_2/\lambda$

Hence,

$$E_{\phi 1} - E_{\phi 2} = -\frac{j}{p^2}\left[\frac{Ajn\beta}{\rho}J_n(u\rho) - B\omega\mu p J_n'(u\rho)\right] - \frac{j}{w^2}\left[\frac{Cjn\beta}{\rho}K_n(w\rho) - D\omega\mu p K_n'(w\rho)\right] = 0 \quad \text{...(5.40)}$$

Similarly, the tangential components of the magnetic field at the core-cladding interface $\rho = a$ (the ϕz plane), the z-component of the field satisfies the condition of continuity. Thus,

$$H_{z1} - H_{z2} = BJ_n(ua) - DK_n(wa) = 0 \quad \text{...(5.41)}$$

and

$$H_{\phi 1} - H_{\phi 2} = -\frac{j}{p^2}\left[\frac{Bjn\beta}{\rho}J_n(u\rho) + B\omega\varepsilon_1 p J_n'(u\rho)\right] - \frac{j}{w^2}\left[\frac{Djn\beta}{\rho}K_n(w\rho) + C\omega\varepsilon_2 w K_n'(w\rho)\right] = 0 \quad \text{...(5.42)}$$

Solution of the four unknown coefficients A, B, C and D requires four equations which are given above. Only discrete values of β restricted to the range given by $n_1\beta = p_1 \geq \beta \geq p_2 = n_2\beta$ will be allowed. We have four unknowns A, B, C and D and four equations in the form of the z- and ϕ-components of the electric and the magnetic fields. A solution to these four unknowns exists only if the determinant formed by these coefficients is zero.

$$\begin{bmatrix} J_n(ua) & 0 & -K_n(wa) & 0 \\ \frac{\beta n}{au^2}J_n(ua) & \frac{j\omega\mu}{u}J_n'(ua) & \frac{\beta n K_n(wa)}{aw^2} & \frac{j\omega\mu K_n'(wa)}{w} \\ 0 & J_n(ua) & 0 & -K_n(wa) \\ -\frac{j\omega\varepsilon_1}{u}J_n'(ua) & \frac{\beta n}{au^2}J_n(ua) & -\frac{j\omega\varepsilon_2 K_n'(wa)}{w} & \frac{\beta n K_n(wa)}{aw^2} \end{bmatrix} = 0 \quad \text{...(5.43)}$$

To evaluate (5.43), let us do it the following way:

Let

$$J_n(ua)=a_1,\ -K_n(wa)=a_2,\ \frac{j\omega\mu}{u}J_n'(ua)=b_1$$

and $$\frac{j\omega\mu K_n'(wa)}{w}=b_2 \qquad \text{...(5.44a)}$$

The above determinant takes the form of the following

$$\begin{bmatrix} a_1 & 0 & a_2 & 0 \\ \frac{\beta n}{au^2}a_1 & b_1 & -\frac{\beta n}{aw^2}a_2 & b_2 \\ 0 & a_1 & 0 & a_2 \\ -\frac{\varepsilon_1}{\mu}b_1 & \frac{\beta n}{au^2}a_1 & -\frac{\varepsilon_2}{\mu}b_2 & -\frac{\beta n}{aw^2}a_2 \end{bmatrix}=0 \qquad \text{...(5.44b)}$$

The first cofactor is evaluated by considering the following

$$\begin{bmatrix} b_1 & -\frac{\beta n}{aw^2}a_2 & b_2 \\ a_1 & 0 & a_2 \\ \frac{\beta n}{au^2}a_1 & -\frac{\varepsilon_2}{\mu}b_2 & -\frac{\beta n}{aw^2}a_2 \end{bmatrix} \qquad \text{...(5.44c)}$$

$$= b_1\left(\frac{a_2 b_2 \varepsilon_2}{\mu}\right)+\frac{\beta n}{aw^2}a_2\left[-\frac{a_1a_2}{aw^2}\beta n-\frac{a_1a_2}{au^2}\beta n\right]+b_2\left(-\frac{a_1b_2\varepsilon_2}{\mu}\right)$$

Hence, the first cofactor of (5.43b) becomes

$$a_1\left\{b_1\left(\frac{a_2b_2\varepsilon_2}{\mu}\right)+\frac{\beta n}{aw^2}a_2\left[-\frac{a_1a_2}{aw^2}\beta n-\frac{a_1a_2}{au^2}\beta n\right]+b_2\left(-\frac{a_1b_2\varepsilon_2}{\mu}\right)\right\} \qquad \text{...(5.44d)}$$

$$=\frac{a_1a_2b_1b_2\varepsilon_2}{\mu}-(a_1a_2)^2\left(\frac{\beta n}{aw^2}\right)^2-(a_1a_2)^2\left(\frac{\beta n}{a}\right)^2\frac{1}{u^2w^2}-(a_1b_2)^2\frac{\varepsilon_2}{\mu}$$

The second and the fourth cofactors vanish. To evaluate the third cofactor of (5.44b), we have

$$\begin{bmatrix} \frac{\beta n}{au^2}a_1 & b_1 & b_2 \\ 0 & a_1 & a_2 \\ -\frac{\varepsilon_1}{\mu}b_1 & \frac{\beta n}{au^2}a_1 & -\frac{\beta n}{aw^2}a_2 \end{bmatrix} \qquad \text{...(5.44e)}$$

$$= \frac{\beta n}{au^2} a_1 \left(-a_1 a_2 \frac{\beta n}{aw^2} - a_1 a_2 \frac{\beta n}{au^2} \right) - b_1 \left[-\left(-a_2 b_1 \frac{\varepsilon_1}{\mu} \right) \right] + b_2 \left(a_1 b_1 \frac{\varepsilon_1}{\mu} \right)$$

Hence, the third cofactor of (5.44b) becomes

$$a_2 \left\{ \frac{\beta n}{au^2} a_1 \left(-a_1 a_2 \frac{\beta n}{aw^2} - a_1 a_2 \frac{\beta n}{au^2} \right) - b_1 \left(a_2 b_1 \frac{\varepsilon_1}{\mu} \right) + b_2 \left(a_1 b_1 \frac{\varepsilon_1}{\mu} \right) \right\}$$

$$= -(a_1 a_2)^2 \left(\frac{\beta n}{a} \right)^2 \frac{1}{u^2 w^2} - (a_1 a_2)^2 \left(\frac{\beta n}{au^2} \right)^2 - (a_2 b_1)^2 \frac{\varepsilon_1}{\mu} + \frac{a_1 a_2 b_1 b_2 \varepsilon_1}{\mu} \quad \text{...(5.44f)}$$

Combining these two, we obtain

$$\frac{a_1 a_2 b_1 b_2 \varepsilon_2}{\mu} - (a_1 a_2)^2 \left(\frac{\beta n}{aw^2} \right)^2 - (a_1 a_2)^2 \left(\frac{\beta n}{a} \right)^2 \frac{1}{u^2 w^2} - (a_1 b_2)^2 \frac{\varepsilon_2}{\mu}$$

$$- (a_1 a_2)^2 \left(\frac{\beta n}{a} \right)^2 \frac{1}{u^2 w^2} - (a_1 a_2)^2 \left(\frac{\beta n}{au^2} \right)^2 - (a_2 b_1)^2 \frac{\varepsilon_1}{\mu} + \frac{a_1 a_2 b_1 b_2 \varepsilon_1}{\mu}$$

$$= \frac{a_1 a_2 b_1 b_2}{\mu} (\varepsilon_1 + \varepsilon_2) - (a_1 a_2)^2 \left[\left(\frac{\beta n}{aw^2} \right)^2 + \left(\frac{\beta n}{au^2} \right)^2 + 2 \left(\frac{\beta n}{a} \right)^2 \frac{1}{u^2 w^2} \right]$$

$$- \left[\frac{(a_1 b_2)^2 \varepsilon_1 + (a_2 b_1)^2 \varepsilon_2}{\mu} \right] \quad \text{...(5.44g)}$$

This should be set equal to zero in order to find out a solution to (5.43). Thus, we have

$$\frac{a_1 a_2 b_1 b_2}{\mu} (\varepsilon_1 + \varepsilon_2) - \left[\frac{(a_1 b_2)^2 \varepsilon_1 + (a_2 b_1)^2 \varepsilon_2}{\mu} \right] \quad \text{...(5.45)}$$

$$= (a_1 a_2)^2 \left[\left(\frac{\beta n}{aw^2} \right)^2 + \left(\frac{\beta n}{au^2} \right)^2 + 2 \left(\frac{\beta n}{a} \right)^2 \frac{1}{u^2 w^2} \right]$$

Dividing both the sides of (5.45) by $(a_1 a_2)^2$, we obtain

$$\frac{b_1 b_2}{a_1 a_2 \mu} (\varepsilon_1 + \varepsilon_2) - \left[\frac{(b_1 / a_2)^2 \varepsilon_2 + (b_1 / a_1)^2 \varepsilon_1}{\mu} \right]$$

$$= \left[\left(\frac{\beta n}{aw^2} \right)^2 + \left(\frac{\beta n}{au^2} \right)^2 + 2 \left(\frac{\beta n}{a} \right)^2 \frac{1}{u^2 w^2} \right] \quad \text{...(5.46)}$$

$$= \left(\frac{\beta n}{a}\right)^2 \left(\frac{1}{u^2} + \frac{1}{w^2}\right)^2$$

To evaluate $\frac{b_1 b_2}{a_1 a_2}$, we substitute the actual terms from (5.44a) in (5.46) yielding the following.

$$\frac{b_1 b_2}{a_1 a_2} = \frac{\frac{j\omega\mu}{u} J_n'(ua) \frac{j\omega\mu}{w} K_n'(wa)}{-J_n(ua)K_n(wa)}$$

$$= \frac{\omega^2\mu^2}{uw} \frac{J_n'(ua)K_n'(wa)}{J_n(ua)K_n(wa)} \qquad \text{...(5.47)}$$

$$= \omega^2\mu^2 \frac{J_n'(ua)}{uJ_n(ua)} \frac{K_n'(wa)}{wK_n(wa)}$$

$$= \omega^2\mu^2 (I_n \mathrm{K}_n)$$

where
$$I_n = \frac{J_n'(ua)}{uJ_n(ua)} \qquad \text{...(5.48a)}$$

and
$$K_n = \frac{K_n'(wa)}{wK_n(wa)} \qquad \text{...(5.48b)}$$

Use of (5.47) in (5.46) gives the following. The first term in the LHS of (5.46) becomes

$$\frac{b_1 b_2}{a_1 a_2 \mu}(\varepsilon_1 + \varepsilon_2) = \frac{\omega^2\mu^2}{\mu}(I_n \mathrm{K}_n)(\varepsilon_1 + \varepsilon_2) = \omega^2\mu(I_n \mathrm{K}_n)(\varepsilon_1 + \varepsilon_2) \qquad \text{...(5.49)}$$

$$= (p_1^2 + p_2^2)(I_n \mathrm{K}_n)$$

The second term in the LHS of (5.46) becomes, similarly

$$\left[\frac{(b_2/a_2)^2 \varepsilon_2 + (b_1/a_1)^2 \varepsilon_1}{\mu}\right] = \frac{1}{\mu}\left[\frac{\frac{j\omega\mu}{w} K_n'(wa)}{-K_n(wa)}\right]^2 \varepsilon_2 + \left[\frac{\frac{j\omega\mu}{u} J_n'(ua)}{J_n(wa)}\right]^2 \varepsilon_1$$

$$= \frac{1}{\mu}\left[-\omega^2\mu^2 \left(\frac{K_n'(wa)}{wK_n(wa)}\right)^2 \varepsilon_2 - \omega^2\mu^2 \left(\frac{J_n'(ua)}{uJ_n(ua)}\right)^2 \varepsilon_1\right] \qquad \text{...(5.50)}$$

$$= -\omega^2\mu\left[\left(\frac{K_n'(wa)}{wK_n(wa)}\right)^2 \varepsilon_2 + \left(\frac{J_n'(ua)}{uJ_n(ua)}\right)^2 \varepsilon_1\right]$$

Therefore,

$$-\left[\frac{(b_2/a_2)^2\varepsilon_2 + (b_1/a_1)^2\varepsilon_1}{\mu}\right]$$

$$= \omega^2\mu\left[\left(\frac{K_n'(wa)}{wK_n(wa)}\right)^2 \varepsilon_2 + \left(\frac{J_n'(ua)}{uJ_n(ua)}\right)^2 \varepsilon_1\right] \quad ...(5.51)$$

$$= \omega^2\mu\varepsilon_2 K_n^2 + \omega^2\mu\varepsilon_1 I_n^2$$

$$= p_2^2 K_n^2 + p_1^2 I_n^2$$

corresponding to the second term of the LHS of (5.46).

Thus, combining (5.49) and (5.51), we have

$$(p_1^2 + p_2^2)(I_n K_n) + p_2^2 K_n^2 + p_1^2 I_n^2 \quad ...(5.52)$$

$$= (I_n + K_n)(p_1^2 I_n + p_2^2 K_n)$$

The evaluation of the determinant of (5.43) through the use of (5.52) yields the following eigenvalue equation for β

$$(I_n + K_n)(p_1^2 I_n + p_2^2 K_n) = \left(\frac{\beta n}{a}\right)^2 \left(\frac{1}{u^2} + \frac{1}{w^2}\right)^2 \quad ...(5.53)$$

Solution of (5.53) for β reveals that only discrete values restricted to the range given by are allowed. The above equation is usually solved by numerical methods corresponding to a particular mode. Let us consider this equation in some detail for a couple of lower order modes. The actual substitution of I_n and K_n from (5.48) and (5.49) respectively into (5.53) gives

$$(I_n + K_n)(p_1^2 I_n + p_2^2 K_n) = \left(\frac{\beta n}{a}\right)^2 \left(\frac{1}{u^2} + \frac{1}{w^2}\right)^2 \quad ...(5.54)$$

$$\Rightarrow \quad \left[\frac{J_n'(ua)}{uJ_n(ua)} + \frac{K_n'(ua)}{wK_n(ua)}\right]\left[\frac{\omega^2\mu\varepsilon_1}{u}\frac{J_n'(ua)}{J_n(ua)} + \frac{\omega^2\mu\varepsilon_2}{w}\frac{K_n'(ua)}{K_n(ua)}\right]$$

$$= \left(\frac{\beta n}{a}\right)^2 \left(\frac{1}{u^2} + \frac{1}{w^2}\right)^2$$

This is a general equation for any value of E_z and H_z including zero. Substitution of $n = 0$ makes the RHS of (5.54) equal to zero and the equation is split into two equations.

$$\frac{J_0'(ua)}{uJ_0(ua)} + \frac{K_0'(ua)}{wK_0(ua)} = 0 \qquad \text{...(5.55)}$$

and

$$\frac{\omega^2\mu\varepsilon_1}{u}\frac{J_0'(ua)}{J_0(ua)} + \frac{\omega^2\mu\varepsilon_2}{w}\frac{K_0'(ua)}{K_0(ua)} = 0 \qquad \text{...(5.56)}$$

$$\Rightarrow \qquad \frac{\varepsilon_1}{u}\frac{J_0'(ua)}{J_0(ua)} + \frac{\varepsilon_2}{w}\frac{K_0'(ua)}{K_0(ua)} = 0$$

We make use of the following recursions in subsequent derivations.

$$\pm\, uJ_n'(u) = nJ_n(u) - uJ_{n+1}(u)$$
$$\pm\, wK_n'(w) = nK_n(w) \mp wK_{n\pm1}(w)$$
$$J_{n+1}(u) = \left(\frac{2n}{u}\right)J_n(u) - J_{n-1}(u) \qquad \text{...(5.57)}$$
$$K_{n+1}(w) = \left(\frac{2n}{w}\right)K_n(w) - K_{n-1}(w)$$

5.3.1. Modes in Step Index Fibers

The J functions exhibit oscillatory behavior for real values of p and hence they are similar to harmonic functions. An illustrative nature of the Bessel's function of first kind, first order is presented in Fig. 5.2. Because of this oscillatory behavior, the equations satisfying the above contain roots for a given value of n. The roots being designated by β_{nm} give rise to the corresponding modes as TE_{nm}, TM_{nm}, EH_{nm} and HE_{nm}. For an optical fiber waveguide, all modes are hybrid modes except those for which $n = 0$. When this happens, the RHS of (5.54) become zero. We obtain two different eigen values for β corresponding to this case. These are

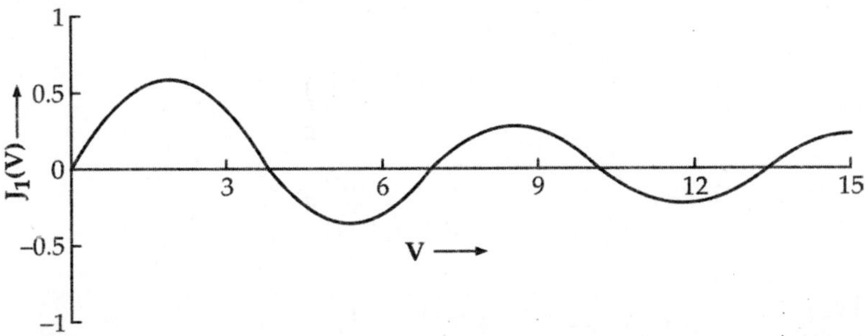

Fig. 5.2: A sketch of the Bessel function of first kind, first order

$$I_0 = \frac{J_0'(ua)}{uJ_0(ua)} = \frac{-J_1(ua)}{uJ_0(ua)} \quad \because \; J_0'(x) = -J_1(x) \text{ and similarly,} \qquad ...(5.58)$$

$$K_0 = \frac{K_0'(wa)}{wK_0(wa)} = \frac{-K_1(wa)}{wK_0(wa)} \quad \because \; K_0'(wa) = -K_1(wa) \qquad ...(5.59)$$

From the continuity of the electric field at the boundary, we get from (5.38)

$$E_{z1} - E_{z2} = AJ_n(u\rho) - CK_n(w\rho) = 0 \qquad ...(5.60)$$

$$\Rightarrow \qquad AJ_n(ua) = CK_n(wa)\Big|_{\rho=a}$$

Similarly, the component $\frac{\partial E_z}{\partial \rho}$ must also be continuous at the boundary otherwise the field $\frac{\partial^2 E_z}{\partial \rho^2}$ would contain an impulsive term that would make $\frac{\partial^2 E_z}{\partial \rho^2} + \frac{1}{\rho}\frac{\partial E_z}{\partial \rho} + \frac{1}{\rho^2}\frac{\partial^2 E_z}{\partial \varphi^2} + p^2 E_z = 0$ invalid.

Carrying out the differentiation with respect to the radius of the fiber, we get

$$\frac{\partial E_{z1}}{\partial \rho} - \frac{\partial E_{z2}}{\partial \rho} = 0$$

$$\Rightarrow \qquad AuJ_n'(u\rho) - CwK_n'(w\rho) = 0 \qquad ...(5.61)$$

$$\Rightarrow \qquad AuJ_n'(ua) = CwK_n'(wa)\Big|_{\rho=a}$$

Dividing (5.60) by (5.61), we get,

$$\frac{AJ_n(ua)}{AuJ_n'(ua)} = \frac{CK_n(wa)}{CwK_n'(wa)}$$

$$\Rightarrow \qquad \frac{J_n(ua)}{uJ_n'(ua)} = \frac{K_n(wa)}{wK_n'(wa)} \qquad ...(5.62)$$

$$\Rightarrow \qquad \frac{uJ_n'(ua)}{J_n(ua)} = \frac{wK_n'(wa)}{K_n(wa)}$$

We make use of the identities given in (5.57) in describing the transverse fields. (5.62) can be written in either of the following two forms

$$\frac{uJ_{n+1}(u)}{J_n(u)} = -\frac{wK_{n+1}(w)}{K_n(w)} \qquad ...(5.63)$$

or

$$\frac{uJ_{n-1}(u)}{J_n(u)} = -\frac{wK_{n-1}(w)}{K_n(w)} \qquad ...(5.64)$$

(5.63) and (5.64) represent two important classes of modal equations.

Multiplying both the sides of (5.64) by $\frac{1}{uw}$, we get

$$\frac{J_{n-1}(u)}{wJ_n(u)} = -\frac{K_{n-1}(w)}{uK_n(w)} \qquad ...(5.65)$$

Rearranging (5.65) we obtain

$$\frac{uJ_{n-1}(u)}{J_n(u)} = -\frac{wK_{n-1}(w)}{K_n(w)} \qquad ...(5.66)$$

Multiplying both the sides of (5.62) by uw, we get

$$\frac{J_n(ua)}{uJ'_n(ua)}.uw = \frac{K_n(wa)}{wK'_n(wa)}.uw \qquad ...(5.67)$$

$$\Rightarrow \qquad \frac{J_n(ua)}{wJ'_n(ua)} = \frac{K_n(wa)}{uK'_n(wa)}$$

Rearranging (5.67), we obtain

$$\frac{uJ_n(ua)}{J'_n(ua)} = \frac{wK_n(wa)}{K'_n(wa)} \qquad ...(5.68)$$

Taking the inverse of both the sides of (5.68),

$$\frac{J'_n(ua)}{uJ_n(ua)} = \frac{K'_n(wa)}{wK_n(wa)} \qquad ...(5.69)$$

$$\Rightarrow \qquad \frac{J'_n(ua)}{uJ_n(ua)} - \frac{K'_n(wa)}{wK_n(wa)} = 0$$

As $J'_0(x) = -J_1(x)$ and similarly, $K'_0(x) = -K_1(x)$ ...(5.70)

Substitution of this in (5.69) gives us

$$\frac{-J_1(ua)}{uJ_0(ua)} + \frac{K_1(wa)}{wK_0(wa)} = 0 \qquad ...(5.71)$$

We understand that $p^2 = \omega^2\mu\varepsilon - \beta^2$

An important parameter connected with the cutoff condition is the normalized frequency V (also called the V number or V-parameter) defined by

$$V_2 = (u^2 + w^2)a^2$$

$$= (p_1^2 - \beta^2 + \beta^2 - p_2^2)a^2$$

$$\Rightarrow \quad \left[\left(\frac{2\pi}{\lambda}n_1\right)^2 - \beta^2 + \beta^2 - \left(\frac{2\pi}{\lambda}n_2\right)^2\right]a^2 \qquad \text{...(5.72)}$$

$$\Rightarrow \quad \left(\frac{2\pi a}{\lambda}\right)^2 (n_1^2 - n_2^2)$$

$$= \left(\frac{2\pi a}{\lambda}\right)^2 .n_1^2.2\Delta$$

We have substituted the expressions of u and w from (5.39a) and (5.39b) in (5.72). Hence, the V number is

$$V = \left(\frac{2\pi a}{\lambda}\right) n_1\sqrt{2\Delta} \qquad \text{...(5.73)}$$

$$= \left(\frac{2\pi a}{\lambda}\right).\text{NA}$$

This is a dimensionless number which indicates how many modes a fiber can support. The number of modes that can exist in a waveguide as a function of V may be represented in terms of a normalized propagation constant defined as

$$b = \frac{a^2w^2}{V^2} = \frac{a^2(\beta^2 - p_2^2)}{\left(\frac{2\pi a}{\lambda}\right)^2 (n_1^2 - n_1^2)} = \frac{\left(\frac{\beta}{p}\right)^2 - n_2^2}{n_1^2 - n_2^2} \qquad \text{...(5.74)}$$

It is found that each mode can exist only for values of V that exceed a certain limiting value. The modes are then cut off when $\frac{\beta}{p} = n_2$. The HE_{11} mode has no cutoff and ceases to exist only when the core diameter is zero. This is the principle on which a single mode fiber is based. By appropriately choosing the core diameter and the refractive indices of the core and the cladding so that

$$\left(\frac{2\pi}{\lambda}\right)a\sqrt{n_1^2 - n_1^2} \leq 2.405 \qquad ...(5.75)$$

which is the value at which the lowest-order Bessel function J_0 is zero.

5.3.2. TE Modes

For the TE modes, as we understand, the electric field is entirely transverse to the direction of propagation; the z-axis; all of the components of the electric field are in the $\rho - \phi$ plane only. However, there exists a longitudinal field H_z along the direction of propagation. As $E_z = 0$, we have the following fields defined as follows

$$E_\rho = \frac{-j}{p^2}\left(\beta\frac{\partial E_z}{\partial \rho} + \frac{\mu\omega}{\rho}\frac{\partial H_z}{\partial \phi}\right)$$
$$= \frac{-j}{p^2}\frac{\mu\omega}{\rho}\frac{\partial H_z}{\partial \phi} \qquad ...(5.76)$$

Inside the case $H_z(\rho < a) = J_n(\rho u)(C_n \cos n\phi + D_n \sin n\phi)e^{-j\beta z}$

$$\frac{\partial H_z}{\partial \varphi} = J_n(\rho u)(-C_n n \sin n\phi + D_n n \cos n\phi)$$

We find that,

$$\Rightarrow \qquad \frac{\partial H_z}{\partial \phi} = 0 \quad \text{when } n = 0 \qquad ...(5.77)$$

Here, n is an integer. (5.76) tells us that the radial component of the electric field vanishes for the particular case of $n = 0$. Hence forth, we consider $n = 0$ in order to simplify the analysis.

Similarly, it can be shown that, the field H_ϕ is also zero. Let us investigate the remaining two fields to find out if they are non zero; $E_\phi \neq 0$ and $H_\rho \neq 0$. The magnetic field is

$$H_\rho = -\frac{j}{p^2}\left(\beta\frac{\partial H_z}{\partial \rho} - \frac{\omega\varepsilon}{\rho}\frac{\partial E_z}{\partial \phi}\right)$$
$$= -\frac{j}{p^2}\beta\frac{\partial H_z}{\partial \rho} \qquad ...(5.78)$$

As we see from (5.78), the radial magnetic field is nonzero. To proceed further, let us make use of the continuity of the boundary values in setting up a new way of understanding the TE modes. We write,

$$H_{\rho 1} = H_{\rho 2}$$

$$= -\frac{j}{u^2}\beta A J_0'(\rho u).u = \frac{j}{w^2}\beta C K_0'(\rho w).w$$

$$= -\frac{A J_0'(\rho u)}{u} = \frac{C K_0'(\rho w)}{w}\Bigg|_{\rho=a} \quad \text{...(5.79)}$$

$$= -\frac{A}{C}\frac{J_0'(au)}{u} = \frac{K_0'(aw)}{w}$$

In the above expression, we have replaced the denominator of both the sides (p^2) by $u^2 = k_1{}^2 - \beta^2$ in the LHS and $w^2 = \beta^2 - k_2{}^2$ in the RHS. However, the term $p^2 = w^2\mu\varepsilon - \beta^2$ is positive corresponding to u^2 in the core and is negative corresponding to w^2 in the cladding; $p^2 = -w^2$. Now, the continuity of the longitudinal components of the magnetic field give us; at the core-cladding interface,

$$A J_0(\rho u) = C K_0(\rho w)\Big|_{\rho=a}$$

$$= \frac{A}{C} = \frac{K_0(aw)}{J_0(au)} \quad \text{...(5.80)}$$

Making use of (5.80) in (5.79), we get

$$-\frac{K_0(aw)}{J_0(au)}\frac{J_0'(au)}{u} = \frac{K_0'(aw)}{w} \quad \text{...(5.81)}$$

$$= -\frac{J_0'(au)}{uJ_0(au)} = \frac{K_0'(aw)}{wK_0(aw)}$$

$J_0'(x) = -J_1(x)$ and similarly, $K_0'(x) = -K_1(x)$

Substituting these two in (5.81), we have

$$\frac{J_1(au)}{uJ_0(au)} = -\frac{K_1(aw)}{wK_0(aw)} \quad \text{...(5.82)}$$

$$\Rightarrow \quad \frac{J_1(au)}{uJ_0(au)} + \frac{K_1(aw)}{wK_0(aw)} = 0$$

This represents the TE mode. We may designate this field as the TE_{0m} field as this has been obtained for the case of $n = 0$ and there are m roots of the Bessel's function. We therefore, have, through the use of (5.58) and (5.59)

$$I_0 + K_0 = 0 \quad \text{...(5.83)}$$

In other words, the TE_{0m} modes are characterized by (5.83) which represents the first term in the LHS of (5.53) corresponding to $n = 0$.

5.3.3. TM Modes

For this mode, the magnetic field is entirely transverse to the direction of propagation. The longitudinal component of the magnetic field is $H_z = 0$. However, we have a finite value of the longitudinal electric field. Therefore, $E_z \neq 0$. All the other components would be lying in the $\rho - \phi$ plane. As we may note, the radial component of the magnetic field

$$H_\rho = -\frac{j}{p^2}\left(\beta\frac{\partial H_z}{\partial \rho} - \frac{\omega\varepsilon}{\rho}\frac{\partial E_z}{\partial \varphi}\right) = 0 \qquad ...(5.84)$$

corresponding to the case of $n = 0$. In a similar manner, it can be shown that the electric field $E_\phi = 0$. We, therefore, have E_ρ and H_ϕ which are nonzero. We make use of the continuity of the boundary conditions of these fields in order to have more insight into the field behavior. Let us consider the H_ϕ component at the core cladding interface.

$$H_\phi = -\frac{j}{p^2}\left(\omega\varepsilon\frac{\partial E_z}{\partial \rho} + \frac{\beta}{\rho}\frac{\partial H_z}{\partial \phi}\right)$$

$$\Rightarrow \qquad H_{\phi 1} = H_{\phi 2}\Big|_{\rho=a}$$

$$\Rightarrow \qquad -\frac{j}{u^2}\omega\varepsilon_1\frac{\partial E_z}{\partial \rho} = \frac{j}{w^2}\omega\varepsilon_2\frac{\partial E_z}{\partial \rho}\Big|_{\rho=a} \qquad ...(5.85)$$

$$\Rightarrow \qquad -\frac{\omega\varepsilon_1}{u^2}AJ_n'(ua).u = \frac{\omega\varepsilon_2}{w^2}CK_n'(aw).w$$

$$\Rightarrow \qquad -\frac{\omega\varepsilon_1}{u}AJ_n'(ua) = \frac{\omega\varepsilon_2}{w}CK_n'(aw)$$

Let us multiply both sides of (5.85) by $\omega\mu$ which gives us

$$-\frac{\omega^2\mu\varepsilon_1}{u}AJ_0'(ua) = \frac{\omega^2\mu\varepsilon_2}{w}CK_0'(aw) \qquad ...(5.86)$$

$$\frac{p_1^2}{u}\frac{A}{C}J_1(ua) = -\frac{p_2^2}{w}K_1(aw)$$

$$E_{z1} = E_{z2}$$

$$\Rightarrow \qquad AJ_0(ua) = CK_0(wa) \qquad ...(5.87)$$

$$\Rightarrow \qquad \frac{A}{C} = \frac{K_0(wa)}{J_0(ua)}$$

Therefore, (5.86) becomes through the use of (5.87),

$$\frac{p_1^2}{u}\frac{K_0(wa)}{J_0(ua)}J_1(ua) = -\frac{p_2^2}{w}K_1(aw)$$

$$\Rightarrow \qquad \frac{p_1^2 J_1(ua)}{uJ_0(ua)} = -\frac{p_2^2 K_1(aw)}{wK_0(wa)} \qquad \text{...(5.88)}$$

$$\Rightarrow \qquad \frac{p_1^2 J_1(ua)}{uJ_0(ua)} + \frac{p_2^2 K_1(aw)}{wK_0(wa)} = 0$$

Hence, we understand that the TM modes are characterized by (5.88). This is the second term of in the LHS of the eigen value equation as given in (5.53). The particular mode as described above corresponding to the case of $n = 0$ is known as the TM_{0m} mode. We get complex modes when $n \neq 0$ and numerical methods are followed to solve the eigen value equation. However, simplified and highly accurate approximations based on the principle that the core and the cladding refractive indices are nearly the same are found in [1]. We note that when $w \to 0$, we have

$$\frac{K_1(w)}{wK_0(w)} \to -\frac{1}{w^2}\ln w \to \infty \qquad \text{...(5.89)}$$

Hence, (5.88) approximates to

$$\frac{J_1(au)}{uJ_0(au)} = -\frac{K_1(aw)}{wK_0(aw)}$$

$$\Rightarrow \qquad \frac{J_1(au)}{uJ_0(au)} \to \infty \qquad \text{...(5.90)}$$

$$\Rightarrow \qquad J_0(au) \to \infty$$

$$\Rightarrow \qquad u \to 0$$

However, $u = 0$ is not a solution as the Bessel's function gives us

$$\lim_{u \to 0}\left[\frac{J_1(u)}{uJ_0(u)}\right] = \frac{1}{2} \qquad \text{...(5.91)}$$

The limiting values u_{c0} of the TE_{0m} modes are the roots of $J_0(u)$.

5.3.4. Hybrid Modes

From the foregoing discussion we understand that for the TE and the TM modes inside the core of an optical fiber which acts as a dielectric, we have either $E_z = 0$ and $H_z = 0$ respectively. However, for modes satisfying the condition $E_z \neq 0$, and $H_z \neq 0$, we have the hybrid modes. There are basically two types of hybrid modes; (a) HE hybrid modes and (b) EH hybrid modes.

(a) If $H_t > E_t$ then we get the HE mode and

(b) If $E_t > H_t$ then we get the EH mode

The magnetic field has its longitudinal component along the positive z-axis whereas its transverse or tangential component is along the positive x-axis. The resultant of the transverse component $H_t = H_x$ and the longitudinal component H_z gives the resultant H field as shown in Fig. 5.3(a). We may observe further from this diagram that the tangential component of the magnetic field $H_t = H_x$ is higher than the tangential electric field $E_t = E_y$ in strength. Therefore, a mode satisfying this condition is known as an HE mode.

Similarly, if the transverse component E_t of the electric field is greater than the corresponding transverse component of the magnetic field H_t, then the mode is called the EH mode. In Fig. 5.3(b) the transverse component of the electric field is aligned along the positive y-axis whereas the tangential component lies along the direction of propagation; the positive z-axis. The vector sum of E_y and E_z constitutes the E vector.

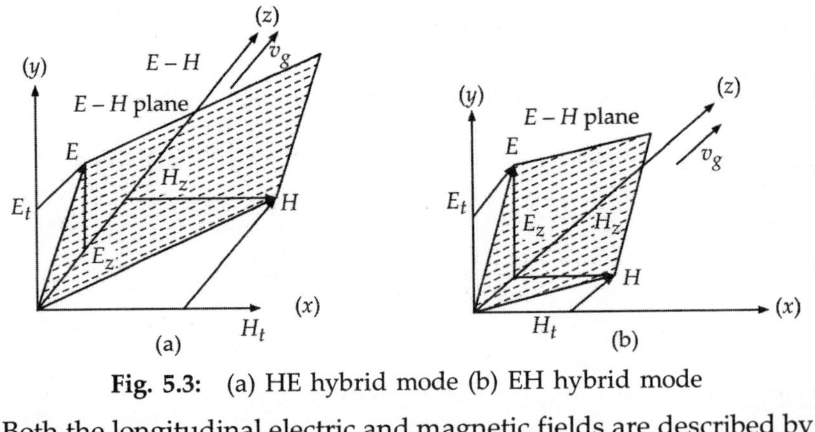

Fig. 5.3: (a) HE hybrid mode (b) EH hybrid mode

Both the longitudinal electric and magnetic fields are described by the Bessel's function of n-th kind. The discussion of cutoff conditions of the hybrid modes dictate different treatments of these modes corresponding to $n = 1$ and the other $n > 1$.

For higher order modes, we note that if the electric field is having a higher strength compared to the corresponding magnetic field, then the cutoff condition for the EH_{nm}, $n \geq 2$ would be dictated by the following condition.

$$J_n(ua) = 0 \qquad \text{...(5.92)}$$

The continuity of the tangential components at the core cladding interface gives us

$$E_{\phi 1} = E_{\phi 2}\Big|_{\rho=a}$$

$$\Rightarrow \quad -\frac{j}{u^2}\left[\frac{Ajn\beta}{\rho}J_n(ua) - B\omega\mu u J_n'(ua)\right]$$

$$= \frac{j}{w^2}\left[\frac{Cjn\beta}{\rho}K_n(w\rho) - D\omega\mu w K_n'(wa)\right]$$

$$\Rightarrow \quad -\frac{j}{u^2}[-B\omega\mu u J_n'(ua)] = \frac{j}{w^2}[-D\omega\mu w K_n'(wa)] \qquad \text{...(5.93)}$$

$$\Rightarrow \quad \frac{B}{D}\frac{J_n'(ua)}{u} = -\frac{K_n'(wa)}{w}$$

$$\Rightarrow \quad \frac{K_n(wa)}{J_n(ua)}\frac{J_n'(ua)}{u} = -\frac{K_n'(wa)}{w}$$

$$\Rightarrow \quad \frac{J_n'(ua)}{uJ_n(ua)} = -\frac{K_n'(wa)}{wK_n(wa)}$$

The hybrid modes are characterized by the condition of $n \neq 0$. The eigen value equation corresponding to this case may be expressed in a quadratic form that would produce two families of fields. This is obtained by expanding (5.53) through the use of (5.48) and (5.49). We, therefore, have

$$(I_n + K_n)(p_1^2 I_n + p_2^2 K_n) = \left(\frac{\beta n}{a}\right)^2\left(\frac{1}{u^2} + \frac{1}{w^2}\right)^2$$

$$\Rightarrow \quad p_1^2 I_n + p_2^2 I_n K_n + p_1^2 I_n K_n + p_2^2 K_n^2 = n^2\left(\frac{\beta}{a}\right)^2\left(\frac{1}{u^2} + \frac{1}{w^2}\right)^2 \qquad \text{...(5.94)}$$

$$= p_1^2 I_n + (p_1^2 + p_2^2) I_n K_n + p_2^2 K_n^2 = n^2\left(\frac{\beta}{a}\right)^2\left(\frac{1}{u^2} + \frac{1}{w^2}\right)^2$$

For simplification of (5.94), let us assume that

$$\frac{J_n'(ua)}{uJ_n(ua)} = x, \frac{K_n'(wa)}{wK_n(wa)} = b \text{ and } \left(\frac{\beta}{a}\right)^2\left(\frac{1}{u^2}+\frac{1}{w^2}\right)^2 = c_1 \quad ...(5.95)$$

Thus,

$$(x+b)(p_1^2 x + p_2^2 b) = n^2 c_1$$

$$\Rightarrow \quad p_1^2 x^2 + (p_1^2 + p_2^2)bx + p_2^2 b^2 = n^2 c_1$$

$$\Rightarrow \quad p_1^2 x^2 + (p_1^2 + p_2^2)bx + (p_2^2 b^2 - c_1) = 0 \quad ...(5.96)$$

$$\Rightarrow \quad x^2 + \left(1+\frac{p_2^2}{p_1^2}\right)bx + \left(\frac{p_2^2 b^2}{p_1^2} - \frac{c_1}{p_1^2}\right) = 0$$

The term,

$$\frac{p_2^2}{p_1^2} = \frac{(2\pi/\lambda)^2 n_2^2}{(2\pi/\lambda)^2 n_1^2} = \frac{n_2^2}{n_1^2}$$

$$\Rightarrow \quad 1+\frac{p_2^2}{p_1^2} = 1+\frac{n_2^2}{n_1^2} = 1+1-2\Delta \quad ...(5.97)$$

$$= 2(1-\Delta)$$

This is because

$$n_1^2 - n_2^2 = n_1^2 . 2\Delta$$

$$\Rightarrow \quad \frac{n_1^2}{n_1^2} - \frac{n_2^2}{n_1^2} = 2\Delta \quad ...(5.98)$$

$$\Rightarrow \quad 1 - 2\Delta = \frac{n_2^2}{n_1^2}$$

The term $n^2 c_1$ is expanded as follows

$$\left(\frac{\beta n}{a}\right)^2\left(\frac{1}{u^2}+\frac{1}{w^2}\right)^2$$

$$= \left(\frac{\beta n}{a}\right)^2\left(\frac{u^2+w^2}{u^2 w^2}\right)^2 \quad ...(5.99)$$

$$= \left(\frac{\beta n}{a}\right)^2 \left(\frac{V}{a}\right)^4 \frac{1}{u^4 w^4}$$

$$= \left(\frac{\beta n}{a}\right)^2 \left(\frac{V}{auw}\right)^4$$

Therefore, the last term in the LHS of (5.96) becomes

$$\frac{c_1}{p_1^2} = \frac{1}{p_1^2}\left(\frac{\beta n}{a}\right)^2 \left(\frac{V}{auw}\right)^4 \qquad \text{...(5.100)}$$

$$= n^2\left(\frac{n_{eff}}{n_1}\right)^2 \left(\frac{V}{auw}\right)^4$$

In the above, we have taken

$$\frac{\beta}{(2\pi/\lambda)} = n_{eff} \qquad \text{...(5.101)}$$

Hence, the quadratic given by (5.96) in x is expressed as, after all the manipulations carried out in (5.97) to (5.101)

$$x^2 + 2(1-\Delta)bx + \left[(1-2\Delta)b^2 - \frac{c_1}{p_1^2}\right] = 0$$

$$\Rightarrow \quad x = \frac{-2(1-\Delta)b \pm \left\{[2(1-\Delta)b]^2 - 4\left[(1-2\Delta)b^2 - \frac{c_1}{p_1^2}\right]\right\}^{\frac{1}{2}}}{2} \qquad \text{...(5.102)}$$

$$\Rightarrow \quad x = -b(1-\Delta) \pm \left\{(1+\Delta^2 - 2\Delta)b^2 - \left[(1-2\Delta)b^2 - \frac{c_1}{p_1^2}\right]\right\}^{\frac{1}{2}}$$

$$\Rightarrow \quad x = -b(1-\Delta) \pm \left(b^2\Delta^2 + \frac{c_1}{p_1^2}\right)^{\frac{1}{2}}$$

We can write x in the following way also by multiplying and dividing (5.102) by w^2

$$x = \frac{1}{w^2}\left\{-bw^2(1-\Delta) \pm \left[(bw^2)^2\Delta^2 + \frac{c_1 w^4}{p_1^2}\right]^{\frac{1}{2}}\right\} \quad ...(5.103)$$

The ± sign in the discriminator of the RHS in (5.103) refers to the mode type. A + sign means an EH mode or HE mode otherwise.

From (5.72) we get,

$$\left(\frac{V}{u}\right)^2 = 1 + \frac{w^2}{u^2}$$

$$\Rightarrow \quad \frac{V}{u} = \sqrt{1 + \frac{w^2}{u^2}} \to 1 + \frac{w^2}{2u^2} \text{ for } \frac{w}{u} << 1 \quad ...(5.104)$$

$$\Rightarrow \quad \left(\frac{V}{u}\right)^4 \to 1 + \frac{2w^2}{u^2}$$

Therefore, (5.95) gives us

$$bw^2 = w^2 \frac{K_n'(wa)}{wK_n(wa)}$$

$$= \frac{K_n'(wa)}{K_n(wa)} \to \begin{cases} -1 + w^2 \ln w & \text{for } n = 1 \\ -n - \dfrac{w^2}{2(n-1)} & \text{for } n > 1 \end{cases} \quad ...(5.105)$$

Similarly,

$$(bw^2)^2 \to \begin{cases} 1 - 2w^2 \ln w & \text{for } n = 1 \\ n^2 + \dfrac{nw^2}{n-1} & \text{for } n > 1 \end{cases} \quad ...(5.106)$$

In both (5.105), and (5.106) the higher powers have been neglected. Further,

$$\frac{c_1}{p_1^2} = \frac{1}{p_1^2}\left(\frac{\beta n}{a}\right)^2 \left(\frac{V}{auw}\right)^4$$

$$= n^2 \left(\frac{n_{eff}}{n_1}\right)^2 \left(\frac{V}{auw}\right)^4 \quad ...(5.107)$$

We may observe from the above discussion that when the fields are not entirely transverse; that either $E_z \neq 0$, all the radial fields are nonzero. Similarly, the ϕ-components of both the electric and the magnetic fields are also nonzero. Similar statements can be made for the case of $H_z \neq 0$. Hence, the nonzero longitudinal component of one field would make the longitudinal component of the other field also nonzero. Such modes are referred to as EH or HE modes depending on the field having a higher strength. The roots of the Bessel function of first kind would determine the cutoff conditions of the modes referred to as HE_{1m} or EH_{1m} modes.

5.3.4.1. *Determination of Cutoff Condition for the hybrid modes corresponding to n = 1*

We show that, the cutoff condition of these modes is determined by the following condition. We have, for this, from (5.95),

$$J_1(ua) = 0 \qquad \text{...(5.108)}$$

For $n = 1$ in (5.103), we get

$$x = \frac{1}{w^2}\left\{-bw^2(1-\Delta) \pm \left[(bw^2)^2\Delta^2 + \frac{c_1 w^4}{p_1^2}\right]^{\frac{1}{2}}\right\}$$

$$\Rightarrow \quad \frac{1}{w^2}\left\{(1-w^2\ln w^2)(1-\Delta) \pm \left[\begin{matrix}(1-2w^2\ln w)\Delta^2 \\ +\left(1-2\Delta+2\Delta\dfrac{w^2}{u^2+w^2}\right)\left(1+\dfrac{2w^2}{u^2}\right)\end{matrix}\right]^{\frac{1}{2}}\right\}$$

$$\Rightarrow \quad \frac{1}{w^2}\left\{\begin{matrix}(1-w^2\ln w^2)(1-\Delta)\pm \\ \left[\begin{matrix}(1-2w^2\ln w)\Delta^2 + \\ \left(1+\dfrac{2w^2}{u^2}-2\Delta-4\Delta\dfrac{w^2}{u^2}+2\Delta\dfrac{w^2}{u^2}+4\Delta\dfrac{w^4}{u^4}\right)\end{matrix}\right]^{\frac{1}{2}}\end{matrix}\right\}$$

$\because \quad w \to 0$

$$\Rightarrow \quad \frac{1}{w^2}\left\{\begin{matrix}(1-w^2\ln w^2)(1-\Delta)\pm(1-\Delta) \\ \left[1-\dfrac{2w^2}{(1-\Delta)^2}\left[\Delta^2\ln w+\left(\dfrac{1-\Delta}{u^2}\right)\right]^{\frac{1}{2}}\right]\end{matrix}\right\} \qquad \text{...(5.109)}$$

(5.109) is further equal to

$$x \to \frac{1}{w^2}\left\{\begin{aligned}&(1-w^2\ln w^2)(1-\Delta)\\&\pm\left[(1-2w^2\ln w)\Delta^2+\left(1+\frac{2w^2}{u^2}-2\Delta-2\Delta\frac{w^2}{u^2}\right)\right]^{\frac{1}{2}}\end{aligned}\right\}$$

$$\Rightarrow \frac{1}{w^2}\left\{\begin{aligned}&(1-w^2\ln w^2)(1-\Delta)\\&\pm\left(\Delta^2-2w^2\ln w\Delta^2+1+\frac{2w^2}{u^2}-2\Delta-2\Delta\frac{w^2}{u^2}\right)^{\frac{1}{2}}\end{aligned}\right\}$$

$$\Rightarrow \frac{1}{w^2}\left\{\begin{aligned}&(1-w^2\ln w^2)(1-\Delta)\\&\pm\left(1+\Delta^2-2\Delta-2w^2\ln w\Delta^2+\frac{2w^2}{u^2}-2\Delta\frac{w^2}{u^2}\right)^{\frac{1}{2}}\end{aligned}\right\}$$

$$\Rightarrow \frac{1}{w^2}\left\{\begin{aligned}&(1-w^2\ln w^2)(1-\Delta)\\&\pm\left[(1-\Delta)^2-2w^2\left(\Delta^2\ln w+\frac{1}{u^2}-\frac{\Delta}{u^2}\right)\right]^{\frac{1}{2}}\end{aligned}\right\}$$

$$\Rightarrow \frac{1}{w^2}\left\{\begin{aligned}&(1-w^2\ln w^2)(1-\Delta)\\&\pm\left[(1-\Delta)^2-2w^2\left[\Delta^2\ln w+\left(\frac{1-\Delta}{u^2}\right)\right]\right]^{\frac{1}{2}}\end{aligned}\right\}$$

$$\Rightarrow \frac{1}{w^2}\left\{\begin{aligned}&(1-w^2\ln w^2)(1-\Delta)\\&\pm(1-\Delta)\left[1-\frac{2w^2}{(1-\Delta)^2}\left[\Delta^2\ln w+\left(\frac{1-\Delta}{u^2}\right)\right]\right]^{\frac{1}{2}}\end{aligned}\right\} \qquad \text{...(5.110)}$$

By developing the square root of the second term in the RHS of (5.110) to the first term, (5.110) becomes

$$x \to \frac{1}{w^2}\left\{\begin{aligned}&(1-w^2\ln w^2)(1-\Delta)\pm(1-\Delta)\\&\left[1-\frac{2w^2}{(1-\Delta)^2}\left[\Delta^2\ln w+\left(\frac{1-\Delta}{u^2}\right)\right]\right]^{\frac{1}{2}}\end{aligned}\right\}$$

$$\Rightarrow \quad \frac{1}{w^2}\left\{\begin{aligned}&(1-w^2\ln w^2)(1-\Delta)\pm(1-\Delta)\\&\left[1-\frac{2w^2}{2(1-\Delta)^2}\left[\Delta^2\ln w+\left(\frac{1-\Delta}{u^2}\right)\right]\right]\end{aligned}\right\}$$

$$\Rightarrow \quad \frac{1}{w^2}\left\{\begin{aligned}&(1-w^2\ln w^2)(1-\Delta)\\&\pm\left[(1-\Delta)-\frac{w^2}{(1-\Delta)}\left[\Delta^2\ln w+\left(\frac{1-\Delta}{u^2}\right)\right]\right]\end{aligned}\right\}$$

$$\Rightarrow \quad \frac{1}{w^2}\left\{\begin{aligned}&(1-w^2\ln w^2)(1-\Delta)\\&\pm\left[(1-\Delta)-\frac{w^2\Delta^2\ln w}{(1-\Delta)}+\frac{w^2}{u^2}\right]\end{aligned}\right\} \qquad \text{...(5.111)}$$

The EH modes correspond to the positive sign of the discriminant in (5.11). For such a mode, we obtain

$$x \to \frac{1}{w^2}\left\{(1-w^2\ln w^2)(1-\Delta)+\left[(1-\Delta)-\frac{w^2\Delta^2\ln w}{(1-\Delta)}+\frac{w^2}{u^2}\right]\right\}$$

$$\to \frac{1}{u^2}+\frac{2(1-\Delta)}{w^2}-(1-\Delta)\ln w^2-\frac{\Delta^2\ln w}{(1-\Delta)} \qquad \text{...(5.112)}$$

$$\to \frac{1}{u^2}+\frac{2(1-\Delta)}{w^2}-\left[\frac{(1-\Delta)^2+\Delta^2}{(1-\Delta)}\right]\ln w$$

$$\to \frac{1}{u^2}+\frac{2(1-\Delta)}{w^2}-\left[\frac{1-2\Delta+2\Delta^2}{(1-\Delta)}\right]\ln w$$

We can write $\dfrac{J_n'(ua)}{uJ_n(ua)}=x$ as the following

$$x=\frac{J_n'(ua)}{uJ_n(ua)}=\begin{cases}\dfrac{n}{u^2}-\dfrac{J_{n+1}(ua)}{uJ_n(ua)}\\[2ex]\dfrac{J_{n-1}(ua)}{uJ_n(ua)}-\dfrac{n}{u^2}\end{cases} \qquad \text{...(5.113)}$$

(5.112), hence becomes, for $n = 1$,

$$x=\frac{J_1'(ua)}{uJ_1(ua)}=\begin{cases}\dfrac{1}{u^2}-\dfrac{J_2(ua)}{uJ_1(ua)}\\[2ex]\dfrac{J_0(ua)}{uJ_1(ua)}-\dfrac{1}{u^2}\end{cases} \qquad \text{...(5.114)}$$

Therefore, substitution of (5.114) in (5.112) gives

$$\frac{J_2(ua)}{uJ_1(ua)} \to -\frac{2(1-\Delta)}{w^2} + \left[\frac{1-2\Delta+2\Delta^2}{(1-\Delta)}\right]\ln w \to -\infty \qquad \text{...(5.115)}$$

$\because \quad w \to 0$

This means that

$$J_1(ua) = 0 \qquad \text{...(5.116)}$$

However, (5.16) is not a solution to (5.115) as

$$\lim_{u\to 0} \frac{J_2(ua)}{uJ_1(ua)} = \frac{1}{4} \qquad \text{...(5.117)}$$

We, therefore are led to the conclusion that the limit values of the EH_{1m} modes are the roots of $J_1(u) = 0$, except for $u = 0$.

For the HE modes corresponding to the negative sign of the discriminant, we obtain, in a similar manner,

$$x \to \frac{1}{w^2}\left\{(1 - w^2 \ln w^2)(1-\Delta) - \left[(1-\Delta) - \frac{w^2\Delta^2 \ln w}{(1-\Delta)} + \frac{w^2}{u^2}\right]\right\}$$

$$\to \frac{(1-\Delta)}{w^2} - \frac{(1-\Delta)}{w^2} - (1-\Delta)\ln w^2 + \frac{\Delta^2 \ln w}{(1-\Delta)} - \frac{1}{u^2}$$

$$\to -\frac{1}{u^2} - \left[\frac{(1-\Delta)^2 - \Delta^2}{(1-\Delta)}\right]\ln w \qquad \text{...(5.118)}$$

$$\to -\frac{1}{u^2} - \left[\frac{1-2\Delta}{(1-\Delta)}\right]\ln w$$

We make use of (5.113) in (5.118) to obtain

$$\frac{J_0(ua)}{uJ_1(ua)} \to -\ln w\left[\frac{1-2\Delta}{(1-\Delta)}\right] \to \infty \text{ as } w \to 0 \qquad \text{...(5.119)}$$

$$\Rightarrow \qquad J_1(ua) \to 0$$

For (5.119), $u = 0$ is a solution as

$$\lim_{u\to 0} \frac{J_0(ua)}{uJ_1(ua)} = \frac{2}{u^2} \to \infty \qquad \text{...(5.120)}$$

We, therefore are led to the conclusion that the limit values of the HE_{1m} modes are the roots of $J_1(u) = 0$ including $u = 0$. The first root of

$J_1(u) = 0$ occurs at $u = 0$. A mode characterized by this is known as the HE_{11} mode. As we may observe, the HE_{11} mode has no cutoff and ceases to exist only when the core diameter is zero. This is the principle on which the single-mode fiber is based. By appropriately choosing a, n_1, n_2 such that,

$$V = \frac{2\pi a}{\lambda}\sqrt{(n_1 - n_2)} \leq 2.405$$

which is the value at which the lowest-order Bessel function J_0 is zero, all modes except the HE_{11} mode are cut off. This particular mode ceases to exist only when the core diameter is zero. Hence, the single mode fibers propagate light through the HE_{11} mode.

5.3.4.2. *Determination of Cutoff Condition for the hybrid modes corresponding to n > 1*

For $n > 1$, we adopt a somewhat different treatment that is outlined as described next. First, we show that

$$\left(\frac{n_{eff}}{n_1}\right)^2 = \left(\frac{\beta}{2\pi/\lambda}\right)^2\left(\frac{1}{n_1}\right)^2 \qquad \text{...(5.121)}$$

To show the validity of (5.121), we start by writing the following

$$\begin{aligned}\left(1 - 2\Delta\frac{u^2a^2}{V^2}\right) &= 1 - \frac{n_1^2 - n_2^2}{n_1^2}\frac{u^2a^2}{(2\pi/\lambda)^2(n_1^2 - n_2^2)a^2} \\ &= 1 - \frac{u^2}{(2\pi/\lambda)^2 n_1^2} \\ &= 1 - \frac{(2\pi/\lambda)^2 n_1^2 - \beta^2}{(2\pi/\lambda)^2 n_1^2} \\ &= 1 - 1 + \left[\frac{\beta}{(2\pi/\lambda)n_1}\right]^2 \qquad \text{...(5.122)} \\ &= \left(\frac{n_{eff}}{n_1}\right)^2\end{aligned}$$

Hence, the quantity $\frac{c_1 w^4}{p_1^2}$ becomes,

$$\frac{c_1 w^4}{p_1^2} = n^2\left(\frac{n_{eff}}{n_1}\right)^2\left(\frac{V}{au}\right)^4$$

$$\Rightarrow \quad n^2\left(1-2\Delta\frac{u^2a^2}{V^2}\right)\left(\frac{V}{au}\right)^4$$

$$\Rightarrow \quad n^2\left(1-2\Delta\frac{V^2-w^2}{V^2}\right)\left(\frac{V}{au}\right)^4 \quad \text{...(5.123)}$$

$$\Rightarrow \quad n^2\left(1-2\Delta+2\Delta\frac{w^2}{V^2}\right)\left(\frac{V}{au}\right)^4$$

$$\Rightarrow \quad n^2\left(1-2\Delta+2\Delta\frac{w^2}{u^2+w^2}\right)\left(\frac{V}{au}\right)^4$$

$$\Rightarrow \quad n^2\left(1-2\Delta+2\Delta\frac{w^2}{u^2}\right)\left(\frac{V}{au}\right)^4 \quad \because w \to 0$$

We note that

$$\left(\frac{V}{u}\right)^2 = 1+\frac{w^2}{u^2}$$

$$\Rightarrow \quad \frac{V}{u} = \sqrt{1+\frac{w^2}{u^2}} \quad \text{...(5.124)}$$

$$\to \quad 1+\frac{w^2}{2u^2}$$

as a first approximation. Therefore,

$$\left(\frac{V}{u}\right)^4 \to \left(1+\frac{w^2}{2u^2}\right)^4$$

$$\Rightarrow \quad \left(1+\frac{w^2}{u^2}+\frac{w^4}{4u^4}\right)\left(1+\frac{w^2}{u^2}+\frac{w^4}{4u^4}\right) \quad \text{...(5.125)}$$

$$\Rightarrow \quad 1+2\frac{w^2}{u^2} \text{ neglecting powers of } w > 2 \text{ as } w \to 0$$

Hence (5.103) becomes after retaining the terms containing only

$$x = \frac{1}{w^2}\left\{\begin{array}{l} -\left(n^2 + \dfrac{nw^2}{n-1}\right)^{\frac{1}{2}}(1-\Delta) \\ \pm\left[\left(n^2 + \dfrac{nw^2}{n-1}\right)\Delta^2 + n^2\left(1 - 2\Delta\dfrac{u^2a^2}{V^2}\right)\left(\dfrac{V}{au}\right)^4\right]^{\frac{1}{2}} \end{array}\right\}$$

$$= \frac{1}{w^2}\left\{\begin{array}{l} -\left(n^2 + \dfrac{nw^2}{n-1}\right)^{\frac{1}{2}}(1-\Delta) \\ \pm n\left\{\left[1 + \dfrac{w^2}{n(n-1)}\right]\Delta^2 + \left(1 - 2\Delta + 2\Delta\dfrac{w^2}{u^2}\right)\left(1 + \dfrac{2w^2}{u^2}\right)^{\frac{1}{2}}\right\} \end{array}\right\} \quad ...(5.126)$$

This quadratic has two real solutions as the square root term is positive. The positive and the negative real roots correspond to two different families of modes. By convention, the EH modes are associated with the positive root and the HE modes are associated with the minus sign. As we see, these modes have nonzero longitudinal components E_z and H_z; hence they are called hybrids. For the EH modes, we have seen previously that the longitudinal electric field E_z has a higher magnitude compared to the magnetic field H_z. Hence the roots of these modes would naturally correspond to the zeros of the Bessel's function of first kind and order n. The first term in the square root of (5.126) becomes

$$\left(n^2 + \frac{nw^2}{n-1}\right)^{\frac{1}{2}}(1-\Delta)$$

$$= n\left(1 + \frac{w^2}{n(n-1)}\right)^{\frac{1}{2}}(1-\Delta)$$

$$= n\left[1 + \frac{w^2}{2n(n-1)}\right](1-\Delta) \quad ...(5.127)$$

$$= \left[n + \frac{w^2}{2(n-1)}\right](1-\Delta)$$

The discriminant of (5.126) becomes, with the substitution of (5.127)

$$n\left\{\left[1+\frac{w^2}{n(n-1)}\right]\Delta^2+\left(1-2\Delta+2\Delta\frac{w^2}{u^2}\right)\left(1+\frac{2w^2}{u^2}\right)\right\}^{\frac{1}{2}}$$

$$=n\left(\Delta^2+\frac{\Delta^2 w^2}{n(n-1)}+1+\frac{2w^2}{u^2}-2\Delta-4\Delta\frac{w^2}{u^2}+2\Delta\frac{w^2}{u^2}+4\Delta\frac{w^4}{u^4}\right)^{\frac{1}{2}}$$

$$=n\left(1+\Delta-2\Delta+\frac{\Delta^2 w^2}{n(n-1)}+\frac{2w^2}{u^2}-2\Delta\frac{w^2}{u^2}\right)^{\frac{1}{2}} \text{ neglecting } \frac{w^4}{u^4}$$

$$=n\left[(1-\Delta)^2+\frac{\Delta^2 w^2}{n(n-1)}+\frac{2w^2}{u^2}-2\Delta\frac{w^2}{u^2}\right]^{\frac{1}{2}}$$

$$=n(1-\Delta)\left\{1+\frac{1}{(1-\Delta)^2}\left[\frac{\Delta^2 w^2}{n(n-1)}+\frac{2w^2}{u^2}(1-\Delta)\right]\right\}^{\frac{1}{2}} \quad ...(5.128)$$

$$=n(1-\Delta)\left\{1+\frac{1}{2(1-\Delta)^2}\left[\frac{\Delta^2 w^2}{n(n-1)}+\frac{2w^2}{u^2}(1-\Delta)\right]\right\}$$

$$=\left[n(1-\Delta)+\frac{\Delta^2 w^2}{2(1-\Delta)(n-1)}+n\frac{w^2}{u^2}\right]$$

Hence, (5.126) becomes, through the use of (5.127) and (5.128),

$$x\to\frac{1}{w^2}\left[n+\frac{w^2}{2(n-1)}\right](1-\Delta)\pm\left[n(1-\Delta)+\frac{\Delta^2 w^2}{2(1-\Delta)(n-1)}+n\frac{w^2}{u^2}\right] \quad ...(5.129)$$

For the EH modes, corresponding to the positive sign, in (5.129) we get

$$x=\frac{1}{w^2}\left\{\left[n+\frac{w^2}{2(n-1)}\right](1-\Delta)+\left[n(1-\Delta)+\frac{\Delta^2 w^2}{2(1-\Delta)(n-1)}+n\frac{w^2}{u^2}\right]\right\}$$

$$=\frac{n(1-\Delta)}{w^2}+\frac{n(1-\Delta)}{w^2}+\frac{n}{u^2} \quad ...(5.130)$$

$$=\frac{n}{u^2}+2\frac{n(1-\Delta)}{w^2}$$

This is because as $w \to 0$, the terms containing w^2 in the numerator have been made equal to zero. Therefore, making use of (5.113), we get

$$\frac{J_{n+1}(u)}{uJ_n(u)} \to -2\frac{n(1-\Delta)}{w^2} \to -\infty \qquad ...(5.131)$$

$$\Rightarrow \qquad J_n(u) \to 0$$

However, $u = 0$ is not a solution as we have the following limit

$$\lim_{u\to 0}\left[\frac{J_{n+1}(u)}{uJ_n(u)}\right] \to \frac{1}{2(n+1)} \neq 0 \qquad ...(5.132)$$

Hence, we have $J_n(ua) = 0$ satisfied as the roots when $n = 2$. This corresponds to the EH_{nm} modes. Therefore, the limit values of the EH_{nm} modes corresponding to $n > 1$ are the roots of $J_n(u)$ except for $u = 0$.

Similarly, for the HE modes corresponding to the negative sign, we write

$$x \to \frac{1}{w^2}\left[n + \frac{w^2}{2(n-1)}\right](1-\Delta) - \left[n(1-\Delta) + \frac{\Delta^2 w^2}{2(1-\Delta)(n-1)} + n\frac{w^2}{u^2}\right]$$

$$= \frac{n(1-\Delta)}{w^2} - \frac{n(1-\Delta)}{w^2} + \frac{(1-\Delta)}{2(n-1)} - \frac{\Delta^2}{2(n-1)(n-1)} - \frac{n}{u^2}$$

$$= \frac{1}{2(n-1)}\left[(1-\Delta) - \frac{\Delta^2}{(1-\Delta)}\right] - \frac{n}{u^2} \qquad ...(5.133)$$

$$= -\frac{n}{u^2} + \frac{1-2\Delta}{2(n-1)(1-\Delta)}$$

As in the case of EH modes, we write

$$x \to \frac{J_{n-1}(u)}{uJ_n(u)} - \frac{n}{u^2}$$

$$\frac{J_{n-1}(u)}{uJ_n(u)} \to \frac{1-2\Delta}{2(n-1)(1-\Delta)} \qquad ...(5.134)$$

We may expand (5.134) as follows

$$\frac{J_{n-1}(u)}{uJ_n(u)} \to \frac{1-2\Delta}{2(n-1)(1-\Delta)}$$

$$\Rightarrow \qquad (1-\Delta)J_{n-1}(u) = \frac{1-2\Delta}{2(n-1)}uJ_n(u)$$

$$\Rightarrow \qquad \frac{n_1^2+n_2^2}{2n_1^2} J_{n-1}(u) = \frac{n_2^2}{n_1^2 2(n-1)} u J_n(u) \qquad \text{...(5.135)}$$

$$\Rightarrow \qquad \frac{n_1^2+n_2^2}{n_2^2} J_{n-1}(u) = \frac{uJ_n(u)}{(n-1)}$$

$$\Rightarrow \qquad \left(1+\frac{n_1^2}{n_2^2}\right) J_{n-1}(u) = \frac{uJ_n(u)}{(n-1)}$$

This limit is finite. This expression may be written in an alternative form by inverting the fractions of the previous expression and by using the recurrence relation

$$uJ_n(u) = 2(n-1)J_{n-1}(u) - uJ_{n-2}(u) \qquad \text{...(5.136)}$$

P. Verify (5.136).

Therefore,

$$\frac{2(n-1)J_{n-1}(u) - uJ_{n-2}(u)}{J_{n-1}(u)} \to \frac{2(n-1)(1-\Delta)}{1-2\Delta}$$

$$\Rightarrow \qquad -\frac{uJ_{n-2}(u)}{(n-1)J_{n-1}(u)} \to \frac{2(1-\Delta)}{1-2\Delta} - 2 \qquad \text{...(5.137)}$$

$$\Rightarrow \qquad -\frac{uJ_{n-2}(u)}{(n-1)J_{n-1}(u)} \to \frac{2\Delta}{1-2\Delta}$$

$$\Rightarrow \qquad \frac{uJ_{n-2}(u)}{(n-1)J_{n-1}(u)} \to -\frac{2\Delta}{1-2\Delta}$$

The limit values of the HE_{nm} modes ($n > 1$) are the roots to (5.137), the right hand side of which will tend towards zero in the weakly guiding regime, *i.e.*, when $\Delta \to 0$. The lowest four order modes are illustrated in Fig. 5.4.

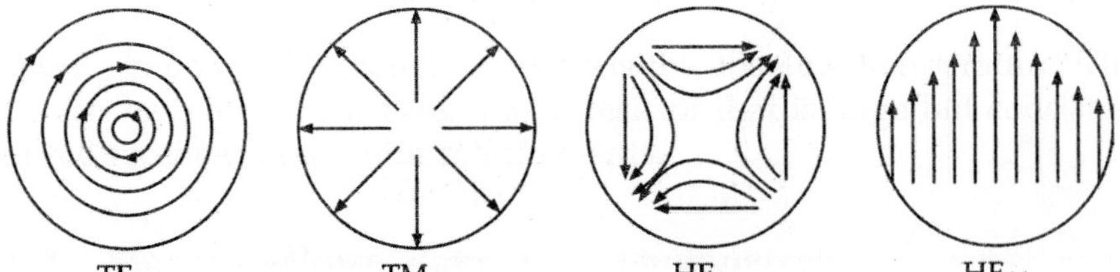

Fig. 5.4: Schematic cross sections showing the transverse electric field vectors in the fiber core for the four lowest order modes in an SI fiber

The discussion from (5.38) to (5.137) is summarized in Table 5.2 that lists the cutoff conditions for various modes.

Table 5.2: Cutoff conditions for some lower order modes

n	*Mode*	*Cutoff condition*
0	TE_{0m} TM_{0m}	$J_0(ua) = 0$
1	HE_{1m} EH_{1m}	$J_1(ua) = 0$
2 and higher	EH_{2m}	$J_n(ua) = 0$
2 and higher	HE_{2m}	$\left(\frac{n_1^2}{n_2^2}+1\right) J_{n-1}(ua) = \frac{ua}{n-1} J_n(ua)$

The roots of (5.103) corresponding to the cutoff conditions of various modes can also be written in the following way.

$$b = 1 - \left(\frac{ua}{V}\right)^2$$

$$\Rightarrow \qquad \left(\frac{ua}{V}\right)^2 = 1 - b \qquad \text{...(5.138)}$$

$$\Rightarrow \qquad 1 - \frac{\beta^2 - \beta_2^2}{\beta_1^2 - \beta_2^2}$$

$$= \frac{\beta_1^2 - \beta_2^2 - (\beta^2 - \beta_2^2)}{\beta_1^2 - \beta_2^2}$$

$$= \frac{n_1^2 - 1}{n_1^2 - n_2^2}$$

We understand that

$$\left(1 - 2\Delta \frac{u^2 a^2}{V^2}\right)$$

$$= 1 - \frac{n_1^2 - n_2^2}{n_1^2} \cdot \frac{n_1^2 - 1}{n_1^2 - n_2^2}$$

$$= 1 - \frac{n_1^2 - 1}{n_1^2} \qquad \text{...(5.139)}$$

$$= \frac{1}{n_1^2}$$

Hence,

$$\frac{c_1}{p_1^2} = \frac{1}{p_1^2}\left(\frac{\beta n}{a}\right)^2\left(\frac{V}{auw}\right)^4$$

$$= n^2\left(\frac{\beta}{p_1}\right)^2\left(\frac{V}{auw}\right)^4$$

$$= n^2\left(\frac{1}{n_1}\right)^2\left(\frac{V}{auw}\right)^4 \quad \because \frac{\beta}{p_1} = \frac{(2\pi/\lambda)}{(2\pi/\lambda)n_1} \qquad \text{...(5.140)}$$

$$= n^2\left(1 - 2\Delta\frac{u^2a^2}{V^2}\right)\left(\frac{V}{auw}\right)^4$$

$$= n^2h^2$$

The typical ray paths have been illustrated in Figs. 5.5 and Fig. 5.6. We see from Fig. 5.5 that light propagates by many paths in a multimode fiber. The modes are responsible for the different paths taken by a light beam coupled to the fiber at its input end. In a single mode fiber, light beam propagates by only one path which is the axial ray and the presence of just one mode is responsible for the single path taken by the light beam as shown in Fig. 5.6.

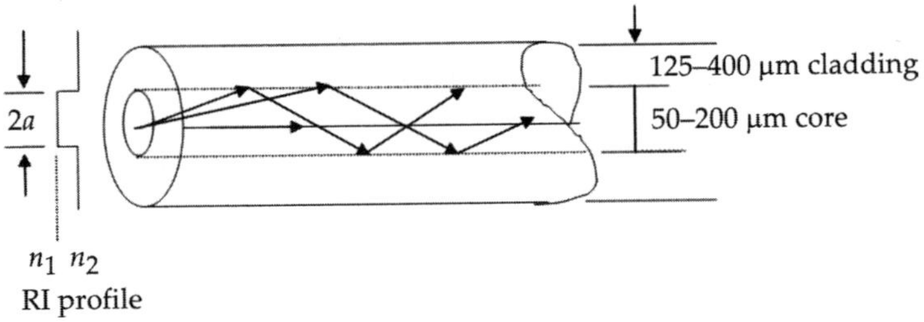

Fig. 5.5: Typical ray paths and dimensions in a multimode step index waveguide

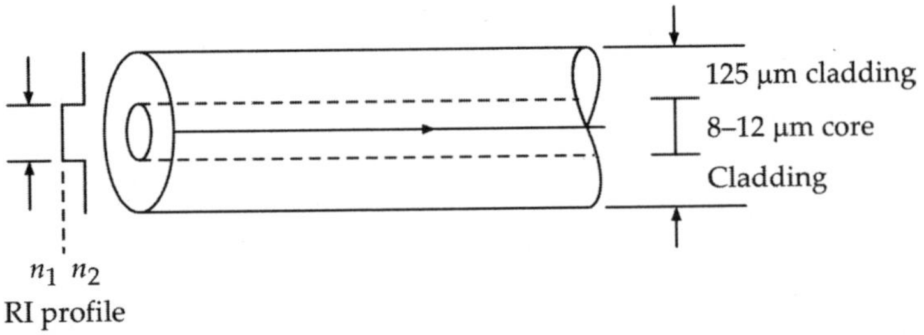

Fig. 5.6: Typical ray paths and dimensions in a monomode step index waveguide

5.4. LINEARLY POLARIZED MODES

As we might have observed from the previous discussion, the exact analysis of the modes in a fiber is mathematically very complex. However, a simple but highly accurate approximation may be used, based on the principle that in a typical step-index fiber the difference between the indices of refraction of the core and cladding is very small, that is $\Delta << 1$. This is the basis of the weakly guiding fiber approximation [2]. In this approximation, the electromagnetic field patterns and the propagation constants of the mode pairs HE_{n+1}, and $EH_{n-1,m}$ are very similar. This holds likewise for the three modes TE_{0m}, and HE_{2m}. Corresponding to the case of $\Delta << 1$, we have $p_1^2 \approx p_2^2 = \beta^2$. Hence the eigen value equation (5.94) is modified as

$$(I_n + K_n)(p_1^2 I_n + p_2^2 K_n) = \left(\frac{\beta n}{a}\right)^2 \left(\frac{1}{u^2} + \frac{1}{w^2}\right)^2$$

$$\Rightarrow \quad p_1^2 (I_n + K_n)^2 = \left(\frac{\beta n}{a}\right)^2 \left(\frac{1}{u^2} + \frac{1}{w^2}\right)^2 \qquad ...(5.141)$$

$$\Rightarrow \quad (I_n + K_n)^2 = \left(\frac{n}{a}\right)^2 \left(\frac{1}{u^2} + \frac{1}{w^2}\right)^2$$

$$\Rightarrow \quad (I_n + K_n) = \pm\left(\frac{n}{a}\right) \left(\frac{1}{u^2} + \frac{1}{w^2}\right)$$

This gives us two solutions corresponding to the positive and the negative signs contained in the RHS of (5.141). Use of recurrence relations in Bessel's function gives us, for the positive sign,

$$\frac{J_n'(ua)}{uJ_n(ua)} + \frac{K_n'(wa)}{wK_n(wa)} = \left(\frac{n}{a}\right)\left(\frac{1}{u^2} + \frac{1}{w^2}\right)$$

$$\Rightarrow \quad \frac{\frac{n}{ua}J_n(ua) - J_{n+1}(ua)}{uJ_n(ua)} + \frac{\frac{n}{wa}K_n(wa) - K_{n+1}(wa)}{wK_n(wa)} = \left(\frac{n}{a}\right)\left(\frac{1}{u^2} + \frac{1}{w^2}\right)$$

$$\Rightarrow \quad \frac{n}{au^2} - \frac{J_{n+1}(ua)}{uJ_n(ua)} + \frac{n}{aw^2} - \frac{K_{n+1}(wa)}{wK_n(wa)} = \left(\frac{n}{a}\right)\left(\frac{1}{u^2} + \frac{1}{w^2}\right) \qquad ...(5.142)$$

$$\Rightarrow \quad -\frac{J_{n+1}(ua)}{uJ_n(ua)} - \frac{K_{n+1}(wa)}{wK_n(wa)} = 0$$

$$\Rightarrow \quad \frac{J_{n+1}(ua)}{uJ_n(ua)} + \frac{K_{n+1}(wa)}{wK_n(wa)} = 0$$

The solution of (5.142) gives a set of modes called the EH modes. Similarly, for the negative sign in (5.141), we have

$$\frac{J_n'(ua)}{uJ_n(ua)} + \frac{K_n'(wa)}{wK_n(wa)} = -\left(\frac{n}{a}\right)\left(\frac{1}{u^2} + \frac{1}{w^2}\right)$$

$$\Rightarrow \quad \frac{-J_n'(ua)}{uJ_n(ua)} + \frac{-K_n'(wa)}{wK_n(wa)} = \left(\frac{n}{a}\right)\left(\frac{1}{u^2} + \frac{1}{w^2}\right)$$

$$\Rightarrow \quad \frac{\frac{n}{ua}J_n(ua) - J_{n-1}(ua)}{uJ_n(ua)} + \frac{\frac{n}{wa}K_n(wa) + K_{n-1}(wa)}{wK_n(wa)} = \left(\frac{n}{a}\right)\left(\frac{1}{u^2} + \frac{1}{w^2}\right)$$

$$\Rightarrow \quad \frac{n}{au^2} - \frac{J_{n-1}(ua)}{uJ_n(ua)} + \frac{n}{aw^2} + \frac{K_{n-1}(wa)}{wK_n(wa)} = \left(\frac{n}{a}\right)\left(\frac{1}{u^2} + \frac{1}{w^2}\right) \quad ...(5.143)$$

$$\Rightarrow \quad -\frac{J_{n-1}(ua)}{uJ_n(ua)} + \frac{K_{n-1}(wa)}{wK_n(wa)} = 0$$

$$\Rightarrow \quad \frac{J_{n-1}(ua)}{uJ_n(ua)} - \frac{K_{n-1}(wa)}{wK_n(wa)} = 0$$

Inverting (5.143), we get

$$\frac{uJ_n(ua)}{J_{n-1}(ua)} = \frac{wK_n(wa)}{K_{n-1}(wa)} \quad ...(5.144)$$

An alternate expression can be derived by making use of the following recurrence relation.

$$uJ_n(ua) = 2(n-1)J_{n-1}(u) - uJ_{n-2}(u) \quad ...(5.145a)$$

and similarly,

$$wK_n(wa) = 2(n-1)K_{n-1}(w) + wK_{n-2}(w) \quad ...(5.145b)$$

We prove the identities in (5.145) as follows:

$$J_{n+1}(x) = \frac{2n}{x}J_n(x) - J_{n-1}(x)$$

$$\Rightarrow \quad xJ_{n+1}(x) = 2nJ_n(x) - xJ_{n-1}(x) \quad \text{(Proved)}$$

$$\Rightarrow \quad xJ_m(x) = 2(x)J_{m-1}(x) - xJ_{n-2}(x) \because m = n+1$$

Similarly,

$$K_{n+1}(x) = \frac{2n}{x}K_n(x) + K_{n-1}(x)$$

$$\Rightarrow \qquad xK_{n+1}(x) = 2nK_n(x) + xK_{n-1}(x)$$

$$\Rightarrow \qquad xJ_m(x) = 2(m-1)K_{m-1}(x) + xK_{n-2}(x) \because m = n+1$$

Substitution of (5.145) into (5.144) gives us

$$\frac{2(n-1)J_{n-1}(ua) - uJ_{n-2}(ua)}{J_{n-1}(ua)} = \frac{2(n-1)K_{n-1}(wa) + wK_{n-2}(wa)}{K_{n-1}(wa)}$$

$$-\frac{uJ_{n-2}(ua)}{J_{n-1}(ua)} = \frac{wK_{n-2}(wa)}{K_{n-1}(wa)} \qquad ...(5.146)$$

These modes are known as HE modes.

We can have another set of modes corresponding to the inverse of (5.145). To do so, we carry out the manipulation of (5.96) as following.

$$\frac{c_1}{p_1^2} = \frac{1}{p_1^2}\left(\frac{\beta}{a}\right)^2\left(\frac{1}{u^2} + \frac{1}{w^2}\right)^2$$

$$= \frac{1}{p_1^2}\left(\frac{\beta}{a}\right)^2\left(\frac{w^2 + u^2}{u^2 w^2}\right)$$

$$= \frac{1}{p_1^2}\left(\frac{\beta}{a}\right)^2\left(\frac{n_1^2 - n_2^2}{u^2 w^2}\right) \qquad ...(5.147)$$

$$= \frac{1}{(\beta n_1)^2}\left(\frac{\beta}{a}\right)^2\left(\frac{n_1^2 2\Delta}{u^2 w^2}\right)$$

$$= \frac{2\Delta}{(auw)^2}$$

For the condition of $\Delta << 1$, we have $\frac{c_1}{p_1^2} << 1$ and hence it may be neglected in (5.103) which becomes

$$x = -b(1-\Delta) \pm \left(b^2\Delta^2 + \frac{c_1}{p_1^2}\right)^{\frac{1}{2}}$$

$$\Rightarrow \qquad x = -b \pm \frac{\sqrt{c_1}}{p_1} \qquad ...(5.148)$$

$$\Rightarrow \qquad x = -b$$

We define a new parameter

$$\xi = \begin{cases} 1 & \text{for TE and TM modes} \\ n+1 & \text{for EH modes} \\ n-1 & \text{for HE modes} \end{cases} \qquad \text{...(5.149)}$$

Then all the modes can be expressed as

$$\frac{uJ_{\xi-1}(ua)}{J_\xi(ua)} = -\frac{wK_{\xi-1}(wa)}{K_\xi(wa)} \qquad \text{...(5.150)}$$

These equations show that within the weakly guiding approximation all modes are characterized by a common set of ξ and m that satisfy the same characteristic equation. This means that these modes are degenerate. Thus, if an $HE_{n+1,m}$ mode is degenerate with an $EH_{n-1,m}$ mode (that is if HE and EH nodes of corresponding radial order m and equal circumferential order n form degenerate pairs), then any combination of an $H_{En+1,m}$ mode with an $EH_{n-1,m}$ mode will likewise constitute a guided mode of the fiber. Such degenerate modes may be called linearly polarized modes designated as $LP_{\xi m}$ modes regardless of their TE, TM, EH or HE field configuration. In general we have the following

(i) Each LP_{0m} mode is derived from an HE_{1m} mode
(ii) Each LP_{1m} mode comes from TE_{0m}, TM_{0m}, HE_{2m} modes
(iii) Each LP_{nm} mode ($n \geq 2$) is from a $HE_{n+1'm}$ and $HE_{n+1,m}$ mode.

The correspondence between five numbers of lowest LP modes (that is, those having the lowest cutoff frequencies) and the traditional TE, TM, EH and HE modes is given in Table 5.3 as an illustration. This table also shows the number of degenerate modes.

The reader is strongly recommended to verify any entry of Table 5.3.

Table 5.3: Composition of lower order LP modes

LP-mode type	*Traditional mode type and Number of modes*	*Number of degenerate modes*
LP_{01}	$HE_{11} \times 2$	2
LP_{11}	TE_{01}, TM_{01}, $HE_{21} \times 2$	4
LP_{21}	$EH_{11} \times 2$, $HE_{31} \times 2$	4
LP_{02}	$HE_{12} \times 2$	2
LP_{31}	$HE_{21} \times 2$, $HE_{41} \times 2$	4

A very useful feature of the LP-mode designation is the ability to readily visualize a mode. In a complete set of modes only one electric and one magnetic field component are significant. The electric field vector **E** may be chosen to lie along an arbitrary axis, with the magnetic field vector **H** being perpendicular to it. In addition there are two equivalent solutions with the field polarities reversed. Since each of the two possible polarization directions can be coupled with either a cos $\xi\phi$ or sin $\xi\phi$ a azimutal dependence, four discrete mode patterns can be obtained from a single $LP_{\xi m}$ label. As an example, the four possible electric and magnetic field directions and the corresponding intensity distributions for the LP_{11} mode are shown in Fig. 5.4. Figures illustrate how two LP_{11} modes are composed from the exact HE_{21} plus TE_{01} and the exact HE_{21} plus TM_{01} modes, respectively.

In summary, we may state the following:

(a) The cutoff condition for TE_{0m} and TM_{0m} modes is obtained by solving for $J_0(ua) = 0$ for $n = 0$.

(b) The cutoff condition for EH_{1m} and HE_{1m} modes is obtained by solving for $J_1(ua) = 0$ for $n = 1$.

(c) The cutoff condition for higher order EH_{nm} modes is obtained by solving for $J_{n1}(ua) = 0$ for $n \geq 0$.

(d) The cutoff condition for higher order HEn_m modes is obtained by solving for for $\left[1+\left(\frac{n_1}{n_2}\right)^2\right] J_{n-1}(ua) = 0.$

5.5. FRACTIONAL MODAL POWER IN THE CORE

One of the important parameters associated with a fiber optic waveguide is the fractional power carried in the core. Now, the power in the core of the fiber is given by

$$\begin{aligned} P_{core} &= C\int_0^a\int_0^{2\pi} |\Psi|^2 \rho d\rho d\phi \\ &= 2C\int_0^a J_n^2\left(\frac{u\rho}{a}\right)\rho d\rho \int_0^{2\pi}\cos^2 n\phi d\phi \\ &= \frac{2C\pi a^2}{u^2}\int_0^u J_n^2(x)x dx; \quad x = \frac{\rho}{a} \qquad \text{...(5.151)} \\ &= C\pi a^2\left[1 - \frac{J_{n-1}(u)J_{n+1}(u)}{J_n^2(u)}\right] \end{aligned}$$

where C is a constant and standard integrals associated with Bessel functions have been used. In a similar manner, we can write down an expression for the power in the cladding.

$$
\begin{aligned}
P_{\text{cladding}} &= C\int_0^a\int_0^{2\pi} |\Psi|^2 \rho d\rho d\phi \\
&= 2C\int_a^{\infty} K_n^2\left(\frac{u\rho}{a}\right)\rho d\rho \int_0^{2\pi}\cos^2 n\phi d\phi \\
&= C\pi a^2\left[\frac{K_{n-1}(u)K_{n+1}(u)}{K_n^2(u)} - 1\right]
\end{aligned}
\qquad \text{...(5.152)}
$$

The total power carried by the optical fiber is the sum of the powers of the wave in the core and the cladding. Using (5.151) and (5.152), we have

$$
\begin{aligned}
P_{\text{tot}} &= P_{\text{core}} + P_{\text{cladding}} \\
&= C\pi a^2\left[1 - \frac{J_{n-1}(u)J_{n+1}(u)}{J_n^2(u)}\right] + C\pi a^2\left[\frac{K_{n-1}(w)K_{n+1}(w)}{K_n^2(w)} - 1\right] \\
&= C\pi a^2\left[1 - \frac{J_{n-1}(u)J_{n+1}(u)}{J_n^2(u)} + \frac{K_{n-1}(w)K_{n+1}(w)}{K_n^2(w)} - 1\right] \\
&= C\pi a^2\left[\frac{K_{n-1}(w)K_{n+1}(w)}{K_n^2(w)} + \frac{w^2}{u^2}\frac{K_{n-1}(w)K_{n+1}(w)}{K_n^2(w)}\right] \\
&= C\pi a^2\frac{V^2}{u^2}\frac{K_{n+1}(w)K_{n-1}(w)}{K_n^2(w)}
\end{aligned}
\qquad \text{...(5.153)}
$$

where we have used the following relation

$$
\frac{u^2 J_{n+1}(u)J_{n-1}(u)}{J_n^2(u)} = -\frac{w^2 K_{n+1}(w)K_{n-1}(w)}{K_n^2(w)}
$$

The fractional power propagating in the core is thus given by

$$
\eta = \frac{P_{core}}{P_{tot}} = \frac{C\pi a^2\left[1 - \dfrac{J_{n-1}(u)J_{n+1}(u)}{J_n^2(u)}\right]}{C\pi a^2\dfrac{V^2}{u^2}\dfrac{K_{n+1}(w)K_{n-1}(w)}{K_n^2(w)}}
$$

$$= \frac{1 + \dfrac{w^2}{u^2} \dfrac{K_{n+1}(w)K_{n-1}(w)}{K_n^2(w)}}{\dfrac{V^2}{u^2} \dfrac{K_{n+1}(w)K_{n-1}(w)}{K_n^2(w)}} \qquad ...(5.154)$$

$$= \frac{u^2 + w^2 \dfrac{K_{n+1}(w)K_{n-1}(w)}{K_n^2(w)}}{V^2 \dfrac{K_{n+1}(w)K_{n-1}(w)}{K_n^2(w)}} = \frac{w^2}{V^2} + \frac{u^2}{V^2} \frac{K_n^2(w)}{K_{n+1}(w)K_{n-1}(w)}$$

We can find from the above that as the mode approaches cutoff

$$V \to V_c;\ w \to 0 \text{ and } u \to V_c \qquad ...(5.155)$$

The limiting forms of $K_n(w)$ give us

$$K_0(w) \underset{w \to 0}{\to} -\ln\left(\frac{w}{2}\right) \qquad ...(5.156)$$

and
$$K_n(w) \underset{w \to 0}{\to} \frac{1}{2}\Gamma(n)\left(\frac{2}{w}\right)^n; \quad n > 0$$

We note that as $w \to 0$, the function

$$\frac{w^2}{V^2} + \frac{u^2}{V^2} \frac{K_n^2(w)}{K_{n+1}(w)K_{n-1}(w)}$$

$$\to 0 \text{ for } n = 0,1$$

$$\to \frac{K_n^2(w)}{K_{n+1}(w)K_{n-1}(w)} \qquad \because \frac{u^2}{V^2} = \frac{u^2}{u^2 + w^2} = 1$$

$$= \frac{\frac{1}{2}\Gamma(n)\left(\frac{2}{w}\right)^n \frac{1}{2}\Gamma(n)\left(\frac{2}{w}\right)^n}{\frac{1}{2}\Gamma(n+1)\left(\frac{2}{w}\right)^{n+1} \frac{1}{2}\Gamma(n-1)\left(\frac{2}{w}\right)^{n-1}}$$

$$= \frac{\Gamma(n)\Gamma(n)}{\Gamma(n+1)\Gamma(n-1)} = \frac{\Gamma(n)}{n\Gamma(n)} \frac{(n-1)\Gamma(n)}{\Gamma(n-1)} = \frac{n-1}{n} \geq 2 \qquad ...(5.157)$$

The power associated with a particular mode is concentrated in the core for large values of that is, far from cutoff.

Example 5.1: Consider an interface of a dielectric and a metal, that is

$$n^2 = \begin{cases} n_1^2; x > 0 \\ n_m^2; x < 0 \end{cases}$$

where n_m^2 is complex having a large negative part. Show that such a structure can support a TM mode. Find out the propagation constant of such a mode.

Solution: For a step index fibre

$$\mu_1 \varepsilon_1 = \frac{n_1^2}{c^2}$$

in one medium and similarly

$$\mu_2 \varepsilon_2 = \frac{n_2^2}{c^2}$$

in another medium

For a dielectric media,

$$\mu \varepsilon_1 = \frac{n_1^2}{c^2}$$

and similarly,

$$\mu \varepsilon_2 = \frac{n_2^2}{c^2}$$

When an interface consists of a boundary between a conducting surface and a dielectric, we note that the requirement of a TM mode dictates a loop current in the *xy*-plane which is not possible here as one axis has a dielectric. As there is no loop current in the *xy*-plane, no magnetic field would exist along the unit normal to the loop; *i.e.*, the *z*-axis, which means that $H_z = 0$. The fields would be exponentially decaying in both the regions. Let us denote the propagation constants in the two regions by k_1 and k_2 respectively. The fields in the two regions are

$$H_y(x) = A_1 \exp(j\beta x)\exp(-k_1 z) \text{ for } z > 0$$

And similarly,

$$H_y(x) = A_2 \exp(j\beta x)\exp(k_2 z) \text{ for } z < 0$$

The associated electric fields are given by

$$E_z = -\frac{j}{\omega\varepsilon_0 n_1^2}\frac{\partial H_y}{\partial x}$$

$$\Rightarrow \qquad E_z = -\frac{j}{\omega\varepsilon_0 n_1^2}(-k_1)A_1 \exp(j\beta x)\exp(-k_1 z)$$

for $z > 0$

Similarly,

$$E_z = -\frac{j}{\omega\varepsilon_0 n_2^2}(k_2)A_1 \exp(j\beta x)\exp(k_2 z)$$

for $z < 0$

The continuity of the tangential components require that

$$-\frac{j}{\omega\varepsilon_0 n_1^2}(-k_1)A_1 \exp(j\beta x)\exp(-k_1 z)$$

$$= -\frac{j}{\omega\varepsilon_0 n_2^2}(k_2)A_1 \exp(j\beta x)\exp(k_2 z)$$

$$\Rightarrow \qquad -\frac{k_1}{k_2} = \frac{n_1^2}{n_2^2}$$

We observe that the real part of n_1 must be a negative quantity in order to satisfy the continuity of the tangential components across the boundary. Further, we have

$$k_1^2 = \beta^2 - k_0^2 n_1^2 \text{ and } k_2^2 = \beta^2 - k_0^2 n_2^2$$

We have, dividing these two

$$\frac{k_1^2}{k_2^2} = \frac{\beta^2 - k_0^2 n_1^2}{\beta^2 - k_0^2 n_2^2} = \frac{n_1^4}{n_2^4}$$

$$\Rightarrow \qquad n_2^4(\beta^2 - k_0^2 n_1^2) = n_1^4(\beta^2 - k_0^2 n_2^2)$$

$$\Rightarrow \qquad \beta^2(n_2^4 - n_1^4) = k_0^2 n_2^4 n_1^2 - k_0^2 n_1^4 n_2^2$$

$$\Rightarrow \qquad \beta^2(n_2^2 + n_1^2)(n_2^2 - n_1^2) = k_0^2 n_1^2 n_2^2 (n_2^2 - n_1^2)$$

$$\Rightarrow \qquad \frac{\beta^2}{k_0^2} = \frac{n_1^2 n_2^2}{n_2^2 + n_1^2} = \frac{n_1^2 n_m^2}{n_1^2 + n_m^2}$$

when the second medium is a metal having a refractive index given as

$$z_2^2 = n_m^2$$

5.6. CHAPTER SUMMARY

The step index fiber propagates light using the principle of total internal reflection through various modes resulting in different field patterns. The fields obey Bessel's function of first kind of order *n* in the core while those in the cladding follow the modified Bessel's function of the second kind of order *n*. The choice of the core and cladding refractive indices and the core diameter determines the type of mode conveying light. TE, TM and hybrid modes are established each characterized by different cutoff conditions. The cutoff condition refers to the zeros of the modal fields set up in the core and the cladding by solving the scalar wave equation subject to appropriate boundary conditions. This is the condition at which no light propagation is possible. We saw that the HE_{11} mode does not have any cutoff condition and this is the mode by which light is propagated in single mode fibers. The power is usually carried by the higher order modes inside the core.

SUGGESTED READING

[1] Kaiser, G. "Optical Fiber Communications", 2nd Edn., McGraw Hill.
[2] Gowar, J. "Optical Communication Systems", 2nd Edn., PHI, 1995.
[3] Gloge, D. "Weakly guiding fibers", Applied Optics, vol. 10, pp. 2252-2258, Oct 1971.

❑❑❑

6

Waves in Graded Index (GI) Fibres

6.1. INTRODUCTION

In the graded index fibre design the core refractive index decreases continuously with increasing radial distance ρ from the centre of the fibre but is generally constant in the cladding. The most commonly used construction for the refractive index variation in the core is the power law relationship

$$n(\rho) = \begin{cases} n_1\left[1-2\Delta\left(\dfrac{\rho}{a}\right)^{\alpha}\right]^{1/2} & \text{for } 0 \le \rho \le a \\ n_1(1-2\Delta)^{1/2} \approx n_1(1-\Delta) = n_2 \text{ for } \rho \ge a & \end{cases} \quad ...(6.1)$$

Here ρ is the radial distance from the fibre axis, a is the core radius, n_1 is the refractive index at the core axis, n_2 is the refractive index of the cladding, and the dimensionless parameter α defines the shape of the refractive index profile. The index difference Δ for the graded index fibre is given by

$$\Delta = \frac{n_1^2 - n_2^2}{2n_1^2} \approx \frac{n_1 - n_2}{n_1} \quad ...(6.2)$$

The approximation on the right-hand side of (6.2) reduces the expression for Δ to that of the step index fibre. The same symbol is used in both the cases. For, $\alpha = \infty$, (6.1) reduces to the step index profile, $n(\rho) = n_1$.

6.2. GRADED INDEX NUMERICAL APERTURE (NA)

The determination of the NA for the GI fibre is more complex than for the step index fibres. In a GI fibre, the NA is a function of position across the core end face. This is in contrast to the step index fibre where the NA is constant across the core. Geometrical optics considerations show that light incident on the fibre at position ρ will propagate as a guided mode only if it is within the local numerical aperture NA (ρ) at that point. The local numerical aperture is defined as

$$\text{NA}(\rho) = \begin{cases} [n^2(\rho) - n_2^2]^{1/2} \approx NA(0)\sqrt{1-(\rho/a)^{\alpha}} & \text{for } \rho \le a \\ 0 & \text{for } \rho \ge a \end{cases} \quad \text{...(6.3)}$$

where the axial numerical aperture is defined as

$$\text{NA}(0) = [n^2(0) - n_2^2]^{1/2} = (n_1^2 - n_2^2)^{1/2} \approx n_1\sqrt{2\Delta} \quad \text{...(6.4)}$$

The details are presented in Appendix-6.A. It is clear from (6.3) that the NA of a GI fibre decreases from NA(0) to zero as ρ moves from the fibre axis to the core-cladding boundary.

6.3. MODES IN GI FIBRES

A modal analysis of an optical fibre based on solving Maxwell's equations can only be carried out rigorously if the core refractive index is uniform, that is for step index fibres. In other cases, such as GI fibres, approximation methods are needed. The most widely used analysis of modes in a GI fibre is an approximation based on the WKB (Wenzel, Kramers and Brillouin), which is commonly used in quantum mechanics. The purpose of the WKB method is to obtain an asymptotic representation for the solution of a differential equation containing a parameter that varies slowly over the desired range of the equation. That parameter in this case is the refractive index profile $n(\rho)$, which varies only slightly over distances on the order of an optical wavelength. Similar to the SI fibres as in (5.31), we have here the corresponding scalar wave function expressed as

$$\frac{d^2P}{d\rho^2} + \frac{1}{\rho}\frac{dP}{d\rho} + \left(p^2 - \frac{n^2}{\rho^2}\right)P = 0 \quad \text{...(6.5)}$$

where $$p^2 = \omega^2\mu\varepsilon - \beta^2 \quad \text{...(6.6)}$$

However, as we are considering the GI fibres, we have

$$\sqrt{\mu\varepsilon} = \frac{n(\rho)}{c} \quad \text{...(6.7)}$$

representing the wave velocity in the optical fibre as the refractive index is now a function of the radial distance and is not a constant. Thus, the wave equation becomes

$$\frac{d^2P}{d\rho^2}+\frac{1}{\rho}\frac{dP}{d\rho}+\left[\frac{\omega^2}{c^2}n^2(\rho)-\beta^2-\frac{n^2}{\rho^2}\right]P=0 \qquad ...(6.8)$$

Let us make the substitution

$$U = P\sqrt{\rho} \qquad ...(6.9)$$

This gives us further,

$$P = \frac{U}{\sqrt{\rho}} \qquad ...(6.10)$$

$$\Rightarrow \qquad \frac{dP}{d\rho} = \frac{d}{d\rho}\left(\frac{U}{\sqrt{\rho}}\right) = \left(\frac{dU}{d\rho\sqrt{\rho}}-\frac{U}{2\rho^{3/2}}\right)$$

$$\Rightarrow \qquad \frac{d^2P}{d\rho^2} = \frac{d}{d\rho}\left(\frac{dU}{d\rho\sqrt{\rho}}-\frac{U}{2\rho^{3/2}}\right)$$

$$\Rightarrow \qquad \left(\frac{d^2U}{d\rho^2\sqrt{\rho}}-\frac{1}{2\rho^{3/2}}\frac{dU}{d\rho}-\frac{1}{2\rho^{3/2}}\frac{dU}{d\rho}-\frac{1}{2}\times-\frac{3}{2}\frac{U}{\rho^{5/2}}\right)$$

$$\Rightarrow \qquad \frac{d^2U}{d\rho^2\sqrt{\rho}}-\frac{dU}{\rho^{3/2}d\rho}+\frac{3}{4}\frac{U}{\rho^{5/2}}$$

Therefore, (6.5) becomes

$$\frac{d^2U}{d\rho^2\sqrt{\rho}}-\frac{dU}{\rho^{3/2}d\rho}+\frac{3}{4}\frac{U}{\rho^{5/2}}+\frac{1}{\rho}\left(\frac{dU}{\sqrt{\rho}d\rho}-\frac{U}{2\rho^{3/2}}\right)$$
$$+\left[\frac{\omega^2}{c^2}n^2(\rho)-\beta^2-\frac{n^2}{\rho^2}\right]\frac{U}{\sqrt{\rho}}=0$$

$$\Rightarrow \qquad \frac{1}{\sqrt{\rho}}\left(\frac{d^2U}{d\rho^2}+\frac{U}{4\rho^2}\right)+\left[\frac{\omega^2}{c^2}n^2(\rho)-\beta^2-\frac{n^2}{\rho^2}\right]\frac{U}{\sqrt{\rho}}=0 \qquad ...(6.11)$$

$$\Rightarrow \qquad \frac{d^2U}{d\rho^2}+\left[\frac{\omega^2}{c^2}n^2(\rho)-\beta^2-\frac{\left(n^2-\frac{1}{4}\right)}{\rho^2}\right]U=0$$

Let us write the bracketed terms in (6.11) as

$$\frac{\omega^2}{c^2}n^2(\rho)-\beta^2-\frac{\left(n^2-\frac{1}{4}\right)}{\rho^2}=P^2 \qquad ...(6.12)$$

A close look at (6.12) reveals that for light of a particular frequency ω, propagating in a mode having a given propagation constant β, there are some regions of the fibre where P is real and some where it is imaginary. The real value of P is obtained if the following condition is satisfied

$$\frac{\omega^2}{c^2}n^2(\rho)-\beta^2\geq\frac{\left(n^2-\frac{1}{4}\right)}{\rho^2} \qquad ...(6.13)$$

Within this region the field distributions vary periodically with radial distance, whereas outside these radii, P is imaginary and the fields decay exponentially. We associate periodic field distributions with guided modes that are constrained within the cylindrical shell bounded by ρ_1 and ρ_2, and fringed by decaying evanescent fields inside and outside these radii. When $n = 0$, there is no inner boundary, ρ_1, and the periodic nature of the guided mode fields extends to the fibre axis. Such modes may be thought to correspond to meridional rays. In this case the fields have radial symmetry.

The periodic field distributions between ρ_1 and ρ_2 are required to match the decaying fields outside, and in consequence only certain discrete values of propagation constant, β, satisfy the boundary conditions and give rise to allowed modes of propagation. These permitted values of β are designated as β_{nm}. The integer n is the azimuthal mode number. The integer m represents the number of half-periods of P between ρ_1 and ρ_2.

Exact analytical solutions for $P(\rho)$ can be obtained in closed form in only a few special cases. One is the step-index profile considered in detail in chapter 5. Another is the parabolic index profile given as

$$n^2(\rho)=n_1^2\left[1-2\Delta\left(\frac{\rho}{a}\right)^2\right] \qquad ...(6.14)$$

However, approximate solutions for P and β_{nm} in a multimode GI fibre may be obtained, provided that the fractional variation in $n(\rho)$ across the fibre diameter is small $n(\rho)$. We still expect the solutions to be periodic and closely related to the case where n is uniform. With independent of ρ in (6.11), we obtain the sinusoidal solutions for the modes:

$$U = U_0\exp(jP\rho) \qquad ...(6.15)$$

As a first approximation, when $n(\rho)$ is slowly varying, we may expect solutions for the higher-order modes to be of the form as given below.

$$U = U_0 \exp\left(j\int P\,d\rho\right) \qquad ...(6.16)$$

Then,

$$P = \frac{U}{\sqrt{\rho}} = \frac{U_0 \exp\left(j\int P\,d\rho\right)}{\sqrt{\rho}} \qquad ...(6.17)$$

As we may observe, each half-period of P corresponds to $\int P\,d\rho = \pi$. With m such half-periods between the roots of $P^2 = 0$ (ρ_1 and ρ_2),

$$\left(m - \frac{\pi}{2}\right) \approx \int_{\rho_1}^{\rho_2} P\,d\rho = \int_{\rho_1}^{\rho_2} \left[\frac{\omega^2}{c^2} n^2(\rho) - \beta^2 - \frac{\left(n^2 - \frac{1}{4}\right)}{\rho^2}\right]^{1/2} d\rho \qquad ...(6.18)$$

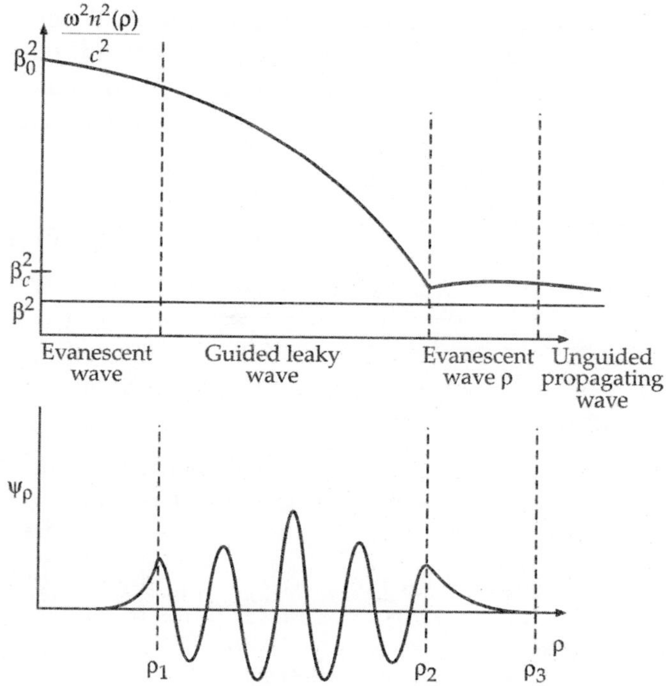

Fig. 6.1: Leaky modes in GI fibre (a) various waves in different regions that satisfy (6.13) (b) oscillatory waves in the region between ρ_1 and ρ_2.

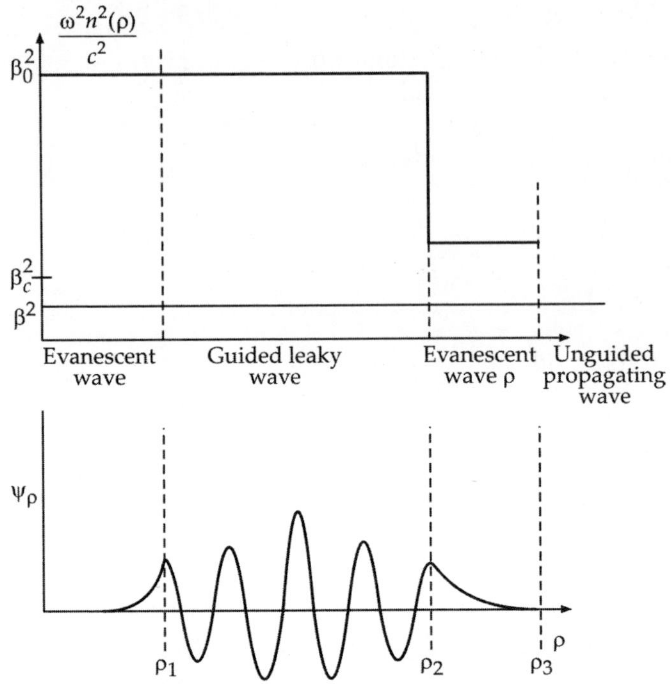

Fig. 6.2: Leaky modes in SI fibre (a) various waves in different regions that satisfy (6.13) (b) oscillatory waves in the region between ρ_1 and ρ_2.

When $n > 0$, (6.13) can be satisfied over a range of radii, even though $\beta < \beta_c$, so the wave does not meet the condition for guidance. Such waves are known as leaky modes. The situation is illustrated for both SI and GI profiles in Figs. 6.1 and 6.2 respectively. It may be observed from these diagrams that the fields of the leaky wave between ρ_1 and ρ_2 do not couple directly with those of the unguided, radiating wave in the cladding beyond ρ_3.

They are separated by the evanescent fields between ρ_2 and ρ_3. Leaky modes, although more highly attenuated than the normal fully guided modes, can propagate considerable distances before they radiate away.

An alternative treatment for the modes set up in a GI fibre having a square law profile is developed next. The GI fibres having a square law profile as given by (6.3) can also be expressed in Cartesian coordinates,

$$n^2 = n_1^2\left[1 - 2\Delta\left(\frac{x^2 + y^2}{a^2}\right)\right] \qquad ...(6.19)$$

The wave velocity in a medium having a refractive index satisfying (6.19) is

$$v = \frac{c}{n} \quad \text{...(6.20)}$$

The scalar wave equation

$$\nabla^2 \Psi = \mu\varepsilon \frac{\partial^2 \Psi}{\partial t^2} = \frac{1}{v^2}\frac{\partial^2 \Psi}{\partial t^2} \quad \text{...(6.21)}$$

Becomes

$$\nabla^2 \Psi = \frac{n_1^2}{c^2}\left[1 - 2\Delta\left(\frac{x^2 + y^2}{a^2}\right)\right]\frac{\partial^2 \Psi}{\partial t^2} \quad \text{...(6.22)}$$

Let us assume a solution of the form of

$$\Psi(x, y, z, t) = \Psi(x, y) \exp[j(\omega t - \beta z)] \quad \text{...(6.23)}$$

Hence, (6.22) becomes

$$\frac{\partial^2 \Psi}{\partial x^2} + \frac{\partial^2 \Psi}{\partial y^2} + \frac{\partial^2 \Psi}{\partial z^2} = -\frac{n_1^2}{c^2}\omega^2\left[1 - 2\Delta\left(\frac{x^2 + y^2}{a^2}\right)\right]\Psi$$

$$\Rightarrow \quad \frac{\partial^2 \Psi}{\partial x^2} + \frac{\partial^2 \Psi}{\partial y^2} - \beta^2\Psi + \frac{n_1^2}{c^2}\omega^2\left[1 - 2\Delta\left(\frac{x^2 + y^2}{a^2}\right)\right]\Psi = 0 \quad \text{...(6.24)}$$

$$\Rightarrow \quad \frac{\partial^2 \Psi}{\partial x^2} + \frac{\partial^2 \Psi}{\partial y^2} + \left\{k_0^2 n_1^2\left[1 - 2\Delta\left(\frac{x^2 + y^2}{a^2}\right)\right] - \beta^2\right\}\Psi = 0$$

where the free space wave propagation constant is

$$k_0 = \frac{\omega}{c} = \frac{2\pi f}{c} = \frac{2\pi}{\lambda} \quad \text{...(6.25)}$$

We use the method of separation of variables to solve (6.24) for which we write

$$\Psi(x, y) = X(x)\, Y(y) \quad \text{...(6.26)}$$

We substitute (6.26) in (6.25) and divide it by XY to obtain

$$\left(\frac{1}{X}\frac{d^2X}{dx^2} - k_0^2 n_1^2 \frac{2\Delta}{a^2}x^2\right) + \left(\frac{1}{Y}\frac{d^2Y}{dy^2} - k_0^2 n_1^2 \frac{2\Delta}{a^2}y^2\right) + \left(k_0^2 n_1^2 - \beta^2\right) = 0 \quad \text{...(6.27)}$$

We separate the variables and equate each of the terms in the LHS of (6.27) to

$$\left(\frac{1}{X}\frac{d^2X}{dx^2} - k_0^2 n_1^2 \frac{2\Delta}{a^2}x^2\right) = -C_1 \qquad \text{...(6.28)}$$

and

$$\left(\frac{1}{Y}\frac{d^2Y}{dy^2} - k_0^2 n_1^2 \frac{2\Delta}{a^2}y^2\right) = -C_2 \qquad \text{...(6.29)}$$

Of course, in (6.28) and (6.29) the terms C_1 and C_2 are constants.

Further, we get

$$-C_1 - C_2 + (k_0^2 n_1^2 - \beta^2) = 0 \qquad \text{...(6.30)}$$

$$\Rightarrow \qquad \beta^2 = k_0^2 n_1^2 - C_1 - C_2$$

Let us further substitute

$$\xi = \vartheta x \text{ and } \chi = \vartheta y \qquad \text{...(6.31)}$$

In (6.31), we have used

$$\vartheta = \left(\frac{k_0 n_1 \sqrt{2\Delta}}{a}\right)^{1/2} = \frac{\sqrt{V}}{a} \qquad \text{...(6.32)}$$

In (6.32), we define

$$V = k_0 n_1 a\sqrt{2\Delta} \qquad \text{...(6.33)}$$

represents the waveguide parameter.

Hence, (6.28) and (6.29) become, with the substitution of ...(6.32)

$$\left(\frac{1}{X}\frac{d^2X}{dx^2} - \vartheta^4 x^2\right) = -C_1$$

$$\Rightarrow \qquad \vartheta^2 \frac{d^2}{d\zeta^2}(X) - \vartheta^2.\vartheta^2 x^2 X = -C_1 X$$

$$\Rightarrow \qquad \vartheta^2 \frac{d^2}{d\zeta^2}(X) - \vartheta^2 \zeta^2 X + C_1 X = 0 \quad \because \zeta^2 = \vartheta^2 x^2$$

$$\Rightarrow \qquad \frac{d^2X}{d\zeta^2} + \left(\frac{C_1}{\vartheta^2} - \zeta^2\right) X(\zeta) = 0$$

$$\Rightarrow \qquad \frac{d^2X}{d\zeta^2} + (\Lambda_1 - \zeta^2)X(\zeta) = 0 \qquad \text{...(6.34)}$$

In a similar manner, we can write

$$\frac{d^2Y}{d\chi^2} + (\Lambda_2 - \chi^2)Y(\chi) = 0 \qquad \text{...(6.35)}$$

where

$$\Lambda_1 = \frac{C_1}{\vartheta^2} \qquad \text{...(6.36)}$$

and

$$\Lambda_2 = \frac{C_2}{\vartheta^2} \qquad \text{...(6.37)}$$

We observe that $X(\zeta)$ and $Y(\chi)$ must tend toward zero as $\zeta, \chi \to \pm\infty$ ($x, y, \to \pm\infty$) for bounded solutions. This means that

$$\Lambda_1 = \frac{C_1}{\vartheta^2} = 2m + 1; \; m = 0, 1, 2, .. \qquad \text{...(6.38)}$$

and

$$\Lambda_2 = \frac{C_2}{\vartheta^2} = 2n + 1; \; n = 0, 1, 2, ... \qquad \text{...(6.39)}$$

The corresponding solutions to $X(x)$ and $Y(y)$ are the Hermite-Gauss functions. Hence the mode profiles are expressed as

$$\Psi_{mn}(x, y) = \left[N_m H_m(\zeta)\exp\left(-\frac{1}{2}\zeta^2\right)\right]\left[N_n H_n(\chi)\exp\left(-\frac{1}{2}\chi^2\right)\right] \qquad \text{...(6.40)}$$

and the propagation constant becomes

$$\beta_{mn} = k_0 n_1 \left[1 - \frac{2(m+n+1)}{k_0 n_1}\sqrt{\frac{2\Delta}{a^2}}\right]^{1/2} \qquad \text{...(6.41)}$$

Hermite functions are polynomial solutions of

$$\frac{d^2y}{dx^2} - 2x\frac{dy}{dx} + 2my(x) = 0$$

The functions $H_m(\zeta)$ are the well-known Hermite polynomials given by

$$H_0(\zeta) = 1, \; H_1(\zeta) = 2\zeta$$

$$H_2(\zeta) = 4\zeta^2 - 2, \; H_3(\zeta) = 8\zeta^3 - 12\zeta$$

and in general

$$H_{m+1}(\zeta) = 2\zeta H_m(\zeta) - 2mH_{m-1}(\zeta) \qquad ...(6.42)$$

Corresponding to the values of m, n = 0, 1, 2, ... and

$$N_m = \left[\frac{\vartheta}{2^m m!\sqrt{\pi}}\right]^{\frac{1}{2}} = \left[\frac{\sqrt{V}/a}{2^m m!\sqrt{\pi}}\right]^{\frac{1}{2}} \qquad ...(6.43)$$

represents the normalization constant. Different values of m and n correspond to different modes of the fibre. An analytic solution of the propagation constant of the (m, n)th mode in such a fibre is given by (6.41). For a typical multimode parabolic index fibre,

$$n_1 = 1.46,\ \Delta = 0.01 \text{ and } a = 25\ \mu\text{m}$$

Corresponding to a free space wavelength $\lambda_0 \approx 0.8\ \mu$m

$$\frac{2\sqrt{2\Delta}}{k_0 n_1 a} \approx 10^{-3}$$

Hence, for $m + n << 10^3$, we are justified in making a binomial expansion in (6.41) and retaining the first-order term, we obtain

$$\begin{aligned}\beta_{nm} &\approx \frac{2\pi}{\lambda} - (m+n+1)\frac{\sqrt{2\Delta}}{a} \\ &= \frac{2\pi}{(c/n_1)/f} - (m+n+1)\frac{\sqrt{2\Delta}}{a} \\ &= \frac{\omega}{c}n_1 - (m+n+1)\frac{\sqrt{2\Delta}}{a}\end{aligned} \qquad ...(6.44)$$

If we neglect the wavelength dependence of n_1 and Δ-that is if we neglect material dispersion, the following expression for the group velocity is obtained.

$$\frac{1}{v_g} = \frac{d\beta_{nm}}{d\omega} \approx \frac{n_1}{c} \qquad ...(6.45)$$

This implies that all the modes set up in the fibre travel with almost the same group velocity. Translated to the language of ray optics, this corresponds to the fact that all rays take the same amount of time to propagate through the fibre. Typical ray paths in a GI fibre are shown in Fig. 6.3.

In an actual parabolic index fibre, the refractive index given by (6.1) is valid beyond which it has a constant value equal to

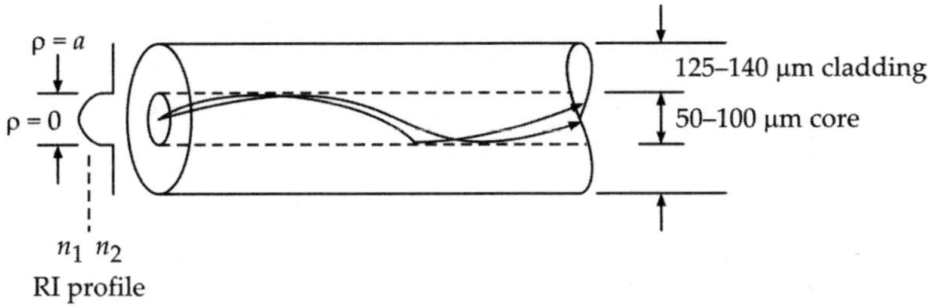

Fig. 6.3: Typical ray paths and dimensions in a multimode graded index fiber

$$n = n_1\sqrt{1-2\Delta} = n_2 \text{ for } (x^2+y^2) > a^2$$

And the mode solutions given by (6.40) and (6.41) are then not exact solutions for such a profile. However, even for realistic refractive index profile, the solutions given by (6.40) and (6.41) are reasonably accurate for lower order modes in a multimoded fibre. The guided modes correspond to only those values of m and n for which

$$k_0 n_2 < \beta_{nm} < k_0 n_1 \qquad \text{...(6.46)}$$

The lowest order mode obviously corresponds to $m = n = 0$ and hence

$$\Psi_{00} = \frac{\vartheta}{\sqrt{\pi}}\exp\left[-\frac{1}{2}\vartheta^2(x^2+y^2)\right] = \frac{\vartheta}{\sqrt{\pi}}\exp\left(-\frac{1}{2}\vartheta^2\rho^2\right)$$

$$= \frac{1}{a}\sqrt{\frac{V}{\pi}}\exp\left[-\frac{1}{2}V\left(\frac{\rho}{a}\right)^2\right] \qquad \text{...(6.47)}$$

which has a Guassian distribution.

The next mode is two-fold degenerate ($m = 0$, $n = 1$ and $m = 1$, $n = 0$). Hence the field distributions are given by

$$\Psi_{01} = \sqrt{\frac{2}{\pi}}\vartheta^2 y \exp\left(-\frac{1}{2}\vartheta^2\rho^2\right) \qquad \text{...(6.48)}$$

and

$$\Psi_{10} = \sqrt{\frac{2}{\pi}}\vartheta^2 x \exp\left(-\frac{1}{2}\vartheta^2\rho^2\right) \qquad \text{...(6.49)}$$

Let us mention here that the modal fields as represented by Ψ can either be an electric field or a magnetic field. Hence, we may substitute Ψ by E_y corresponding to the TE modes or H_y for the TM modes. We have the y-directed fields due to the variation of the refractive index with respect to x. In a similar manner, we may obtain the fields E_x and E_y corresponding to the case of the refractive index profile as given by (6.1).

Example 6.1: Show that if $n^2(x) = n^2(-x)$, then the modal field patterns are either symmetric or antisymmetric functions of x.

Solution: The wave equation corresponding to the TE modes is

$$\frac{d^2E_y(x)}{dx^2} + k_0 n^2(x) E_y(x) = \beta^2 E_y(x)$$

Making the transformation $x \to -x$, we have

$$\frac{d^2E_y(-x)}{dx^2} + k_0 n^2(x) E_y(-x) = \beta^2 E_y(-x)$$

A close look at the field expressions $E_y(x)$ and $E_y(-x)$ reveals that both of them satisfy the same wave equation and therefore they are eigen functions belonging to the same value of β^2. Thus, if the modes are nondenerate, then $E_y(-x)$ must be a multiple of x which means that

$$E_y(-x) = \alpha E_y(x)$$

$$\Rightarrow \qquad E_y(x) = \alpha E_y(-x)$$

$$\Rightarrow \qquad E_y(x) = \alpha \cdot \alpha E_y(-x) = \alpha^2 E_y(x)$$

which means that $\lambda^2 = 1$ or $\lambda = \pm 1$. Hence

$$E_y(-x) = \pm E_y(x)$$

Modal fields belonging to the class $\lambda = 1$ and $\lambda = -1$ are symmetric and antisymmetric functions of x respectively.

Example 6.2: Using the condition that for guided modes $\Psi_m(x) \to 0$ as $x \to \pm\infty$, show that the mode fields are orthogonal to each other.

Solution: We start by writing the one dimensional wave equation as

$$\frac{d^2\Psi_m}{dx^2} + [k_0 n^2(x) - \beta_m^2]\Psi_m(x) = 0 \qquad \text{...(A)}$$

Let us write this as an eigenvalue equation

$$\frac{d^2\Psi_m}{dx^2} + [k_0 n^2(x) - \beta_m^2]\Psi_m(x) = 0 \qquad \text{...(B)}$$

where $\lambda_m = \beta^2$ represents the eigen value of the operator $\left[\frac{d^2}{dx^2} + k_0 n^2(x)\right]$.

We write the complex conjugate of the eigenvalue equation corresponding to the eigen value λ_k

$$\frac{d^2\Psi_k^*}{dx^2} + k_0 n^2(x)\Psi_k^*(x) = \lambda_k^* \Psi_k^*(x) \qquad \text{...(C)}$$

Let us multiply (A) by Ψ_k^* and (B) by ψ_k and subtract these from each other following which we obtain

$$\Psi_k^* \frac{d^2\Psi_m}{dx^2} - \Psi_m \frac{d^2\Psi_k^*}{dx^2} = (\lambda_m - \lambda_k^*)\Psi_m(x)\Psi_k^*(x) \qquad \text{...(D)}$$

The LHS of (D) becomes

$$\frac{d}{dx}\left(\Psi_k^* \frac{d\Psi_m}{dx} - \Psi_m \frac{d\Psi_k^*}{dx}\right) = \Psi_k^* \frac{d^2\Psi_m}{dx^2} - \Psi_m \frac{d^2\Psi_k^*}{dx^2} \qquad \text{...(E)}$$

Integrating (E) within from to , we obtain

$$(\lambda_m - \lambda_k^*) \int_{-\infty}^{\infty} \Psi_m(x)\Psi_k^*(x)dx = \int_{-\infty}^{\infty} \frac{d}{dx}\left(\Psi_k^* \frac{d\Psi_m}{dx} - \Psi_m \frac{d\Psi_k^*}{dx}\right)$$

$$= \left[\Psi_k^* \frac{d\Psi_m}{dx} - \Psi_m \frac{d\Psi_k^*}{dx}\right]_{-\infty}^{\infty} \qquad \text{...(F)}$$

The RHS of (F) vanishes as for guided modes the fields vanish at $x = \pm\infty$. Hence, we obtain

$$(\lambda_m - \lambda_k^*) \int_{-\infty}^{\infty} \Psi_m(x)\Psi_k^*(x)dx = 0 \qquad \text{...(G)}$$

For $k = m$, the integral $\int_{-\infty}^{\infty} |\Psi_m(x)|^2 dx$ is positive definite and we therefore must have $\lambda_m = \lambda_m^*$ proving that all eigenvalues β_m^2 must be real. Further, for $\lambda_m \neq \lambda_k$, we must have

$$\int_{-\infty}^{\infty} \Psi_k^*(x)\Psi_m(x)dx = 0 \text{ for } \lambda_m \neq \lambda_k$$

This proves that the modal fields are orthogonal to each other.

6.4. CHAPTER SUMMARY

In a step index fibre, the modal fields exhibit a Bessel function behaviour inside the core and a modified Bessel function of second kind in the cladding.

In a graded index fibre, the modal fields normally exhibit a Gaussian behaviour.

In both the fibres, the modal fields are orthogonal to each other.

APPENDIX-6.A

$$n(\rho) = \begin{cases} n_1\left[1 - 2\Delta\left(\dfrac{\rho}{a}\right)^{\alpha}\right]^{1/2} & \text{for } 0 \le \rho \le a \\ n_1(1-2\Delta)^{1/2} \approx n_1(1-\Delta) = n_2 & \text{for } \rho \ge a \end{cases} \qquad \text{...(6.A-1)}$$

From this, immediately we can see that

$$n^2(0) = n_1^2 \qquad \text{...(6.A-2)}$$

The quantity $NA(0)$ has been defined as

$$NA_2(0) = [n^2(0) - n_2^2] = (n_1^2 - n_2^2)$$

From (3.146), we get $NA \approx n_1\sqrt{2\Delta}$. As the refractive index difference between the core and the cladding is very small, we may approximate $NA(0) = (n_1^2 - n_2^2)^{1/2} \approx n_1\sqrt{2\Delta}$. Use of this in (6.A-1) gives us the following

$$NA(\rho) = \left\{n_1^2\left[1 - 2\Delta\left(\frac{\rho}{a}\right)^{\alpha}\right] - n_2^2\right\}^{1/2} = \left\{n_1^2 - n_2^2 - n_1^2 2\Delta\left(\frac{\rho}{a}\right)^{\alpha}\right\}^{1/2}$$

$$= \left\{[NA(0)]^2 - n_1^2 2\Delta\left(\frac{\rho}{a}\right)^{\alpha}\right\}^{1/2} = NA(0)\left\{1 - \frac{n_1^2 2\Delta\left(\frac{\rho}{a}\right)^{\alpha}}{NA^2(0)}\right\}^{1/2}$$

$$= NA(0)\left\{1 - \frac{n_1^2 2\Delta\left(\frac{\rho}{a}\right)^{\alpha}}{n_1^2 2\Delta}\right\}^{1/2} = NA(0)\left[1 - \left(\frac{\rho}{a}\right)^{\alpha}\right]^{1/2} \qquad \text{...(6.A-3)}$$

❑❑❑

7

Pulse Dispersion Mechanisms

7.1. INTRODUCTION

As a light wave propagates along an optical fibre, it gets distorted due to a number of factors. The intramodal dispersion is pulse spreading in the time domain (also known as the temporal spreading) that occurs within a single mode. This is due to that fact that the group velocity is a function of the wavelength λ. Since intramodal dispersion depends on the wavelength, its effect on signal distortion increases with the spectral width of the optical source. This spectral width is the band of wavelengths over which the source emits light. It is normally characterized by the root-mean square (rms) spectral width. For the LEDs, the rms spectral width is approximately 5% of a central wavelength. For example, if the peak emission wavelength of an LED source is 850 nm, a typical source width would be 40 nm; that is, the source emits most of its optical power in the 830-870-nm wavelength band. Laser diode optical sources have much narrower spectral widths, typical values being 1 to 2 nm.

7.2. DISPERSION OF PULSES IN AN OPTICAL FIBRE

A very elementary optical transmitter consists of a source of light (as discussed in Chapter 4) that needs to be biased properly by a driver circuit as shown in Fig. 7.1. Information sent in the form of optical pulses down a length of an optical fibre suffers from a couple of dispersion effects. These dispersion phenomena result in widening an optical pulse; the temporal spreading of an optical pulse thereby placing a limit on the maximum amount of data that can be transferred over the fibre so that the receiver can distinguish between any two consecutive pulses. The receiver, for this case consists of a photodetector coupled to an amplifier with a view to strengthen the weak electrical output as produced by the detector. In a fibre, two distinct types of dispersion are observed as discussed in the following.

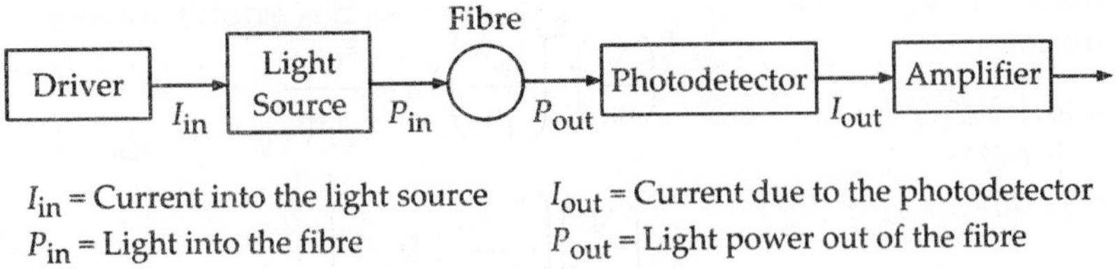

Fig. 7.1: Elements of a typical elementary optical communication system

(a) Intermodal dispersion (or simply modal dispersion or between different modes)

(b) Intramodal despension (within a single mode)

Two main causes of intramodal dispersion are:

(a) Material dispersion, which arises from the variation of the refractive index of the core material as a function of wavelength. (Material dispersion is also called chromatic dispersion or spectral dispersion, since this is the same effect by which a prism spreads out a spectrum). This causes a wavelength dependence of the group velocity of any given mode; that is, pulse spreading occurs even when different wavelengths follow the same path.

(b) Waveguide dispersion, which occurs because a single-mode fibre only confines about 80% of the optical power to the core. Dispersion thus arises, since the remaining 20% of the light propagating in the cladding travels faster than the light confined to the core (due to the smaller refractive index of the cladding). The amount of waveguide dispersion depends on the fibre design, since the modal propagation β constant is a function of a/λ (the optical fibre dimension relative to the wavelength λ.

The other factor responsible for pulse spreading is intermodal delay, which is a result of each mode having a different value of the group velocity at a single frequency.

The waveguide dispersion is usually ignored in multimode fibres. This effect, however, is significant in single mode fibres. The SMFs exhibit a dispersion behaviour that limits the long-distance and very high speed transmission. The dispersion of a single mode silica fibre is lowest at 1300 nm whereas its attenuation is a minimum at 1550 nm, where the dispersion is higher. Hence, for achieving a maximum transmission distance of a high-capacity link, the dispersion null should be at the wavelength of minimum attenuation. To achieve this, one can adjust the basic fibre parameters to shift the zero dispersion minimum to longer wavelengths. An alternative to reduce fibre dispersion is by spreading the dispersion minimum out over a broader range. This approach is known as dispersion flattening.

Material dispersion tends to be the dominant pulse-distorting effect in graded index fibres. Intermodal delay distortion can be made very small by carful tailoring of the core refractive-index profile.

A light-pulse will broaden as it travels along a fibre. This pulse broadening eventually causes a pulse to overlap with neighbouring pulses. After a certain amount of overlap, has occurred, adjacent pulses can no longer be individually distinguished at the receiver and errors will occur. Thus the dispersive properties determine the limit of the information capacity of the fibre.

Electrical bandwidth is defined as the frequency at which the voltage gain of the circuit falls to 0.707. The optical bandwidth is defined as the frequency at which power gain of the circuit falls to half of the maximum value. In case of optical systems, the power is directly proportional to current and not to the square of the current. Hence, optical bandwidth is more than the electrical bandwidth. In case of an optical fibre, the bandwidth depends upon its length. The fibre bandwidth is normally expressed in terms of the bandwidth times kilometre product. So a 100 MHz.km fibre can usually operate with a 200 MHz bandwidth for a 5 kilometre of optical fibre. A longer fibre results in more pulse spreading and hence leads to a lower bandwidth as a result.

The information-carrying capacity can be determined by examining the deformation of short light pulses propagating along the fibre. A measure of the information capacity of an optical fibre is usually specified by the bandwidth-distance product in MHz.km. For a step index fibre, the various distortion effects as outlined above tend to limit the bandwidth-distance product to about 20 MHz.km. In graded index fibres, the radial refractive index profile can be carefully selected so that pulse broadening is minimized at a specific operating wavelength. This has led to bandwidth-distance product as high as 2-5 GHz.km. Single-mode fibres can have capacities well in excess of this. The flatness of the attenuation curves for the fibres extend up to the microwave spectrum.

7.3. RAY OPTICS AND PULSE DISPERSION

Pulse dispersion is an important characteristic of an optical fiber as it determines the maximum information carrying capacity in terms of bits per second. In digital communication system, information to be sent is first coded in the form of pulses and these pulses of light are transmitted from the transmitter to the receiver where the information is decoded. The higher the number of pulses that can be transmitted per second and still be resolvable at the receiver, the larger is the information carrying capacity of the system. A pulse of light launched onto an optical fibre broadens in time as it propagates through it due to the reasons mentioned below.

(a) Different rays take different times to propagate through a given length of the fiber.

(b) Any given source emits radiation over a range of wavelengths and because of the intrinsic property of the material, different wavelengths take different amounts of time to propagate along the same path (also referred to as material dispersion)

This spreading out of a pulse while propagating in the fibre is referred to as pulse dispersion and sets the upper limit on the maximum information carrying capacity. The simplest type of optical fibre consists of a thin cylindrical structure of a transparent glassy material of uniform refractive index n_1 surrounded by a cladding of another material but slightly lower refractive index n_2. These fibres are referred to as step index fibres because of the step discontinuity of the refractive index profile at the core-cladding interface.

We note from Fig. 7.2 that rays making larger angles with the axis have to traverse a longer optical path length and they take a longer time to reach the output end. Consequently, the pulse broadens as it propagates through the fiber. Hence, even though two pulses may be well resolved at the input end of the fiber, they may not be uniquely identified at the output end because of this broadening. As we might understand, if the pulses are not resolved then no information can be retrieved from the incoming data stream at the receiver. Thus, the smaller the pulse dispersion, the greater is the information carrying capacity of the system.

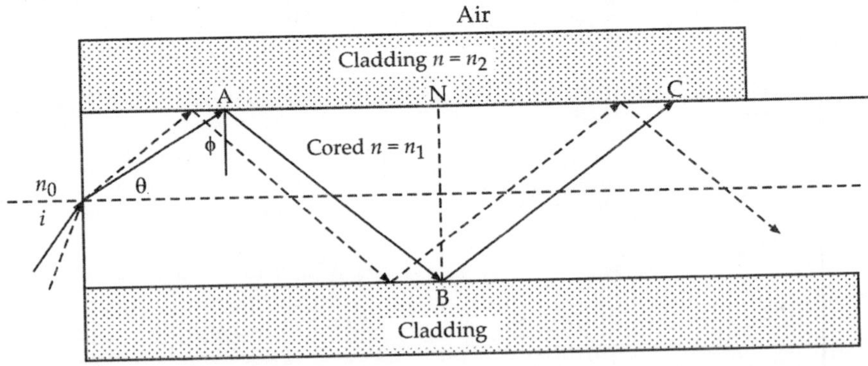

Fig. 7.2: Light rays striking the core-cladding interface at an angle greater than the critical angle are trapped inside the core (total internal reflection)

We have shown wave propagation inside the core of an optical fiber to take place in the form of rays in Fig. 7.2. However, we have seen from chapters 5 and 6 that inside a core the light waves are a class of guided waves that trace out different electric and magnetic field patterns known as modes decided by their respective cut off frequencies. Here we are describing dispersion of light waves by showing ray diagrams. Is the concept of modes and rays the same? Well, to answer that, we note that the physical concept of a ray is a representation of a pencil of plane electromagnetic waves in the limiting case of $\lambda \to 0$ that may be realized in practice by a very thin laser beam. A guided wave travelling along z-axis can be decomposed into a family of superimposed plane waves that collectively form a standing wave pattern along a direction transverse to the fiber axis. With any plane wave we associate a light ray that is perpendicular to the phase front of the wave, a family of plane waves corresponding to a particular mode forms a set of rays known as a "ray congruence". Each ray of this particular set travels at the same angle relative to the fiber axis in the fiber. Ray tracing approach, moreover provides good approximations to light acceptance and guiding characteristics of optical fibers when the fiber radius is large compared to the wavelength of light. Moreover, it is often practical and advantageous to understand various optical phenomena like reflection and refraction through the ray tracing approach than the drawing of various modal patterns.

To derive the ray equation, we consider a medium in which the refractive index depends only on the x coordinate as given by $n = n(x)$. A medium with continuously varying refractive index describe as above may be thought of as a limiting case of a medium consisting of a set of thin layers, each characterized by a specific value of the refractive index as shown in Fig. 7.3. To trace out the rays through such a stack of thin layers Snell's law is used which gives us

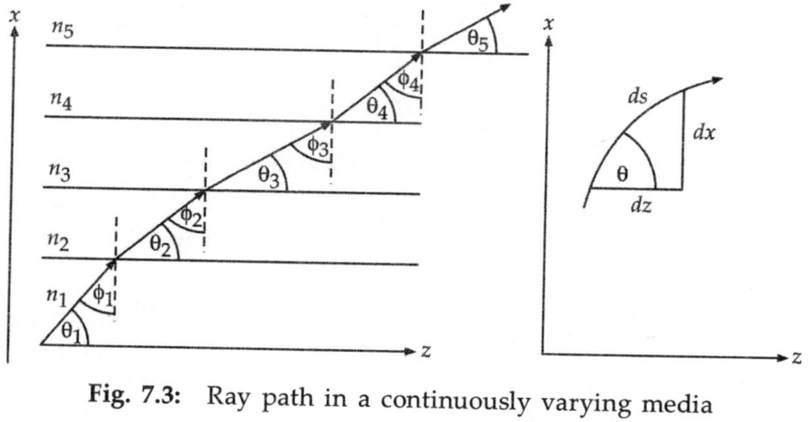

Fig. 7.3: Ray path in a continuously varying media

$n_1 \sin\phi_1 = n_2 \sin\phi_2 = n_3 \sin\phi_3 = ...$ constant where $\phi_1, \phi_2, \phi_{3,...}$ are the angles of incidence at various boundaries of the layers of different refractive indices. If $\theta_1, \theta_2, \theta_{3,...}$ are the corresponding angles that the rays make with the axis, then

$$n_1 \sin(90 - \theta_1) = n_2 \sin(90 - \theta_2) = n_3 \sin(90 - \theta_3) = ... \quad ...(7.1)$$

which is $n_1 \cos\theta_1 = n_2 = n_3 \cos\theta_3 =$ constant α (say) ...(7.2)

The ray has been chosen to lie in the *xz*-plane. When the refractive index variation is continuous, the thickness of each layer becomes infinitesimally small and the piecewise straight line form a continuous curve. The rays bend in such a way that the product $n(x)\cos\theta(x)$ remains constant which we have denoted by α. Therefore,

$$n(x)\cos\theta(x) = \alpha \text{ (invariant of the ray path)} \quad ...(7.3)$$

The infinitesimal arc length along the ray path is given by

$$(ds)^2 = (dx)^2 + (dz)^2$$

$$\Rightarrow \qquad \left(\frac{ds}{dz}\right)^2 = \left(\frac{dx}{dz}\right)^2 + 1 \quad ...(7.4)$$

From Fig. 7.3,

$$\cos\theta = \frac{dz}{ds} \quad ...(7.5)$$

Thus,

$$\frac{ds}{dz} = \frac{1}{\cos\theta(x)} = \frac{n(x)}{\beta} \quad ...(7.6)$$

Substituting (7.6) in (7.4), we have

$$\left(\frac{dx}{dz}\right)^2 = \frac{n^2(x)}{\alpha^2} - 1 \quad ...(7.7)$$

which is known as the ray equation.

The expression in (7.7) represents the rigorously ray equation for n^2 depending only on the x coordinate. We note that ray paths are possible only in regions having $\alpha \le n(x)$ and hence the ray would move in such a way so as to satisfy this condition.

For a given refractive index profile, a ray becomes parallel to the *z*-axis and turn back toward the axis when $\frac{dx}{dz} = 0$, that is at the value $x = x,$

satisfying $n(x_t) = \alpha$. The point $x = x_t$ is known as the turning point of the ray. By differentiating (7.7) with respect to z, we get

$$2\frac{dx}{dz}\frac{d^2x}{dz^2} = \frac{1}{\alpha^2}\frac{dn^2}{dx}\frac{dx}{dz} \qquad \text{...(7.8)}$$

$$\Rightarrow \qquad \frac{d^2x}{dz^2} = \frac{1}{2\alpha^2}\frac{dn^2}{dx}$$

is also another form of ray equation.

7.3.1. Ray Paths in a Homogeneous Medium

In a homogeneous medium, n^2 is a constant and the ray equation (7.7) is simplified to

$$\frac{d^2x}{dz^2} = \frac{1}{2\alpha^2}\frac{dn^2}{dx} = 0$$

The solution to this is

$$x(z) = A + Bz \qquad \text{...(7.9)}$$

which represents a straight line. The result is that the ray paths in a homogeneous medium are straight lines.

7.3.2. Ray Paths in Square Law Media

A square law medium is characterized by the following refractive index profile

$$n^2(x) = n_1^2\left[1 - 2\Delta\left(\frac{x}{a}\right)^2\right], \quad |x| < a \quad \text{core} \qquad \text{...(7.10)}$$

$$= n_1^2[1 - 2\Delta] = n_2^2, \quad |x| > a \quad \text{clading}$$

Ray paths in these media are important as they represent the optical fibers with a parabolic refractive index profile. Such types of fibers are extensively used in optical communication systems.

We consider the ray paths in the core of such waveguides. The ray equation then becomes

$$\frac{d^2x}{dz^2} = \frac{1}{2\alpha^2}\frac{dn^2}{dx} \qquad \text{...(7.11)}$$

$$\Rightarrow \qquad \frac{d^2x}{dz^2} + \xi^2 x(z) = 0$$

where $\xi = \dfrac{n_1\sqrt{2\Delta}}{\alpha a}$

The general solution to this kind of differential equation given by (7.11) is

$$x(z) = A \sin \xi z + B \cos \xi z \qquad ...(7.12)$$

rigorously correct raypaths in square law media.

which represents the general path. In a parabolic index fibre the meridional ray paths are rigorously given by the above equation and without any loss of generality, we further write $x(0) = 0$ which means that $B = 0$. Hence,

$$x(z) = A \sin \xi z \qquad ...(7.13)$$

If the ray makes an angle of θ_1 with the z-axis at $z = 0$, then

$$\tan \theta_1 = \left.\frac{dx}{dz}\right|_{z=0} = A\xi$$

$$\Rightarrow \qquad A = \frac{\alpha a \tan\theta_1}{n_1\sqrt{2\Delta}} = \frac{n_1 \cos\theta_1 a \tan\theta_1}{n_1}$$

$$= \frac{a \sin\theta_1}{\sqrt{2\Delta}} = \frac{a}{\sqrt{2\Delta}}\left[1-\left(\frac{\alpha}{n_1}\right)^2\right]^{1/2} \qquad ...(7.14)$$

Hence, the ray equation becomes

$$x(z) = \frac{a\sin\theta_1}{\sqrt{2\Delta}}\sin \xi z = \frac{a\sin\theta_1}{\sqrt{2\Delta}}\sin\left(\frac{n_1\sqrt{2\Delta}}{a\alpha}z\right) \qquad ...(7.15)$$

through the use of the expression for ξ. The periodic length of the sinusoidal ray path is, therefore given by

$$z_p = \frac{2\pi}{\zeta} = \frac{2\pi\alpha a}{n_1\sqrt{2\Delta}} \qquad ...(7.16)$$

Typical ray paths are shown in Fig. 7.4 with each ray corresponding to a specific value of α and bends in such a way that the product $n(x)$ $\cos\theta(x)$ remains a constant. The ray paths given by the above equation are valid only for $|x| < a$ since the refractive index profile is parabolic only in the region $|x| < a$. If α is such that the turning point $x = x_t$ lies between $x = +a$ and $x = -a$, then the ray will propagate by periodically oscillating around the z-axis as shown in Fig. Such rays form guided rays. For this to take place, α must be greater than n_2 since the turning point is determined by $n(x_t) = \alpha$ and the refractive index reduces monotonically from the axis. The maximum value of the refractive index is n_1 which gives us $\alpha \leq n_1$. Thus, for guided waves we should satisfy

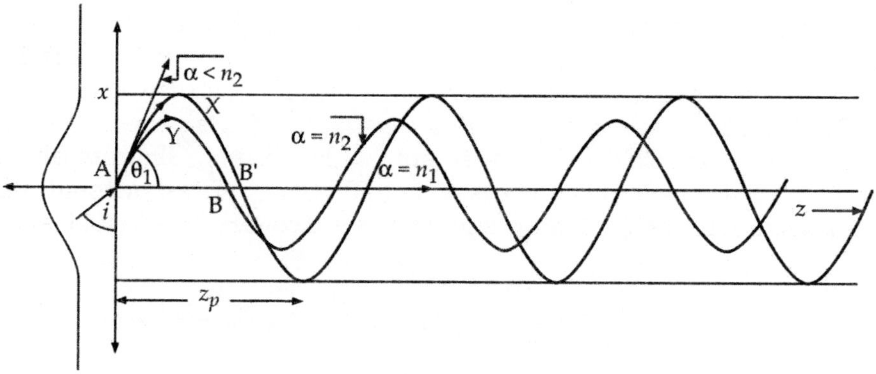

Fig. 7.4: Ray paths in a fiber having a parabolic refractive index profile

$$n_2 < \alpha < n_1 \text{ for guided rays} \qquad \text{...(7.17)}$$

If $\alpha < n_2$, then any ray will intersect the core-cladding interface at a finite angle and will be transmitted into the cladding. Further, since the cladding has uniform refractive index, the rays travel in a straight line. Hence, such type of rays are called refracting rays and these are not guided. Hence, we have $0 < \alpha < n_2$ for the refracting rays.

The rays shown in Fig. 7.4 have slightly different periods. However, if θ_1 is such that $\cos \theta_1 \approx 1$, (as is indeed the case for usual launch conditions), then all the rays have the same periodic length given by

$$z_p \approx \frac{2\pi a}{\sqrt{2\Delta}} \qquad \text{...(7.18)}$$

which is independent of the launch angle θ_1. Physically, although the ray AXB′, covers a larger path as compared to the ray AYB, it does so in a medium of relatively lower refractive index thus, the greater the path length is compensated by a greater "average speed" and hence, all rays take virtually the same amount of time meaning thereby all the modes have the same group velocity.

Let PXQ denote a curved surface separating the two media of refractive indices n_1 and n_2. Let I be the paraxial image point of the object point O. Let us consider a point Q on the left of I. The straight line path OXY connects the points O and Y. This is the only allowed path connecting O and Y. This path corresponds to L_{op} which is the optical path length being a minimum as in Fig. 7.5(a) and all nearby paths like OPY take a longer time.

$$n_1.OP + n_2.PY > n_1.OX + n_2.XY$$

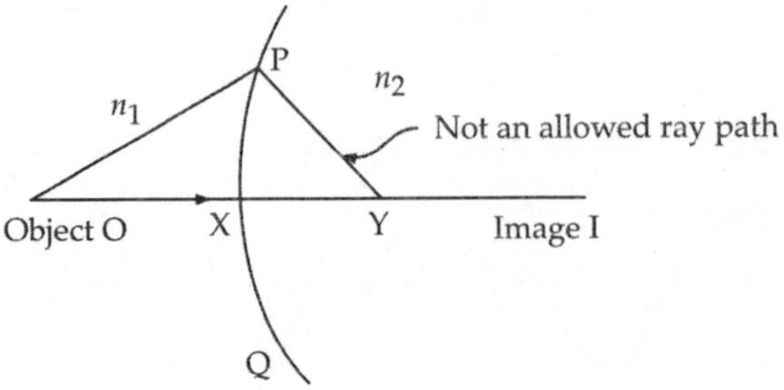

Fig. 7.5: (a) Non allowable paths at the interface of two different media

Similarly, as shown in Fig. 7.5(b), for a point Z on the right of the point I, the only allowed ray path (connecting O and Z) is the straight line path OXZ, which corresponds to L_{op} being maximum and all nearby paths like OPZ take less time

$$n_1.OP + n_2.PZ < n_1.OX + n_2.XZ \quad ...(7.19)$$

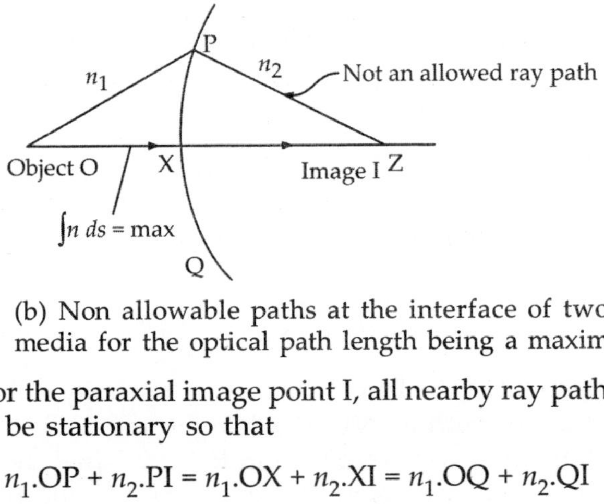

Fig. 7.5: (b) Non allowable paths at the interface of two different media for the optical path length being a maximum

Finally, for the paraxial image point I, all nearby ray paths are allowed and L_{op} will be stationary so that

$$n_1.OP + n_2.PI = n_1.OX + n_2.XI = n_1.OQ + n_2.QI \quad ...(7.20)$$

and all such paraxial rays take the same amount of time. Returning to Fig. 7.5(c), rays connecting P and Q correspond to

$$\int nds = \text{stationary} \quad ...(7.21)$$

and all paraxial rays take the same amount of time. This would lead to extremely small pulse dispersion. The transit time corresponding to the calculation of pulse dispersion is carried out next.

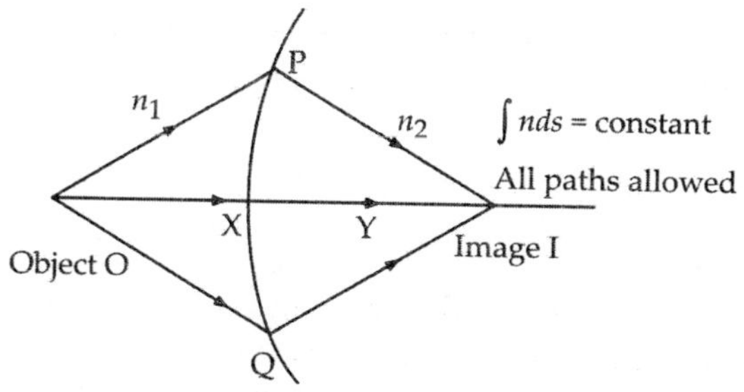

Fig. 7.5: (c) All allowed paths

7.4. TRANSIT TIME CALCULATIONS (Intermodal Dispersion)

One of the mechanisms leading to pulse broadening in time domain is the difference in time taken by different rays to traverse a given length. If all rays are launched simultaneously in a waveguide, at the output the rays arrive at different times and correspond to a time domain spreading of the optical pulse. To calculate the pulse dispersion, we need to compute the time taken by a ray to traverse a given length of the fiber. Here, we discuss intermodal dispersion.

Let τ_p represent the time taken for the ray to traverse a length z_p as shown in Fig. 7.4. The time taken to cover an arc of length ds along the ray is

$$d\tau = \frac{ds}{v(x)} = \frac{1}{c}n(x)ds \qquad ...(7.22)$$

We have assumed a refractive index profile to be a function of x. As we may note, $v(x) = \frac{c}{n(x)}$ is the velocity of the ray along the ray path. Hence, the time taken for one period is

$$\tau_p = \int \frac{n(x)}{c} ds \qquad ...(7.23)$$

where the integral is over one period. Thus,

$$ds = \sqrt{(dx)^2 + (dz)^2} = \sqrt{1 + \left(\frac{dz}{dx}\right)^2}\, dx \qquad ...(7.24)$$

which is further expressed as using the ray equation,

$$ds = \frac{n(x)}{\sqrt{[n^2(x) - \alpha^2]}} dx \qquad ...(7.25)$$

Hence, the time taken to travel over a quarter of a period (from the axis to the turning point) is $\tau_p/4$, which is given by

$$\frac{1}{4}\tau_p = \frac{1}{c}\int_0^{x_t} \frac{n^2(x)}{\sqrt{[n^2(x)-\alpha^2]}}dx \qquad ...(7.26)$$

where x_t is the x coordinate at the turning point. The above ray equation is valid for any arbitrary profile. Let us consider some specific cases.

7.4.1. Pulse Dispersion in a Parabolic Index Medium

For a medium having a parabolic refractive index profile, we have

$$n_1^2\left[1-2\Delta\left(\frac{x_t}{a}\right)^2\right] = \alpha^2$$

$$\Rightarrow \qquad x_t = \frac{a}{n_1\sqrt{2\Delta}}\sqrt{n_1^2-\alpha^2}$$

$$\Rightarrow \qquad x_t^2 = \frac{a^2}{n_1^2 2\Delta}(n_1^2-\alpha^2) \qquad ...(7.27)$$

$$\Rightarrow \qquad n_1^2 = \frac{\alpha^2}{\left[1-2\Delta\left(\frac{x_t}{a}\right)^2\right]}$$

Thus, we have

$$\tau_p = \frac{4}{c}\int_0^{x_t} \frac{n_1^2\left[1-2\Delta\left(\frac{x}{a}\right)^2\right]}{\sqrt{\left\{n_1^2\left[1-2\Delta\left(\frac{x}{a}\right)^2\right]-\alpha^2\right\}}}dx$$

$$= \frac{4}{c}\int_0^{x_t} \frac{n_1^2-2\Delta n_1^2\left(\frac{x}{a}\right)^2-\alpha^2+\alpha^2}{\sqrt{\left\{n_1^2\left[1-2\Delta\left(\frac{x}{a}\right)^2\right]-\alpha^2\right\}}}dx;$$

adding and subtracting α^2

$$= \frac{4}{c}\int_0^{x_t}\left[\sqrt{\left\{n_1^2\left[1-2\Delta\left(\frac{x}{a}\right)^2\right]-\alpha^2\right\}} + \frac{\alpha^2}{\sqrt{\left\{n_1^2\left[1-2\Delta\left(\frac{x}{a}\right)^2\right]-\alpha^2\right\}}}\right]dx \qquad ...(7.28)$$

Thus,

$$\left\{n_1^2\left[1-2\Delta\left(\frac{x}{a}\right)^2\right]-\alpha^2\right\}$$

$$= \frac{n_1^2 2\Delta}{a^2}x_t^2 - n_1^2 2\Delta\left(\frac{x}{a}\right)^2 = \frac{n_1^2 2\Delta}{a^2}(x_t^2 - x^2) \qquad ...(7.29)$$

Substitution of the above relation in the expression for the transit time gives us

$$\tau_P = \frac{4}{c}\int_0^{x_t}\left[\frac{n_1\sqrt{2\Delta}}{a}\sqrt{x_t^2-x^2}\,dx + \frac{\alpha^2}{\frac{n_1\sqrt{2\Delta}}{a}\sqrt{x_t^2-x^2}}dx\right]$$

$$= \frac{4}{c}\left[\frac{n_1\sqrt{2\Delta}}{a}\int_0^{x_t}\sqrt{x_t^2-x^2}\,dx + \frac{\alpha^2 a}{n_1\sqrt{2\Delta}}\int_0^{x_t}\frac{dx}{\sqrt{x_t^2-x^2}}\right] \qquad ...(7.30)$$

Integration of the first term in RHS of (7.30) gives us

$$\frac{n_1\sqrt{2\Delta}}{a}\int_0^{x_t}\sqrt{x_t^2-x^2}\,dx = \frac{n_1\sqrt{2\Delta}}{a}\left[\frac{x\left(\sqrt{x_t^2-x^2}\right)}{2} + \frac{x_t^2}{2}\sin^{-1}\frac{x}{x_t}\right]_0^{x_t}$$

$$= \frac{n_1\sqrt{2\Delta}}{a}\left(\frac{x_t^2}{2}\sin^{-1}\frac{x_t}{x_t}\right) = \frac{n_1\sqrt{2\Delta}}{a}\frac{x_t^2}{2}\sin^{-1}1 = \frac{n_1\sqrt{2\Delta}}{a}\frac{x_t^2}{2}\frac{\pi}{2} \qquad ...(7.31)$$

Integration of the second term in RHS of (7.30) gives us

$$\frac{\alpha^2 a}{n_1\sqrt{2\Delta}}\int_0^{x_t}\frac{dx}{\sqrt{x_t^2-x^2}} = \frac{\alpha^2 a}{n_1\sqrt{2\Delta}}\sin^{-1}\frac{x}{x_t}\Bigg|_0^{x_t} = \frac{\alpha^2 a}{n_1\sqrt{2\Delta}}\frac{\pi}{2} \qquad ...(7.32)$$

Substitution of (7.31) and (7.32) in (7.30) gives us

$$\tau_p = \frac{4}{c}\left[\frac{n_1\sqrt{2\Delta}}{a}\frac{x_t^2}{2}\frac{\pi}{2}+\frac{\alpha^2 a}{n_1\sqrt{2\Delta}}\frac{\pi}{2}\right]$$

$$= \frac{4}{c}\frac{\pi}{2}\frac{a}{n_1\sqrt{2\Delta}}\left[\frac{n_1^2 2\Delta}{a^2}\cdot\frac{a^2}{n_1^2 4\Delta}(n_1^2-\alpha^2)+\alpha^2\right]$$

$$= \frac{4}{c}\frac{\pi}{2}\frac{a}{n_1\sqrt{2\Delta}}\left(\frac{n_1^2-\alpha^2}{2}+\alpha^2\right) \qquad ...(7.33)$$

$$= \frac{4}{c}\frac{\pi}{2}\frac{a}{n_1\sqrt{2\Delta}}\frac{1}{2}(n_1^2+\alpha^2)$$

$$= \frac{\pi}{c}\frac{a}{n_1\sqrt{2\Delta}}(n_1^2+\alpha^2)$$

We note from (7.33) that the turning point corresponds to $z = z_p/4$ and therefore,

$$\frac{1}{4}\Gamma z_p = \frac{1}{4}\frac{2\pi}{z_p}.z_p = \frac{\pi}{2} \qquad ...(7.34)$$

from which we obtain

$$z_p = \frac{2\pi}{\Gamma} = \frac{2\pi a\alpha}{n_1\sqrt{2\Delta}} \qquad ...(7.35)$$

Hence if $\tau(z)$ corresponds to the time taken by the ray to traverse the distance z then

$$\frac{\tau_p}{z_p} = \frac{\dfrac{\pi}{c}\dfrac{a}{n_1\sqrt{2\Delta}}(n_1^2+\alpha^2)}{\dfrac{2\pi a\alpha}{n_1\sqrt{2\Delta}}} = \frac{1}{2c}\left(\frac{n_1^2+\alpha^2}{\alpha}\right) = \frac{1}{2c}\left(\alpha+\frac{n_1^2}{\alpha}\right) \qquad ...(7.36)$$

Since $n_2 < (x) < n_1$ we have for guided rays $n_2 < \alpha < n_1$.

A ray with $\alpha = n_1$ takes the minimum time given by

$$\tau_{min}(z) = \frac{1}{2c}\left(n_1+\frac{n_1^2}{n_1}\right)z = \frac{n_1}{c}z \qquad ...(7.37)$$

We observe that the minimum ray traverse time is obtained for the ray aligned with the core for which $\theta = 0$ (in the core) which is the axial ray. For a ray with $\alpha = n_2$ we get the maximum time for ray traverse given by

$$\tau_{max}(z) = \frac{1}{2c}\left(n_2 + \frac{n_1^2}{n_2}\right)z \qquad ...(7.38)$$

The maximum time is obtained for a ray becoming parallel to the z-axis at $x = a$ (corresponding to $\alpha = n_2$). Therefore, the pulse dispersion is obtained as

$$\Delta\tau = \tau_{max} - \tau_{min} = \frac{1}{2c}\left(n_2 + \frac{n_1^2}{n_2}\right)z - \frac{n_1}{c}z$$

$$= \frac{z}{2c}\left(\frac{n_2^2 + n_1^2}{n_2}\right) - \frac{n_1}{c}z$$

$$= \frac{z}{2c}\left(\frac{n_2^2 + n_1^2 - 2n_1n_2}{n_2}\right) = \frac{z}{2cn_2}(n_1 - n_2)^2 \qquad ...(7.39)$$

$$= \frac{z}{2cn_2}\Delta^2 n_1^2 \approx \frac{n_1}{2c}\Delta^2 z$$

Where we have assumed $n_1 \approx n_2$.

7.4.2. Pulse Dispersion in a Planer Step Index Waveguide

The refractive index profile of a step index fibre is given as

$$n_2(x) = n_1^2 \; |x|, \; a \text{ (core)}$$

$$= n_2^2 = n_1^2(1 - 2\Delta) \; |x| > a \text{ (cladding)} \qquad ...(7.40)$$

In a step index fiber, the refractive index changes abruptly (in a step like manner) as we move from the core to the cladding. The core has a higher refractive index than that of the cladding. As we may infer, both the core and the cladding are homogeneous media as the refractive index remains constant throughout the transverse dimension of the material. Thus the ray paths are straight lines in both the core as well as the cladding. For a ray to be guided by the core, it must undergo total internal reflection at the core-cladding interfaces at $x = \pm a$. Thus, if the angle made by the ray with the z-axis is θ, then for a ray to be guided we should ensure that $0 < \theta < \theta_c$; the ray should make an angle less than the critical angle

$$\Rightarrow \qquad 0 < \theta < \cos^{-1}\left(\frac{n_2}{n_1}\right); \sin\varphi = \cos\theta$$

$$\Rightarrow \qquad \cos\theta_c < \cos\theta < 1$$

$$\Rightarrow \qquad \frac{n_2}{n_1} < \cos\theta < 1 \qquad ...(7.41)$$

$$\Rightarrow \qquad n_2 < n_1 \cos\theta < n_1$$

$$\Rightarrow \qquad n_1 < n_1 \cos\theta > n_2$$

Since $\alpha = n(x)\cos\theta\,(x)$, for the present case, and we obtain

$$n_2 < \alpha < n_1 \text{ for guided rays}$$

and

$$\alpha < n_2 \text{ for refracting rays} \qquad ...(7.42)$$

The pulse dispersion is evaluated by finding out the difference of the minimum and the maximum values of the transit time as a ray of light traverses a given length of the optical fiber. To do so, we refer to Fig. 7.2 and we first find out the ray period z_P

$$z_P = AC = 2AN = 2.\frac{2a}{\tan\theta} \qquad ...(7.43)$$

The ray undergoes the first total internal reflection at point *A* whereupon it travels a distance of AB before suffering from the second total internal reflection at the core-cladding interface marked by point *B*. Therefore, the ray spends a time equal to the two step journey made at the interfaces at *A* and *B*. The time taken by the ray τ_P spent in the two step journey is in covering the distance *AB* and *BC*.

This may be further expressed as

$$\tau_P = \frac{AB + BC}{v} = \frac{2AB}{(c/n_1)} = \frac{2n_1}{c}AB$$

$$\frac{2n_1}{c}.\frac{2a}{\sin\theta} \qquad \because AB = \frac{2a}{\sin\theta} \qquad ...(7.44)$$

The time taken by the ray to cover one period, therefore becomes

$$\tau_P = \frac{2n_1}{c}.\frac{2a}{\sin\theta} = \frac{4n_1}{c}\frac{a\cos\theta}{\sin\theta.\cos\theta}$$

$$= \frac{4an_1}{c\cos\theta}\frac{1}{(\sin\theta/\cos\theta)} \qquad ...(7.45)$$

$$= \frac{n_1 z_p}{c(\alpha/n_1)} = \frac{n_1^2}{c\alpha} z_p \quad \because n_1 \cos\theta = \alpha$$

It is obvious that, the time taken to cover a distance of z is evaluated as follows

$$\frac{z}{z_p} = \frac{\tau}{\tau_p}$$

$$\Rightarrow \qquad \tau = \tau_p . \frac{z}{z_p} = \frac{n_1^2}{c\alpha} z_p . \frac{z}{z_p} \qquad ...(7.46)$$

$$= \frac{n_1^2}{c\alpha} z$$

We observe from the above that,

$$\tau \propto \frac{1}{\alpha} \qquad ...(7.47)$$

The minimum transit time would correspond to the core as it has the highest refractive index.

Hence,
$$\tau_{min} = \frac{n_1^2}{cn_1} z = \frac{n_1}{c} z \qquad ...(7.48)$$

And the maximum time

$$\tau_{max} = \frac{n_1^2}{cn_2} z \qquad ...(7.49)$$

Therefore, the pulse spreading in the temporal domain is

$$\Delta\tau = \tau_{max} - \tau_{min}$$

$$\Rightarrow \qquad \frac{n_1^2}{cn_2} z - \frac{n_1}{c} z = \frac{n_1}{cn_2}(n_1 - n_2) z \qquad ...(7.50)$$

$$\approx \frac{n_1}{c} \Delta z$$

Here we have used $n_1 - n_2 \approx n_1 \Delta$ and $n_1 \approx n_2$. We thus find that, the pulse spreading is proportional to the refractive index of the core and the numerical aperture of the fibre. Hence, use of a small refractive index for the core and a lower NA would reduce the pulse spreading.

7.5. WAVEGUIDE DISPERSION

The intermodal and material dispersion disappears for a single mode fibre operating at a wavelength near 1300 nm. However, a different type of dispersion known as waveguide dispersion appears. At the condition of minimal intermodal and material dispersion, the waveguide dispersion predominates. It is difficult to build a pure monochromatic source of light. However, a narrow bandwidth may be considered around λ_0 so as to result in an optimum dispersion. This would cause a corresponding shift in group velocity, which varies in a complicated way with the wavelength and subsequently results in pulse broadening. This effect is only the result of the guiding properties of the fiber and hence the name.

7.5.1. Dispersion Power Penalty

The received pulse is broadened due to the redistribution of light energy in time. This reduced optical power can be compensated by boosting the transmitter power. The added power is referred to as dispersion power penalty (P_{dp}). The dispersion power penalty in dB, for a digital transmission system including intersymbol interference is expressed as, for a fiber of length,

$$P_{dp} = 10 \log g_d + 10 \log h_d \qquad ...(7.51)$$

In the above expression

$$g_d = 1 - \frac{1}{2}(\omega \tau L)^2 \qquad ...(7.52)$$

and

$$h_d 2 + \left[\frac{0.1}{(\omega/\omega_c)^2} \right] \qquad ...(7.53)$$

τ is the pulse broadening, $\omega = 2\pi f$, $f = 2\times$ and f_c represents the fiber bandwidth

The power penalty for different bandwidth-length-bit rate products is shown in Figs 7.6(a) and (b).

7.5.2. Total Dispersion Delay

The pulse dispersion is a consequence of different kinds of dispersion. This depends upon the type of fiber and the operating conditions. For example, in a multimode step index fiber, intermodal dispersion

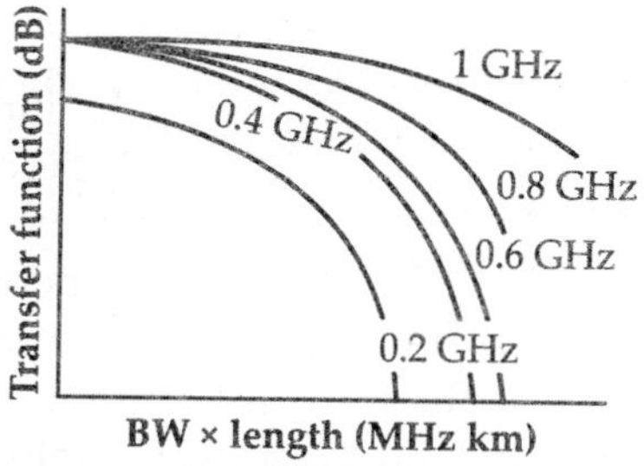

(a)

Fig. 7.6 (a) Bandwidth and transfer function

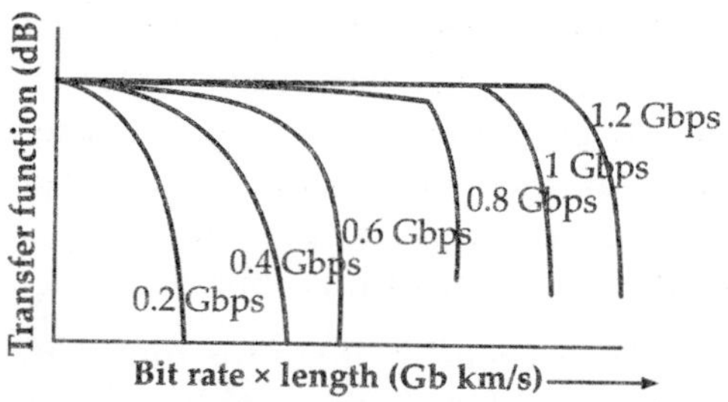

(b)

Fig. 7.6 (b) Transfer function as a function of bit rate

predominates while in single mode fiber operating at 1300 nm, the waveguide dispersion dominates. The total dispersion at any wavelength is expressed as

$$\Delta t_T = \sqrt{(\Delta t_{imd})^2 + (\Delta t_{md})^2 + (\Delta t_{wgd})^2} \qquad ...(7.54)$$

These three dispersions are computed on a theoretical basis. Practically, these values are slightly higher and hence the total is also higher. However, this gives a satisfactory estimate of the fiber behavior.

7.5.3. Maximum Transmission Rate

We have seen previously how dispersion affects the maximum rate of information transmission. To see it more clearly, let us consider three pulses A, B, C with different delays that have reached the receiver and are added to form a broadened or dispersed pulse. The pulse B experiences an average delay while pulse A has the minimum delay and pulse C has the maximum delay. The thick waveform represented by W is the overall received pulse and is the summation of all the pulse components between the limits of A and C. Here, the transmitted pulse width is represented by Δt_w, whereas the received pulse width is represented by Δt_r. Of course, we have

$$\Delta t_r = \Delta t_w + \Delta t_t \qquad ...(7.55)$$

If we want to detect the next pulse properly, it must be transmitted after an interval of time equal to Δt_r as this provides the necessary time margin for the pulse spreading to decay to a significant amount. The bit rate under the ideal condition is given by

$$B = \frac{1}{\Delta t_r} = \frac{1}{\Delta t_w + \Delta t_T} \qquad ...(7.56)$$

However, the value of dispersion appearing in the above expression is a theoretical value which is less than the actual values obtained in practice. Moreover, the pulse tails tend to spread beyond the average pulse limits. From all these it is evident that a safety factor of about 5 times on the received pulse width Δt_r has to be allowed to obtain minimum desired error rates in transmission.

To get an estimate of the maximum bit rates, the transmitted pulse is to be considered reasonably smaller than the total dispersion; when $\Delta t_w << \Delta t_T$.

$$B_r(\max) \cong \frac{1}{5\Delta t_T} \qquad \text{...(7.57)}$$

B_r (max) is the maximum bit rate

Some more relations may be written as

$$\Delta t_w > 4\,\Delta t_r \Rightarrow B_r < \frac{1}{4\Delta t_T} \qquad \text{...(7.58)}$$

7.6. DISPERSION SHIFTED FIBERS

In optical fiber communications there are three low loss windows as given in the following

(i) 820-880 nm where the loss is 2.2 dB/km

(ii) 1200-1320 nm where the loss is 0.6 dB/km

(iii) 1550-1610 nm where the loss is 2.2 dB/km

Single losses are low in these windows. However, it has been observed that the two intermodal dispersion wavelength of around 1300 nm is most suitable whereas from the point of view of loss, the wavelength range of 1550 to 1610 nm is most suitable. Hence 1550 to 1610 nm range is particularly promising in terms of losses. That is why in recent years, sources and detectors in this range have been developed. The choice of wavelength is determined for getting minimum loss and dispersion. Special fibers have been developed to minimize dispersion in this range (1550 to 1610 nm). Such special fibers are called dispersion shifted fibers.

7.7. LOSSES IN FIBERS

The other most important impairment meted out to an optical pulse traveling through an optical fiber is attenuation that contributes to signal loss. The attenuation and dispersion combined together decide the repeater spacing in a long haul fiber optic communication link. Attenuation of an optical pulse is measured in terms of dB. The loss in dB corresponding to an input power and output power of P_{in} and P_{out} respectively is expressed as

$$\alpha = 10\log_{10}\frac{P_{in}}{P_{out}} \qquad ...(7.59)$$

In communication systems, it is customary to talk about 3 dB power loss which represents a loss of input power by a factor of one half. Similarly, an attenuation of one-tenth would give rise to a loss of 10 dB. The losses in glassy materials have been steadily declining over the years. For example, starting with the Egyptian glass that incurs a loss of about 10^7 dB/km, The various loss mechanism are (i) absorptive type and (ii) radiative type. Absorptive losses can be further classified as intrinsic and extrinsic type. The intrinsic losses are caused by interaction of the propagating lightwave with one or more major components of glass that constitute the fiber's material composition. Extrinsic absorption is due to the presence of minute quantities of materials like transition metal ions and hydroxyl ions dissolved in glass. The presence of these hydroxyl ions leads to absorption peaks at 0.72, 0.88, 0.95, 1.13, 1.24 and 1.38 μm. Advances made in glass manufacturing have led to the development of very low loss glasses having losses of about 0.2 dB/km at 1.55μm.

Glass, by its very manufacturing process tends to have microscopic variations around the average material density and local microscopic variations in the composition Each of these gives rise to fluctuations in the refractive index on a scale that is small compared to optical wavelength. This is fundamental to any glassy material, however carefully it might be fabricated and is responsible for light to be scattered in a manner known as Rayleigh scattering. This also accounts for signal loss in an optical fiber. In general, a Rayleigh scattering loss coefficient α_R is defined as

$$\alpha_R = \alpha_0\left(\frac{\lambda_0}{\lambda}\right)^4 \qquad ...(7.60)$$

In (7.60), α_0 = 1.7 dB/km and λ_0 = 0.85 μm. We may have another expression for α_R expressed as

$$\alpha_R = A_R\left(\frac{1}{\lambda}\right)^4 \qquad ...(7.61)$$

The constant A_R is determined by the material type in (7.61).

Besides these two types of losses, a signal may also be attenuated due to the presence of tight bends that cause some of the light not to be internally reflected but to propagate into the cladding eventually to be lost. Theoretically, the optical power scattered out of a given fiber at a major bend is an exponential function of the bend radius. The bending loss, then is proportional to exp (R/R_c) where R is the bend radius and the critical radius R_c (for multimode fibers) is function of the fiber NA expressed as $R_c \approx a/(\text{NA})^2$ where is the core radius. At a bend radius equal to the critical radius, the signal loss is significant. However, the loss is kept to acceptable limits for less tight bends because of the exponential relation. The loss in single mode fibers is decided by the extent to which the electromagnetic waves penetrate into the cladding that is a function of the refractive index profile.

A succession of very small bends in a fiber may also lead to signal loss known as microbending loss. This is particularly important when a fiber is wound in tension on to a drum. The microbends arise as the fiber is distorted by the variable pressure exerted on it by the irregularities on the drum's surface. Continuous, small variations in core diameter or its circularity appearing during the manufacturing process may also result in a scattering mechanism thereby resulting in signal loss.

However, as we shall see later, the bends would be utilized in a constructive an effective manner to form the basis of many an optical sensors using what is known as intensity modulated sensors.

7.8. CHAPTER SUMMARY

One of the primary factors determining the maximum data rate that can be supported by a typical optical fiber is pulse dispersion; the widening of an impulse in time domain as it travels a given length of it. This also places an upper limit on the repeater placements in a given optical communication system. The nature of pulse dispersion varies from a step index fiber to a graded index fiber. Important causes for pulse dispersion are intermodal, material and intramodal dispersion. By using a small numerical aperture, the dispersion can be reduced in a graded index fiber. The two basic loss mechanisms meted out to an optical pulse while it travels along the core are absorptive, radiative (scattering) types. The loss together with the various dispersions set the upper limit on the maximum amount of data that can be carried by a given optical fiber reliably over a certain distance and hence the number of repeaters needed in a given fiber optic link.

❑❑❑

8

Wave Phenomena of Light

8.1. INTRODUCTION

The wave nature of light has been amply discussed in terms of reflection and refraction in the previous chapters. These are wave phenomena that are described by the nature of the media in which light passes taking into account only a single wave. Here we discuss two different categories of phenomena that are due to the wave nature of light and which are also concerned with multiple light waves (at least two). However, the nature of the medium does not come into picture here. We are more concerned about the phenomena that arise when a number of different light beams interact with each other at a given point. Interference is an application of the principle of superposition of waves according to which the resultant displacement (at a particular point) produced by a number of waves is the vector sum of the displacements produced by each one of the disturbances.This can be observed on a stretched string at the two ends of which two waves of equal amplitude travel down the length the string with equal but opposite velocities. At a given point on the string, the two waves interfere with each other and give rise to a pattern that is stationary. The pattern is characterized by points of higher strength (called antinodes) and very small (ideally zero) strengths (called nodes) repeated throughout the length of the string. In a similar manner, two light waves coming from two different sources combine to produce an interference pattern on a screen. The interference between two light waves being emanated from two independent source cannot be observed as these do not maintain a constant phase relationship with each other. Hence the interfering waves are derived from a single wave so that a fixed phase relationship between these two can be maintained. There are basically two methods for achieving this. A beam is allowed to fall on two closely spaced holes and the two beams emanating from the holes interfere. This method is known as division of wave front. In the second method, a beam is divided at two or more reflecting surfaces and the reflected beams interfere. This method is known as interference by division of amplitude. The algebraic combination or superposition of waves by division of wave front is considered first. This chapter also discusses diffraction of light waves: another wave phenomenon.

Study of interference and diffraction phenomena is important from the view point of optical fibre based sensors and instrumentation. A number of optical sensors used to sense and produce a measurable output for a good number of physical parameters do rely upon interference and diffraction of light waves as they pass through narrow openings. Hence we study these wave phenomena here.

8.2. THE INTERFERENCE PATTERN: DIVISION OF WAVE FRONTS

Let S_1 and S_2 represent the two pinholes of the Young's interference experiment. The objective is to determine the positions of maxima and of minima on the line LL' which is parallel to the y-axis and lies in the xz-plane containing the two sources and the point S. We would prove that the interference pattern (around the point O) consists of a series of dark and bright lines perpendicular to the plane; O being the foot of the perpendicular from the point S on the screen. For an arbitrary point P (on the line LL') to correspond to a maximum we must have

$$S_2P - S_1P = n\lambda;\ n = 0, 1, 2, \qquad \text{...(8.1)}$$

Now, where

$$(S_2P)^2 - (S_1P)^2 = \left[D^2 + \left(y_n + \frac{d}{2}\right)^2\right] - \left[D^2 + \left(y_n - \frac{d}{2}\right)^2\right] = 2y_n d \qquad \text{...(8.2)}$$

where $S_1S_2 = d$ and $OP = y_n$ ...(8.3)

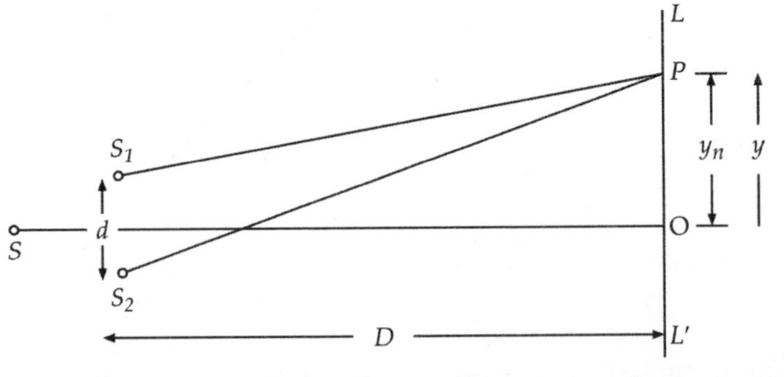

Fig. 8.1: Arrangement for producing Young's interference pattern

Thus,

$$S_2P - S_1P = \frac{2y_n d}{S_2P + S_1P} \qquad ...(8.4)$$

For values of y_n, $d << D$, we may approximate $S_2P + S_1P \approx 2D$ resulting in negligible error. For example, with a value of $d = 0.02$ cm, $D = 50$ cm, $OP = 0.5$ cm (which corresponds to typical values for a light interference experiment) we note that

$$[50^2 + 0.51^2]^{1/2} + [50^2 + 0.49^2]^{1/2} = 100.005 \text{ cm} \approx 2D \qquad ...(8.5)$$

Thus, the error involved in replacing $S_2P + S_1P$ by $2D$ is about 0.005%. In this approximation, we further have

$$S_2P - S_1P \approx \frac{2y_n d}{2D} = \frac{y_n d}{D} \qquad ...(8.6)$$

and

$$y_n = \frac{D}{d}(S_2P - S_1P) \qquad ...(8.7)$$

If we wish to have a bright spot at P, then the path difference must satisfy

$$S_2P - S_1P = n\lambda;\ n = 0, 1, 2,... \qquad ...(8.8)$$

and thus,

$$y_n = \frac{D}{d}n\lambda \qquad ...(8.9)$$

The dark and bright fringes are equally spaced and the distance between two consecutive dark (or bright) fringes is given by

$$\beta = y_{n+1} - y_n = \frac{(n+1)\lambda D}{d} - \frac{n\lambda D}{d} = \frac{\lambda D}{d} \qquad ...(8.10)$$

As we may observe, β in (8.10) gives the expression for the fringe width. Next, we wish to determine the shape of the interference pattern for which the locus of the point P satisfies the following

$$S_2P - S_1P = \Delta \qquad ...(8.11)$$

Let us assume the origin to be at O and the z-axis perpendicular to the plane of the screen as shown in Fig. 8.2. The y-axis is assumed to be parallel to S_2S_1. We consider an arbitrary point Q on the plane of the screen ($z = 0$). Let the coordinates be $(x, y, 0)$.

The coordinates of the points S_1 and S_2 are $\left(0, \frac{d}{2}, D\right)$ and $\left(0, -\frac{d}{2}, D\right)$ respectively. Thus,

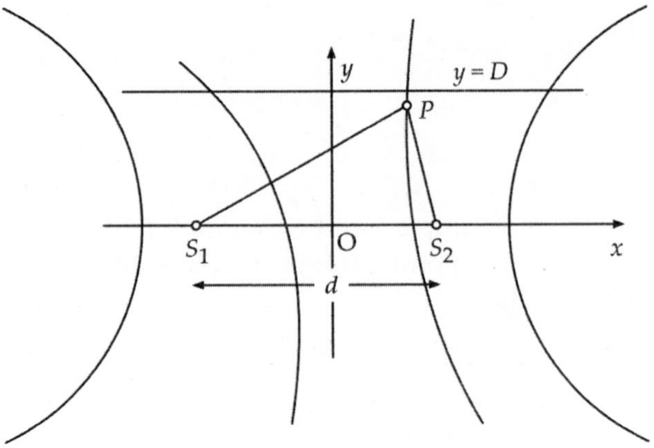

Fig. 8.2: The nodal curves

$$S_2P - S_1P \quad \left[x^2 + \left(y + \frac{d}{2}\right)^2 + D^2\right]^{1/2} \qquad \text{(say)} \qquad ...(8.12)$$

$$- \left[x^2 + \left(y - \frac{d}{2}\right)^2 + D^2\right]^{1/2} = \Delta$$

Squaring both the sides

$$\left[x^2 + \left(y + \frac{d}{2}\right)^2 + D^2\right] = \left\{\Delta + \left[x^2 + \left(y - \frac{d}{2}\right)^2 + D^2\right]^{1/2}\right\}^2 \qquad ...(8.13)$$

Manipulation of (8.13) gives

$$[2yd - \Delta^2]^2 = (2\Delta)^2 \left[\Delta + \left[x^2 + \left(y - \frac{d}{2}\right)^2 + D^2\right]\right] \qquad ...(8.14)$$

Thus,

$$(d^2 - \Delta^2)\, y^2 - \Delta^2 x^2 = \Delta^2 \left[D^2 + \frac{1}{4}(d^2 - \Delta^2)\right] \qquad ...(8.15)$$

$$\frac{(d^2 - \Delta^2)}{\Delta^2 \left[D^2 + \frac{1}{4}(d^2 - \Delta^2)\right]} y^2 - \frac{x^2}{D^2 + \frac{1}{4}(d^2 - \Delta^2)} = 1 \qquad ...(8.16)$$

Rearranging the terms we have

$$\frac{y^2}{\Delta^2\left(\frac{1}{4}+\frac{D^2}{(d^2-\Delta^2)}\right)}-\frac{x^2}{D^2+\frac{1}{4}(d^2-\Delta^2)}=1 \qquad ...(8.17)$$

This represents the equation of a hyperbola. Thus, the shape of the fringes is hyperbolic. On rearranging we get

$$y=\pm\left(\frac{\Delta^2}{d^2-\Delta^2}\right)^{1/2}\left[x^2+D^2+\frac{1}{4}(d^2-\Delta^2)\right]^{1/2} \qquad ...(8.18)$$

For values of such that $x^2 << D^2$, the loci are straight lines parallel to the x-axis. Hence we obtain approximately straight line fringes on the screen. It should be emphasized that the fringes are straight lines although the sources S_1 and S_2 are point sources. It is easy to see that if we had slits instead of the point sources we would have obtained again straight line fringes with increased intensities. The fringes so produced are said to be non-localized; they can be photographed by just placing a film on the screen; they can also be seen through the eyepiece.

Example 8.1: The locus of points which correspond to minima are known as nodal lines. Show that the equation of a nodal line is a hyperbola. Also obtain the locus of points which correspond to maxima. It is given that $S_1P - S_2P = \Delta$.

Solution: Let the coordinates of point P be (x, y). From Fig. 8.2, we do find that, and similarly, $S_1P = \left[\left(x+\frac{d}{2}\right)^2+y^2\right]^{1/2}$ and similarly

$$S_2P=\left[\left(x-\frac{d}{2}\right)^2+y^2\right]^{1/2}$$

Substitution of these two expressions for the path difference, we obtain

$$\left[\left(x+\frac{d}{2}\right)^2+y^2\right]^{1/2}-\left[\left(x+\frac{d}{2}\right)^2+y^2\right]^{1/2}=\Delta$$

$$\Rightarrow \qquad \left[\left(x+\frac{d}{2}\right)^2+y^2\right]^{1/2}=\left[\left(x+\frac{d}{2}\right)^2+y^2\right]^{1/2}+\Delta$$

Squaring both the sides, we have

$$x^2 + xd + \frac{d^2}{4} + y^2 = x^2 - xd + \frac{d^2}{4} + y^2 + \Delta^2 + 2\Delta\left[\left(x - \frac{d}{2}\right)^2 + y^2\right]^{1/2}$$

$$\Rightarrow \qquad 2xd - \Delta^2 = 2\Delta\left[\left(x - \frac{d}{2}\right)^2 + y^2\right]^{1/2}$$

Again squaring both the sides, we obtain

$$(2xd - \Delta^2) = 4\Delta^2\left[\left(x - \frac{d}{2}\right)^2 + y^2\right]$$

$$\Rightarrow \qquad 4x^2d^2 + \Delta^4 - 4xd\Delta^2 = \left(x^2 - xd + \frac{d^2}{4} + y^2\right)4\Delta^2$$

$$\Rightarrow \qquad 4(d^2 - \Delta^2)x^2 = \Delta^2 d^2 - \Delta^4 + 4\Delta^2y^2$$

$$\Rightarrow \qquad 4(d^2 - \Delta^2)x^2 = \Delta^2(d^2 - \Delta^2) + 4\Delta^2y^2$$

Dividing both the sides by $\Delta^2(d^2 - \Delta^2)$, we obtain

$$\frac{4x^2}{\Delta^2} = 1 + \frac{4y^2}{d^2 - \Delta^2}$$

$$\Rightarrow \qquad \frac{x^2}{\frac{1}{4}\Delta^2} - \frac{y^2}{\frac{1}{4}(d^2 - \Delta^2)} = 1$$

This represents the equation of a hyperbola. The minima are obtained corresponding to those points where the phase difference of the two interfering waves is equal to $(2n+1)\pi$ which corresponds to a path difference of $\left(n + \frac{1}{2}\right)\lambda$ where n is an integer. Similarly, points of maxima are obtained corresponding to a path difference of $n\lambda$. We further note that, for large values of x and y the curves asymptotically tend to straight lines which are expressed as

$$y = \pm\left(\frac{d^2 - \Delta^2}{\Delta^2}\right)^{1/2} x$$

Example 8.2: Find out the positions of maxima and minima when there exist an initial phase difference of π radians between two interfering waves.

Solution: We understand that, a phase difference of $2n\pi$ between two interfering waves at a given point results in a constructive interference or the positions of maxima. As we already have an initial phase difference of π, we require an additional π radians between the two waves to result in maxima. A path difference of $\frac{\lambda}{2}$ gives a phase difference of π. Hence, the required path differences should correspond to $\left(n+\frac{1}{2}\right)\lambda$ for obtaining positions of maxima. Similarly, minima positions will be obtained corresponding to a path difference of $n\lambda$.

8.3. SUPERPOSITION OF SINUSOIDAL WAVES

Let the displacement produced by the sources at S_1 and S_2 be given by

$$y_1 = a \cos \omega t \qquad ...(8.19)$$

and

$$y_2 = a \cos (\omega t + \phi) \qquad ...(8.20)$$

The phase difference between the two waves is, therefore ϕ. We wish to find the resultant displacement. To do so, we note that

$$y_2 = a(\cos \omega t \,.\, \cos \phi - \sin \omega t \,.\, \sin \phi) \qquad ...(8.21)$$

Thus, the total horizontal displacement

$$y_h = a + a \cos \phi = a(1 + \cos \phi) \qquad ...(8.22)$$

and the total vertical displacement

$$y_v = -a \sin \phi \qquad ...(8.23)$$

Hence the resultant is given as

$$y = (y_h^2 + y_v^2)^{1/2} = a[1+\cos^2\phi + 2\cos\phi + 1 + \sin^2\phi]^{1/2}$$

$$= a[2(1+\cos\phi)]^{1/2} = a\left(2.2\cos^2\left(\frac{\phi}{2}\right)\right)^{1/2} = 2a\cos\left(\frac{\phi}{2}\right) \qquad ...(8.24)$$

The angle that the resultant displacement makes with the horizontal is given as

$$\tan\theta = \frac{-a\sin\phi}{a(1+\cos\phi)} = -\frac{2\sin\left(\frac{\phi}{2}\right)\cos\left(\frac{\phi}{2}\right)}{2\cos^2\left(\frac{\phi}{2}\right)} = -\tan\left(\frac{\phi}{2}\right) \quad \text{...(8.25)}$$

Hence,
$$\theta = -\frac{\phi}{2} \quad \text{...(8.26)}$$

The resultant is hence expressed as

$$y = 2a\cos\left(\frac{\phi}{2}\right)\cos\left(\omega t + \frac{\phi}{2}\right) \quad \text{...(8.27)}$$

The amplitude of the resultant is observed to vary as $2a\cos\left(\frac{\phi}{2}\right)$.

Let us generalize the superposition of the two sinusoidal waves defined as follows

$$y_1 = a_1\cos(\omega t + \theta_1) \text{ and } y_2 = a_2\cos(\omega t + \theta_2) \quad \text{...(8.28)}$$

$$y_1 = a_1(\cos\omega t \,.\, \cos\theta_1 - \sin\omega t \,.\, \sin\theta_1) \text{ and similarly} \quad \text{...(8.29)}$$

$$y_2 = a_2(\cos\omega t \,.\, \cos\theta_2 - \sin\omega t \,.\, \sin\theta_2) \quad \text{...(8.30)}$$

The total horizontal component of these two phasors

$$y_h = a_1\cos\theta_1 + a_2\cos\theta_2 \quad \text{...(8.31)}$$

and similarly the total vertical component becomes

$$y_v = a_1\sin\theta_1 + a_2\sin\theta_2 \quad \text{...(8.32)}$$

Hence the resultant is given as

$$y_2 = y_h^2 + y_v^2 = a_1^2\cos_2\theta_1 + a_2^2\cos_2\theta_2 + 2a_1a_2\cos\theta_1\cos\theta_2$$
$$+ a_1^2\sin\theta_1 + a_2^2\sin^2\theta_2 + 2a_1a_2\sin\theta_1\sin\theta_2$$
$$= a_1^2 + a_2^2 + 2a_1a_2(\cos\theta_1\cos\theta_2 + \sin\theta_1\sin\theta_2)$$
$$= a_1^2 + a_2^2 + 2a_1a_2\cos(\theta_1 - \theta_2) \quad \text{...(8.33)}$$

Therefore, the amplitude of the resultant is given as

$$y_2 = (y_h^2 + y_v^2) = [a_1^2 + a_2^2 + 2a_1a_2\cos(\theta_1 - \theta_2)]^{1/2} \quad \text{...(8.34)}$$

The angle that the resultant makes with the horizontal is

$$\tan\phi = \frac{a_1\sin\theta_1 + a_2\sin\theta_2}{a_1\cos\theta_1 + a_2\cos\theta_2} \quad \text{...(8.35)}$$

We may note that ϕ is not uniquely determined from the above equation. However, if we assume the a part to be positive, then cos θ and sin θ can be determined from (8.34) and (8.35) which would uniquely determine. ϕ We note that, if

$$\theta_1 - \theta_2 = 2m\pi \quad \text{then} \quad y = a_1 + a_2 \qquad ...(8.36)$$

The two waves are in phase and this is known as constructive interference. However, if $\theta_1 \sim \theta_2 = (2m + 1)\pi$ then $y = a_1 - a_2$. The resultant amplitude is the difference of the two amplitudes. This is known as destructive interference. It may be mentioned that when constructive and destructive interference occur, there is no violation of conservation of energy; the energy is merely redistributed. In general, let us consider the superposition N of phasors each of different amplitude, same frequency but different phase shifts. Thus, if

$$y_1 = a_1 \cos(\omega t + \theta_1);$$

$$y_2 = a_2 \cos(\omega t + \theta_2); \; ... \;;$$

$$y_N = a_N \cos(\omega t + \theta_N) \qquad ...(8.37)$$

Then as before, the total horizontal component is

$$y_v = a_1 \cos\theta_1 + a_2 \cos\theta_2 + ... + a_N \cos\theta_N = a \cos\phi \qquad ...(8.38)$$

and the total vertical component is

$$y_v = a_1 \sin\theta_1 + a_2 \sin\theta_2 + ... + a_N \sin\theta_N = a \sin\phi \qquad ...(8.39)$$

The resultant is expressed as

$$y = y_1 + y_2 + ... + y_N = a \cos(\omega t + \phi) \qquad ...(8.40)$$

Let us consider a simplified case of the above as the following

$$a_1 = a_2 = a_3 = ... = a_N = a$$

and further

$$\theta_1 = 0 \text{ and } \theta_2 = \theta_0,\ \theta_3 = 2\theta_0 \text{ ... and } \theta_N = (N-1)\,\theta_0 \qquad ...(8.41)$$

Therefore,

$$\tan\phi = \frac{a \sin\theta_0 + ... + a \sin(N-1)\,\theta_0}{1 + a\cos\theta_0 + ... + a\cos(N-1)\,\theta_0} \qquad ...(8.42)$$

In simplifying (8.42), we make use of the following identity

$$\sum_{n=0}^{N} \exp[j(\theta + n\phi)] = \frac{\sin\left[\frac{(N+1)\phi}{2}\right]}{\sin\left(\frac{\phi}{2}\right)} \exp\left[j\left(\theta + \frac{N\phi}{2}\right)\right] \qquad ...(8.43)$$

For the case of $\theta = 0$, the above reduces to

$$\sum_{n=0}^{N} \exp[jn\phi] = \frac{\sin\left[\frac{(N+1)\phi}{2}\right]}{\sin\left(\frac{\phi}{2}\right)} \exp\left(j\left(\frac{N\phi}{2}\right)\right) \qquad ...(8.44)$$

Expansion of (8.44) gives us

$$1 + \cos\phi + \cos 2\phi + ... + \cos N\phi + j(\sin\phi + \sin 2\phi + ... + \sin N\phi)$$

$$= \frac{\sin\left[\frac{(N+1)\phi}{2}\right]}{\sin\left(\frac{\phi}{2}\right)} \left[\cos\left(\frac{N\phi}{2}\right) + j\sin\left(\frac{N\phi}{2}\right)\right] \qquad ...(8.45)$$

By equating the real and imaginary parts respectively of (8.45), we have

$$1 + \cos\phi + \cos 2\phi + ... + \cos N\phi = \frac{\sin\left[\frac{(N+1)\phi}{2}\right]}{\sin\left(\frac{\phi}{2}\right)} \cos\left(\frac{N\phi}{2}\right) \qquad ...(8.46)$$

and similarly,

$$\sin\phi + \sin 2\phi + ... + \sin N\phi = \frac{\sin\left[\frac{(N+1)\phi}{2}\right]}{\sin\left(\frac{\phi}{2}\right)} \sin\left(\frac{N\phi}{2}\right) \qquad ...(8.47)$$

Hence, our resultant angle becomes

$$\tan\phi = \frac{a\sin\theta_0 + ... + a\sin(N-1)\theta_0}{a + a\cos\theta_0 + ... + a\cos(N-1)\theta_0}$$

$$= \frac{\dfrac{\sin\left(\frac{N\theta_0}{2}\right)}{\sin\left(\frac{\theta_0}{2}\right)} \sin\left[\frac{(N-1)\theta_0}{2}\right]}{\dfrac{\sin\left(\frac{N\theta_0}{2}\right)}{\sin\left(\frac{\theta_0}{2}\right)} \cos\left[\frac{(N-1)\theta_0}{2}\right]} \qquad ...(8.48)$$

$$= \tan\left[(N-1)\frac{\theta_0}{2}\right]$$

Therefore, $$\phi = \frac{(N-1)\theta_0}{2} \qquad \text{...(8.49)}$$

As we might observe, the amplitude of the resultant would vary as the following

$$y^2 = (y_h^2 + y_v^2) = [a + a\cos\theta_0 + ... + a\cos(N-1)\,\theta_0]^2$$
$$+ [a\sin\theta_0 + ... + a\sin(N-1)\,\theta_0]^2$$

$$= a^2\left[\frac{\sin\left(\frac{N\theta_0}{2}\right)}{\sin\left(\frac{\theta_0}{2}\right)}\right]^2 \qquad \text{...(8.50)}$$

The resultant of N phasors each of different amplitude, same frequency but different phase shifts related to each other harmonically is given as

$$y = a\frac{\sin\left(\frac{N\theta_0}{2}\right)}{\sin\left(\frac{\theta_0}{2}\right)}\cos\left[\omega t + \frac{(N-1)\theta_0}{2}\right] \qquad \text{...(8.51)}$$

We thus observe that, the superposition of N number of sinusoids follows a pattern proportional to

$$\frac{\sin\left(\frac{N\theta_0}{2}\right)}{\sin\left(\frac{\theta_0}{2}\right)} \qquad \text{...(8.52)}$$

8.4. THE INTENSITY DISTRIBUTION

Let $\mathbf{E}_1$ and $\mathbf{E}_2$ represent the electric fields produced at the point P by the two sources S_1 and S_2 respectively. The electric fields $\mathbf{E}_1$ and $\mathbf{E}_2$ have, in general different directions and different magnitudes. However, if the distances S_1P and S_1P are very large in comparison to the distance S_1S_2, the two fields are almost point in the same direction. Hence, we write

$$\mathbf{E}_1 = E_{01}\cos\left(\frac{2\pi}{\lambda}S_1P - \omega t\right)\hat{x} \qquad \text{...(8.53)}$$

and $$\mathbf{E}_2 = E_{02}\cos\left(\frac{2\pi}{\lambda}S_2P - \omega t\right)\hat{x} \qquad ...(8.54)$$

In these two expressions, $\hat{x}$ represents the unit vector along the direction of either of the electric fields. The resultant field is given by

$$\mathbf{E} = \mathbf{E}_1 + \mathbf{E}_2$$

$$= \left[E_{01}\cos\left(\frac{2\pi}{\lambda}S_1P - \omega t\right) + E_{02}\cos\left(\frac{2\pi}{\lambda}S_2P - \omega t\right)\right]\hat{x} \qquad ...(8.55)$$

The intensity I is proportional to the square of the electric field and is given by

$$I = KE^2 \qquad ...(8.56)$$

where K is a proportionality constant.

This becomes

$$I = K\begin{bmatrix} E_{01}^2\cos^2\left(\frac{2\pi}{\lambda}S_1P - \omega t\right) + E_{02}^2\cos^2\left(\frac{2\pi}{\lambda}S_2P - \omega t\right) \\ +2E_{01}E_{02}\cos\left(\frac{2\pi}{\lambda}S_1P - \omega t\right)\cos\left(\frac{2\pi}{\lambda}S_2P - \omega t\right)\end{bmatrix} \qquad ...(8.57)$$

This is simplified as, through the use of trigonometric identities,

$$I = K\begin{bmatrix} E_{01}^2\cos^2\left(\frac{2\pi}{\lambda}S_1P - \omega t\right) + E_{02}^2\cos^2\left(\frac{2\pi}{\lambda}S_2P - \omega t\right) \\ +E_{01}E_{02}\cos\left\{\frac{2\pi}{\lambda}(S_1P - S_2P)\right\}\cos\left\{2\omega t + \frac{2\pi}{\lambda}(S_1P + S_2P)\right\}\end{bmatrix} \qquad ...(8.58)$$

For a light wave, the frequency represented by ω is very large, ($\omega \approx 10^{15}$/sec) and all the terms depending on ωt vary too fast to be recorded by a detector that reads only the average values. The time average value of the cosine squared terms in the above expression become

$$\left\langle\cos^2(\omega t - \theta)\right\rangle = \frac{1}{2\tau}\int_{-\tau}^{\tau}\frac{1}{2}[1 + \cos 2(\omega t - \theta)]dt$$

$$= \frac{1}{2} + \frac{1}{16\pi}\frac{T}{\tau}\left\{[\sin 2(\omega t - \theta)]_{-\tau}^{\tau}\right\} \qquad ...(8.59)$$

where $T = \frac{2\pi}{\omega}(\approx 2\pi \times 10^{-15})$ sec for a light wave. Here, $<x>$ denotes the time average of a variable x. For any practical detector, $\frac{T}{\tau} <<< 1$, and the quantity between the curly brackets vary between –2 and +2, (the bracketed term is neglected in comparison to the constant value of ½). we write

$$\left\langle \cos^2(\omega t - \theta) \right\rangle = \frac{1}{2} \qquad ...(8.60)$$

The factor cos $(2\omega t - \theta)$ oscillates between +1 and –1 and its average value is hence equal to zero. Thus, the intensity as recorded by a detector is given by

$$I = I_1 + I_2 + 2\sqrt{I_1 I_2} \cos\delta \qquad ...(8.61)$$

where

$$\delta = \frac{2\pi}{\lambda}(S_2P - S_1P) \qquad ...(8.62)$$

represents the phase difference between the displacements reaching the point P from S_1 and S_2. Further,

$$I_1 = \frac{1}{2}KE_{01}^2 \qquad ...(8.63)$$

represents the intensity produced by the source S_1 if no light from S_2 is allowed to fall on the screen; similarly, $I_2 = \frac{1}{2}KE_{02}^2$ represents the intensity produced by the source S_2 if no light from S_1 is allowed to fall on the screen. From the above, we infer that

(a) The maximum and the minimum values of cos δ are +1 and –1 respectively; as such the maximum and the minimum values of the intensity are given as

$$I_{\max} = \left(\sqrt{I_1} + \sqrt{I_2}\right)^2 \qquad ...(8.64)$$

and

$$I_{\min} = \left(\sqrt{I_1} - \sqrt{I_2}\right)^2$$

(b) The maximum intensity occurs when $\delta = 2m\pi$ which corresponds to the case of

$$S_2P - S_1P = m\lambda$$

(c) The minimum intensity occurs for the case of $\delta = (2m + 1)\pi$ which corresponds to the case of $S_2P - S_1P = \left(m + \frac{1}{2}\right)\lambda$. In these, is a positive integer.

We further notice that, when the amplitudes of the interfering light waves are equal, their maximum values of individual intensities are equal and the resultant minimum intensity becomes zero. For the more general case of $I_1 \neq I_2$, the minimum intensity is not zero.

(d) If the holes S_1 and S_2 are illuminated by different light sources, then the phase difference δ remains constant for about 10^{-9} seconds and thus δ varies with time in a random way. Averaging this phase difference over time scales which are of the order of 10^{-9}, we obtain

$$< \cos \delta > = 0$$

and
$$I = I_1 + I_2 \qquad ...(8.65)$$

$I_2 \approx I_1 \approx I_0$ (say). Hence, for incoherent sources, the resultant intensity is the sum of the individual intensities produced by each one of the sources independently and no interference pattern can be observed. In the experimental set up shown in Fig. 8.1, if the distances S_1P and S_2P are extremely large in comparison with d, then we obtain

Then,
$$I = 2I_0 + 2I_0 \cos \delta = 4I_0 \cos^2 (\delta/2) \qquad ...(8.66)$$

The intensity distribution is often termed as $\cos^2$ pattern. This is illustrated in Fig. 8.3.

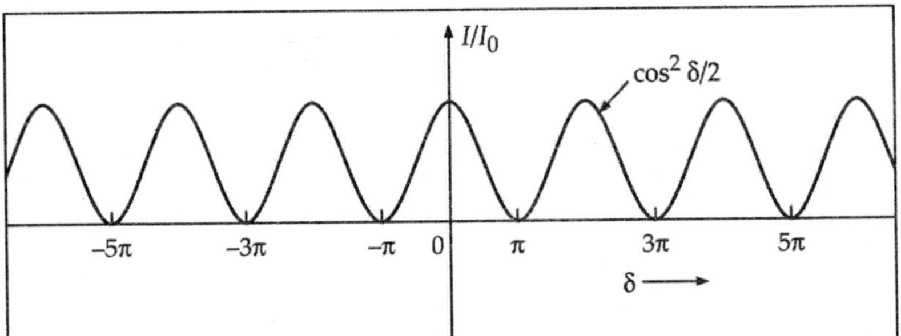

Fig. 8.3: The normalized intensity pattern following a cosine squared law

8.5. INTERFERENCE BY DIVISION OF AMPLITUDE

In the previous section, we considered the interference pattern produced by division of a wave front; for example, light coming out of a pinhole is allowed to fall on two holes, and spherical waves emanating from these two holes interfere with each other to result in an interference pattern consisting of alternate bright and dark fringes. Here we consider the formation of interference pattern produced by division of amplitude; for example, if a plane wave falls on a thin film then the wave reflected from the upper surface interferes with the wave reflected from the lower surface. Such studies account for the display of beautiful colours by a soap or oil film illuminated by white light.

Let us consider a plane wave incident normally on a thin film of uniform thickness d. The waves reflected from the upper surface interfere with the waves reflected from the lower surface. In order to observe this interference pattern without obstructing the incident beam, we use a partially reflecting plate P as shown in Fig. 8.4. Such an arrangement also enables us to eliminate the direct beam from reaching the photographic plate P or the eye. The plane wave may be produced by placing an illuminated pinhole at the focal point of a corrected lens; alternatively, it may just be a beam coming out of a laser.

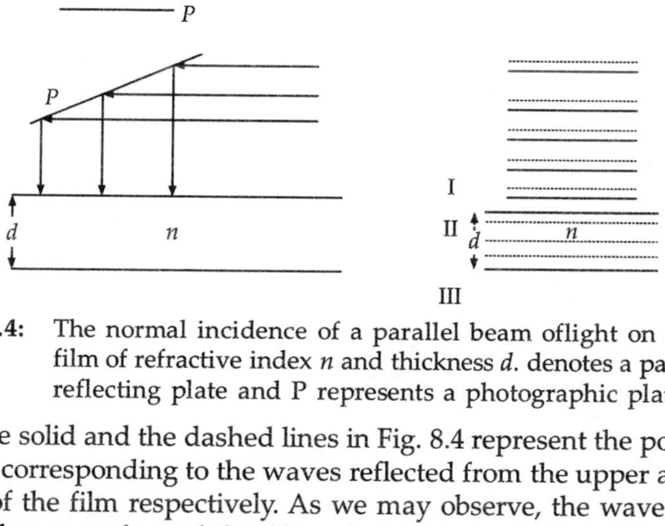

Fig. 8.4: The normal incidence of a parallel beam oflight on a thin film of refractive index n and thickness d. denotes a partially reflecting plate and P represents a photographic plate.

Let the solid and the dashed lines in Fig. 8.4 represent the positions of the crests corresponding to the waves reflected from the upper and lower surfaces of the film respectively. As we may observe, the wave reflected from the lower surface of the film traverses an additional path of $2nd$, where is the refractive index of the film. Further, if the film is placed in air, then the wave reflected from the upper surface of the film will undergo a sudden change in phase of π and as such the conditions for destructive or constructive interference is given by

$$2nd = m\lambda \text{ for destructive interference and} \quad ...(8.67)$$

$$2nd = \left(m + \frac{1}{2}\right)\lambda \quad ...(8.68)$$

for constructive interference.

In the above, λ is the free space wavelength and m is a positive integer including zero. A photographic plate placed at P receives uniform illumination. It becomes dark when $2nd = m\lambda$ and bright when

$2nd = \left(m + \frac{1}{2}\right)\lambda$. Instead of placing the photographic plate, if we try to view the film from the top with naked eye, then the film appears to be uniformly illuminated. We may observe that the amplitudes of the waves reflected from the upper and lower surfaces, in general are different and as such the interference is not completely destructive. However, with appropriate choice of the refractive indices of media II and III, the two amplitudes can be made very nearly equal.

For an air film between two glass plates, no phase change occurs on reflection at the glass-air interface, but a phase change of π does occur on reflection at the air-glass interface and the conditions for maxima and minima remain the same. On the other hand, if the medium I is a medium of refractive index 1.47 and medium II is another medium with a refractive index of 1.55 and the third medium has a refractive index of 1.60, then a phase change of π occurs at both reflections and the conditions for maxima and minima would be

$$2nd = \left[m + \frac{1}{2}\right]\lambda \qquad \text{...(8.69)}$$

for destructive interference and

$$2nd = m\lambda$$

for constructive interference (8.70)

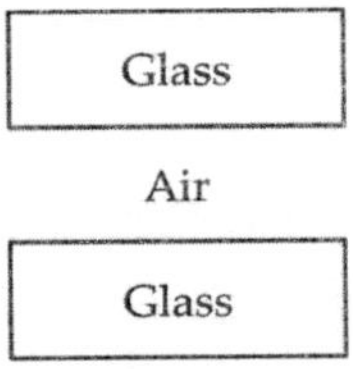

Fig. 8.5: Thin film of air formed between two glass plates

In general, whenever the refractive index of medium II lies in between the refractive indices of media I and III, then the conditions of maxima and minima are given by (8.69) and (8.70).

Let us next consider the oblique incidence of the plane wave on the thin film as shown in Fig. 8.5. Here also, the wave reflected from the upper surface of the film interferes with the wave reflected from the lower surface of the film. The latter traverses an additional optical path length of Δ, which is given by

$$\Delta = n_2\,(BD + DF) - n_1\,BC \qquad \text{...(8.71)}$$

where C is the foot of the perpendicular from the point F on BG. Let us show that

$$\Delta = 2n_2 d \cos\theta' \quad ...(8.72)$$

where θ' is the angle of refraction.

Outline of the proof for the additional path length traversed by the ray reflected from the lower surface of the film

We refer to Fig. 8.6. Let θ and θ' denote the angles of incidence and refraction respectively. We have constructed a perpendicular BJ from the point B on the lower surface LL′ and extended BJ and FD to the point B' where they meet. We see that,

$$\angle JBD = \angle BDN = \angle NDF = \theta' \quad ...(8.73)$$

where N is the foot of the perpendicular drawn from the pint D on BF. As we may observe,

$$\angle BDJ = \frac{\pi}{2} - \theta' \quad ...(8.74)$$

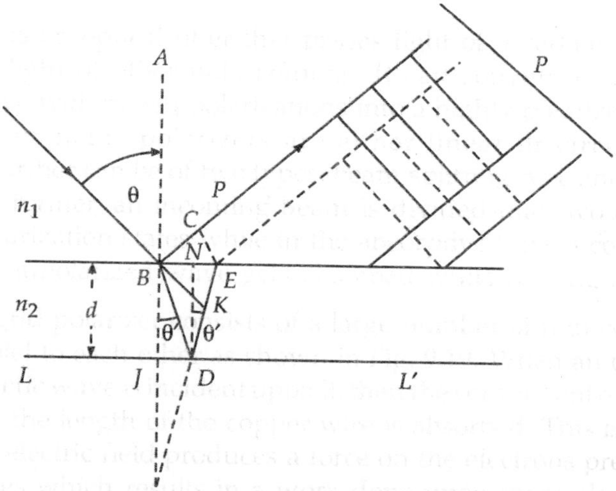

Fig. 8.6: Calculation of optical path difference between the waves reflected from the upper surface of the film and from the lower surface of the film. The solid and the dashed lines represent the corresponding positions of crests. P denotes the photographic plate.

Moreover,

$$\angle B'DJ + \angle BDJ + \angle BDN + \angle NDF = \pi \quad ...(8.75)$$

$$\angle B'DJ = \pi - \left[\left(\frac{\pi}{2} - \theta'\right) + \theta' + \theta'\right] = \frac{\pi}{2} - \theta' \qquad ...(8.76)$$

This is due to the fact that

$$\angle B'DF = \pi,\ \angle BDN = \theta' \text{ and } \angle NDF = \theta' \qquad ...(8.77)$$

Hence,
$$BD = BD' \text{ and } BJ = B'J = d \qquad ...(8.78)$$

or,
$$BD + DF = B''D + DF = B'F \qquad ...(8.79)$$

$$\Delta = n_2\ D'F - n_1\ BC \qquad ...(8.80)$$

$$\angle CFB = \angle CBA = \theta \qquad ...(8.81)$$

Hence,

$$BC = BF \sin\theta = \frac{KF}{\sin\theta'} \sin\theta \qquad ...(8.82)$$

This is because and
$$\angle BFK = \frac{\pi}{2} - \theta'$$

$$BF \cos\left(\frac{\pi}{2} - \theta'\right) = KF$$

$\Rightarrow$
$$BF = \frac{KF}{\sin\theta'} \qquad ...(8.83)$$

Therefore,

$$BC = KF\frac{n_2}{n_1} \qquad ...(8.84)$$

where K is the foot of the perpendicular from B on $B'F$. Substitution of this expression gives us

$$\Delta = n_2\ B'F - n_1\ BC = n_2\ D'F - n_2\ KF = n_2\ B'K \qquad ...(8.85)$$

This is known as the cosine law.

The last step in (8.85) follows from the fact that in the triangle $BB'K$, we have

$$BB' = 2d \text{ and } \angle BB'\ K = \theta' \qquad ...(8.86)$$

Hence,
$$B'K = BB' \cos\theta' = 2d\cos\theta' \qquad ...(8.87)$$

For a film placed in air, a phase change of π occurs when reflection takes place at the point B and as such, the conditions of destructive and constructive interference is given by

$$\Delta = 2n_2 d \cos\theta' = m\lambda \text{ for minima} \qquad ...(8.88)$$

and, for maxima we should have

$$\Delta = \left(m + \frac{1}{2}\right)\lambda \qquad ...(8.89)$$

8.6. DIFFRACTION

Let us consider a plane wave incident on a long narrow slit of width b as shown in Fig. 8.7. According to geometrical optics we expect the region XY of the screen S to be illuminated and the remaining portion (known as the geometrical shadow) to be absolutely dark. However, careful observation reveals that if the width of the slit is not very large compared to the wavelength, then the light intensity in the region XY is not uniform and there is also some intensity inside the geometrical shadow. Further, if the width of the slit is made smaller, larger amounts of energy reach the geometrical shadow. This spreading out of a wave when it passes through a narrow opening is known as the phenomenon of diffraction and the intensity distribution on the screen is known as the diffraction pattern. We show subsequently that the spreading out decreases with decrease in wavelength. Indeed, since the light wavelengths are very small ($\lambda \sim 5 \times 10^{-5}$) cm, the effects due to diffraction are not readily observed.

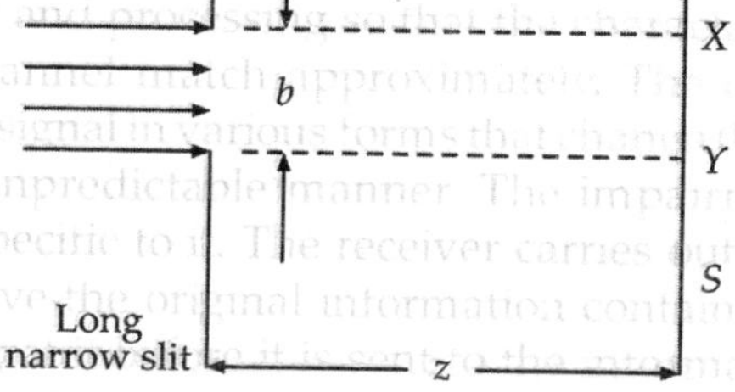

Fig. 8.7: A plane wave is incident normally on a long narrow slit of width b. Only the region is XY supposed to be illuminated according to geometrical optics

It may be mentioned here that there is not much of a difference between the phenomena of interference and diffraction. Interference results due to the superposition of waves coming out from a number of point or line sources and diffraction corresponds to the situation when we have waves coming out from an area source like a circular or rectangular aperture or even a large number of rectangular apertures (like the diffraction grating).

The diffraction phenomena are usually divided into two categories: (a) Fresnel diffraction and (b) Fraunhofer diffraction.

In the Fresnel diffraction, the source of light and the screen are, in general, at a finite distance from the diffracting aperture as shown in Fig. 8.8(a).

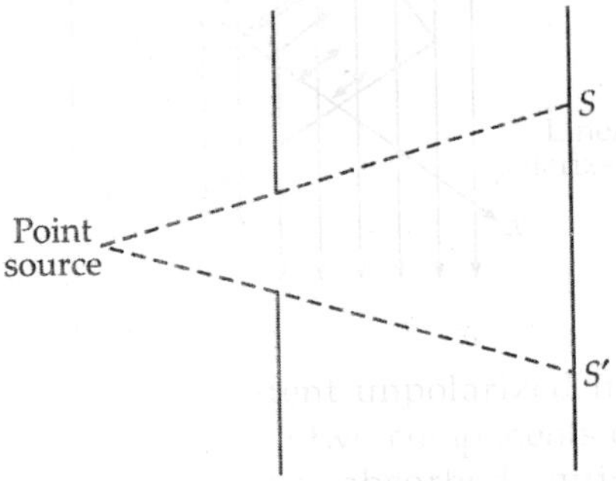

Fig. 8.8: (a) When either the source or the screen (or both) are at finite distances from the aperture, the diffraction pattern corresponds to the Fresnel class

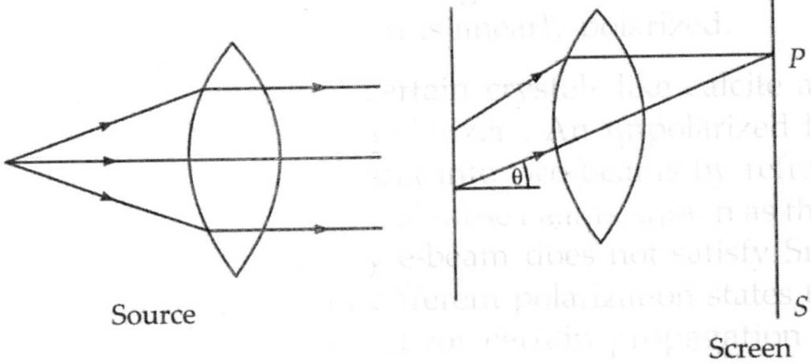

Fig. 8.8: (b) The source and the screen are placed at the focal planes of two lenses. The beam diffracted along the direction θ is observed at point P resulting in Fraunhofer kind of diffraction.

In the second category of diffraction, the source and the screen are at infinite distances from the aperture; this is easily achieved by placing the source on the focal plane of a convex lens and placing the screen on the focal plane of another convex lens Fig. 8.8(b). The two lenses effectively move the source and the screen to infinity because the first lens makes the light beam parallel and the second lens effectively makes the screen receive a parallel beam of light. It turns out that it is much easier to calculate the intensity distribution of a Fraunhofer class of diffraction pattern. This kind of diffraction is also easy to observe which is done by an ordinary laboratory spectrometer; the collimator renders a parallel beam of light and the telescope receives parallel beams of light on its focal plane. The diffracting aperture is placed on the prism table.

8.6.1. Single Slit Diffraction Pattern

We consider the Fraunhofer class of diffraction produced by an infinitely long slit of width b. A plane wave is assumed to fall normally on the slit for which the intensity distribution is calculated on the focal plane of the lens L as shown in Fig. 8.9.

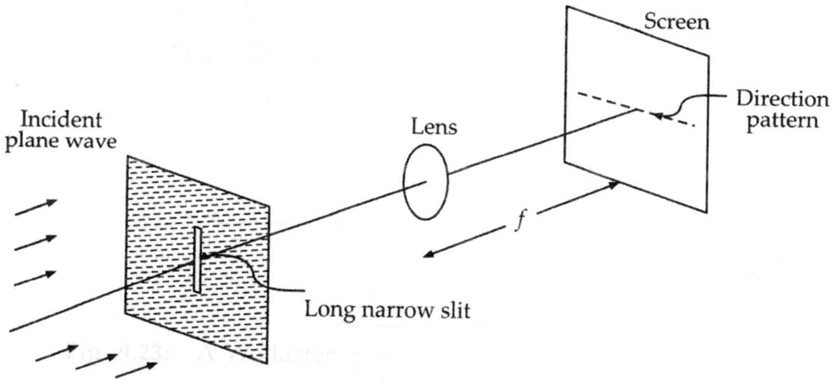

Fig. 8.9: Diffraction of a plane wave incident normally on a long narrow slit of width b

The slit is assumed to consist of a large number of equally spaced point sources and each point on the slit acts as a source of Huygen's secondary wavelets that interfere with the wavelets originating from other points. Let the point sources be at $A_1, A_2, \ldots A_n$ and let the distance between any two consecutive points be Δ. Hence, the width of the slit is expressed in terms of the point sources as (the total number of point sources being n)

$$b = (n - 1)\,\Delta \qquad \ldots(8.90)$$

We now calculate the resultant field produced at an arbitrary point P produced by these n number of point sources. The receiving parallel rays make an angle of θ with the normal to the slit. Since the slit has been assumed to consist o a continuous distribution of sources, we let n approach infinity in the general case of field calculation, thereby making Δ go to zero such that $n\Delta \to b$.

We make use of Fig. 8.10 to calculate the intensity distribution at point P on the screen. Let the amplitudes of the disturbances reaching point P be $A_1, A_2, \ldots A_n$ that are nearly the same as the point P is situated at a distance which is very large in comparison to b. However, because of even slightly different path lengths to the point P, the field produced by

A_1 differs from that produced by A_2 by a phase difference that obviously, is related to their path difference.

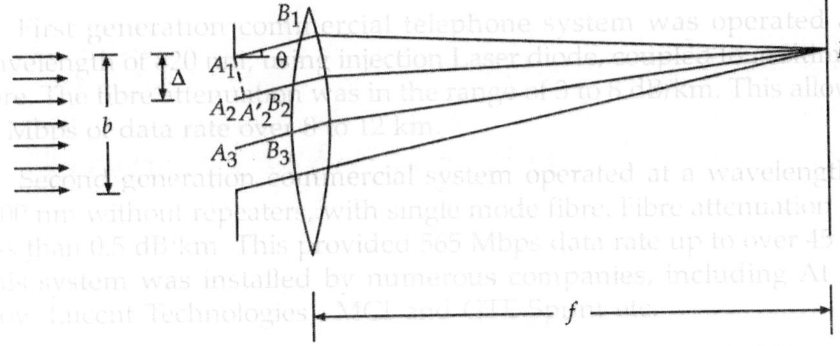

Fig. 8.10: The slit of width b is assumed to consist of a large number of equally spaced points

For an incident plane wave, the points A_1, A_2, ... are in phase and therefore, the additional path traversed by the disturbance emanating from the point A_2 is A_2, A_2 where is the foot of the perpendicular drawn from on . This follows from the fact that the optical paths A_1B_1P and $A_2'B_2P$ are the same. If the diffracted rays make an angle θ with the normal to the slit then the path difference would be

$$A_2A_2' = \Delta \sin\theta \qquad ...(8.91)$$

The corresponding phase difference, ϕ, is therefore, given by

$$\phi = \frac{2\pi}{\lambda}\Delta \sin\theta \qquad ...(8.92)$$

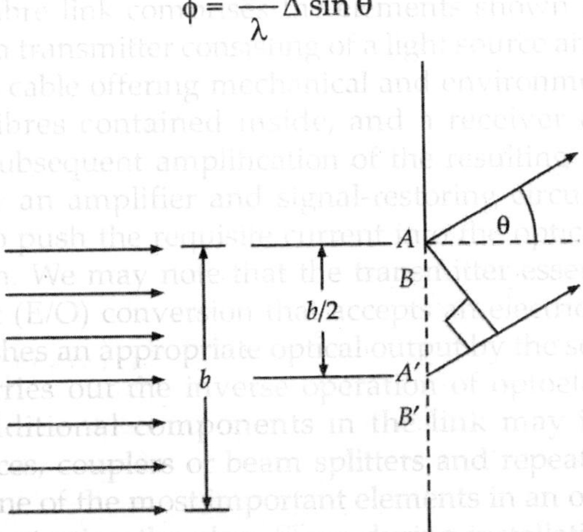

Fig. 8.11: The slit is divided into two halves for deriving the condition for the first minimum

Hence, if the field at the point P due to the disturbance coming from point A_1 is $A \cos \omega t$, then the field produced due to the disturbance coming from A_2 would be $A \cos (\omega t - \phi)$. The field due to the source A_2 lags behind that due to the point A_1 as it has to traverse an extra path length of A_2A_2' as compared tothe field due to A_1. Hence, the difference in the phases of the disturbance reaching from the points A_2 and A_3 are also ϕ and thus the resultant field at point P is given by

$$E = A\,[\cos \omega t + \cos (\omega t - \phi) + \cos (\omega t - 2\phi) + \ldots + \cos \{\omega t - (n-1)\phi\}] \quad \text{...(8.93)}$$

We have seen previously that

$$\cos \omega t + \cos(\omega t - \phi) + \cos (\omega t - 2\phi) + \ldots + \cos \{\omega t - (n-1)\,\phi\}$$

$$= \frac{\sin (n\phi/2)}{\sin (\phi/2)} \cos [\omega t - (n-1)\phi] \quad \text{...(8.94)}$$

Thus,

$$E = E_A \cos [\omega t - (n-1)\,\phi] \quad \text{...(8.95)}$$

where the amplitude of the resultant field is given by

$$E_A = A\frac{\sin (n\phi/2)}{\sin (\phi/2)} \quad \text{...(8.96)}$$

In the limiting case, as we have noted previously, $n \to \infty$ and $\Delta \to 0$ such that $n\Delta \to b$, we have

$$\frac{n\phi}{2} = \frac{\pi}{\lambda} n\Delta \sin\theta \to \frac{\pi}{\lambda} b \sin\theta \quad \text{...(8.97)}$$

Further, the phase difference ϕ becomes

$$\phi = \frac{2\pi}{\lambda} \Delta \sin\theta = \frac{2\pi}{\lambda}\frac{b\sin\theta}{n} \text{ which would tend to zero} \quad \text{...(8.98)}$$

Therefore, we write

$$E_A \approx \frac{A\sin(n\phi/2)}{(\phi/2)} = nA\frac{\left(\dfrac{\pi b\sin\theta}{\lambda}\right)}{\left(\dfrac{\pi b\sin\theta}{\lambda}\right)} = A_R\frac{\sin\beta}{\beta} \quad \text{...(8.99)}$$

In the above,

$$A_R = nA \text{ and } \beta = \frac{\pi b\sin\theta}{\lambda} \quad \text{...(8.100)}$$

Hence, the resultant field becomes, in the limiting case of $n \to \infty$,

$$E_A = A_R \frac{\sin\beta}{\beta} \cos(\omega t - \beta) \qquad ...(8.101)$$

The corresponding intensity distribution is given by

$$I = \frac{\sin^2\beta}{\beta^2} \qquad ...(8.102)$$

where I_0 represents the intensity at $\theta = 0$.

Positions of Maxima and Minima

The variation of the intensity with β is shown in Fig. 8.3. We note that the intensity is zero when

$$\beta = m\pi,\ m \neq 0 \qquad ...(8.103)$$

We further observe that, corresponding to $\beta = 0$, the factor $\frac{\sin\beta}{\beta} = 1$ and $I = I_\theta$ which corresponds to the maximum of the intensity. Substituting the value of we obtain

$$b \sin\theta = m\lambda;\ m = 1, \pm 2, \pm 3 \qquad ...(8.104)$$

for minima.

The first minimum occurs at $\theta = \pm \sin^{-1}\left(\frac{\lambda}{b}\right)$; the second minimum at $\theta = \pm \sin^{-1}\left(\frac{2\lambda}{b}\right)$, etc. Since the maximum value of is unity, the maximum value of is the integer which is less than and closest to $\left(\frac{b}{\lambda}\right)$.

The positions of minima are obtained by first considering the case of $m = 1$. The angle θ satisfies the equation

$$b \sin\theta = \lambda \qquad ...(8.105)$$

The slit is divided into two halves as shown in Fig. 8.11. Let us consider two points A and A' separated by a distance of $b/2$. The path difference between the waves reaching the point P from A and B is $\frac{b}{2}\sin\theta$ which in this case is $\lambda/2$. The corresponding phase difference is, thus $\frac{\pi}{2}$ resulting in a zero output at this point. Similarly, the wave originating from point

B is cancelled by the wave originating from the corresponding point *B* in the bottom half. Thus the resultant disturbance due to the upper half of the slit gets cancelled by the disturbances starting from the lower half that gives rise to zero intensity at point P. In a similar manner when

$$b \sin \theta = 2\lambda \qquad \text{...(8.106)}$$

The slit has been divided into four equal parts. The first and second quarters cancel each other while the third and the fourth quarters similarly cancel each other. Similarly, when $m = 3$, the slit is divided into six equal parts and so on.

In order to determine the positions of maxima, we differentiate (8.102) with respect to β and set it equal to zero. Hence,

$$\frac{dI}{d\beta} = I_0 \left[\frac{\sin\beta \cos\beta}{\beta^2} - \frac{2\sin^2\beta}{\beta^3} \right] = 0 \qquad \text{...(8.107)}$$

Therefore,

$$\sin\beta\,(\beta - \tan\beta) = 0 \qquad \text{...(8.108)}$$

This condition of $\sin\beta = 0$, or $\beta = m\pi (m \neq 0)$ corresponds to minima. The conditions for maxima correspond to the roots of the equation formed by the second part of (8.108).

$$\tan\beta = \beta \text{ (for maxima)} \qquad \text{...(8.109)}$$

The root $\beta = 0$ corresponds to the central maximum. The other roots can be found by determining the points of intersection of the curves for $y = \beta$ and $y = \tan\beta$. The intersections occur at $\beta = 1.34\pi$, 2.46π etc. And are known as the first maximum, second maximum, etc. Since $\left[\frac{\sin(1.43\pi)}{1.43\pi}\right]^2$ is about 0.0496, the intensity of the first maximum is about 4.96% of the central maximum. Similarly, the intensities of the second and third maxima are about 1.68% and 0.83% of the central maximum respectively.

8.6.2. Diffraction by a Circular Aperture

We have seen previously that a plane wave incident on a long narrow slit of b width spreads out along its width with an angular divergence of $\sim\lambda/b$. In a similar manner we discuss the diffraction of a plane wave by a circular aperture. An experimental set up for doing this is shown in Fig. 8.12. A plane wave is shown to be incident normally on a circular aperture and a lens whose diameter is much larger than that of the aperture is placed close to the aperture and the Fraunhofer diffraction pattern is observed on the focal plane of the lens. The diffraction pattern is observed to consist of concentric dark and bright rings owing to the rotational symmetry of the circular aperture. This kind of concentric dark and bright rings known as the Airy pattern has an intensity distribution given by

$$I = I_0\left[\frac{2J_1(\phi)}{\phi}\right] \qquad ...(8.110)$$

where
$$\phi = \frac{2\pi}{\lambda} a \sin\theta \qquad ...(8.111)$$

In this case, a is the radius of aperture, λ is the wavelength and θ is the angle of diffraction, I_0 being the intensity corresponding to $\theta = 0$ that represents the central maximum. $j_L(\phi)$ is the Bessel function of the first order. As we have seen previously, $J_1(0) = 0$, however,

$$\lim_{x \to 0} \frac{2J_1(x)}{x} = 1 \qquad ...(8.112)$$

Other zeros of $J_1(x)$ occur at

$$x = 3.832, 5.136, 7.016, ...$$

that correspond to the successive dark circles in the Airy pattern. The first dark ring appears when

$$\sin\theta = \frac{3.832\lambda}{2\pi a} \sim \frac{1.22\lambda}{D} \qquad ...(8.113)$$

where $D = 2a$ is the diameter of the aperture. About 84% of the energy is contained within the first dark ring; thus we may infer that the angular spread is approximately given by

$$\Delta\theta \sim \frac{\lambda}{D} \qquad ...(8.114)$$

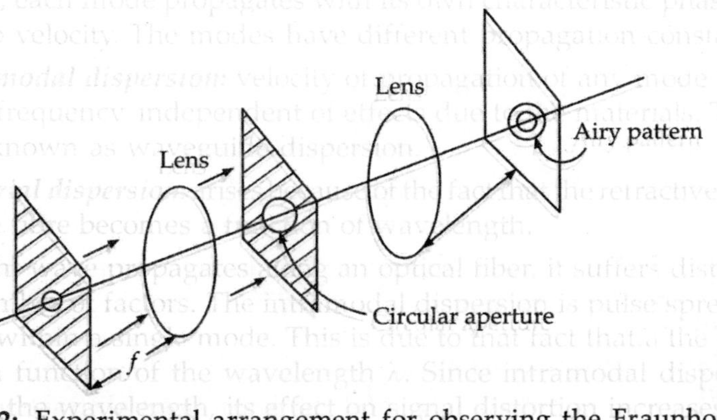

Fig. 8.12: Experimental arrangement for observing the Fraunhofer diffraction pattern by a circular aperture

We may say that the angular divergence associated with the diffraction pattern may be written in the following general form as

$$\Delta\theta \sim \frac{\lambda}{\text{linear dimension of the aperture}}$$

An interesting application of the above phenomenon is shown in Fig. 8.13. A layman would expect that to obtain more directionality of sound waves, one should use a loudspeaker of small aperture as shown in Fig. 8.13. However, thisresults in a greater diffraction divergence and only a small fraction of energy will reach the observer. On the other hand, if one uses a loudspeaker of larger diameter, greater directionality is achieved.

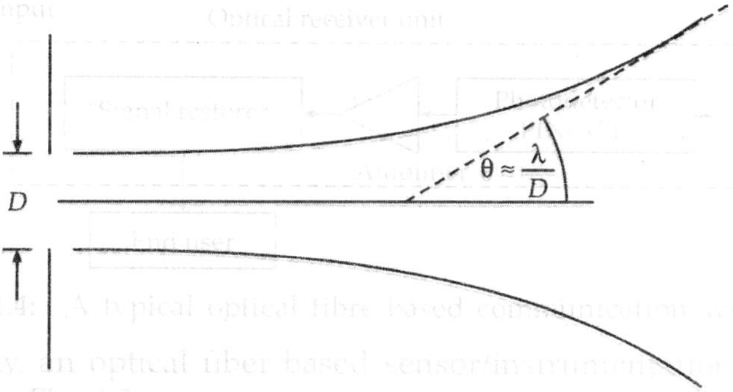

Fig. 8.13: Diffraction divergence of a plane wave incident normally on a circular aperture of diameter D

An interesting application of diffraction through a circular aperture is illustrated by the observance of large sized speakers in marriage processions, addressing at public gatherings. A large diameter speaker results in greater directionality so that considerable amount of sound energy reaches an observer. A small diameter would result in more diffraction giving rise to higher spread of energy in many directions.

8.7 DIFFRACTION GRATING

An arrangement of a large number of equidistant slits to produce a diffraction pattern is known as a diffraction grating. The pattern so produced is known as a grating spectrum. Since the exact positions of the principal maxima in the diffraction pattern depend on the wavelength, the principal maxima corresponding to different spectral lines correspond to different angles of diffraction. The grating spectrum, thus, is useful for determination of wavelengths. A large number of parallel slits is required in order to produce a narrow principal maximum. For a good quality grating, the value of N is of the order of 15,000 per inch. This is implemented by cutting grooves with a diamond point on an optically transparent sheet of material so that the grooves act as opaque spaces. Once the cutting of one groove is over, the diamond point moves the sheet forward to produce the next groove. As the distance between the grooves is extremely small, the movement of the sheet is obtained with the help of the rotation of a screw that drives the carriage carrying it. The additional requirement to produce a good quality grating is to make the grooves so cut out equidistant from each other. Hence, the pitch of the screw should be a constant. Commercial gratings are manufactured by taking the cast of an actual grating on a transparent film like that of cellulose acetate. An appropriate strength solution of cellulose acetate is poured on the cut surface and allowed to dry to form a strong thin film, detachable from the parent grating. These impressions of a grating are preserved by mounting the film between two glass sheets. Nowadays, gratings are also produced holographically, where one records the interference pattern between two planes or spherical waves. In contrast to cut gratings, holographic gratings have a much larger number of lines/cm. The positions of the principal maxima are given by

$$d \sin\theta = m\lambda; \quad m = 0, 1, 2, \ldots \tag{8.115}$$

This relation is used to study the dependence of the angle of diffraction θ on the wavelength λ. The zeroth order principal maximum occurs corresponding to $\theta = 0$ irrespective of the wavelength of the light source used. When a polychromatic source of light such as white light is used, the central maximum corresponds to the same color as the source itself. However, for $m \neq 0$, the angles of diffraction are different for different wavelengths and therefore, various spectral components appear at different positions. Hence, by measuring the angles of diffraction for various colors, we can determine the wavelengths for a given value of m. Differentiating θ with respect to λ, we have

$$d \cos\theta = m$$

$$\Rightarrow \frac{\Delta\theta}{\Delta\lambda} = \frac{m}{d\cos\theta} \tag{8.116}$$

From this, we may infer that for very small values of θ such that $\cos\theta \approx 1$, the angle is $\Delta\theta$ directly proportional to the order of the spectrum (m) for a given $\Delta\lambda$ so that for a given m, the quantity $\frac{\Delta\theta}{\Delta\lambda}$ is a constant. Such a spectrum is known as a normal spectrum which means that the difference in angles for two spectral lines is directly proportional to the difference in wavelengths. However, for longer wavelengths such red light, the dispersion tends to be greater (For a longer wavelength, the quantity

$\sin\theta$ tends to be larger for a given value of m and grating spacing d. Hence, $\cos\theta$ tends to be smaller which means a greater value of $\Delta\theta$ or dispersion). Further, as $\Delta\theta$ is inversely proportional to the grating distance, a smaller spacing would result in a larger angular dispersion.

The FBGs are an integral part of modern telecommunications hardware, used in applications such as add/drop filters, fiber lasers, and data multiplexing.

8.8. CHAPTER SUMMARY

Interference and diffraction are wave phenomena related to physical objects of small size. The two interference types are the interference between two light waves coming out of a slit when an incident light beam is divided into two parts. Interference by division of amplitude refers to the interference between light waves reflected from the surfaces of films. Interference results in alternate maxima and minima that are positions of maximum intensity and minimum light intensity. These are known as fringe lines which areapproximately straight lines corresponding to two point sources. The intensity distribution known as Malu's law follows a squared cosine shape. There is not much difference between interference and diffraction. Both are due to the superposition of two or more light waves. However, the mechanism of producing at least two light beams to cause a superposition is a point or line source in case of interference while diffraction is due to the superposition of waves coming out from an area source such as rectangular or circular. A diffraction grating has a large number of rectangular apertures. Further, we can have Fraunhofer type or Fresnel type of diffraction. Here we have studied the former type. The diffraction pattern due to rectangular apertures follow a sinc(.) pattern while that from a circular aperture has a pattern based on Bessel's function. The divergence of waves coming out of a circular aperture is inversely proportional to its diameter which requires the use of large sized objects to result in higher directionality of energy.

❑❑❑

9

Optical Components

9.1. INTRODUCTION

An optical fibre link needs to launch optical power into a particular fibre from a given light source so that it is converted into a corresponding electrical output at the receiving end. Care must be exercised on the ways implemented to couple optical power from one fibre to another. This is because a given optical fibre link cannot be practically implemented by a single fibre. A number of fibres need to be connected to each other in such a manner that the loss due to any coupling is minimized. Launching optical power from a source into a fibre requires considerations such as the numerical aperture, core size, refractive-index profile and core-cladding index difference of the fibre plus the size, radiance and angular distribution of the optical source. Also to distribute optical power from one to many or many to one, use is made of components known as fibre optic couplers. Hence, couplers are defined as components with 'one or many' input fibres and 'one or many' output fibres.

Optical power can be coupled from one fibre to another which is controlled by number of modes that can propagate in each fibre. For example, a fibre in which 500 modes can propagate is connected to another fibre in which 400 modes can propagate, and then at best 80 per cent of optical power from the first fibre can be coupled to the second one provided all the modes are equally excited.

9.2. CLASSIFICATION OF OPTICAL COUPLERS

The optical couplers[1] can be categorized fundamentally as two types: (a) Directional couplers and (b) Distributive or wavelength dependent couplers

Similarly, the directional couplers are broadly divided into the following three categories and may have three or four ports.

(a) Biconically tapered
(b) Offset butt joint and
(c) Beam splitting type

9.2.1. Biconically Tapered Directional Coupler

This type of coupler is shown in Fig. 9.1. This is used to provide low loss couplers with a range of splitting ratio. The tapering length and the cladding thickness determine the splitting ratio. The fabrication procedure of such a coupler is simple. In this coupler, two multimode or single mode fibres are twisted around one another and kept under tension. The junction is heated in order to soften the optical fibres and fuse the cladding. A biconical taper at each of the four ports is made by pulling on the softened fibres. In case of multimode fibres, coupling takes place, as higher order modes do not strike the core-cladding interface beyond the critical angle in the tapered region. We note the presence of two conic sections in this component; one cone marked 1 and 4 representing one fibre and the other marked 2 and 3 marking the other fibre. These two are observed to be joined in the middle by a taper.

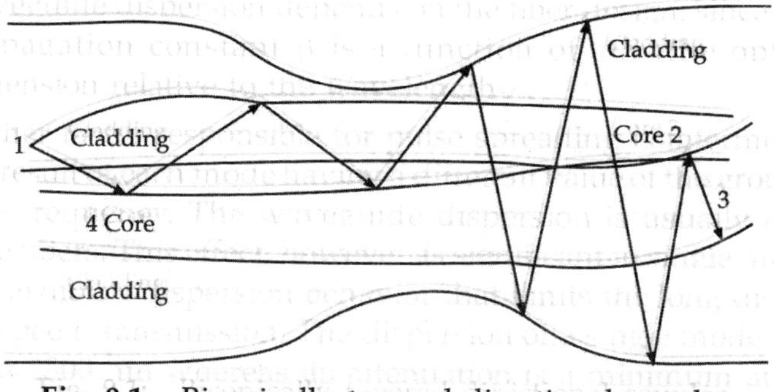

Fig. 9.1: Biconically tapered directional coupler

It is evident from the diagram that higher order modes are trapped by total internal reflection at the outer surface of the cladding and such modes are converted into cladding modes. Hence the optical power contained in the lower order modes is confined to the original fibre. It is also observed from the figure that the central cladding is common to both the cores and hence the power from higher order input modes becomes common to both the fibres. Thus the output tapers of the coupler convert the cladding modes back into the core. This results in a sharing of energy between the two overlapping evanescent waves and energy transfer from one fibre to another. Energy coupling between these two fibres by this technique is known as 'evanescent field coupling'. The tapered and fused coupler for single mode operation is explained by the energy exchange taking place in the overlapping evanescent fields associated with the two fibres.

Advantages of tapering:

(a) The two cores come close to each other due to tapering thereby increasing the evanescent field overlap.
(b) The fibre core diameter is decreased due to tapering causing a decrease in the V-number of the fibre (normalized frequency is lowered)

9.2.2. Offset Butt Joint Directional Coupler

This is a four port directional coupler shown in Fig. 9.2. The input is observed to be provided at port 1 while port 2 collects an amount of power determined by the amount of offset. A fraction of the incident light travels from the joint to the tap (port 3) along a planer curved plastic waveguide.

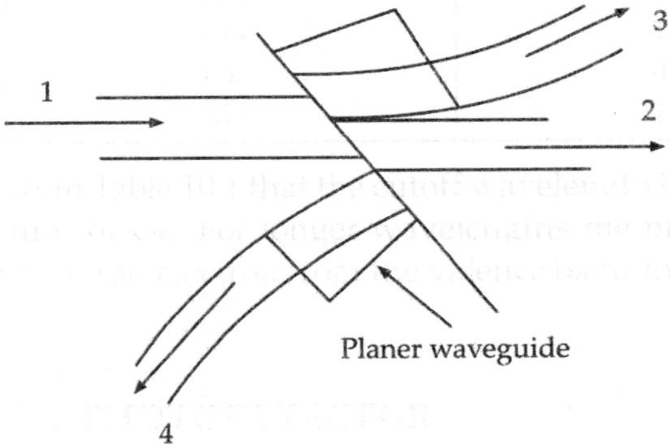

Fig. 9.2: Offset butt joint coupler

9.2.3. Beam Splitting Directional Coupler

A partial reflector, called beam splitter, can be used as a simple directional coupler. Such an arrangement is shown in Fig. 9.3.

A partial reflector, called beam splitter can be used as a simple directional coupler for optic system. Two different types of beam splitters; a plate type and a cube type are illustrated in Figs. 9.3(a) and (b). The plate type has a thin partially reflective coating on a transparent substrate. The beam splitting ratio is dependent on the thickness and composition of the coating. It displaces the transmitted beam laterally with reference to the incident beam. The cube type beam splitter has two prisms separated by a partially reflective coating. It may be noted that a beam splitter on its own cannot be used for power division among optical fibres. The presence of splitter in the path of light is equivalent to a gap which provides large losses, as the diverging rays emitted by the input fibre miss the receiving fibre.

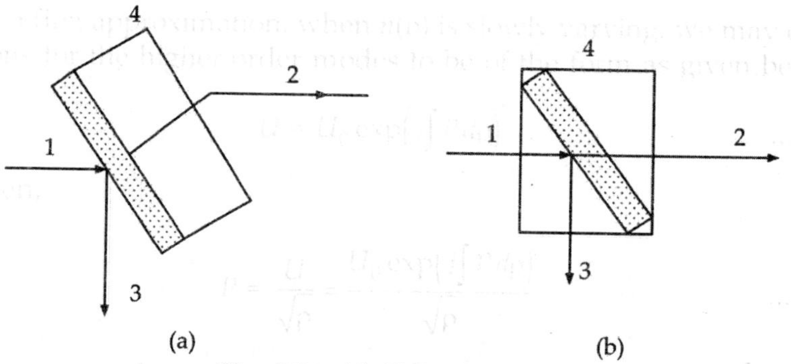

Fig. 9.3: (a) Plate (b) Cube

A small modification to the above mentioned beam splitter is shown in Fig. 9.4. The coupler is made by two quarter pitch GRIN rod lenses, separated by a partially reflective coating. The fibres connected to the coupler are offset from the axis of the lenses. If port 1 is taken as input port, the system of lenses will allow the light to go in the 2nd port. The light reflected by the coating forms an image on port 3. No light reaches port 4. Inputs at the other ports are similarly distributed.

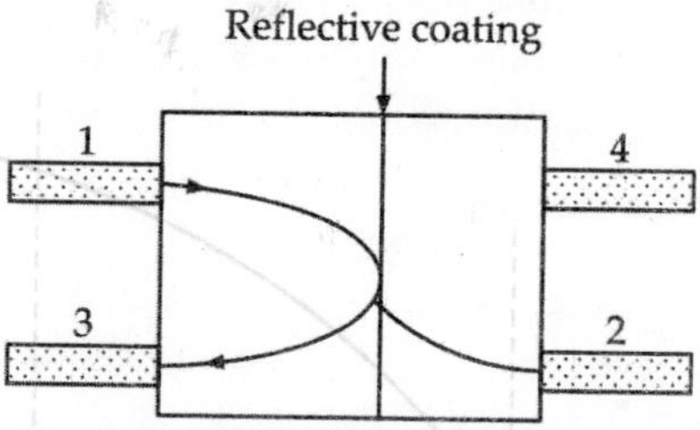

Fig. 9.4: A modified beam splitting coupler

The incident light beam is collimated on the beam splitter and the split light so formed is refocused onto the receiving fibres. This type of directional coupler is shown in Fig. 9.5. The same function can also be realized by replacing the GRIN rod by conventional spherical lenses.

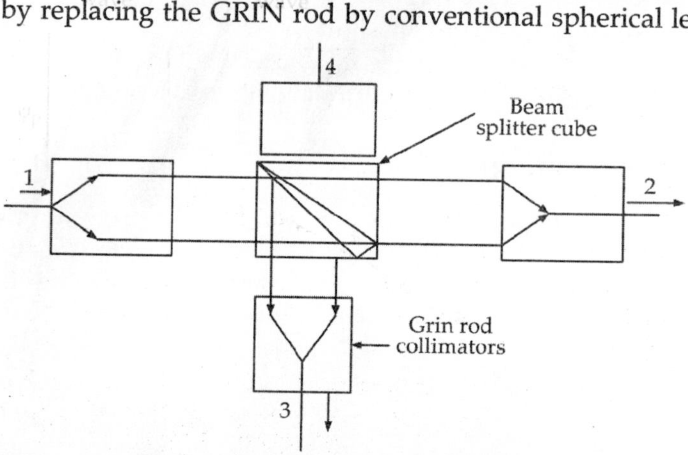

Fig. 9.5: GRID Rod collimating lens based directional coupler

9.4. OTHER DIRECTIONAL COUPLERS

A number of other directional couplers in use is described as follows.

(i) Wavefront-dividing directional coupler: M_1 and M_2 are spherical reflectors of focal length f.

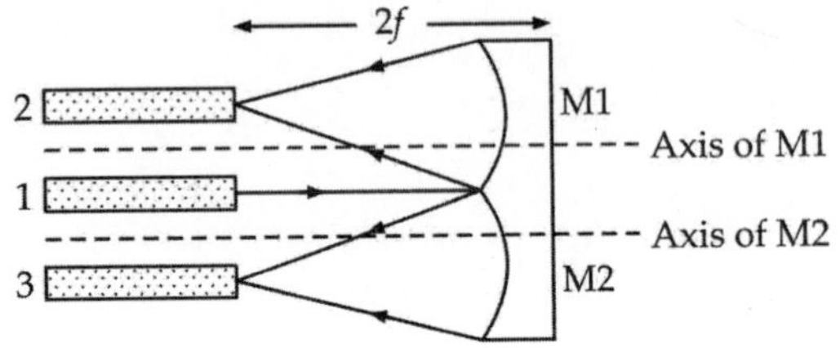

(ii) Four port directional coupler

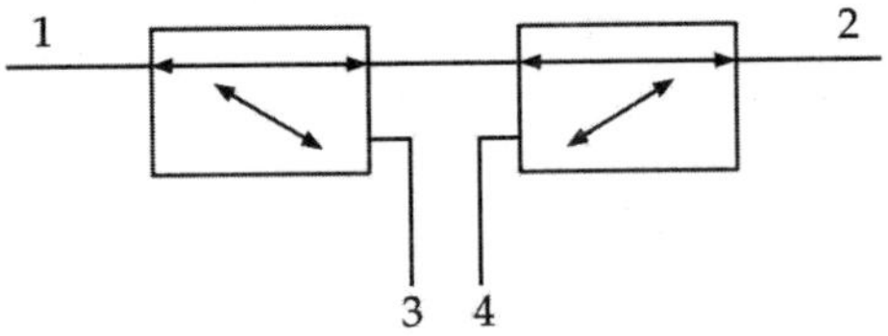

(iii) Radiation bent-fibre directional coupler

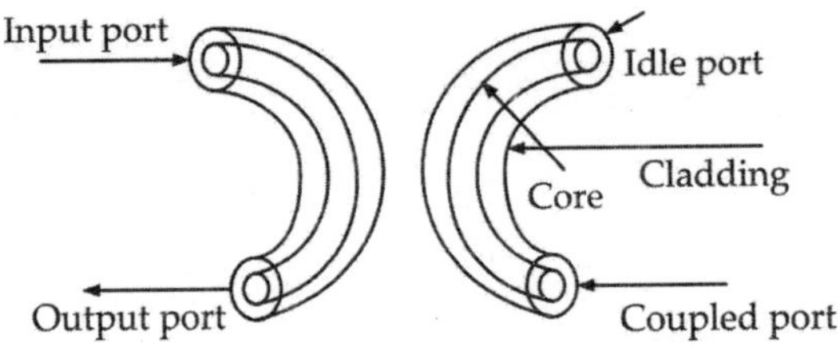

(iii) Bent-fibre type of coupler

Fig. 9.6: Other Directional Couplers

9.5. STAR COUPLERS

These couplers distribute an optical signal from a single input fibre to multi output fibres, as shown in Fig. 9.7. The two main manufacturing techniques to produce multimode fibre star couplers are the mixer-rod and the FBT methods. These are of two types: (i) transmissive type and (ii) reflective type.

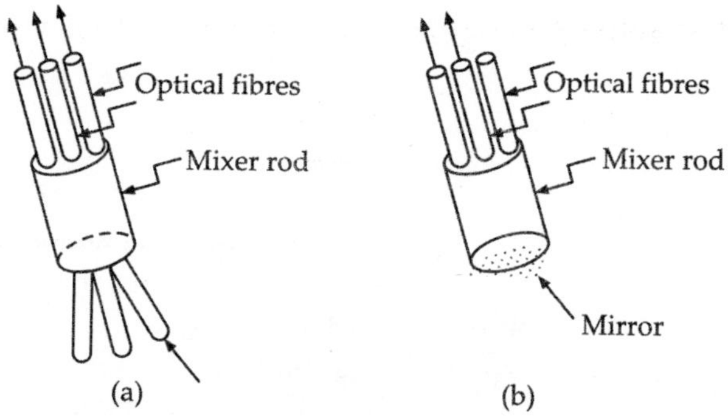

Fig. 9.7: Fibre star couplers using the mixer rod method (a) transmissive type (b) reflective type

As shown in Fig. 9.7, a thin platelet of glass is used which effectively mixes the light from one fibre, dividing it among the outgoing fibres. Such a method can be used to manufacture a transmissive type shown in Fig. 9.7(a) or a reflective type of star coupler as shown in Fig. 9.7(b). The typical insertion loss for an 8 × 8 mixer rod transmissive star coupler with fibre pigtails is 12.5 dB with port to port uniformity of ±0.7 dB. In an ideal star coupler, the optical power from any input fibre is evenly distributed among the output fibres. The total loss associated with the star coupler comprises its theoretical splitting loss together with the excess loss. The splitting loss is related to the number of output ports N by the following expression:

$$\text{Splitting loss (star coupler)} = 10 \log_{10} N \text{ (dB)} \qquad ...(9.1)$$

The basic structure of a transmissive type of coupler is shown in Fig. 9.8. Here the optical signals are sent into the mixing block from any single input fibre which are available at all output fibres. The number of input fibres is the same as that of the output fibres. Power is evenly distributed. For a $N \times N$ star coupler, the power available at each output fibre is $1/N$ th power of any single input fibre.

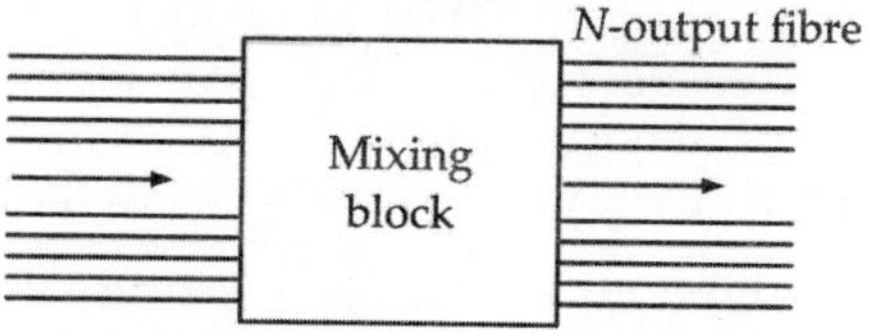

Fig. 9.8: Another transmissive type of coupler

Hence, the power available at a single output port

$$P_0 = \frac{P_{in}}{N} \qquad ...(9.2)$$

The power splitting ratio in dB associated with a star coupler is, therefore,

$$P_{ST} = 10 \log\left(\frac{P_{in}}{P_0}\right) = -10\log\left(\frac{1}{N}\right) \qquad ...(9.3)$$

The negative sign is provided to make the final result positive. The above expression shows that a loss of power takes place due to power splitting. However, this is not so as the total output power on all N output ports may be equal to the input power. The excess power loss is the ratio of the total output power to the power lost from the input.

(ii) Reflective type of star coupler

The basic construction of this type of coupler is shown in Fig. 9.9. The splitting loss of star couplers increases logarithmically as the number of coupler ports increases. As a coupler is not an ideal device, excess loss is an additional loss in these. But this loss occurs only once with a star-coupled network, whereas it occurs at every point in a T network.

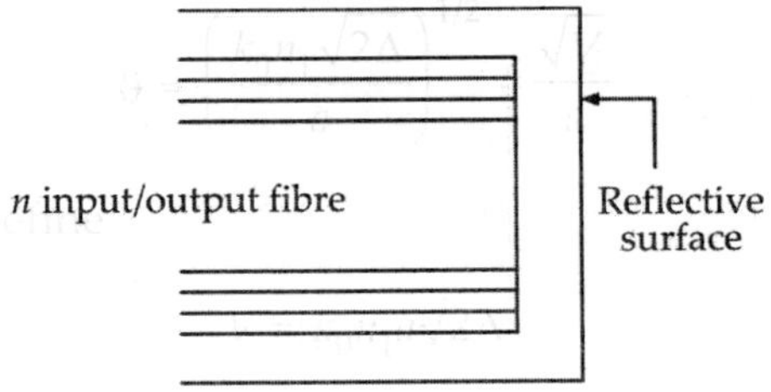

Fig. 9.9: Reflective type of star coupler

For a star coupler, the power into any port is nearly equally divided among all output ports, and loss increases slowly with the number of additional output ports. The penalty paid for using a coupler, more than needed, with ports reserved for future expansion, is balanced against the advantages of easy expansion by simply adding stations. The new station does not affect the power budget for existing stations. If any network involves transreceivers only, the star coupler is preferred. Combinations of star coupler and T coupler are used in a practical scenario.

The transmissive types of star couplers have the disadvantage of cross talk whereas the reflective types of couplers do not have this issue as power from any fibre is distributed to all the fibres.

9.6. T-COUPLERS

A T-coupler is shown in Fig. 9.10. The input port 1 receives the optical power which is split into two parts: port 2 gets P_2 and port 3 gets a power of P_3. The powers P_2 and P_3 need not be equal. The amount of split is decided by the detailed coupler structure. The splitting may be expressed in dB or in terms of a percentage. An important attribute of such a coupler is its directivity. This is defined as the power ratio between one port and another, which is zero for ideal directivity. It is expected that no power be transferred between any two ports on the same side of the coupler. The excess loss associated with a star coupler has the same meaning here also which is defined as the ratio of the total output power to input power.

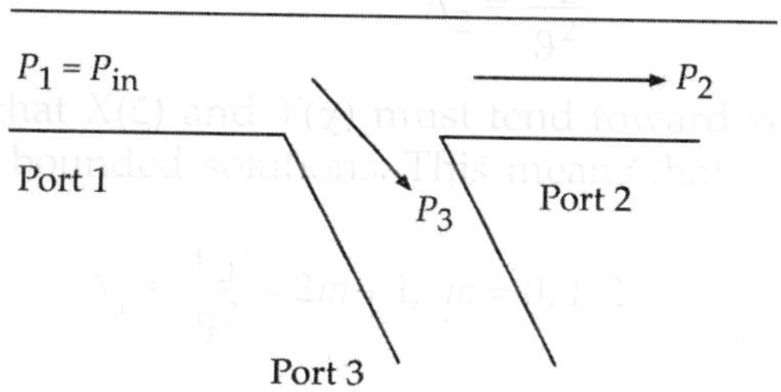

Fig. 9.10: A T-coupler

The T-couplers are categorized as fused, expanded beam and expanded beam with GRIN lenses type.

(a) Fused T-couplers: This type of coupler, construction wise is the same as the transmission type of star coupler. This is illustrated in Fig. 9.11(a).

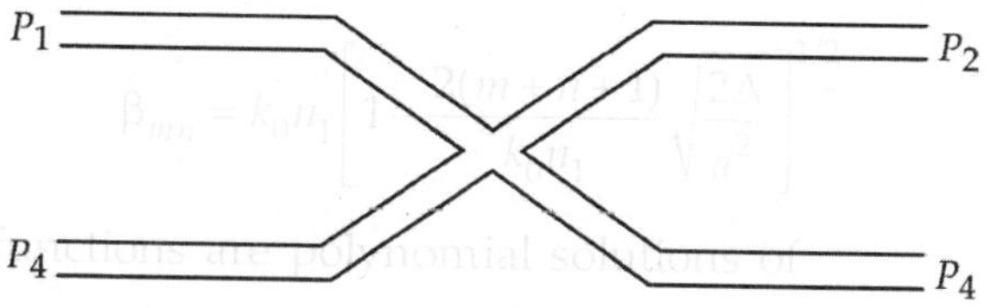

Fig. 9.11: (a) Fused T-coupler

(b) Expanded beam T-coupler: Such a coupler is illustrated in Fig. 9.11(b). It is essentially an expanded beam connector. A partially reflecting mirror is placed between the fibre lenses. The mirror permits a fraction of the incident light to travel from port 1 to port 2 and reflects a part to port 3. This splitting is decided by the extent of the reflectivity of the partially reflecting mirror.

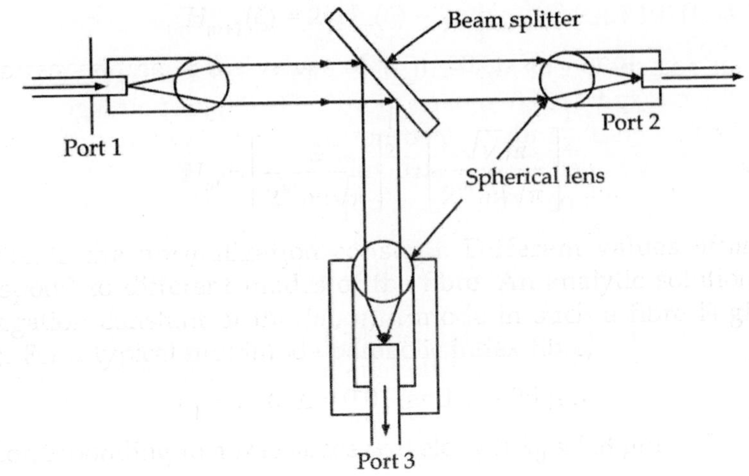

Fig. 9.11: (b) Expanded beam T-coupler

(c) Expanded beam T-coupler with GRIN lens: A schematic of this type of coupler is shown in Fig. 9.11(c). A GRIN lens and a partially reflective surface is used to accomplish the coupling action in this type of coupler. The GRIN lens is sometimes called SELFOC lens. A beam from port 4 is split and is partially reflected to port 3 and the major section is refracted to port 2.

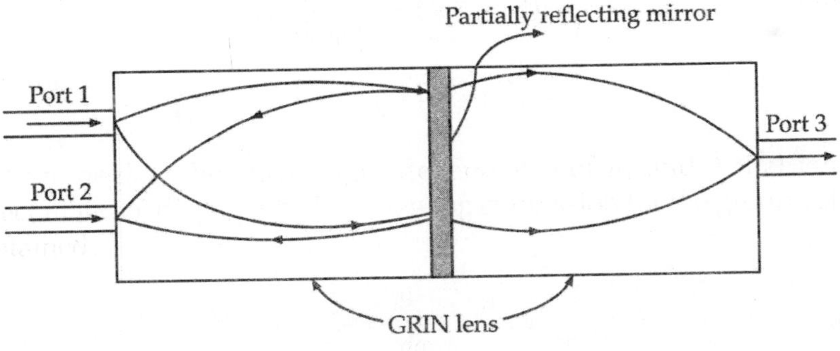

Fig. 9.11: (c) Expanded beam T-coupler with GRIN lens

T-couplers with different splitting ratios are available. However, they introduce significant loss of power from the network. For each mode of network, a T-coupler is essential. As more such couplers are added to the network, the losses increase to a great extent. Moreover, the problem of dynamic range is an added disadvantage. The nodes which are close to the source are subject to larger amounts of power than the nodes that are farther away from the source.

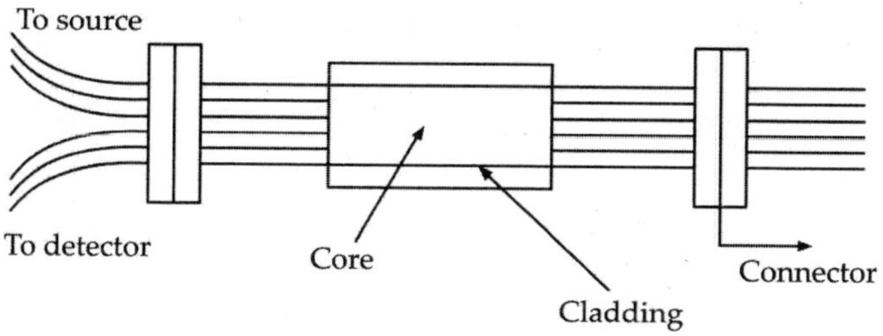

Fig. 9.12: An expanded view of a T-coupler

Butt Coupling: In some cases the fibre is butt coupled to the source. Butt coupling is defined as coupling by placing the flat fibre end directly against the source, without the aid of any lens system. Butt coupling cannot be achieved when the source is mounted in a package with a covering window glass. If the fibre is directly butt coupled to the light source, the ratio of the power accepted by the fibre to the power emitted by the source is expressed as

$$P_i/P_s = \frac{1}{2}(m+1)[\alpha + (\alpha/2)]NA^2 \qquad ...(9.4)$$

where a is the fibre refractive index profile, m is the exponent for source angular distribution for power

As before, the coupling loss in dB is expressed as

$$L_{cop} = -10\log_{10}\left(\frac{P_t}{P_s}\right) \qquad ...(9.5)$$

9.7. CALCULATIONS ON COUPLERS

In order to define some coupler characteristics, let us consider a coupler as shown in Fig. 9.13. The directions of allowed power flow are indicated by the arrows. Let us assume an input power P_1 to be incident on the input port 1 of the coupler. This power is split between port 2 and port 3 according to the required splitting ratio. In the ideal case, no power reaches port 4 and hence it is known as an isolated port. The power emerging from port 2 is P_2 and that at port 3 is P_3. Let us define some loss parameters of the coupler.

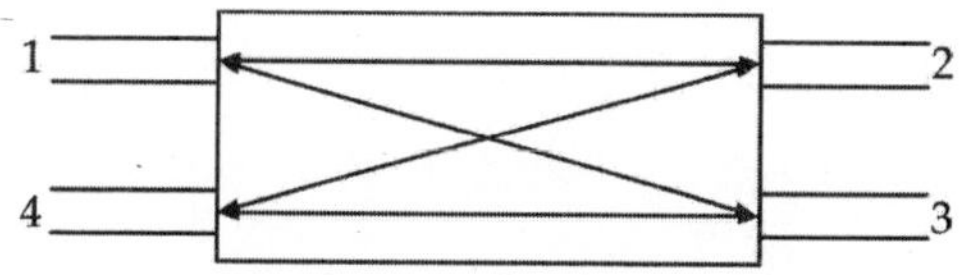

Fig. 9.13: Four port directional coupler

(a) *Tap loss*: The transmission loss between port 1 and port 2 is known as tap loss and is defined as

$$P_{L\,tap} = -10\log\left(\frac{P_3}{P_1}\right) \qquad ...(9.6)$$

(b) *Throughput loss*: The transmission loss incurred between power flow at the input port and the favoured port 2 is expressed as

$$P_{L\,thp} = -10\log\left(\frac{P_3}{P_1}\right) \qquad ...(9.7)$$

(c) *Directionality*: It is defined as the loss between the input port and the isolated port 4 which mathematically is

$$P_{LD} = -10\log\left(\frac{P_4}{P_1}\right) \qquad ...(9.8)$$

(d) *Excess loss*: The power lost within the coupler is termed as the excess loss which is mathematically defined as

$$P_{LE} = -10\log\left(\frac{P_2+P_3}{P_1}\right) \qquad ...(9.9)$$

This loss includes radiation, scattering, absorption and coupling to the isolated port. In (9.6) to (9.9), the base of the logarithm is 10.

9.7.1. Relation between Tap Loss and Throughput Loss

We assume that the isolated port 4 receives no optical power. Further no power is lost in the ideal coupler. Hence,

$$P_1 = P_2 + P_3 \qquad ...(9.10)$$

$$\Rightarrow \qquad P_2 = P_1 - P_3$$

Thus,

$$P_{L\,thp} = -10\log\left(\frac{P_1 - P_3}{P_1}\right) = -10\log\left(1-\frac{P_3}{P_1}\right) \qquad ...(9.11)$$

or, $$P_{L\,thp} = -\,10\log\left(1 - 10^{-P_{L\,thp}/10}\right) \qquad ...(9.12)$$

as $$P_{L\,tap} = -10\log\left(\frac{P_3}{P_1}\right) \quad ...(9.13)$$

$$\frac{P_3}{P_1} = 10^{-P_{L\,tap}}$$

or ⇒ $$-\frac{P_{L\,tap}}{10} = \log_{10}\left(\frac{P_3}{P_1}\right) \quad ...(9.14)$$

Coupling efficiency

$$\eta_c = \frac{P_F}{P_{out}} = \frac{\text{Power into fiber}}{\text{Power out of device}} \quad ...(9.15)$$

This efficiency is a function of the source structure, light intensity distribution and size and NA of the fibre.

Coupling to fibre: An important optical system design parameter is the amount of power coupled into the optical fibre. For an optical fibre the light capturing power is expressed as

$$P_{coup} = \left(\frac{\beta+1}{2}\right)\frac{a_f}{a_e}(NA)^2 P_{emi} \quad ...(9.16)$$

$$\text{Coupling loss in dB} = 10\log_{10}\left(\frac{P_{emi}}{P_{coup}}\right) \quad ...(9.17)$$

In the above expression, β is the LED or laser diode spatial distribution factor, a_f is the fibre core area, a_e is the emitting area and $\frac{a_f}{a_e}$ is the ratio that determines the overlap between the emitter surface and the core area of the fibre. For $a_f > a_e$, the ratio is taken as unity in the expression for light capturing power. We note that doubling the numerical aperture of the optical fibre quadruples the power. The curves in Fig. 8.10 represent the range of input coupling loss for different optical emitter and fibre combinations. The coupling efficiencies for different emitter and fibre combinations are shown in Table 9.1. The fibre core are that overlaps the emitter rectangle is represented by the expression

$$a_0 = d\sqrt{r^2 - \frac{d^2}{4}} + 2d\sin^{-1}\left(\frac{d}{2r}\right) \quad ...(9.18)$$

In the above expression, and represent the thickness and the core radius respectively.

Table 9.1: Coupled power (in dB) for different combinations of optical sources and fibers

Transmitter type	*SMF at 1300 nm*	*SMF at 1550 nm*	*MMF at 850 nm*	*MMF at 1300 nm*
E-LED	–25	–23	–9	–12
S LED	–35	–35	–15	–18
SU LED	–7	–	+0	+0
Laser	+3	+2	+8	+5

We can observe from Table 9.1 that the power coupled from all the three types of LEDs onto the fibres is always negative which means some loss. For example, if an E-LED outputs 1W of power (0 dB), when it is coupled to a single mode fibre operating at 850 nm, a loss of 9 dB occurs and therefore, the effective power that can be coupled to such a fibre is only –9 dB.

9.8. SPLICERS

Splices and connectors are required between individual single fibres or may involve a number of fibres in a multi fibre cable. In the latter case, techniques for terminating and connecting all the fibres simultaneously have been developed. Splices are required during installation and again during service should cables be damaged and fibres break. The typical length that can be drawn into a duct before the drawing tension exceeds the cable weight is about 1-2 km. For undersea cables and those that are directly ploughed into the ground, longer fibre lengths are desirable. The splices may be made in one of two ways. The two fibre ends are either fused together, or glued with a transparent, index-matching adhesive. In each case, a jig is required to hold the ends while they are melted and joined or while the adhesive is applied and cured.

Each splice or connector gives rise to additional attenuation, and the need to minimize such losses sets a maximum tolerance on the permitted and misalignment of fibres. Mismatch may arise because of differences between the two fibres to be joined, differences in numerical aperture, in refractive index profile, in core or cladding diameter, or in their circularity or concentricity. Essentially, the mode spot sizes should be well matched. In the case of multimode fibres, the additional losses are not confined to the immediate vicinity of the splice or connector, because the mismatch transfers some of the optical power into leaky and weakly guided higher-order modes which may propagate for some distance before being radiated out of the core.

Careful preparation of the ends of the fibres is essential for splices and connectors. Low loss is obtained only if the fibre end faces are clean, smooth and perpendicular to the fibre axis, or else at some other precisely controlled angle. Fibres can be cleaved reliably in a controlled manner by bending them in tension over a curved mandrel and scoring the output surface. The fibre then breaks in the required manner as a result of the propagation of a stress crack, a technique that is familiar to anyone who has worked with glass. The precise angle of the cleave varies with any torsion that may be present in the fibre. Torsional strain of 1° per cm gives rise to a cleave angle of approximately 1° measured perpendicular to the axis.

In the most commonly used fusing splicing technique, an electric discharge is struck across the two fibre ends, which melt in the heat generated. As they come into contact, surface tension forces help to self-align the two fibres. Many automated, microprocessor controlled fusion splices are available commercially and allow splices to be made by non-specialist staff in operational environments. Often the splice loss is monitored by local injection of light within the unit, and an average loss per splice of 0.1 dB or less can be obtained for well-matched fibres. Some of the consequences of fibre mismatch and misalignment are shown in Fig. 9.14. It is normal practice to secure the splice inside a glass or polymer sleeve for environmental protection and to carry any subsequent tensile stress. Minimum splice strengths exceeding 2 GPa can be obtained with mean strengths of about 3 PGa.

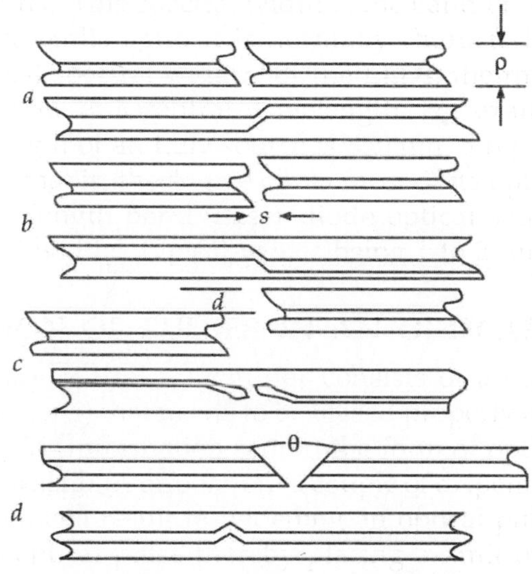

Fig. 9.14: The effect of fibre mismatch and misalignment on a fusion splice: (a) non-concentric core with cladding aligned; (b) non-concentric core with cores aligned; (c) effect of offset; (d) effect of non-parallel end faces. In each case, surface tension forces pull the outer cladding surfaces into alignment.

The amount of radiation loss depends on the degree of misalignment. A mismatch of the core sizes result in some optical signal loss as shown in Fig. 9.14(a).

The longitudinal separation occurs when the fibres have the same axis but have a gap distance between their end faces as shown in Fig. 9.14(b). Due to the longitudinal displacement some of the light spreads beyond the end of the receiving fibre. For the step index fibre we have coupling efficiency η_{s1} defined as

$$\eta_{s1} = \left(\frac{1}{1 + s/\rho \tan \alpha_m}\right)^2 \qquad \text{...(9.19)}$$

In the above, *s* is the separation distance of the fibre ends and α_m is the critical acceptance angle $\sin^{-1}$(NA) and ρ is the radius of the core.

The lateral displacement as shown in (Fig. 9.14c) occurs when the axis of the two fibres are separated by a distance *d*. Mechanical misalignments of the fibres result in losses due to the non overlapping of the fibre areas. In the analysis of the misalignments, the usual assumptions are that the fibres have equal radii, index profile and numerical aperture to isolate the effect of the misalignment. The fibre is assumed to have uniform power distribution across its cross sectional area. For a step index fibre, the coupling efficiency depends on the lateral misalignment distance *d* as

$$\eta_{s1} = \frac{2}{\pi}\cos^{1}\left(\frac{d}{\rho}\right) - \frac{d}{\pi\rho}\sqrt{1 - \left(\frac{d}{2\rho}\right)^2} \qquad \text{...(9.20)}$$

This equation is derived from a calculation of the overlapping areas of two equal circles with centres separated by a distance *d*.

Angular displacement as shown in (Fig. 9.14d) is encountered when the two fibre axes form an angle θ so that the fibres are no longer parallel to each other. The common misalignment encountered in practice is the axial displacement and it is responsible for the greatest power loss. The axial offset reduces the overlap area of the two fibre end faces and consequently reduces the amount of optical power than can be coupled from one end fibre to the other one.

9.8.1. Power Loss in Joining Technique

Each joining technique is subjected to certain conditions which cause the amount of optical power loss at each joint. These losses depend on parameters such as:

(a) The geometrical characteristics of the fibre: For example, if the emitting fibre has a large diameter than the receiving fibre, then the optical power will be lost because the end face area mismatches which has been illustrated in Fig. 9.15.

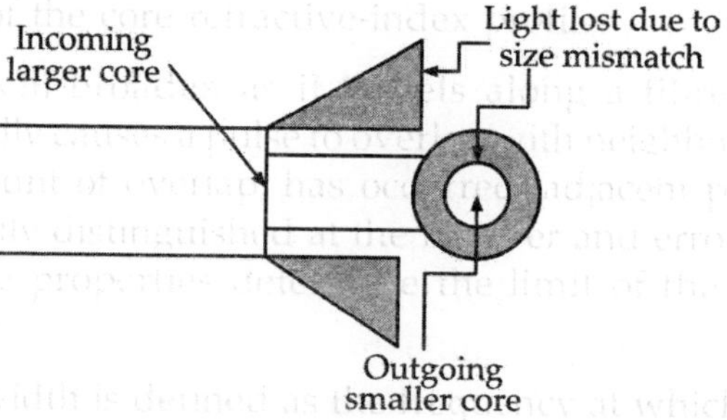

Fig. 9.15: Illustration of optical power being lost because of size mismatch while joining two fibres

(b) *The waveguide characteristics of the fibre*: For example, if an emitting fibre has larger numerical aperture than the receiving fibre, all optical power falling outside the accepting core of the receiving fibre will be lost.

(c) *Fibre end-face quality*: One criterion for low loss joints is that the fibre end faces must be clean, smooth and perpendicular to the axis. End preparation techniques include sawing, grinding, polishing and control fracture. Fig. 9.16(a) shows a fibre is cleaved at an angle to the axis such that when light from the matching sloping surface, some portion of light will be lost by reflection and more by coupling into the leaky modes. Fig. 9.16(b) shows a fibre which was not cleaved properly due to improper application of cleaving stress. The end of the fibre has a scalloped irregular surface which scatters light. These defects arise due to improper fibre end preparation.

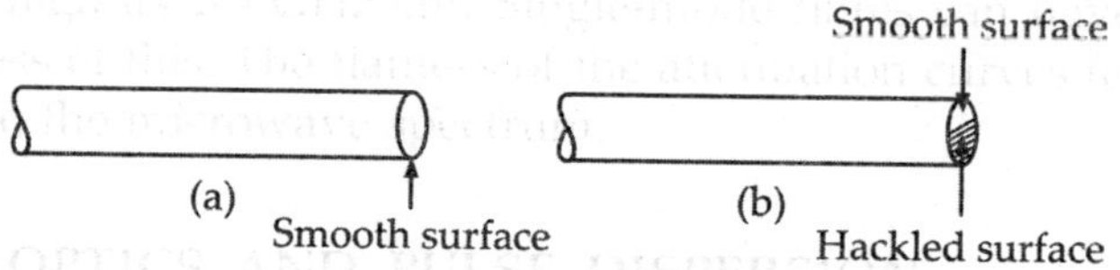

Fig. 9.16: (a) Forking and (b) hackled surface due to improperly cleaved end faces

(d) *Power loss due to dirt*: Any dirt or foreign substance which gets into a connector or splice during or after the assembly may increase its losses or even completely block it. Extreme care should be taken during installation to prevent the inclusion of dirt.

Glued splices are preferred in some applications. A small mandrel typically consists of end guides which accommodate the coated fibres, index-matching gel, and a central insert that maintains the precise alignment of the two fibres. Again, a glass sleeve is bonded over the joint in order to secure and protect the two fibres. Normally, a UV-cured proxy adhesive is used. Average splice losses for identical fibres are in the region of 0.2 dB, and back-reflection can be as small as –40 dB.

It is essential to prepare the end faces of a fibre before the fibres are connected together or spliced with each other. In order to eliminate the deflection or scattering of light at the joint, the fibre end must be flat, smooth and perpendicular to the fibre axis. The commonly followed end preparation technique consists of sawing, grinding, polishing and controlled fracture. Conventional grinding and polishing techniques can produce a smooth surface which is perpendicular to the fibre axis. However, this method is quite time-consuming and requires a good amount of technical skill of the operator.

There are essentially four different methods of fibre splicing. These are discussed as follows.

(i) Fusion Splicing Technique

Fusion splice may be made by thermally bonding together prepared fibre ends as shown in Fig. 9.17(a). The fibre ends are pre aligned and butt together in this method. This is accomplished in a grooved fibre holder or under a microscope with micro manipulators. The butt-joint is then heated with an electric arc or a laser pulse so that the fibre ends are momentarily melted and bonded together. This technique may produce an extremely small splice loss as low as 0.1 to 0.2 dB for identical fibre end. However, extreme care must be taken in this technique. If it is not followed surface damages due to handling and surface defect growth during heating may develop. A change in chemical composition may produce a weak splice. The change in chemical composition arises from the melting of the material.

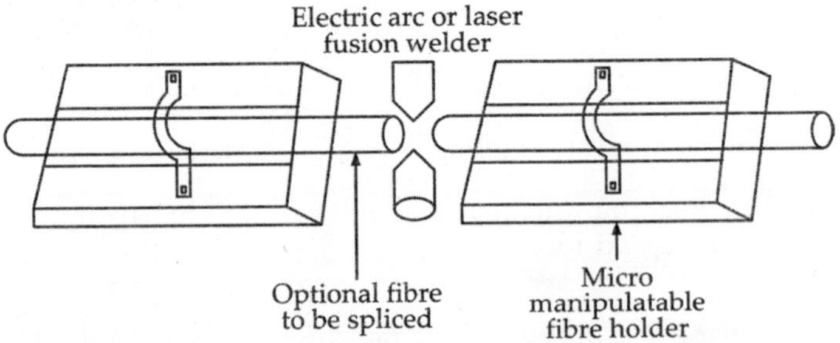

Fig. 9.17: (a) A fusion splice may be made by bonding together prepared fibre ends and the fibre ends are pre aligned and butt together

(ii) V-Groove Splice Technique

This method uses a V-shaped groove as an alignment aid to bring two fibres into mechanical alignment as shown in Fig. 9.17(b) before applying an adhesive such as epoxy or being fastened in place with a cover plate. The V-shaped channel could be either a grooved silicon, plastic ceramic or metal substrate. The loss in the splice in this method depends largely on the fibre size and eccentricity. The outside dimension and core diameter variations determine the fibre size, while the eccentricity depends on the position of the core relative to the centre of the fibre.

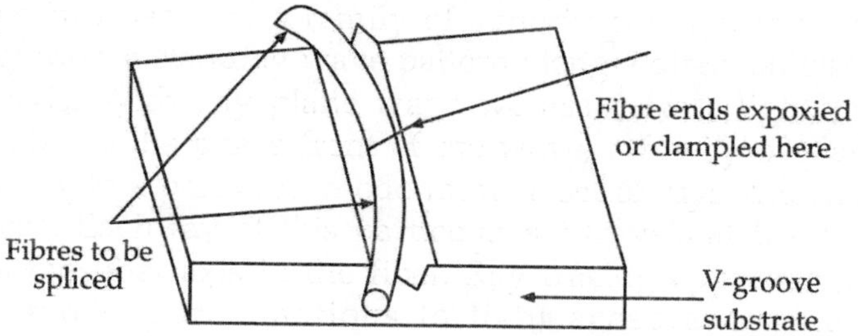

Fig. 9.17: (b) A V-grooved splicing as an alignment aid to bring two fibre into mechanical arrangement

(iii) Loose Tube Splice

The diagram for loose tube splice system is shown in Fig. 9.17(c). This is a little different from V-groove splice. The corner of a rectangular tube is used for alignment in this case. This technique uses the outside surface of the fibre as the reference surface. Hence it is prone to losses due to variations in the core concentricity and core size. The fibre ends require preparation before splicing. Losses as low as several hundredth of a dB have been reported with these techniques. However, splicing fibres of unequal diameters by this technique gives result in unacceptable misalignments.

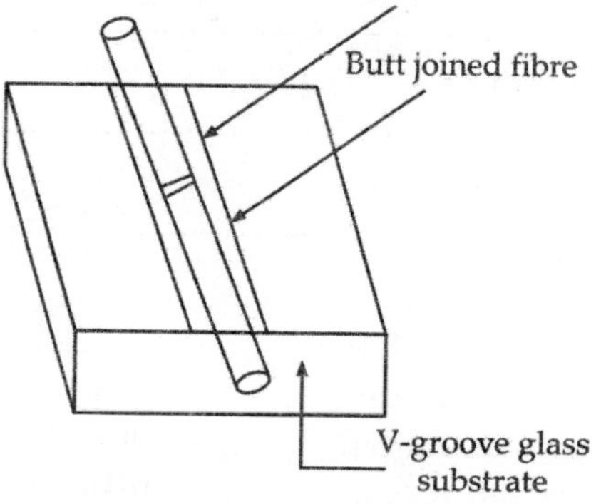

Fig. 9.17: (c) A loose tube splicing system

(iv) Elastic Tube Splice

The cross-sectional view of an elastic tube splice system is shown in Fig. 9.17(d). It is a unique device that can perform lateral, longitudinal and angular alignment automatically. It splices multimode fibres with loss having the same order of values as in commercial fusion splices but less equipments and technical skill are required. The applied mechanism basically consists of a tube made of elastic material. The central hole diameter of the tube is slightly smaller than that of the fibres to be spliced and is tapered in each end of the smooth insertion of the fibre. When the fibre is inserted into the elastic tube, it expands the hole diameter so that the elastic material exerts a symmetrical force on the fibre. The symmetry feature allows an accurate and automatic alignment of the axes of the two joined fibres .A wide range of fibre diameters can be inserted into the elastic tube. Thus, the fibres which are spliced need not be equal in diameter, since each fibre moves into position independently relative to the tube axis. This aspect is a great advantage over the V-groove splice in which fibres of unequal diameter are automatically changed.

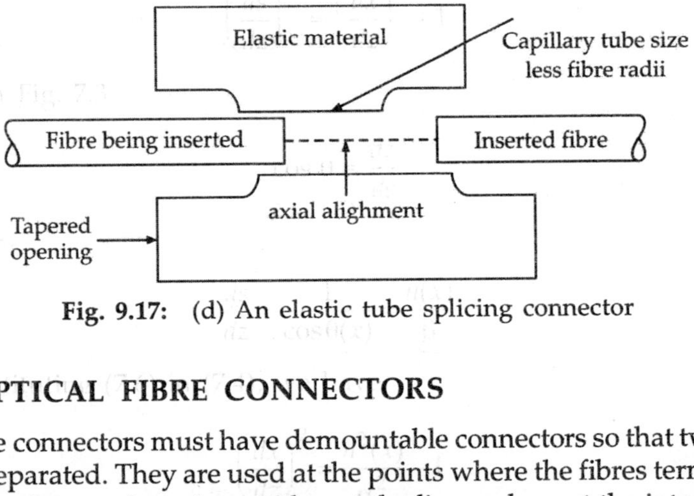

Fig. 9.17: (d) An elastic tube splicing connector

9.9. OPTICAL FIBRE CONNECTORS

The fibre connectors must have demountable connectors so that two fibres can be separated. They are used at the points where the fibres terminate at the transmitter and receivers either at the line ends or at the intermediate repeaters. These connectors must automatically ensure both a minimum loss connector and at the same time it is easy to connect or disconnect.

There are many schemes to build such connectors. Two important and commonly used practices are (a) V-groove connectors and (b) precision-bore hole in a jewel bearing. In the V-groove system lateral and angular alignment is obtained by holding the fibre ends into the bottom of a V-groove cut in a block as shown in Fig. 9.18(a). It is necessary that the fibres must have the same dimensions, otherwise the lateral core misalignment will occur. The fibres must be held in position either by a spring-loading system or by sandwiching them between two V-blocks as shown in Fig. 9.18. The V-blocks are made parts of the mating ferrules which press the fibre ends together when they are made to slide together.

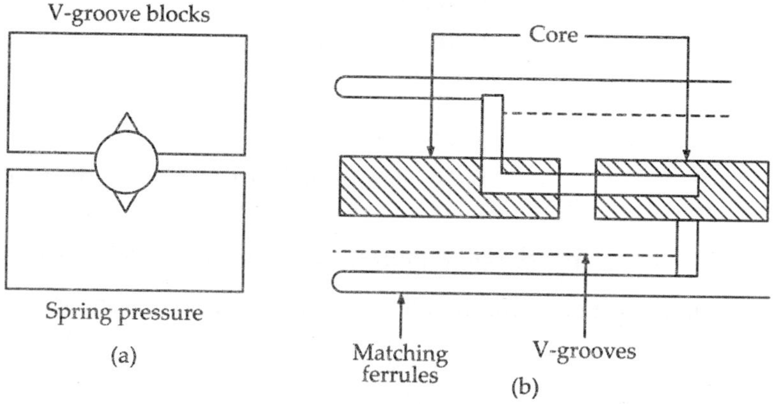

Fig. 9.18: A V-groove system connector

9.10. POLARIZER

A polarizer is an optical filter that passes light of a certain polarization and blocks light of other polarizations. It can convert an unpolarized wave or wave with mixed polarizations into a highly polarized light. The commonly available polarizers are either linear or circular. Linear polarizers, further can be of two types; beam splitting type and absorptive type. In the former, an incoming beam is divided into two beams with opposite polarization states while in the absorptive type, a component of the original unpolarized wave gets absorbed while passing through it.

A wire grid polarizer consists of a large number of thin copper wires placed parallel to each other as shown in Fig. 9.19. When an unpolarized electromagnetic wave is incident upon it, then the component of the electric vector along the length of the copper wire is absorbed. This is due to the fact that the electric field produces a force on the electrons present inside the thin wires which results in a work done upon these electrons. This work results in heating of the wires. Hence energy is lost in this heating loss. However, the component of the electric field perpendicular to the wires does no work on the electrons which makes this component not to lose energy in the form of Joule heating. Thus, the perpendicular component of the electric field passes without much attenuation. We can immediately note that the emergent wave is, therefore polarized (as the electric field now has just one component; the component perpendicular to the wires, which may be along a particular axis, for example). For the system to be effective, the spacing between the wires must be the order of at least one wavelength of the incident light. Fabrication of such a wire-grid polarizer is, therefore an easy task at microwave frequencies (as the wavelength is of the order of 3 cm). The spacing between the wires has to be ≤ 3 cm. However, as the wavelength associated with light waves is very small, the fabrication of this type of wire-grid polarizer at optical frequencies becomes extremely difficult. Instead of thin copper wires, one may place a long chain of polymer molecules that contain atoms (like iodine) giving high conductivity along the length of the chain. These long chains of polymer molecules are arranged in such a manner that they are almost parallel to each other. The component of the electric field parallel to these chains gets absorbed (due to the high conductivity of the atoms). A sheet containing such long chain polymer molecules is known as a Polaroid. When a light beam is incident upon such a Polaroid, the molecules absorb the component of the electric field parallel to the direction of alignment is absorbed while the component perpendicular to the alignment passes through. Thus the aligned conducting chain of molecules acts in a similar manner to the wires present in a wire grid polarizer. Moreover, ad the spacing between the two adjacent long chain of molecules is small compared to the wavelength, this arrangement is very effective in producing polarized light. The aligning of the long chain conducting molecules is not very difficult.

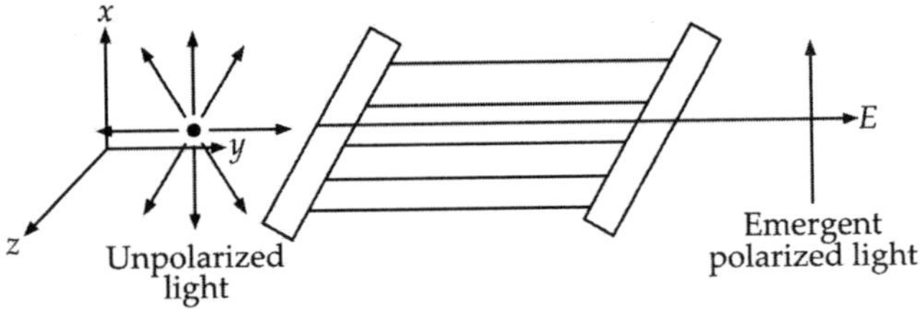

Fig. 9.19: A wire grid polarizer

9.10.1. Polarization by Reflection

Let us consider a plane wave incident on a dielectric. We further assume that the electric field associated with the incident wave lies in the plane of incidence (known as parallel polarization previously). We have seen previously that there occurs no reflection from the dielectric surface corresponding to the Brewster angle which is reproduced here for convenience.

$$\theta = \theta_p = \tan^{-1}\left(\frac{n_2}{n_1}\right) \qquad ...(9.21)$$

This implies that the reflection coefficient is zero. Thus, if an unpolarized light beam is incident at this angle, then the reflected beam will be linearly polarized with its electric vector perpendicular to the plane of incidence. This is shown in Fig. 9.20.

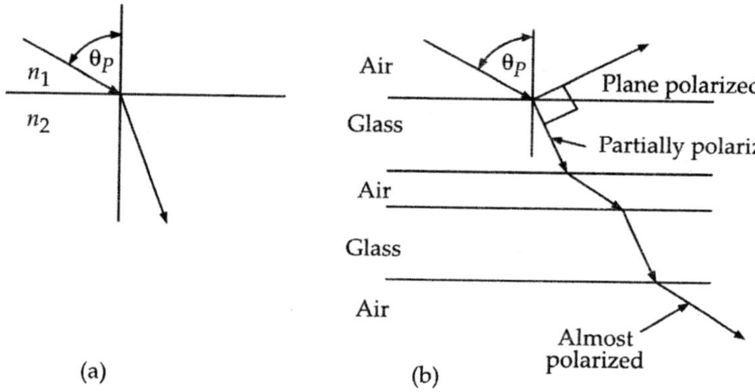

Fig. 9.20: (a) A linearly polarized wave corresponding to parallel polarization is incident on the interface of two dielectrics at the Brewster angle so that there is no reflection and all of light is transmitted (b) An unpolarized light has been incident at this angle, the reflected beam is linearly polarized having its electric field perpendicular to this plane of incidence. The transmitted beam is partially polarized. Upon multiple reflections, the emergent ray becomes almost parallel to the incident ray.

We may note from the above that θ_p is known as the polarizing angle. We may further see that corresponding to an air-glass interface for which $n_1 = 1.0$ and $n_2 = 1.5$, the polarizing angle is $\theta_p \approx 57°$. The transmitted beam is partially polarized and if we use a large number of reflecting surfaces, we obtain a nearly polarized beam.

9.10.2. Polarization by Double Refraction

When an unpolarized wave enters an anisotropic crystal, it splits into two beams, each of them having a certain state of polarization. Elimination one of the beams would lead to the production of a linearly polarized wave. A simple method of elimination is through selective absorption which is otherwise known as dichroism. A tourmaline crystal has two different coefficients of absorption for the two linearly polarized beams into which the incident beam splits up. One of the beams gets absorbed quickly and the other beam passes through without much attenuation. Therefore, when an unpolarized light beam is passed through a tourmaline crystal, the emergent beam is almost polarized. This is illustrated in Fig. 9.21.

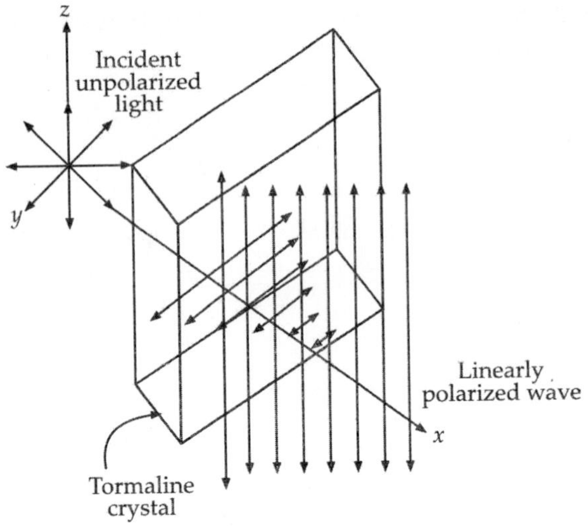

Fig. 9.21: Splitting of the incident unpolarized light beam by the tourmaline crystal into two components one of which (the y-component getting absorbed quickly while the z-component passes through)

Unpolarized light waves can also be polarized by getting scattered. If an unpolarized wave passes through a gaseous medium, then the beam scattered at 90° to the incident beam is linearly polarized.

The birefringent property of certain crystals like calcite and quartz can be exploited to realize linear polarizers. An unpolarized light beam incident at the crystal surface is split into two beams by refraction. The laws of refraction hold good for one of these beams known as the ordinary or o-beam while the extraordinary e-beam does not satisfy Snell's laws. The two beams, in general are in different polarization states though not in linear polarization states except for certain propagation directions relative to the crystal axis. These two rays also experience different refractive indices in the crystal.

An early type of polarizer based on Nicol prism is illustrated in Fig. 9.22. This is made from crystal of calcite that is split and rejoined with Canada balsam. The crystal is so cut that the e- and the o-beams are in orthogonal polarization states. The o-beam is observed to suffer from total internal reflection at the balsam interface while the e-beam passes through the crystal. Thus, the o-beam does not reach the output having undergone total internal reflection whereas the e-beam emerges as a linearly polarized wave. The o-beam suffers total internal reflection as it experiences a larger refractive index in calcite than in balsam. Another birefringent polarizer is the Wollasten prism which is constructed from two triangular pieces of calcite prisms whose orthogonal axes are joined together as shown in Fig. 9.23. An unpolarized light beam is split into two linearly polarized beams at the internal surface. These two beams leave the prism at a divergence angle of 15°-45°. The Rochon and Senarmont prisms are similar to this prism, however with the difference that they use different optical axis orientations.

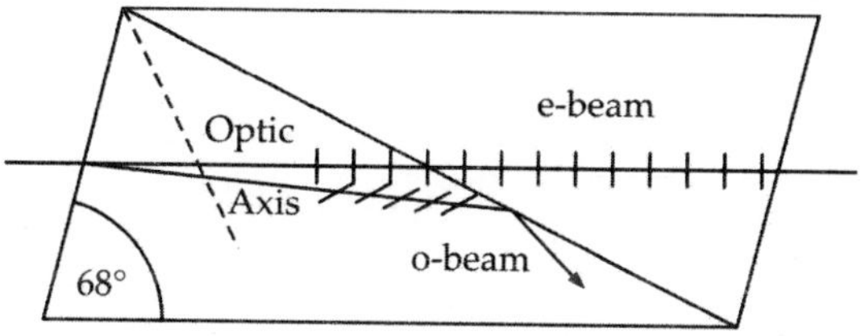

Fig. 9.22: A Nicol prism based polarizer

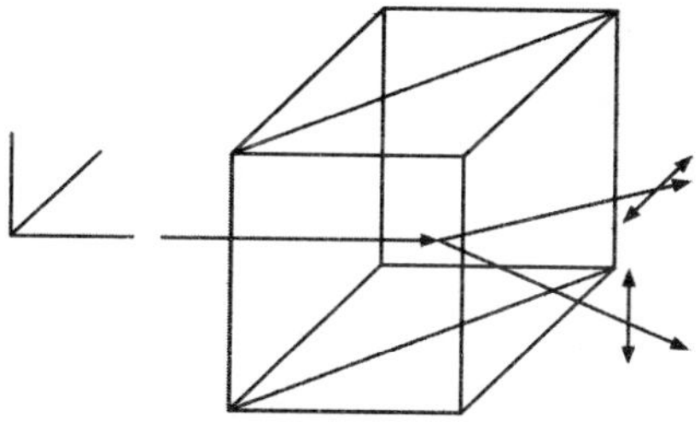

Fig. 9.23: A Wollasten prism based birefringent polarizer

9.11. FIBRE POLARIZERS

In applications such as fibre optic sensors and integrated optics, it is necessary to change the state of polarization of the light entering the device. As circular core fibres do not maintain the polarization state of the propagating medium, it is necessary to be able to polarize light before it enters the fibre device. A few in-line polarizers are discussed below.

Metal-clad polarizers

A metal-dielectric interface can support a TM guided wave (which is polarized perpendicular to the metal dielectric surface) as we saw in Ex 5.1. Such a mode is known as the surface plasmon mode. As a metal has an imaginary component of dielectric constant, such a surface Plasmon mode is lossy. At the same time, the metal-dielectric interface cannot support a truly guided TE wave as the tangential component of the electric field is zero at the metal. However, if a polished fibre half block that is polished almost up to the core of the fibre is coated with a metal, then the fibre mode with polarization perpendicular to the metal surface suffers losses due to the metal, whereas the parallel polarization will suffer much less. Using direct coating of gold or aluminium, one can achieve reasonable losses in the TM polarization.

As the metal film supports a lossy Plasmon mode, the attenuation of the TM polarization can be considerably increased by using an intermediate buffer layer, since in such a case we may consider the metal-coated fibre device as a directional coupler between the fibre mode and the metal-clad Plasmon mode. Changing the buffer refractive index and the thickness leads to changes in the Plasmon mode and fibre mode effective indices. Thus, by varying the buffer refractive index and thickness, phase matching between the Plasmon mode and the fibre mode may be attained. This would lead to an increased attenuation at a specific buffer thickness or refractive index. Compared with the TM polarization, there is no reported case of TE polarization in such metal clad polarizers. The fibre mode with its polarization parallel to the metal interface suffers much less attenuation. This leads to a polarization action of the device.

Thin metal clad polarizers: The thickness of the metal layer is usually very large compared with the field penetration in conventional metal clad polarizers. As a result, the outer metal surface not in contact with the waveguide is not active. In thin metal clad polarizers, the thickness of the metal layer is about 10-15 nm [2]. In such a case, the depth of penetration of the Plasmon mode is larger than the metal thickness and the behaviour of the polarizer is also affected by the refractive index of the overlay medium placed above the polarizer. If the refractive index of the overlay is changed, the loss of the perpendicular polarization component can be very large at some specific overlay refractive index while minimizing the loss in the horizontal component. Such a behaviour leads to a polarizing action of this device.

Birefringent crystal clad polarizer: Such a polarizer is realized by covering a side-polished fibre with a medium such that transmission of the polished fibre depends on the refractive indices of the cover medium. Thus, if the refractive index of the cover is less than the fibre mode refractive index, then total internal reflection continues to take place and there is no additional attenuation. On the other hand, if the refractive index of the cover is higher than that of the core, power from the fibre mode leaks away due to transmission to the high index layer. In case of the fibre core exposure, this leakage is simply due to partial reflection at the core-overlay interface. If a finite thickness of the cladding is still left or a thin low refractive index dielectric buffer layer is deposited on the polished fibre surface, then this is due to what is known as frustrated total internal reflection in which leaky modes are set up responsible for power loss. If a birefringent crystal is used as an overlay and if the effective index of the fibre mode lies between the ordinary and the extraordinary refractive indices of the crystal, then by properly orienting the crystal, one of the polarizations would 'see' an overlay with an index lower than the effective index and hence, would be guided.

9.12. POWER COUPLING TO AN OPTICAL FIBRE

We have seen the importance of the numerical aperture of a given fibre previously. It relates to the light gathering capacity of the fibre. We expect the fibre to gather more light if its NA is on the higher side. However, as we shall see here, not all the light generated by a source is coupled to a fibre that is 100% determined by the NA. Rather only a small fraction of the NA decides the amount of light getting coupled to the fibre. To get a quantitative result, we proceed as follows.

We should have the knowledge about the spatial radiation pattern of the source which is an LED in this case. The light emitted from the LED is incoherent and diffused. It is assumed to be emitted over an area as shown in Fig. 9.24. The centre of the emitting are is assumed to be located at the origin of the spherical coordinate system. Further, the emission is assumed to be fairly uniform over the source area though in practice the power as measured at a given radial distance R is a function of the azimuth angle ϕ and the elevation angle θ. A surface emitting LED has been assumed that is said to follow a lambertian pattern. This is similar to an omnidirectional radiator; one that radiates equally along all directions. Therefore, the power delivered at an angle θ measured relative to a normal (which is represented by the z-axis here) to the emitting surface area as shown varies as $\cos\theta$. Therefore,

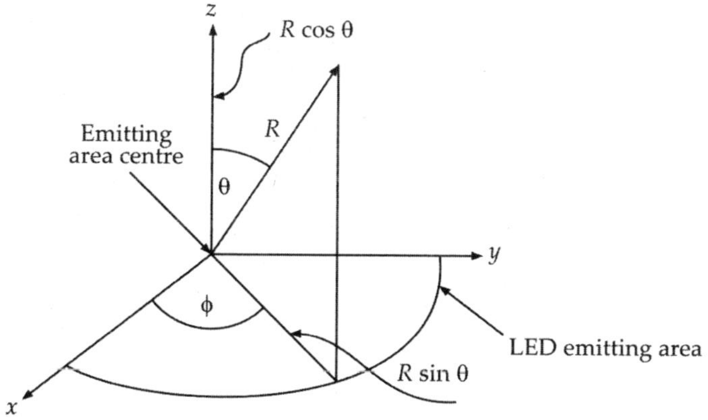

Fig. 9.24: Coordinate system used to compute the amount of light coupled to a fibre from an LED

$$I(\theta, \phi) = I_0 \cos\theta \qquad \text{...(9.22)}$$

where I_0 is the radiance along the normal to the radiating surface.

However, the edge-emitting LEDs and the laser diodes have a more complex optical radiation pattern that is a function of both the angles θ and ϕ as we may note from Fig. 9.25. These optical sources produce a radiation pattern in planes parallel and perpendicular to their emitting-junction planes. An approximation to the radiance pattern for these devices may be put in a general format as follows.

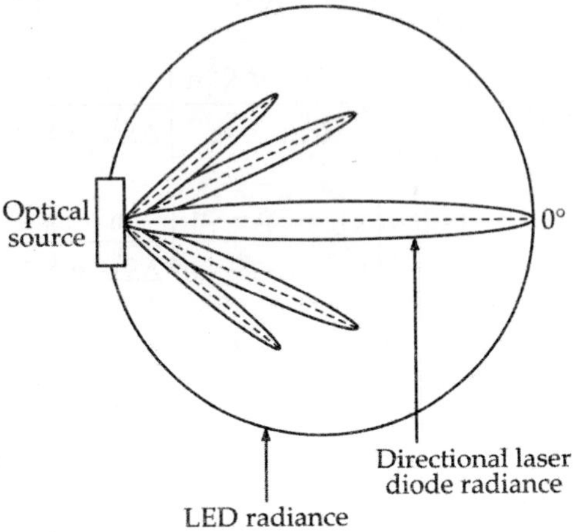

Fig. 9.25: Radiance patterns for LED and laser diodes

$$\frac{1}{I(\theta,\phi)} = \frac{\sin^2\phi}{I_0\cos^T\theta} + \frac{\cos^2\phi}{I_0\cos^L\theta} \quad ...(9.23)$$

The exponents T and L denote the transverse and the lateral power distribution coefficients respectively. For edge-emitters, $L = 1$ and T is much larger. We may interpret this as more optical power radiated along the lateral plane of the LED as compared to the transverse plane of its emitting-edge plane. Similarly, for lasers, it is the other way making the value of $L = 100$. This explains a highly directional optical power from a laser which means more light is coupled to a fibre for identical power ratings of LEDs and lasers. The emitting area of an individual point source may be considered to be $rdrd\theta_s$. For a surface-emitting LED of radius r_s less than the fibre core radius a, the total power emitted becomes

$$P = \int_0^{r_s}\int_0^{2\pi}\left(2\pi I_0 \int_0^{\theta_{max}} \cos\theta\right) r\sin\theta\, d\theta\, dr\, d\theta_s$$

$$= \pi I_0 \int_0^{r_s}\int_0^{2\pi} (NA^2) r dr d\theta_s \quad ...(9.24)$$

For a step index fibre, the NA is independent of both the radial distance from the core and the angle. Hence, (9.24) simplifies to

$$P = \pi I_0 \int_0^{r_s} \int_0^{2\pi} (NA^2) r dr d\theta_s$$

$$I_0 \pi^2 r_s^2 NA^2 = 2\pi^2 r_s^2 I_0 n_1^2 \Delta \qquad ...(9.25)$$

The power contained in $\pi^2 r_s^2 I_0$ is the total optical power radiated from a source of area $\pi^2 r_s^2$ into a hemisphere defined by $(2\pi, sr)$. Therefore, the power available for coupling into the step index fibre becomes

$$P = \begin{cases} P_s n_1^2 \Delta & \text{for } r_s < a \\ \left(\dfrac{a}{r_s}\right)^2 P_s n_1^2 \Delta & \text{for } r_s > a \end{cases} \qquad ...(9.26)$$

For a GI fibre, the NA is dependent on the radial distance r as we have observed in (6.1). Thus, the power coupled from a surface-emitting LED into such a fibre is evaluated as the following.

$$P = \pi I_0 \int_0^{r_s} \int_0^{2\pi} (NA^2)\, r dr\, d\theta_s$$

$$= 2\pi^2 I_0 \int_0^{r_s} [n^2(r) - n_2^2]\, r dr \qquad ...(9.27)$$

$$= 2\pi^2 I_0 \int_0^{r_s} \left[n_1^2 \left(1 - 2\Delta \frac{r}{a}\right)^{\alpha} - n_2^2 \right] r dr$$

$$= 2\pi^2 I_0 \left[\frac{(n_1^2 - n_2^2) r_s^2}{2} - \frac{2\Delta n_1^2}{\alpha + 2} \left(\frac{r_s}{a}\right)^{\alpha+2} \right]$$

This is simplified to

$$P = 2\pi^2 I_0 \left[\frac{(n_1^2 - n_2^2) r_s^2}{2} - \frac{2\Delta n_1^2}{\alpha + 2} \left(\frac{r_s}{a}\right)^{\alpha+2} \right]$$

$$= 2\pi^2 I_0 n_1^2 r_s^2 \left[\frac{(n_1^2 - n_2^2)}{2n_1^2} - \frac{2\Delta}{\alpha + 2} \left(\frac{r_s}{a}\right)^{\varepsilon} \right] \qquad ...(9.28)$$

$$= 2\pi^2 I_0 n_1^2 r_s^2 \left[\Delta - \frac{2\Delta}{\alpha + 2} \left(\frac{r_s}{a}\right)^{\alpha} \right]$$

$$= 2\pi^2 I_0\, n_1^2\, r_s^2 \Delta \left[1 - \frac{2}{\alpha+2}\left(\frac{r_s}{a}\right)^{\alpha}\right]$$

$$= 2P_s\, n_1^2 \Delta \left[1 - \frac{2}{\alpha+2}\left(\frac{r_s}{a}\right)^{\alpha}\right]$$

We may observe from (9.26) and (9.28) that perfect conditions of coupling between the source and the core have been assumed. This is true for the case of perfect matching of the refractive index of the intervening medium between the source and the fibre end with n_1. In case of a mismatch, there will be a loss in the coupled power the amount of loss being proportional to

$$P = \left(\frac{n_1 - n}{n_1 + n}\right)^2 \qquad \text{...(9.29)}$$

9.13. OPTOCOUPLERS

The optocoupler is a component that provides an optical connectivity between a source and a load without any physical or electrical connection. It couples an electrical input to an electrical output by optical means through the aid of a diode and a photodetector thereby providing complete electrical isolation between the input and the output. Its operation is based on light detection. The input to a coupler is connected to a light emitter and the output is a photodetector that may be photovoltaic/ photoconductive type, a photodiode or a phototransistor. Different combinations of LEDs at the input and photodetectors at the output constitute several kinds of optocouplers. An optocoupler has. The basic operation is understood with the aid of Fig. 9.26 that shows a basic diode circuit.

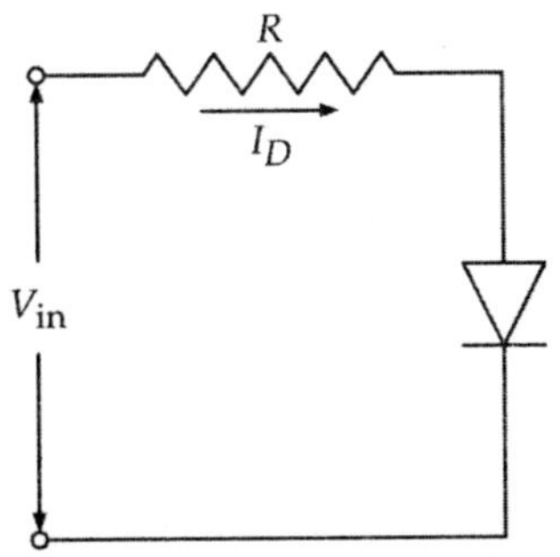

Fig. 9.26: A basic diode circuit

This is the basic circuit that needs a suitable current I_D for its proper operation and usually requires a current limiting resistor in series with the input. The value of the resistor *R*, for a given input voltage V_{in} is computed as

$$R = \frac{V_{in} - V_\gamma}{I_D} \qquad ...(9.30)$$

where V_γ is the cutin potential of the diode. Once a suitable current is established in the circuit, it is used for basically two modes of operation:, pulsed and linear. The first type finds wide applications in digital circuits while the latter is more suitable for analog signal processing applications. For digital circuits, the input LED acts like a switch that is turned on or off by the polarity of an input pulse. The output is another pulse that may be the same pulse or 180° phase with it depending on where it is fed to the load. A very basic circuit is shown in Fig. 9.27.

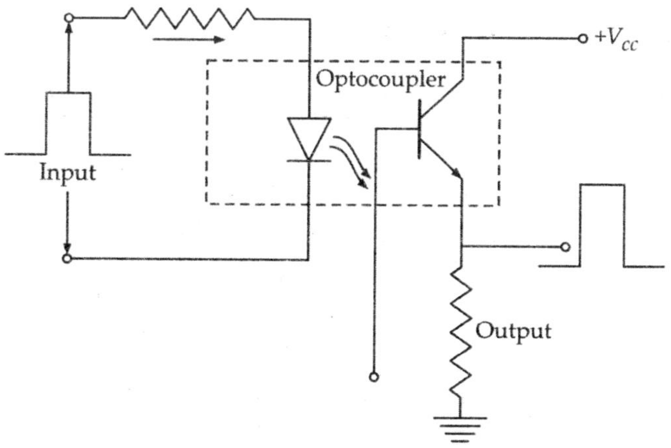

Fig. 9.27: An optocoupler based digital circuit (emitter follower)

An analog circuit realized with an optocoupler is shown in Fig. 9.28.

A look at Fig. 9.28 tells us that this circuit is more suitable for developing an analog signal at its output which is taken across an external resistor connected to the emitter leg of the transistor unlike the digital circuit shown in Fig. 9.27. Another important difference between the digital and analog circuits lies in the use of a proper biasing for the input of the optocoupler for linear mode of operation. This is provided by the use of a constant current source as shown in Fig. 9.28.

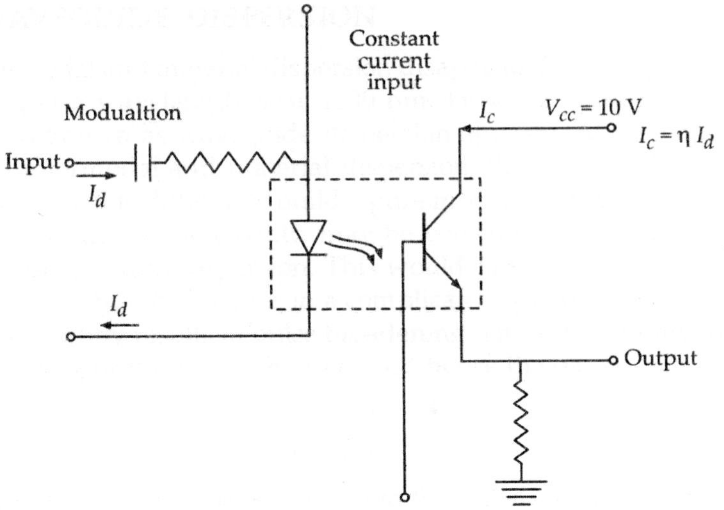

Fig. 9.28: Optocoupler based circuit for linear mode of operation

9.14. CHAPTER SUMMARY

The optical components are used primarily to connect a source and a detector to a fibre. The power available from a source is coupled to the fiber through a directional coupler. All the source power is not coupled to a fibre due to a number of losses taking place. Splicers are used to join two pieces of fibres as on a long-haul link. A misalignment of the two fibers getting connected leads to optical power loss. A polarizer is responsible for producing a polarized light beam by accepting an unpolarized light. The power coupled from a source to a fibre depends upon its numerical aperture. An optocoupler is an optical component that provided optical connectivity between an input and output without any physical or electrical connection, thereby providing an electrical isolation and eliminating the need to provide proper electrical grounding.

SUGGESTED READING

[1] Dorn, S. van. 'Fibre Optic Couplers', Proceedings of SPIE, 574, pp. 2-8, 1985.
[2] Johnstone, W., Stewart, G., Culshaw, B. and Hart, T. "Fibre Optic Polarizers and Polarizing Couplers", Electronics Letters, no. 24, pp. 894, 1988.

□□□

10

Optical Detectors

10.1. INTRODUCTION

We have discussed the LEDs and the laser diodes as the optical sources that generate optical beams in a non coherent or coherent manner. Similarly, at the receiving end, we employ optical detectors to retrieve information contained in the optical form to an equivalent electrical form. The photodetector is an optical detector that senses the luminescent power incident upon it and converts the variation of this optical power into a correspondingly varying current. The optical signal is usually weakened and distorted when it emerges from the end of the fibre. Hence the photodetector must meet very high performance requirements. Among the foremost of these requirements are a high response or sensitivity in the emission wavelength range of the optical source being used, a minimum addition of noise to the system, and a fast response speed or sufficient bandwidth to handle the desired data rate. The photodetector should also be insensitive to variations in temperature, be compatible with the physical dimensions of the optical fibre, have a reasonable cost in relation to the other components of the system, and have a long operating life.

A couple of photodetectors are in existence. Photodetectors, normally are based on two broad optoelecronic conversion principles; (i) the photoconduction property and (ii) the photovoltaic property. In the former, resistivity/conductivity of certain materials change when light of an apppropriate wavelength falls upon it. The light dependent resistor is a classic example of such a detector. These are therefore, passive type of optoelectronic converters. On the other hand, the photovoltaic types generate a dc voltage in response to light of suitable wavelength qualifying therefore as active detectors. A solar cell is a classic example of such an optoelectronic converter.

The photomultipliers, the pyroelectric detectors and semiconductor based photoconductors, phototransistors and photodiodes are noteworthy. However, many of these detectors do not meet one or more of the above requirements. Photomultipliers consist of a photocathode and an electron multiplier packaged in a vacuum tube. They are capable of very high gain and very low noise. However, their large size and high voltage requirements make them unsuitable for optical applications. Pyroelectric photodetectors involve the conversion of photons to heat. Photon absorption results in a temperature change of the detector material. This gives rise to a variation in the dielectric constant which is usually measured as a capacitance change. The response time of this detector is quite flat over a broad spectral band, but its speed is limited by the detector cooling rate after it has been subject to photon excitation. Its chief use is to detect high-speed laser pulses, and it is not suited for optical fibre systems.

The semiconductor based photodetectors are widely used for optical applications because of their small size, suitable material, high sensitivity and fast response time. Two types of photoetectors are available. These are the pin photodetector and the avalanche photodiode (APD).

Both types of photodetectors are based on reverse-biased PN junctions. In the dark, the voltage-current relationship of a typical PN junction diode is expected to follow the following expression

$$I = I_1 + I_2 = I_{s1}[\exp(qV/K_BT) - 1] + I_{s2}[\exp(qV/2K_BT) - 1] \quad ...(10.1)$$

In (10.1), I_s represents the reverse saturation current of the diode. The current I_{s1} is proportional to the squared intrinsic carrier concentration and area of the junction whereas I_{s2} is simply proportional to the intrinsic carrier concentration which becomes important in wide band-gap semiconductors. But when the junction is illuminated with light having a wavelength shorter than some threshold wavelength λ_{th}, the effect is to shift the characteristics by a constant current I_{ph}, as shown in Fig. 10.1.

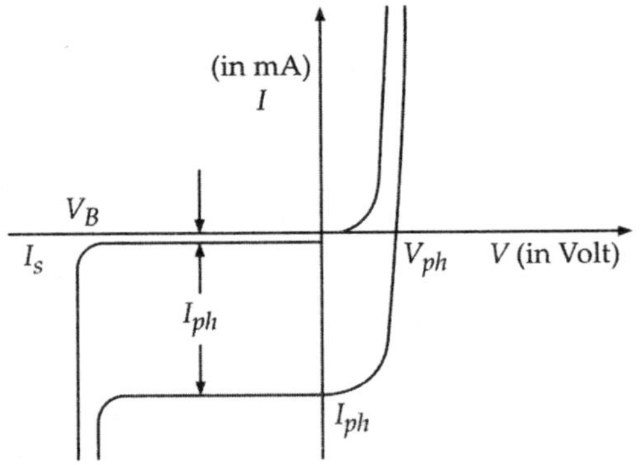

Fig. 10.1: The reverse saturation current (dark current) I_s and the opto generated current I_{ph} in response to light in a *p-n* junction diode

This is the result of a quantum interaction between the radiation and the electrons in the valence band, which causes an electron to be excited across the band-gap, an electron-hole pair to be crated, and a photon of light to be absorbed. The process is illustrated in Fig. 10.2(a).

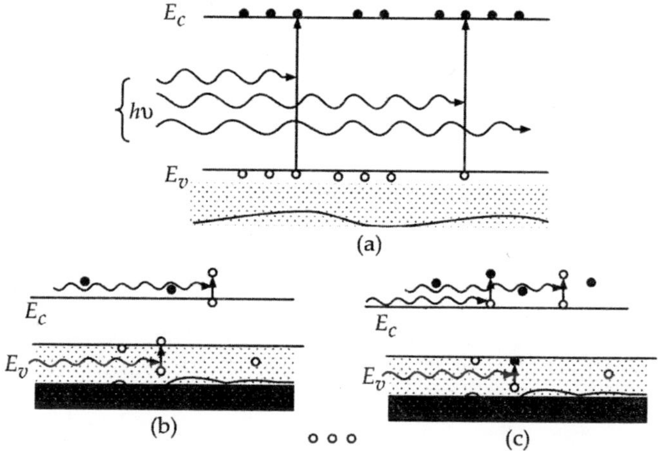

Fig. 10.2: Electron energy band diagrams illustrating optical absorption processes: (a) as a result of band-to-band excitation; (b) following excitation of donor and acceptor impurities; (c) as a result of free particle transitions

The dependence of the optical power absorption coefficient on wavelength is shown in Fig. 10.3 for several photodiode materials. Thus a particular semiconductor material can only be used over a limited wavelength range. The upper wavelength cutoff λ_c is determined by the band-gap energy E_g of the material. If E_g is expressed in eV, then λ_c is given in μm by

$$\lambda_c(\mu m) = \frac{hc}{E_g} = \frac{1.24}{E_g(eV)} \quad ...(10.2)$$

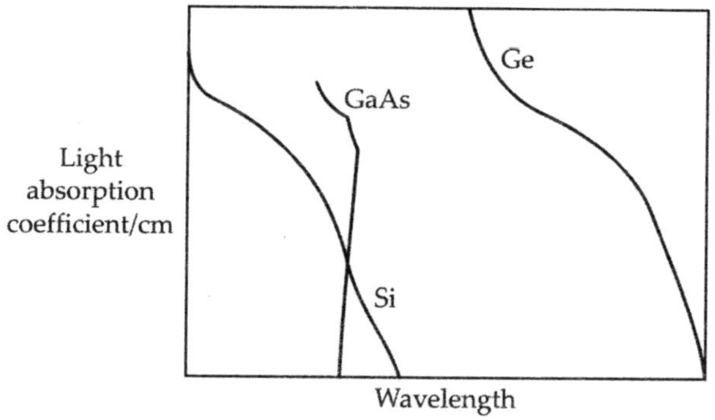

Fig. 10.3: Dependence of optical absorption coefficient on wavelength of operation for three different materials

The cutoff wavelength is about 1.06 μm for Si and 1.6 μm for Ge. For longer wavelengths the photon energy is not sufficient to excite an electron from the valence band to the conduction band.

At the lower-wavelength end, the photoresponse cuts off as a result of the very large values of α_s at the shorter wavelengths. In this case the photons are absorbed very close to the photodetector surface where the recombination time of the generated electron-hole pairs is very short. The generated carriers thus recombine before they can be collected by the photodetector circuitry. Here α_s represents the absorption coefficient and is found to be a function of wavelength.

Table 10.1: Some commonly used materials for use as photodetectors and the corresponding bandgap energy alongwith the cutoff wavelengths

Material type	*Bandgap energy E_g (in eV)*	*Cutoff wavelength λ_c (in μm)*
Si	1.11	1.13
Ge	0.67	1.85
GaAs	1.43	0.87
InGaAs	0.75	1.65

We observe from Table 10.1 that the cutoff wavelength is about 1.16 μm for Si and 1.85 μm for Ge. For longer wavelengths the photon energy is not sufficient to excite an electron from the valence band to the conduction band.

10.2. THE *p-i-n* PHOTODETECTOR

Silicon is the main material used to construct the photodetectors in the wavelength range of 800 to 900 nm region. The semiconductor photodetector is illustrated in Fig. 10.4. The device structure consists of *p* and *n* regions separated by a very lightly *n*-doped intrinsic (i) region. In normal operation, a sufficiently large reverse-bias voltage is applied across

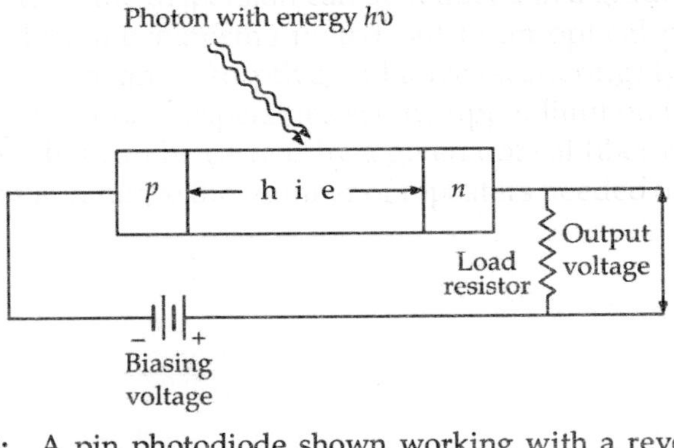

Fig. 10.4: A pin photodiode shown working with a reverse bias

the device so that the intrinsic region is fully depleted of carriers. That is, the intrinsic n and p carrier concentrations are negligibly small in comparison with the impurity concentration in this region.

When an incident photon has energy greater than or equal to the band-gap energy of the semiconductor material, the photon can give up its energy and excite an electron from the valence band to the conduction band. This process generates free electron-hole pairs from the valence which are known as photocarriers, since they are photon generated charge carriers as shown in Fig. 10.5. The photodetector is normally designed so that these carriers are generated mainly in the depletion region where most of the incident light is absorbed. The high electric field present in the depletion region causes the carriers to separate and be collected across the reverse-biased junction. This gives rise to a current flow in an external circuit, with one electron flowing for every carrier pair generated. This current flow is known as photocurrent.

As the charge carriers flow through the material, some electron-hole pairs will recombine and hence disappear. On the average, the charge carriers move a distance L_n or L_p for electrons and holes respectively.

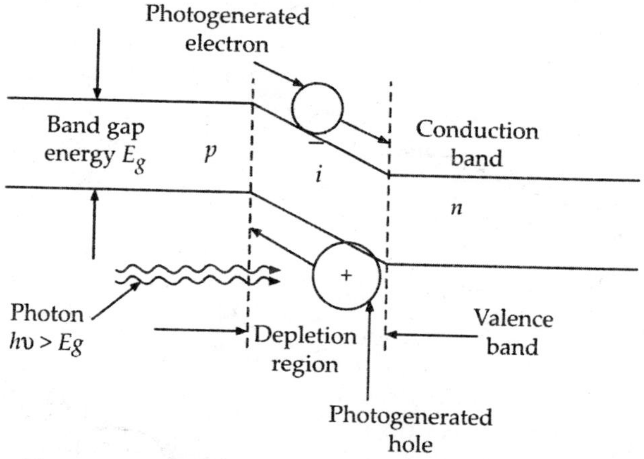

Fig. 10.5: Energy band diagram for a pin photodiode. Photons having energy at least equal to or greater than the band gap energy are observed to release a free electron-hole pair which act as photocurrent carriers

This distance is known as the diffusion length. The time it takes for an electron or hole to recombine is known as the carrier lifetime and is represented by τ_n and τ_p respectively. The lifetimes and the diffusion lengths are related by

$$L_n = \sqrt{D_n \tau_n} \text{ and } L_p = \sqrt{D_p \tau_p} \quad ...(10.3)$$

where D_n and D_p are the electron and hole diffusion coefficients respectively, which are expressed in centimetres squared per second.

Optical radiation is absorbed in the semiconductor material according to the exponential law

$$P(x) = P_0(1 - e^{-\alpha_s(\lambda)x}) \qquad ...(10.4)$$

Here $\alpha_s(\lambda)$ is the absorption coefficient at a wavelength λ, P_0 is the incident optical power level and $P(x)$ is the optical power absorbed in a distance x.

In the case of materials having a direct band-gap, there is no constraint on the carrier-pair creation process, and the absorption coefficient increases sharply for wavelengths just shorter than λ_{th}. With indirect band-gap materials, such as silicon and germanium, a simultaneous interaction with the crystal lattice is required in order that momentum may be conserved in the carrier-generation process. This may involve a lattice vibration, an impurity, or some other form of lattice disorder, but the result is that α_s increases only gradually with wavelength immediately below λ_{th}. At shorter wavelengths such that excitation across the direct band-gap becomes possible (0.8 eV in the case of Ge and 3.4 eV for Si), there is another sharp increase in α_s.

We may deduce from Fig. 10.1 that a junction diode may be used to detect optical radiation in several different ways. The one that is perhaps the simplest is sometimes called photovoltaic or open-circuit mode. The diode is connected directly across the input of a voltage amplifier with a high input impedance, and the variation of V_{ph} is measured. An alternative arrangement is the short-circuit mode, in which the diode current is amplified by a current amplifier having a very low input resistance. The diode voltage is thus maintained near to zero. This, in theory, causes the least random noise to be superimposed on the diode current. In practice, the photodiodes used in optical communication systems employ neither of the modes, but are always reverse biased. This is because the detection efficiency and the bandwidth are then significantly improved for reasons we discuss later. When the reverse-bias voltage is increases to values which approach the breakdown voltage, V_B, the photogenerated current is multiplied as a result of the same carrier avalanche ionization processes that give rise to breakdown. This region is shown in Fig. 10.2.

If the depletion region has a width w, then the total power absorbed in the distance is

$$P(w) = P_0(1 - e^{-\alpha_s w}) \qquad ...(10.5)$$

If we take into account a reflectivity R_f at the entrance face of the photodiode, then the primary photocurrent I_p resulting from the power absorption of the previous equation is

$$I_p = \frac{q}{h\upsilon} P_0 (1 - e^{-\alpha_s w})(1 - R_f) \qquad ...(10.6)$$

where P_0 is the optical power incident on the photodetector, q is the electron charge and $h\upsilon$ is the photon energy.

Example 10.1: Show that (10.6) represents the photogenerated current.

Solution: We note that as a part R_f of the incident optical power is reflected, the power remaining to generate electron hole pairs is $(1 - R_F)P_w$. Taking into account a one-to-one correspondence between a photon of energy $h\upsilon$ and one electron-hole pair, we find that the total number of photons available for carrier generation becomes equal to

$$n = \frac{P_w \times T}{h\upsilon}$$

$$\Rightarrow \qquad I_p = \frac{nq}{T}$$

$$= \frac{P_w T q}{h\upsilon T} = \frac{q P_w}{h\upsilon}$$

$$= \frac{q}{h\upsilon} P_0 (1 - e^{-\alpha_s w})(1 - R_f)$$

Two important characteristics of a photodetector are its quantum efficiency and its response speed. These parameters depend on the material band gap, the operating wavelength, and the doping and thickness of the p, i and n regions of the device.

Here I_p is the average photocurrent generated by a steady-state average optical power P_0 incident on the photodetector. The photogenerated current I_{ph} shown in Fig. 10.1, increases linearly with the incident optical power, P_0, over many orders of magnitude. The ratio of I_{ph} to P_0 is known as the responsivity, $\mathfrak{R}$, of the photodiode. Because of the quantum nature of the interaction, it is customary also to define a quantum efficiency. The quantum efficiency η is the number of electron-hole carrier pairs generated per incident photon of energy $h\upsilon$ and is given by η = number of electron-hole pairs generated/number of incident photons

$$\eta = \frac{I_{ph}/q}{P_0/h\upsilon} \qquad ...(10.7)$$

The excess carriers generated by absorption recombine on average after the mean minority carrier lifetime, τ_n or τ_p which is a function of the type, quality and doping concentration of the semiconductor material. For the incoming optical power to be converted in the detector into an electrical signal, it is necessary that the excess electrons and holes generated should suffer some net separation before they recombine. One of the simplest ways of achieving this is to ensure that some of the carriers are generated within the depletion layer of a reverse-biased PN junction. The electrons and holes are then swept apart and collected at the contacts before they have time to recombine. In consequence a current is induced in the external biasing circuit.

In the ideal photodiode all the incident light would be absorbed in this way in the depletion layer and all the carriers generated would be collected. The quantum efficiency would then be unity, and the responsivity would be $q/E_{ph} = q\lambda/hc$. Of course, in practice some of the incident light is reflected at the semiconductor surface and not all of the light entering the semiconductor is absorbed in the depletion layer. This is illustrated in Fig. 10.6.

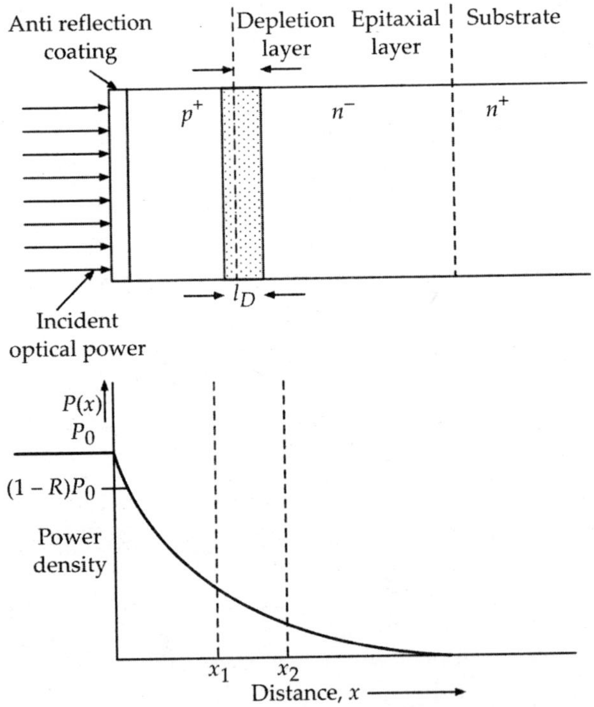

Fig. 10.6: Attenuation of the optical power density with depth below the incident surface. The effect of minimizing surface reflection and the distance x_1 and maximizing the distance $(x_2 - x_1)$ by increasing the depletion layer thickness is illustrated

Obtaining a high value of quantum efficiency depends on:

(i) Minimizing reflection at the incident surface

(ii) Maximizing the absorption within the depletion layer and

(iii) Avoiding recombination before the carriers are collected

It is known that about 33% of the light generated within a semiconductor source suffers Fresnel reflection at the semiconductor-air interface. Similarly, about 33% of the light incident on to semiconductor from the outside is reflected. To minimize reflection, it has become normal practice to coat the semiconductor surface with a film of transparent dielectric about a quarter of a wavelength thick. Ideally, the refractive index of the film should be the square root of the semiconductor refractive index. In practice, thin films of silica ($\mu = 1.46$) are more convenient and give a worthwhile improvement in the quantum efficiency.

We may estimate the fraction of the incident power that is usefully absorbed. This is the power absorbed within about a diffusion length, L_n or L_p on either side of the depletion-layer boundaries. Let these bounds to the 'useful region' be at depths x_1 and x_2 below the surface. Let the incident power density be P_0 and the surface reflection coefficient be R. Then the power density entering the semiconductor is $(1 - R)P_0$ and the power density usefully absorbed is given by the difference between that entering the effective region at x_1 and that leaving at x_2, namely,

$$P = (1-R)P_0[\exp(-\alpha_s x_1) - \exp(-\alpha_s x_2)] \qquad ...(10.8)$$

Thus the quantum efficiency is

$$\eta = P/P_0 = (1-R)\exp(-\alpha_s x_1)[1-\exp\{-\alpha_s\}(x_1 - x_2)] \qquad ...(10.9)$$

At first sight it would seem that x_1 should be as small, and x_2 as large, as possible. The effect of surface recombination means that a compromise has to be reached.

Carrier lifetime τ_n or τ_p, is normally much greater than the carrier transit time across the depletion layer, so that the fraction of carriers generated there that is lost by recombination is negligible. Carriers generated within a diffusion length of the depletion layer tend to diffuse into the high field region over a period of the order of the recombination lifetime. Should these be an appreciable fraction of the total, the quantum efficiency is reduced by the recombination that does occur, and is reduced further at modulation frequencies above $1/2\pi\tau_n$ or $1/2\pi\tau_p$.

In a practical photodiode, 100 photons will create between 30 and 95 electron-hole pairs, thus giving a detector quantum-efficiency ranging from 30 to 95 percent. To achieve a high quantum efficiency, the depletion layer must be thick enough to permit a large fraction of the incident light to be absorbed. However, the thicker the depletion layer, the longer it takes for the photogenerated carriers to drift across the reverse-biased junction. Since the carrier drift time determines the response speed of the photodiode, a compromise has to be made between response speed and quantum efficiency. The performance of a photodiode is often characterized by the responsivity $\mathfrak{R}$. This is related to quantum efficiency by

$$\mathfrak{R} = \frac{I_p}{P_0} = \frac{\eta q}{h\upsilon} \quad \text{...(10.10)}$$

This parameter is quite useful, since it specifies the photocurrent generated per unit optical power. Typical pin photodiode responsivities as a function of wavelength are shown in Fig. Representative values are 0.65 μA/μW for silicon at 900 nm, 0.45 μA/μW for germanium at 1.3 m. and 0.6 μA/μW for InGaAs at 1.3 μm.

In most photodiodes the quantum efficiency is independent of the power level falling on the detector at a given photon energy. Thus the responsivity is a linear function of the optical power. This means that the photocurrent I_p is directly proportional to the optical power P_0 incident upon the photodetector so that the responsivity $\mathfrak{R}$ is constant at a given wavelength. The quantum efficiency is not constant at all wavelengths, since it varies according to the photon energy. Consequently, the responsivity is a function of the wavelength and of the photodiode material (since different materials have different band gap energies). For a given material, as the wavelength of the incident photon becomes longer, the photon energy becomes less than that required to excite an electron from the valence band to the conduction band. The responsivity thus falls off rapidly beyond the cutoff wavelength.

10.2.1. Impulse and Frequency Response of a PIN detector

A normal biasing arrangement for a photodiode is illustrated in Fig. 10.7. The output of the photodiode is observed to be fed as the input to an amplifier. The components R_A and C_A represent the amplifier input impedance. A time-varying optical power $P_0(t)$ produces a time-varying photodiode current $i_p(t)$ and produces a voltage $v_{in}(t)$ at the amplifier input terminals.

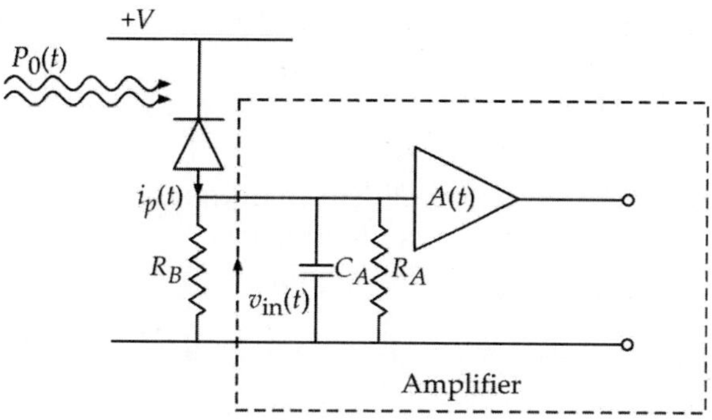

Fig. 10.7: Normal photodiode biasing arrangement

An equivalent circuit for the reverse biased photodiode of Fig. 10.7 is shown in Fig. 10.8. The small shunt conductance G_D represents the slope of the reverse biased characteristic of such a diode. The series resistance R_S is due to the bulk of the semiconductor, the contacts and the leads and is typically less than 10 Ω. The capacitance C_D is due to the junction, the leads and the packaging and can usually be kept below 1 pF. The value this capacitance should be as small as realizable in order to make high frequency operation possible. This in turn determines that the junction area should be no greater than necessary.

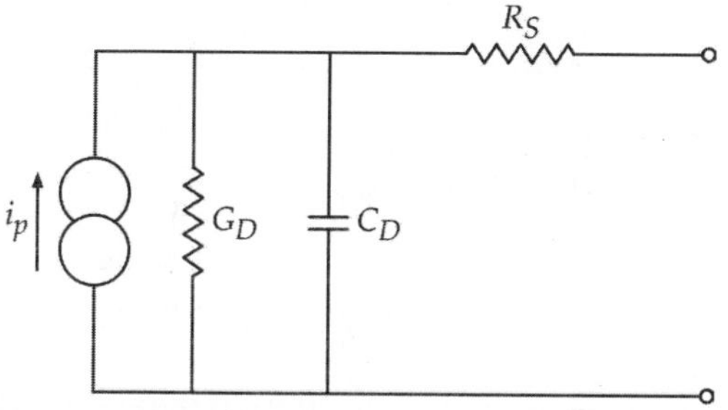

Fig. 10.8: Small-signal equivalent circuit for a reverse-biased photodiode

An equivalent circuit for the biased photodiode connected directly to a voltage amplifier is shown in Fig. 10.9. In addition to the components already described, a distributed stray capacitance C_S is already added. The bias resistance R_B appears across the diode terminals in the equivalent circuit. The circuit of Fig. 10.8 reduces to that of Fig. 10.9 under the assumption of a simplified condition of $R_S \ll R_A$, as the shunt components can be lumped together. Thus,

$$\frac{1}{R_A} + \frac{1}{R_B} + G_D = \frac{1}{R} \qquad \text{...(10.11)}$$

and

$$C = C_D + C_S + C_A \qquad \text{...(10.12)}$$

Usually $1/R_B$ and G_D are much smaller than $1/R_A$, so that $R \approx R_A$.

When the photogenerated current is sinusoidally varying at a frequency f and is represented by the phasor $I_{ph}(f)$, then the response of the load voltage, $V_{in}(f)$ is expressed as

$$\frac{V_{in}(f)}{I_{ph}(f)} = \frac{R}{1 + j2\pi f RC} \qquad ...(10.13)$$

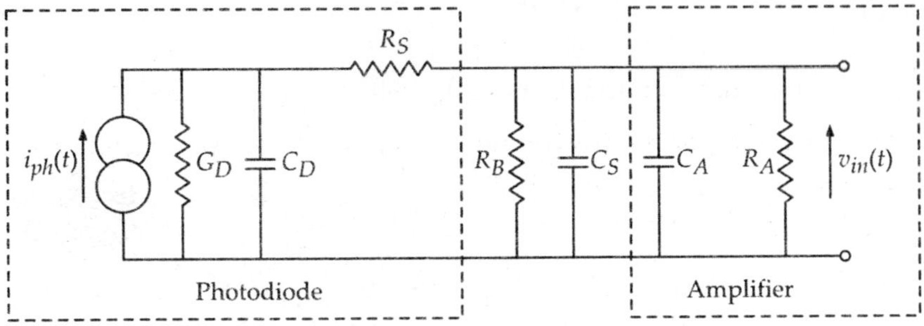

Fig. 10.9: Small-signal equivalent circuit for a normally biased photodiode and amplifier

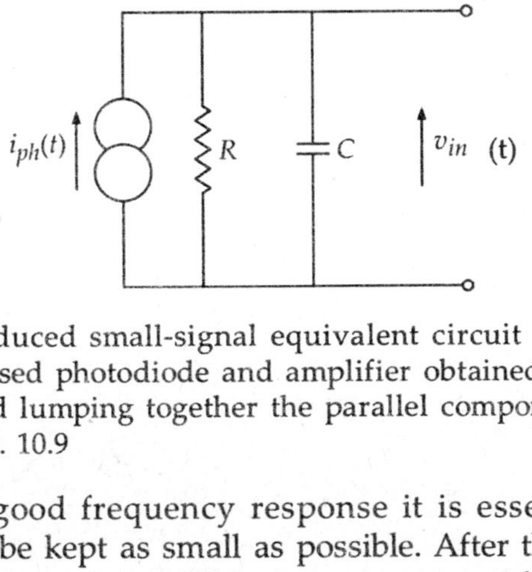

Fig. 10.10: Reduced small-signal equivalent circuit for a normally biased photodiode and amplifier obtained by neglecting and lumping together the parallel components shown in Fig. 10.9

Hence for a good frequency response it is essential that all the components of C be kept as small as possible. After that it is necessary either to reduce R so that $2\pi fRC \ll 1$, or to provide high frequency equalization in the later stages of amplification. The merits of these two techniques have been the subject of much debate.

The response of a photodiode to an optical pulse of some 10-20 ns duration is typically as shown in Fig. 10.11. We expect the voltage, v_{in} to rise or fall exponentially with a time constant RC following a step change in the photogenerated current. However, two further factors affect the response:

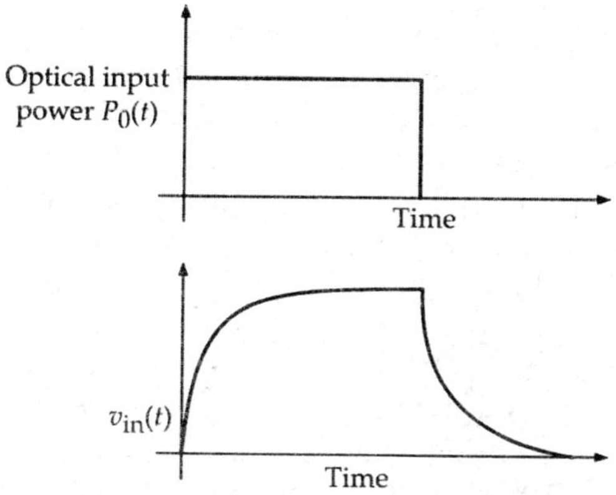

Fig. 10.11: Typical photodiode response to an optical pulse

(i) The transit time resulting from the drift of carriers across the depletion layer

(ii) The delay that results while carriers generated outside the depletion layer diffuse into it

10.3. AVALANCHE PHOTODIODE (APD)

The APDS are high speed, high sensitivity photodiodes that utilize an internal gain mechanism made possible by applying high values of reverse bias in order to enhance the number of carriers. Compared to their pin diode counterparts, they can measure even lower level light and are used in a wide requiring high sensitivity variety of applications such as long distance optical communications and optical distance measurement. These photodiodes internally multiply the primary signal photocurrent before it enters the input circuitry of the following amplifier. This increases receiver sensitivity, since the photocurrent is multiplied before encountering the thermal noise associated with the receiver circuit. In order for carrier multiplication to take place, the photogenerated carriers must traverse a region where a very high electric field is present. In this high-field region a photogenerated electron or hole can gain enough energy so that it ionizes bound electrons in the valence band upon colliding with them. This carrier multiplication mechanism is known as impact ionization. The newly created carriers are also accelerated by the high electric field, thus gaining enough energy to cause further impact ionization. This phenomenon is the avalanche effect. Below the diode breakdown voltage a finite total number of carriers are created, whereas above breakdown the number can be infinite.

When electron-hole pairs are generated in the depletion layer of a photodiode with a reverse bias applied to the PN junction, the electrons drift toward the N^+ side whereas the holes drift toward the P^+ side under the influence of the electric field developed across the junction. As we may see, the drift velocity of these charge carriers is proportional to the electric field strength which in turn is determined by the applied reverse bias. When the electric field is raised to a certain level, the charge carriers are more likely to collide with the crystal lattice so that the drift velocity gets saturated to some average velocity. If the reverse voltage is increased beyond this value, some of the carriers that escaped collision with the crystal lattice are more likely to have a higher amount of energy. New pairs of electron-holes get generated upon collision with the crystal lattice on the part of these highly energized carriers. These electron-hole pairs now generate more pairs leading to an avalanche kind of phenomenon. And the current rises steeply. This phenomenon begins to manifest when the electric field strength reaches 2×10^5 V/cm. The number of electron-hole pairs generated during the time the carriers travel a given distance is referred to as the ionization rate. Usually the ionization of the electrons is referred to as α and that of the holes as β. These ionization rates are important in determining the avalanche multiplication mechanism. The ratio κ of β to α is called the ionization ratio and is used as a parameter to indicate device noise.

A commonly used structure for achieving carrier multiplication with very little excess noise is the reach-through construction. as shown in Fig. 10.12. The reach-through APD is composed of a high-resistivity p-type material deposited as an eptaxial layer on a p^+ (heavily-doped type) substrate. A p-type diffusion or ion implant is then made in the high-resistivity material followed by the construction of an n^+ (heavily doped n-type) layer. For silicon, the dopants used to form these layers are normally

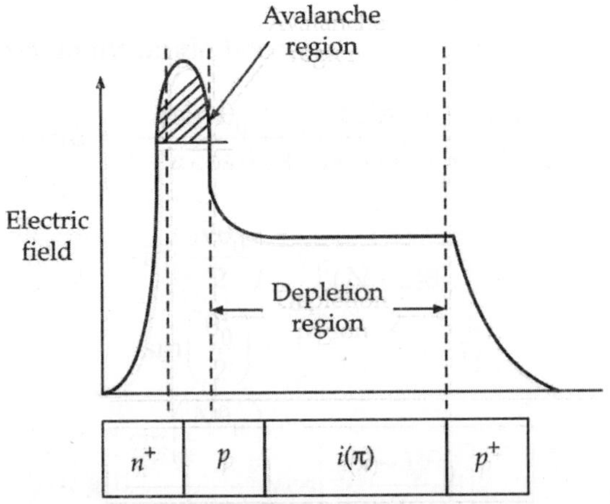

Fig. 10.12: Reach through APD structure and electric fields in the depletion and multiplication region

boron and phosphorus, respectively. This configuration is referred to as $p^+\pi pn^+$ reach-through structure. The π layer is basically an intrinsic material that inadvertently has some p doping because of imperfect purification.

The term reach-through arises from the photodiode operation. When a low reverse-bias is applied, most of the potential drop is across the pn^+ junction. The depletion layer widens with increasing bias until a certain voltage is reached at which the peak electric field at the pn^+ junction is about 5 to 10 percent below that needed to cause avalanche breakdown. At this point the depletion layer just reaches through to the nearly intrinsic π region.

In normal usage the reach-through APD is operated in the fully depleted mode. Light enters the device through the p^+ region and is absorbed in the π material which acts as the collection region for the photogenerated carriers. Upon being absorbed the photon gives up its energy, thereby creating electron-hole pairs, which are then separated by the electric field in the π region. The photogenerated electrons drift through the π region to the pn^+ junction where a high electric field exists. It is in this high-field region that carrier multiplication takes place.

The average number of electron-hole pairs created by a carrier per unit distance is called the ionization rate. Most materials exhibit different electron ionization rates α and hole ionization rates β. The ratio $\kappa = \alpha/\beta$ of the two ionization rates is a measure of the photodetector performance. The APDs constructed of materials in which one type of carrier largely dominates impact ionization exhibit low noise and large gain-bandwidth products. Of all the materials examined so far, only silicon has a significant difference between electron and hole ionization rates.

The multiplication M for all carriers generated in the photodiode is defined by

$$M = \frac{I_M}{I_P} \quad \text{...(10.14)}$$

where I_M is the average value of the total multiplied output current and I_P is the primary unmultiplied photocurrent defined in (10.6). In practice, the avalanche mechanism is a statistical process, since not every carrier pair generated in the diode experiences the same multiplication. Thus the measured value of M is expressed as an average quantity.

The current gain varies with wavelength and is a function of bias voltage for a silicon reach-through APD. The dependence of the gain on the excitation wavelength is attributable to mixed initiation of the avalanche process by electrons and holes when most of the light is absorbed in the

n^+p region close to the detector surface. This is especially noticeable at short wavelengths, where a larger portion of the optical power is absorbed close to the surface than at longer wavelengths. Since the ionization coefficient for holes is smaller than that for electrons in silicon, the total current gain is reduced at the short wavelengths. The responsivity of an APD is characterized by its responsivity

$$\Re_{APD} = \frac{\eta q}{h\upsilon} M = \Re_0 M \qquad ...(10.15)$$

where $\Re_0$ is the unity gain responsivity.

In the 850-nm range, silicon APDs typically require bias voltages of around 250 V to achieve an optimum gain of 100, whereas PIN type of photodetectors require bias voltages of only 10-50 V. At longer wavelengths (1300-1500 nm) the APDs require typical bias of 20-30 V for typical gains of 10-30, whereas PIN devices require only 5-15 V. A typical circuit to achieve this optoelectronic conversion is depicted in Fig. 10.13.

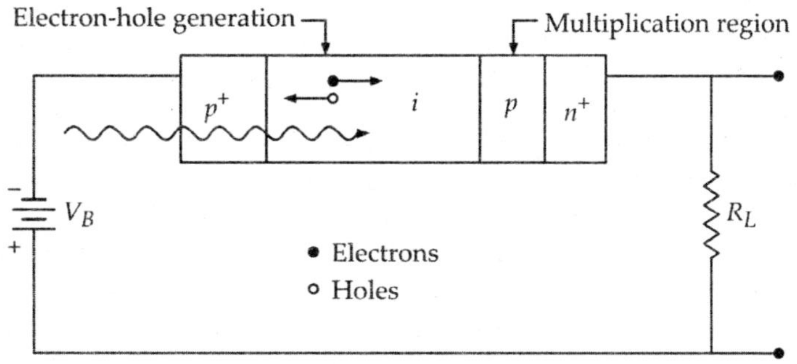

Fig. 10.13: An APD that is reverse biased to a high voltage

10.4. PHOTODETECTOR NOISE

The photodiode is used to detect very weak optical signals. Detection of the weakest possible optical signals requires that the photodetector and its following amplification circuitry be optimized so that a given signal-to-noise ratio (SNR) is maintained. The SNR at the output of an optical receiver is defined as

$$\frac{S}{N} = \frac{\text{Signal power from photocurrent}}{\text{Photodetector noise power} + \text{amplifier noise power}} \qquad ...(10.16)$$

The noise sources in the receiver arise from the photodetector noises resulting from the statistical nature of the photon-to-electron conversion process and the thermal noises associated with the amplifier circuitry.

The following conditions should be satisfied in order to achieve a high SNR.

(a) The photodetector must have a high quantum efficiency to generate a large number of electron-hole pairs thereby giving a large signal power.

(b) The photodetector and the amplifier noise should be kept as small as possible.

For most applications it is the noise currents that determine the minimum optical power that can be detected, since the photodiode quantum efficiency is normally close to its maximum possible value.

The sensitivity of a photodetector in an optical fibre communication system is described in terms of the minimum detectable optical power. This is the optical power necessary to produce a photocurrent of the same magnitude as the root mean square of the total noise current or, equivalently, an SNR of unity. We next discuss the various sources of noise in an optical detector.

10.4.1. Noise Sources

Let us have a look at the simplest receiver model as shown in Fig. 10.14 and its equivalent circuit in order to understand the role of various noises in a photodetector.

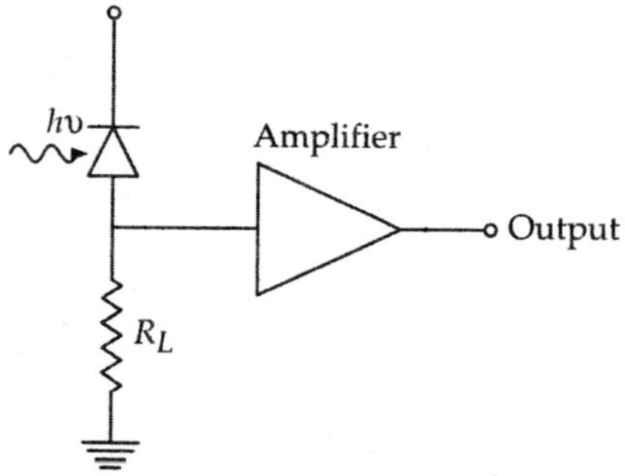

Fig. 10.14: A simple circuit model for photodetector noise

The photodiode has a small series resistance R_s, a total capacitance C_d consisting of junction and packaging capacitances, and a bias or load resistor R_L. The amplifier following the photodiode has an input capacitance C_a and a resistance R_a. For practical purposes, R_s is much smaller than the load resistance R_L and can be neglected.

If a modulated signal of optical power P_o is incident upon the photodiode, then the primary photocurrent $i_p(t)$ generated is given as, as we have seen before

$$i_p(t) = \Re P_o = \frac{\eta q}{h\upsilon} P_o \qquad ...(10.17)$$

We may observe from this that the squared current has an average value given as

$$\left\langle i_p^2(t) \right\rangle = \left\langle \Re P_o \right\rangle^2 = \Re^2 \left\langle P_o^2 \right\rangle \qquad ...(10.18)$$

< . > indicates averaging operaters.

This primary photocurrent consists of a dc value I_P that represents the average photocurrent due to the signal power, and a signal component $i_P(t)$. For PIN photodiodes the mean square signal current is

$$\left\langle i_s^2 \right\rangle = \left\langle i_P^2(t) \right\rangle \qquad ...(10.19)$$

whereas for APD type of photodiodes,

$$\left\langle i_s^2 \right\rangle = M^2 \left\langle i_P^2(t) \right\rangle \qquad ...(10.20)$$

where M is the average of the statistically varying avalanche gain as defined in 10.14. For a sinusoidally varying input signal of modulation index m_a, the signal component takes the form of the following:

$$\left\langle i_P^2(t) \right\rangle = \frac{m_a^2}{2} I_P^2 \qquad ...(10.21)$$

This is proved as follows:

Let the power incident on the photodiode be

$$P_o = P(1 + m_a \cos \omega t)^2 \qquad ...(10.22)$$

(10.22) when expanded, becomes

$$P_o = P^2 (1 + m_a^2 \cos^2 \omega t + 2m_a \cos \omega t) \qquad ...(10.23)$$

This power generates a photocurrent that would now consists of the following three terms in response to the three power terms. The constant power P results in a dc current which has an average value given by

$$I_P = \Re P$$

$$\Rightarrow \qquad I_P^2 = (\Re P)^2 \qquad ...(10.24)$$

The mean square value of the photocurrent generated in response to the second term is found by averaging it over one period assuming the period to be T. The average value of this squared cosine term is 1/2.

$$\frac{1}{T}\int_0^T \cos^2 \omega t = \frac{1}{2T}\int_0^T (1+\cos 2\omega t)dt \qquad ...(10.25)$$

$$= \frac{1}{2}$$

Therefore,

$$\left\langle m_a^2 \cos^2 \omega t \right\rangle = \frac{1}{2} m_a^2 \qquad ...(10.26)$$

and the current generated has a squared average given as

$$i_p^2 = \frac{1}{2} m_a^2 (\Re P_o)^2 \qquad ...(10.27)$$

The contribution of the third term to the squared average current is zero.

The main noises associated with photodetectors having no internal gain are quantum noise, dark current noise generated in the bulk material of the photodiode, and surface leakage current noise. The quantum or shot noise arises from the statistical nature of production and collection of photoelectrons when an optical signal is incident on a photodetector. It has been found that these generations follow a Poisson process. Since the fluctuations in the number of photocarriers created from the photoelectric effect are a fundamental property of the photoelectric process, they set the lower limit on the receiver sensitivity when all other conditions are optimized. The quantum noise current has a mean square value in a bandwidth given as

$$\left\langle i_Q^2 \right\rangle = 2q\, I_p\, BM^2\, F(M) \qquad ...(10.28)$$

where $F(M)$ is a noise figure associated with the random nature of the avalanche process also known as the excess noise factor. It has been found that

$$F(M) \cong M^x \qquad ...(10.29)$$

to a good approximation where $0 \le x \le 1.0$ that depends on the material. The values of both M and $F(M)$ are equal to unity for a PIN type of photodetector.

The photodiode dark current is the current that continues to flow through the bias circuit of the device when no light is incident. This is a combination of bulk and surface currents. The bulk dark current i_{DB} is due to the electrons and/or holes that are thermally generated in the PN junction of the photodiode. These thermally generated carriers get accelerated in an APD by the high electric field present at the junction and hence are subject to a gain known as avalanche multiplication. The mean square value of this current is

$$\left\langle i_{DB}^2 \right\rangle = 2q\, I_D\, M^2\, F(M)B \qquad \text{...(10.30)}$$

where I_D is the primary unmultiplied detector bulk dark current.

The surface dark current is also referred to as a surface leakage current or simply the leakage current. It depends on surface defects, cleanliness, bias voltage and surface area. An effective way of reducing this current is through the use of a guard ring structure that shunts the surface leakage currents away from the load resistor. The mean square value of the surface dark current is

$$\left\langle i_{DS}^2 \right\rangle = 2q\, I_L\, B \qquad \text{...(10.31)}$$

where I_L is the surface leakage current. The surface dark current is not affected by the avalanche gain as it is a bulk effect.

The total value of the mean square noise current in a photodetector is, therefore the sum of all these noise currents as they are statistically uncorrelated. Therefore,

$$\begin{aligned} \left\langle i_N^2 \right\rangle &= \left\langle i_Q^2 \right\rangle + \left\langle i_{DB}^2 \right\rangle + \left\langle i_{DS}^2 \right\rangle \\ &= 2q(I_p + I_D)M^2F(M)B + 2qI_LB \end{aligned} \qquad \text{...(10.32)}$$

There is another source of noise; the thermally agitated carriers give rise to a noise current in the resistors connected to a photodetector. This is known as Johnson or white noise also which has a mean square value equal to

$$\left\langle i_T^2 \right\rangle = \frac{4K_BT}{R_L}B \qquad \text{...(10.33)}$$

where K_B is Boltzman's constant and T is the absolute temperature. This noise can be reduced by using a load resistor that is large but consistent with the receiver bandwidth requirements.

10.4.2. Signal-to-Noise Ratio in a Photodetector

We have the signal current that is a function of the detector's responsivity. Therefore, the signal power to the mean square noise power is given by

$$\frac{S}{N} = \frac{\langle i_p^2 \rangle M^2}{2q(I_p + I_D)M^2F(M)B + 2qI_LB + 4K_BTB/R_L} \quad \text{...(10.34)}$$

This expression is true for an APD type of photodetector. Here photogenerated noise currents dominate the thermal noise currents. For a PIN photodetector, the value of M is unity. The dominant noise sources are thermal current as contributed by the load resistor and the active elements of the amplifier circuitry.

The signal power is multiplied by M^2 and the quantum noise plus bulk dark current is multiplied by a factor of $M^2F(M)$. The surface leakage current is not affected by the avalanche gain mechanism. Since the noise figure $F(M)$ increases with M, there always exists an optimum value of M that maximizes the SNR. The optimum gain at the maximum SNR is found by differentiating (10.34) with respect to M and setting the result equal to zero and then solving for M. For a sinusoidally varying input signal, we have

$$\langle i_p^2 \rangle = \frac{1}{2}(I_p)^2$$

We have

$$\frac{\partial}{\partial M}\left(\frac{S}{N}\right) = (I_p)^2[2q(I_p + I_D)M^{x+2}B + 2qI_LB + 4K_BTB/R_L]M$$

$$-\frac{1}{2}((I_pM)^2)(x+2)2q(I_p + I_D)M^{x+1}B = 0 \quad \text{...(10.35)}$$

$$\Rightarrow \quad B(x+2-2)(I_p + I_D)M^{x+2} = 2qI_LB + 4K_BTB/R_L$$

$$\Rightarrow \quad M_{opt}^{x+2} = \frac{2qI_L + 4K_BT/R_L}{xq(I_p + I_D)}$$

We may observe from (10.35) that this optimum value of the avalanche gain is independent of the detector bandwidth. However, it is a function of the temperature of operation, the dark current and surface leakage current.

A number of figures of merit have been used in order to assess the quality of the photodetectors. These were mainly developed as a guide to the weakest, steady or slowly varying infra red radiation source that could be detected. The three parameters most often used are the noise equivalent power (NEP), the detectivity (D) and the specific detectivity (D^*). These three parameters are related to the wavelength to be detected and the quantum efficiency and the dark current of the photodetector.

The detector is assumed to be working into a very high impedance so that the principal source of noise is the shot noise associated with the dark current and the signal current. The NEP is defined as the optical power at a specified wavelength required to produce a detector current equal to the RMS noise current in a unit bandwidth. Therefore, NEP is evaluated at a fixed wavelength by writing

$$I_p = \frac{\eta q}{h\upsilon} P_o$$

$$\Rightarrow \qquad P_o = \frac{h\upsilon}{\eta q} I_p \qquad \text{...(10.36)}$$

$$\Phi = \frac{I_p}{\eta} \frac{hc}{\lambda q}$$

Upon substitution of $I_p = I_{sh}$, we obtain

$$I_{sh} = \{2q(I_p + I_D)B\}^{1/2} \qquad \text{...(10.37)}$$

$$= \left\{2qI_p B\left(1 + \frac{I_D}{I_p}\right)\right\}^{1/2}$$

$$\Rightarrow \qquad I_{sh}^2 \approx 2qI_p B \quad \because \quad I_D << I_p$$

$$\Rightarrow \qquad I_{sh} \approx 2qB \quad \because \quad I_{sh} = I_p$$

For a unit bandwidth, we get

$$NEP = \Phi = \frac{I_p}{\eta} \frac{hc}{\lambda q}$$

$$= \frac{2q}{\eta} \frac{hc}{\lambda q} = \frac{2hc}{\eta \lambda} \qquad \text{...(10.38)}$$

With a quantum efficiency equal to unity, this represents the NEP of an ideal quantum detector. For $I_D >> I_p$, we get

$$I_{sh} = (2qI_D B)^{1/2}$$

$$\Rightarrow \qquad NEP = \Phi = \frac{I_p}{\eta} \frac{hc}{\lambda q} \qquad \text{...(10.39)}$$

$$= \frac{\sqrt{2qI_D}}{\eta} \frac{hc}{\lambda q} \quad \because \quad B = 1 \text{ Hz}$$

The detectivity is defined as the reciprocal of the NEP as evaluated as above. For a photodetector dominated by dark current and detecting a monochromatic radiation, we write

$$D = D_\lambda = \frac{1}{NEP} = \eta\frac{\lambda q}{hc}\frac{1}{\sqrt{2qI_D}} \qquad ...(10.40)$$

The specific detectivity D^* takes account of the fact that in such photodiodes I_D is often proportional to the area of the detector. This is the case when the background radiation and thermal generation are the primary reasons of I_D rather than surface conduction. Therefore,

$$D^* = D\sqrt{A} = \eta\frac{\lambda q}{hc}\sqrt{\frac{A}{2qI_D}} \qquad ...(10.41)$$

where A is the detector area. The wide bandwidth detectors required for optical communication often work into a low resistance and require a minimum signal current that greatly exceeds the dark current. Then, load resistance, the amplifier and the signal current itself all introduce additional sources of noise that complicate these parameters and render them inappropriate for the present discussion.

10.4.3. Speed of Response

The speed with which a photodetector responds to an incident optical radiation in order to produce an electrical output is determined by three primary factors: the transit time of the photogenerated carriers through the depletion region, the electrical frequency response as determined by the RC time constant, which in turn, depends on the diode capacitance and the slow diffusion of carriers generated outside the depletion region.

The transit time of carriers across the depletion region of width is given by

$$\tau_t = \frac{w}{v_d} \qquad ...(10.42)$$

where v_d is the drift velocity of the carrier. The smaller is w, the smaller is τ_t and the lesser will be the limitation due to transit time. This requirement of smaller w is contrary to that required to achieve larger quantum efficiency. Apart from transit time limitations, photodiode capacitance may also play a significant role. The junction capacitance of a diode that has a layer width of w and area A is given by

$$C_d = \frac{\varepsilon A}{w} \qquad ...(10.43)$$

where ε is the permittivity of the semiconductor. The speed of response in an electrical circuit is given by the *RC* time constant. The rise time of an electrical circuit is dictated by this time constant and is given as

$$\tau_r = 2.2RC = 2.2R_L C_d \quad ...(10.44)$$

$$= 2.2R_L \frac{\varepsilon A}{w}$$

Decreasing the width *w* to reduce transit time would increase the capacitive rise time that needs to be balanced by a corresponding decrease in the load resistance R_L. The bandwidth of the photodiode in terms of this rise time is given as

$$B = \frac{0.35}{t_r} = \frac{0.35}{R_L C_d} \quad ...(10.45)$$

This is due to the fact that the rise time and the bandwidth are related as follows:

$$B = \frac{1}{2\pi RC} = \frac{1}{[2\pi/(t_r/2.2)]} = \frac{2.2}{2\pi t_r} = \frac{0.35}{t_r} \quad ...(10.46)$$

10.5. PHOTOMULTIPLIER TUBES

As the name suggests, this type of photodetector is a glass tube that employs photoelectric emission at the primary level. It does not have semiconductor components. Hence it requires significant amounts of dc voltage for its operation. The multiplication is due to the secondary emission. The photomultiplier tube type of detector finds wide application at wavelengths shorter than 1 μm which is characteristic of unguided optical communication. These types of detectors require a high value of dc voltage of the order of 1 kV for their operation. These are based on the principle of photoelectric effect. The surface of the tube is illuminated with optical radiation of energy of such value that it is sufficient to eject an excited electron from the surface. The energy band diagram corresponding to such a process is shown in Fig. 10.15.

The photodiode shown in Fig. 10.16 is constructed from a *p*-type semiconductor for a number of reasons. We observe that

$$\Phi = h\upsilon = \chi + E_g \quad ...(10.47)$$

Hence, the wavelength of the emitted radiation is

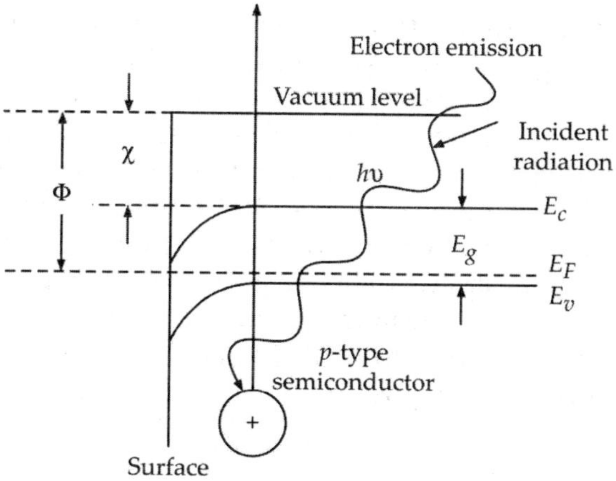

Fig. 10.15: Energy band diagram that shows photoelectric emission from a *p*-type semiconductor

$$\frac{hc}{\lambda} = \chi + E_g \qquad \text{...(10.48)}$$

$$\Rightarrow \qquad \lambda = \frac{hc}{(\chi + E_g)}$$

The quantity χ is known as the electron affinity of the material. A semiconductor is preferred as it has a higher probability of interaction between the incident optical radiation and the valence band electron as compared to a metal where this interaction is between a free electron and the optical radiation. The same applies to an *n*-type semiconductor as well. The work function Φ is increased in the case of a *p*-type semiconductor with the consequence that any thermionic emission is minimized. The emission of an electron from the absorbing surface leaves behind a net positive charge that becomes augmented with the emission of each new electron. Hence, the energy bands bend down as shown in Fig. 10.15. For an *n*-type semiconducting absorbing surface, the energy bands tilt upwards as a result of which the electron affinity increases. This requires a higher threshold photon energy to emit an electron that has a finite probability of having a lesser value of emission velocity.

The photoemissive surface has a quantum efficiency lower than that of the photodetectors discussed earlier. In order to have an appreciable number of emitted electrons, they are made to multiply through the use of what is known as 'secondary-emission multiplier'. This is shown schematically in Fig. 10.16.

The electron multiplication process is carried out through a number of secondary emitters known as dynodes which are maintained at high positive potentials. An electron is accelerated towards these dynodes because of the high positive potential after being emitted from the photoemissive surface which is maintained at 0 V or grounded. The primary electron releases other electrons from the surfaces of these dynodes resulting in secondary emission. This is a cumulative process so that at the end of the final dynode, a large number of electrons are available. These are collected by the collector having the highest potential and connected through a suitable external circuit give rise to a substantial amount of current. The dynodes are usually coated with a material having a high secondary electron emission coefficient κ. The overall gain resulting from this arrangement of N electrodes is

$$M = \kappa^N \qquad ...(10.49)$$

For example, for a value of $N = 10$ and $\kappa = 2$, the overall gain becomes $M = 2^{10}$.

As with an avalanche photodiode, shot noise current is generated in the quantum conversion process at the photocathode. As we may guess, this noise is multiplied together with the signal at each dynode. In addition, some extra noise is added to the signal in the multiplication process and this is taken into account by means of a noise figure F. Therefore, for $\kappa \gg 1$, the noise figure may be approximated as[1-2].

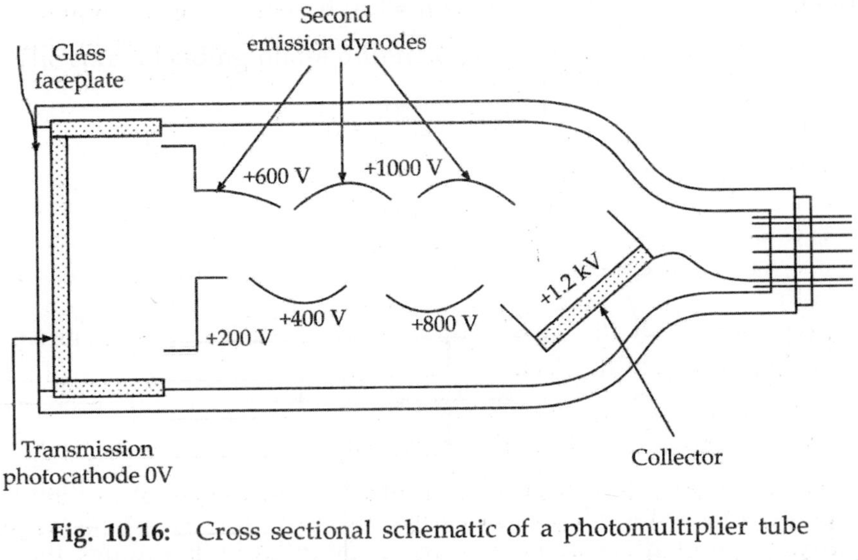

Fig. 10.16: Cross sectional schematic of a photomultiplier tube

$$F \approx \frac{\kappa}{\kappa - 1} \qquad ...(10.50)$$

The gain of such a photodetector is normally sufficient to ensure that the receiver noise is usually dominated by the multiplied shot noise. The bandwidth over which the gain can be maintained depends on the electron transit time through the diode. Careful design in electrostatic photomultiplier tubes such as the one shown in Fig. 10.16 may extend the frequency of operation to 1 GHz.

10.6. PHOTOTRANSISTOR

A phototransistor, as the name suggests accepts light as input and furnishes a current which is an amplified version of the electrical equivalent of the optical input. The transistor provides an internal gain of the photocurrent which is achieved through transistor action rather than avalanche multiplication. The primary difference between a conventional transistor and a phototransistor is shown in Fig. 10.17. As compared to a conventional transistor, a phototransistor does not have a biasing circuit connected to its base. Instead, the base-collector junction is made photosensitive so that when light falls on this junction, a current is generated as the base current. The structure of an *n-p-n* heterojunction phototransistor as shown in Fig. 10.17(b) is grown on an *n*-type InP substrate. The emitter is made from a wide bandgap *n*-type InP material. Radiation falling upon the device passes unattenuated through the wide bandgap emitter and is absorbed in the base, base-collector depletion region and the collector. A large secondary photocurrent between the emitter and collector is obtained as the photogenerated holes are swept into the base, increasing the forward bias on the device. The use of the heterostructure results in a low emitter-base and collector-base junction capacitances together with low base resistance.

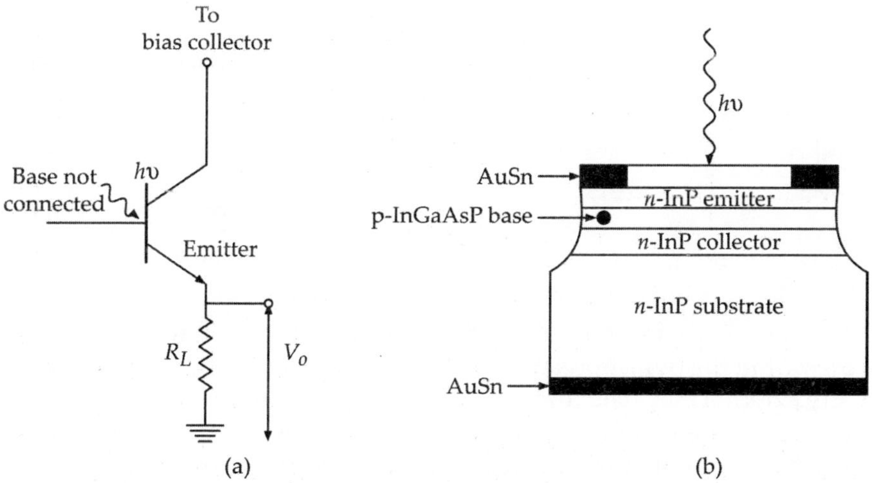

Fig. 10.17: A typical circuit realizing a phototransistor

A large current gain is made possible through low doping levels both at emitter and collector but high doping in the base region. The potential barrier created by the heterojunction at the emitter-base junction effectively eliminates hole injection from the base when the junction is forward biased. The optical gain of such a transistor is given by[3]

$$A_0 = \eta h_{FE} = \frac{h\upsilon}{q} \frac{I_c}{P_0} \qquad ...(10.51)$$

In (10.50), the common emitter current gain is denoted as h_{FE}, P_0 is the incident optical power and I_c is the collector current. Such a transistor operates in the 0.9-1.3 μm wavelength band giving optical gains in excess of 100.

10.7. SOLAR CELL

A solar cell is widely used these days to generate clean, green energy as an alternative energy source. This is an active transducer in the form of an optical detector that utilizes the abundantly available sunlight to generate a dc voltage. As the name indicates, the solar cell takes in the sunlight as the input energy that is used to produce electron-hole pairs, thereby generating electrical energy at its output. This is constructed from a *p-n* junction such that when sunlight falls upon it, electron-hole pairs are generated in the transition region. Such a photodetector works with minority carriers and hence is called a minority carrier device.

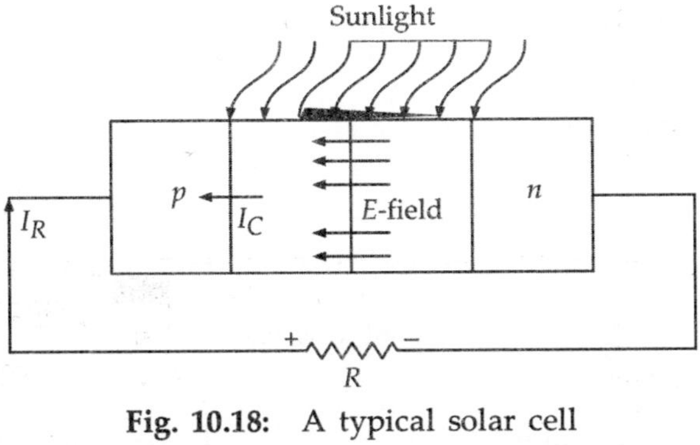

Fig. 10.18: A typical solar cell

We know from elementary semiconductor physics that an electric field is present in the depletion layer of a *p-n* junction diode that is independent of any external bias. The direction of the field is from the *n*-type material to the *p*-type. When sunlight falls upon this junction, it creates electron-hole pairs that are pulled towards the opposite ends to constitute a current. This current flowing through an external resistor produces a voltage. The polarity of the voltage developed across this resistor *R* is so as to forward bias the diode. The net current flowing through the cell is given as

$$I = I_C - I_R = I_C - I_0\left[\exp\left(\frac{qV}{K_BT}\right) - 1\right] \quad ...(10.52)$$

The photocurrent is always in the reverse biased direction as a result of which the solar cell current is along the reverse bias direction. We may note from (10.52) that even though and $R = 0$, the current through the circuit of Fig. 10.17 is not zero. This current is referred to as the short circuit current $I_{SC} = I_C$. This is due to the inherent electric field that exists in the depletion layer of a *p-n* junction as shown in this figure. The equivalent circuit of a solar cell may be constructed as the following as shown in Fig. 10.19.

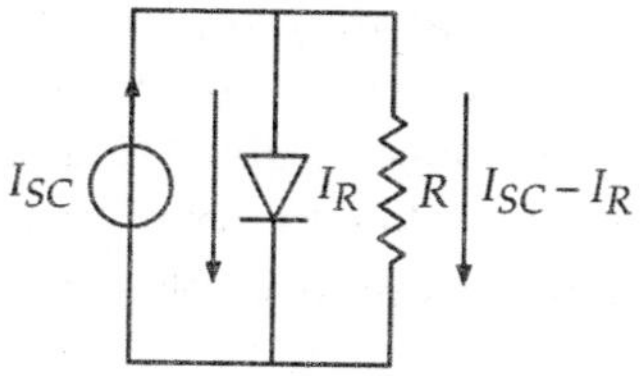

Fig. 10.19: Equivalent circuit of a solar cell

As the solar cell is constructed from a *p-n* junction, it behaves as a semiconductor diode, however with an important difference. While the ordinary *p-n* junction diode requires a forward bias above the threshold voltage for its proper operation, a solar cell is an active diode, meaning thereby it produces a small dc voltage across its terminals when exposed to sunlight. The total cell current goes to zero at a voltage known as the open circuit voltage V_{OC} wherein the forward current becomes equal and opposite to the generated current. From (10.52), we obtain this voltage by setting the net current to zero and working as the following.

$$I = I_C - I_R = I_C - I_0\left[\exp\left(\frac{qV_{OC}}{K_BT}\right) - 1\right] = 0$$

$$\Rightarrow \quad I_C = I_0\left[\exp\left(\frac{qV_{OC}}{K_BT}\right) - 1\right] \quad ...(10.53)$$

$$\Rightarrow \quad \frac{I_C + I_0}{I_0} = \exp\left(\frac{qV_{OC}}{K_BT}\right)$$

$$\Rightarrow \quad V_{OC} = \frac{K_BT}{q}\ln\left(1 + \frac{I_C}{I_0}\right)$$

The V-I characteristics of a solar cell have been shown in Fig. 10.20 in which the open circuit voltage V_{OC} and the device short circuit current

I_{SC} have been shown. The useful voltage seems to be less than the contact potential. This in fact is smaller than the band gap voltage. The operating current is in the range of 10-100 mA for a junction with an area of about 1 cm^2. In order to deliver a significant amount of power, an array of such devices needs to be used. In practice, a number of solar cells are connected in series and commercially available solar panels are made from silicon based *p-n* junctions having a large surface area. A metal grid is attached to the top *n*-type material whereas the bottom *p*-type material is grounded. As obvious, the load is connected between the top and bottom surfaces.

The solar cells are constructed from silicon with homojunctions in which crystalline silicon is used as the main material. They are also made from heterojunction devices that may utilize CdS-CuS cells wherein the two metal sulphides form a heterojunction. GaAs solar cells and amorphous thin film silicon based cells are also available. Design and construction of higher power efficient solar cells is a very active area of research these days because of the impetus given to alternatives of fossil based fuel energies.

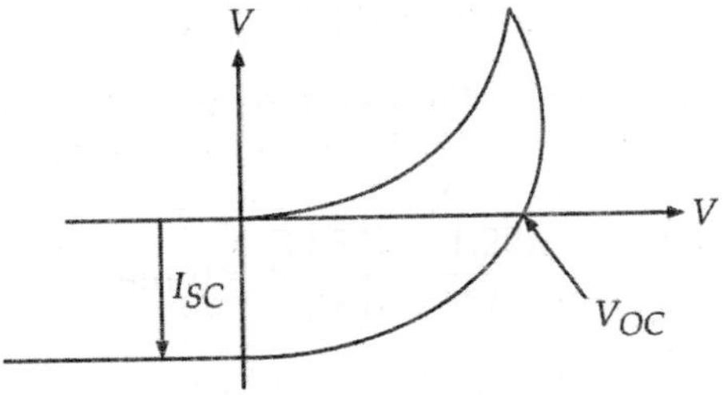

Fig. 10.20: V-I characteristics of a typical solar cell

10.8. CHAPTER SUMMARY

The photodetector is a device that generates an electrical output usually a current in response to an optical power at its input. Both diodes and transistors are used as the opto electric converters. Photoconductive and photovoltaic properties exhibited by certain materials result in passive and active optoelectronic converters. The two primary and widely used diodes are the *p-i-n* type and the APD type. Both have their own carrier generation mechanisms as well as noise. The noise affecting the optoelectric conversion process is shot noise and the avalanche multiplication noise in APD type of receivers. The photomultiplier tube makes use of vacuum tubes employing photoemission and it requires high dc voltages for its operation. The multiplication of charge carriers is different in APD and the photomultiplier tubes. While it is avalanche multiplication in the former, it is secondary emission in the latter. The process of avalanche multiplication gives rise to an additional noise known as multiplication noise. The receiver sensitivity is determined by the inherent noise sources in the receiver that is again given by the noise figure. The two important criteria while selecting a photodetector for a given application are the wavelength of operation, the quantum efficiency and the responsivity. Transistors are also used as optical detectors and heterojunction structures have been used to result in an appreciable current gain. Here also the receiver performance in terms of the photocurrent is decided by shot noise. As we may observe, both shot noise and avalanche noise generation mechanisms are random in nature and require a statistical analysis as both are due to the quantum process of optoelectric conversion. Solar cells, unlike the other photodetectors studied in this chapter are active photodetectors that take sunlight as their input and produce a small dc voltage across a load suitably connected. An array of solar cells having a large surface area in backup with a rechargeable battery can generate significant amount of electrical power that may be connected to a grid for normal use or as a backup source of clean, green power.

SUGGESTED READING

[1] Ziel, A. van der. "Solid State Physical Electronics", Prentice-Hall, 1968.
[2] McIntyre, R.J. "Multiplication noise in uniform avalanche diodes", IEEE Trans. On Electron Devices, ED-18, pp. 164-168, 1966.
[3] Wright, P.D., Nelson, R.J. and Cella, T. "High gain InGaAsP/InP phototransistors for fibre optic communications", *Appl. Phy. Letters,* no. 2, vol. 37, pp. 192-194, 1981.

❑❑❑

11

Optical Sensors

11.1. INTRODUCTION

The mechanism of conversion of optical energy into electrical energy and vice versa constitutes one important class of transducers. In the previous chapters, we have seen lasers, LEDs as optical sources and a variety of detectors that convert light energy into an appropriate electrical signal. These are important transducers from our point of view. The optical fibers, in association with the sources and the detectors have been configured in a number of ways to act as sensors for a host of physical parameters. Our objective here is to discuss sensors that have been developed over the years around light and its various properties. Guided light waves through optical fibers have found prominence into making low cost, robust, light weight sensors. Optical sensors enjoy the benefits of electromagnetic immunity, electrical isolation, wide dynamic range, suitability for multiplexing besides being available in point as well as distributed forms of sensing. We first discuss sensors built around light sources and detectors before discussing optical fiber based sensors. An example is an optical shaft crcerder wed as a position velocity sensor. Optical shaft encoders as shown in Fig. 11.1 are used to provide information about position, velocity etc. They convert angular position of a rotating shaft into an analog or digital data. The optical encoders work by converting mechanical movement into an electrical signal. Light from an LED or IR-LED is made to pass through a rotating high resolution encoded disk that contains the code patterns in the form of binary/grey code/BCD code. Photodetectors scan the opaque disk punctured with slots as it rotates and consequently generate a series of pulses depending on whether they have detected light or not. Accordingly a square wave is produced that is fed to appropriate electronic circuitry to be counted. The count gives an indication of the angular position of the shaft. The angular position can be used to measure

velocity containing the code patterns. The shaft encoders are of two types: absolute position and incremental position. The former provides a unique output for every single position of rotation of the disk containing the position as well as direction. The coded disk consists multiple concentric tracks of alternate light and dark segments. Each track is independent with its own photodetector to read out a unique coded position value for each angle of rotation. The number of tracks on the disk corresponds to the binary bit resolution of the encoder so that an 8-bit absolute encoder has 8 tracks and the same coded value appears once per revolution.

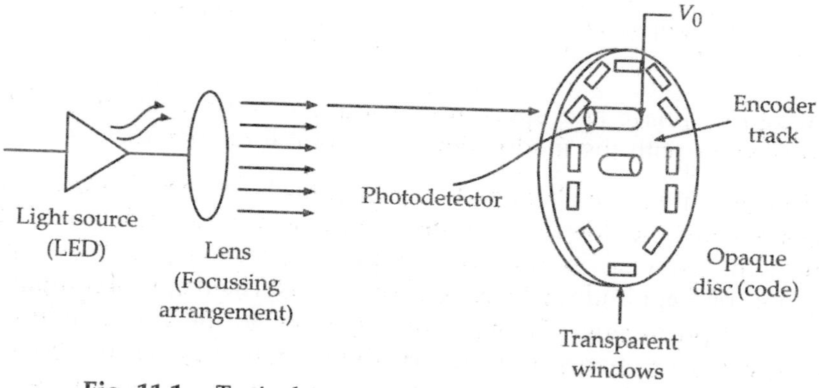

Fig. 11.1: Typical optical shaft encoder arrangement

11.2. CLASSIFICATION OF FIBER BASED SENSORS

Incremental type of shaft encoders measure position with reference to a given reference point.

The advantages of fiber optic sensors lie in the fact that they are of light weight and small size. They are characterized by high sensitivity and a large bandwidth. They can operate in very hazardous environments having high degrees of electromagnetic interference, high temperature, chemical corrosion, high pressure and high voltage etc. These can be configured as point, multiplexed or distributed type of sensors. The measurands that can be sensed and measured using optical fibers are listed in Table 11.1.

In a typical optical fiber sensor, light from a source such as a laser or LED is guided by a piece of optical fiber to the sensing part. Some property of the guided light gets modified or modulated in accordance with the measurands under consideration. The modulated light beam thus obtained is then coupled to another fiber in order to be photodetected and subsequently measured. The parameters of a light wave that can be modulated in accordance with the input baseband are intensity,

Table 11.1: Optical Fiber based Measurands

Displacement	***Electric Field***
Velocity Acceleration Angular velocity Angular rotation Strain Pressure Force Flow Temperature	Magnetic Field Acoustic Field Liquid level pH Chemical species Radiation Electric Current Vibration Humidity

polarization, phase and spectral distribution. Most of the optical fiber sensors work with the modulation of the first three parameters.

Use of multimode fibers for building optical fiber based sensors leads to less sensitive but low cost, simpler solutions. Use of single mode fibers along with lasers as optical source improves sensitivity of these sensors by orders of magnitude at the expense of sophisticated optical components and signal processing. The basic principle lies in the way a light beam gets affected or changed by any of the measurands as mentioned in Table 11.2. Properties that are useful in making optical sensors are studied first in order to gain a comprehensive understanding of the sensors.

Table 11.2: Various effects utilized in optical sensors

Effect	***Use***
Photoelectric	Photocell, photomultiplier
Photosensitive	Acceleration, Strain
Photoelastic	Pressure, Strain
Magneto optic (Faraday)	Current, Voltage

The main component in a optical fiber sensor (OFS) is the modulation zone as depicted in Fig. 11.2. Here modulation (change of certain parameters in accordance with a given physical quantity) of light beam passing through the feed fiber takes place. The block schematic of Fig. 11.2 has been expanded in Fig. 11.3 that shows the physical quantity or the measurand causing one or a combination of changes of light beam parameters. The sensor is typically constructed from a piece of optical fibre so that a given light beam passing through it experiences changes in its parameters when the external stimulus in the form of the mesaurand is applied to it.

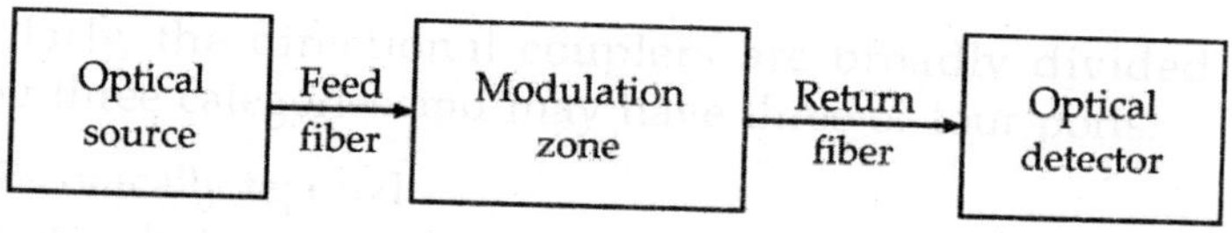

Fig. 11.2: Block schematic of a typical optical fiber based sensor

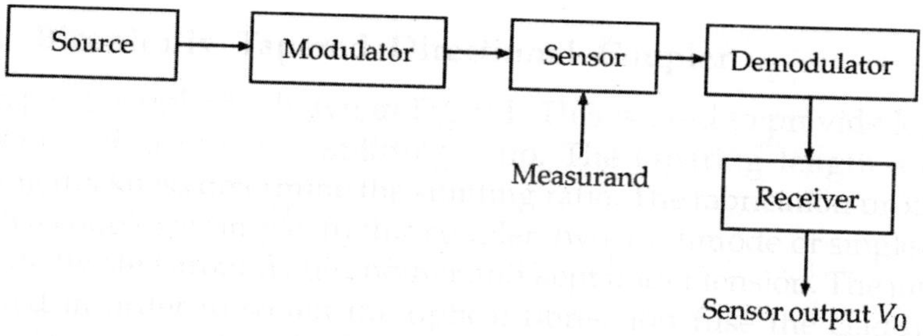

Fig. 11.3: A schematic for an extrinsic type of OFS

Moreover, it is possible to categorize the optical device into two distinct groups based primarily on their use of intensity and phase changes. These are categorized as (a) intrinsic and (ii) extrinsic devices.

In the intrinsic type of optical sensors, the physical parameter acts on the fiber itself so as to produce a change in its transmission characteristics. The latter type of sensor uses the fiber as a guiding medium for light to and from the sensors which is configured to allow the measurands to change to coupling characteristics of feed and the return fiber. A comprehensive comparison of the two broad types of sensors is presented in Table 11.3.

Table 11.3: A brief comparison of intrinsic and extrinsic type of OFSs

Intrinsic	*Extrinsic*
Higher sensitivity Not amenable to multiplexing Less number of connection related issues Requires elaborate signal processing, demodulation More expensive Applications: acceleration, rotation, acoustic pressure, vibration and strain	Reduced sensitivity Amenable to multiplexing More number of issues with connection Signal processing issues less complex Less expensive Temperature, pressure, fluid flow, point velocity, liquid level, electric field, magnetic field Easier to use

The OFSs can be further classified as point type, multiplexed or distributed type. A schematic showing the three configurations is shown in Fig. 11.4. The point type sensor detects the variation of the measurand in the proximity of the sensor. The multiplexed category of OFSs make use of multiple localized sensors placed along the length of a given optical fiber to detect the change in the measurand. For the distributed kind of sensing, as the name suggests we find from Fig. 11.4(c) that that sensing has been distributed throughout the length of the fiber.

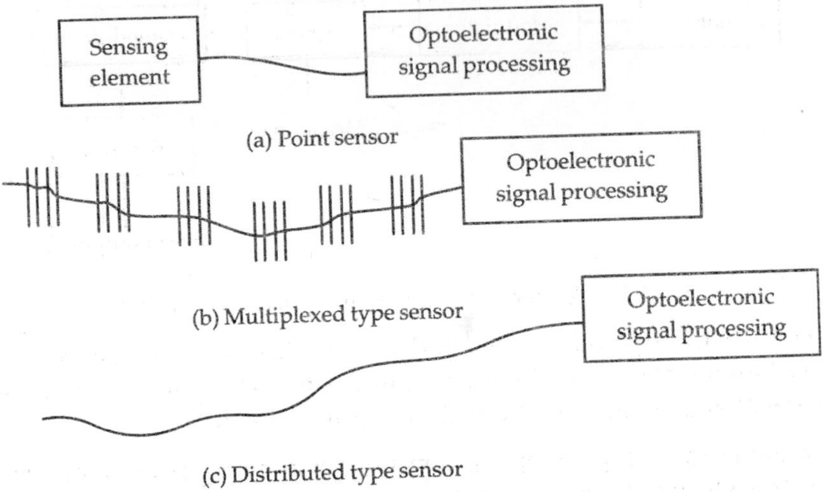

Fig. 11.4: Three different configurations of OFS sensing

Further, the sensing mechanism broadly may be the following two types. This is shown in Fig. 11.5.

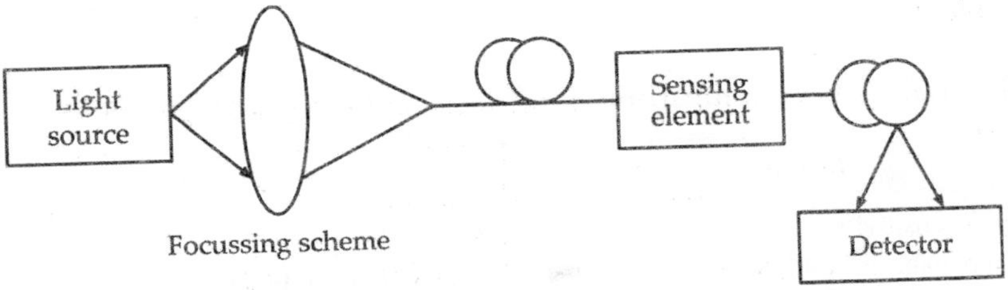

Fig. 11.5: (a) Transmission type of sensing

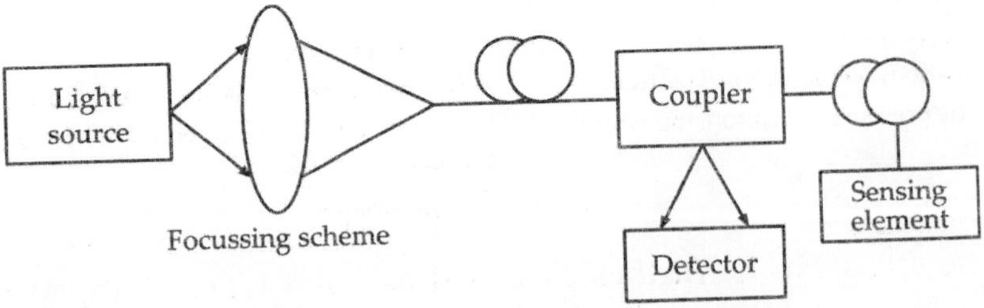

Fig. 11.5: (b) Reflection type of sensing

We observe from Figs. 11.5(a) and (b) that the in the former type, light makes a onward propagation through the fiber and the sensing element changes some attributes of this light which is subsequently detected and gives an indication of the mesaurand. In the reflection type, two light waves interfere at the detector a part of which comes from the direct light and the other after being reflected from the sensing element.

Single mode fibers can conveniently utilize interference phenomena for sensing various fields and are therefore more sensitive. However, multimode fibers carry typically a large number of modes (of the order of 500) so that interference techniques are not suitable for sensing task. The external field to be sensed usually intensity modulates a given light beam propagating in such a fiber which is subsequently detected and analyzed by a photodetector.

11.3. INTENSITY MODULATED (IM) SENSORS

Intensity modulation results through light interruption due to either the displacement or misalignment of one fiber relative to another. This misalignment may be due to three effects; axial (longitudinal) transverse and angular. Transverse misalignment fiber sensor is more sensitive among these three methods. Sensitivity of such fiber optic sensors can be enhanced by cleaving and polishing the two ends at a slant angle and keeping the slant face of the two fibers sufficiently close. This kind of sensor can work with both broadband as well as a monochromatic source of light such as a laser. This is though less sensitive, however requires simple signal processing. The disadvantages of such IM FOSs include susceptibility to power fluctuation of the optical source, variation in modal power distribution as in multimode fibers, receiver noise. The intensity modulation based FoS s make use of the following. The measurand amplitude (intensity) modulates a light beam passing through an optical fiber. For this class of sensors, a normalized modulation index is defined as

$$M_I = \frac{\delta P}{P_0 x} \qquad ...(11.1)$$

where ΔP is the change in optical power as a result of modulation by the measurands, P_0 is the unmodulated optical power reaching the photodetector and x represents the measurands. The limitation of these kinds of sensors is due to the presence of shot noise expressed as

$$i^2_{short} = 2qBI_p \qquad ...(11.2)$$

11.3.1. Macrobend Based IM Sensors

Radiative losses occur whenever an optical fiber undergoes a bend of a finite radius of curvature. Fibers are typically subject to two types of bends: macroscopic bends having radii that are large compared to the fiber diameter that may come into place when a fiber turns a corner and random microscopic bends of the fiber axis that may arise when the fibers are incorporated into cables. One way to induce intensity modulation is through the clever use of macro bends. A macro bend is observed when the fiber is wound around a specimen. These macrobends in an optical fiber induce a signal power loss given as (11.3) holds good under the assumption of $\frac{r}{R\Delta} << 1$.

$$L_{MB} = 10 \log R\left[\frac{\alpha+2}{2\alpha}\frac{r}{R\Delta}\right] \qquad ...(11.3)$$

where r is the core radius and α specifies the refractive index profile of the fiber. For example, $\alpha = 2$ for a square law medium such as a parabolic index as in a GI fiber $\alpha = \infty$ and for an SI fiber, *R* is the radius of curvature of the bend, and Δ is the refractive index difference between the core and the cladding. We may note from the above that, this kind of macrobend loss can be increased by using a larger core radius or using a relatively smaller refractive index. A macro bend loss is observed when a fiber is bent to a radius of several centimetres. Large bending loss occurs when the bend radius approaches a value known as the critical bending radius given as

$$R_c = \frac{3n_1^2}{4\pi(n_1^2 - n_2^2)^{3/2}}\lambda \qquad ...(11.4)$$

The optimum conditions for a large bending radius occurs when either the relative refractive index difference between the core and the cladding is rather small or the fiber works with a light source operating at a longer wavelength. Through the measurement of the resulting diminished light intensity, the measurands can be taken care of by this kind of macro bends. The effect of the misalignment resulting in a diminished light intensity output is shown in Fig. 11.6.

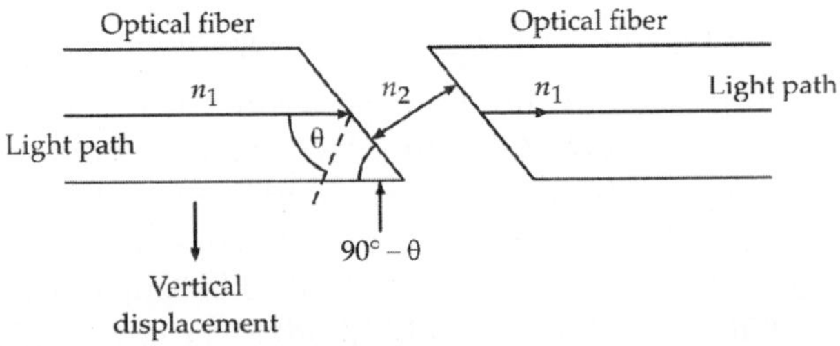

Fig. 11.6: Figure showing misalignment of two optical fibers

A physical parameter is converted into an appropriate change in the amount of light in this type of sensor. Such sensors usually utilize multimode fibers and work on the principle of intensity modulation. Such an arrangement is shown in Fig. 11.7. As we may see, the intensity of light arriving at the detector is related to the displacement of the object. As the displacement Δx increases, the intensity of the detected light decreases correspondingly. Due to frustrated total internal reflection power will leak from the first fibre to the receiving fibre through slant face.

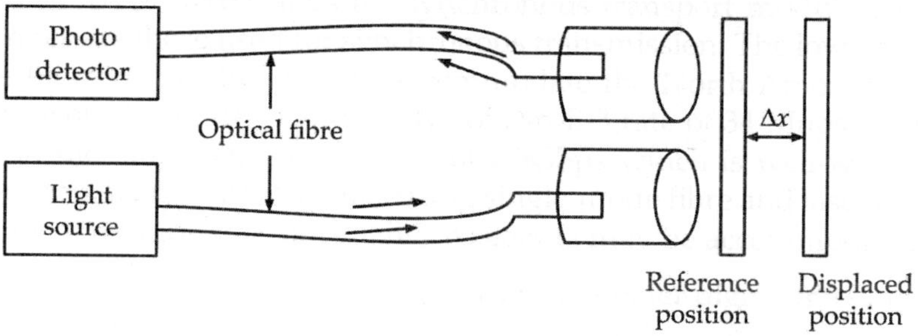

Fig. 11.7: Displacement sensor

In this kind of sensor, spurious light incidence is avoided by (i) illuminating the object with the help of a comparatively narrow line light source and (ii) placing a filter matching the source wavelength in the detector. The second point makes sure that the detector responds to the light coming from the source only and does not respond to lights coming from other sources as they are likely to emit at different wavelengths. Moreover, fiber illumination by other spurious sources of lights is reduced by the use of sleeves extended beyond the end of the fibers. Another class of displacement sensor using microbending based IM is discuss next.

11.3.2. Micro Bending Based IM Sensors

A certain class of intensity modulated fiber optic sensors work on the basis of transmission loss occurring due to microbending of the optical fiber. This technique is widely utilised for the measurement of acoustic pressure, strain, temperature, displacement etc. Here the fiber may be sandwiched between a pair of toothed or serrated plates to induce microbending as shown in Fig. 11.8. Sharp curvatures involving local axial displacements of a few micrometers and spatial wavelengths of a few millimetres can cause these kind of micro bends. These result in significant radiative loss and mode coupling as shown in Fig. 11.9. The lower order modes are converted into higher order modes and eventually get lost in radiation due to the multiple bends.

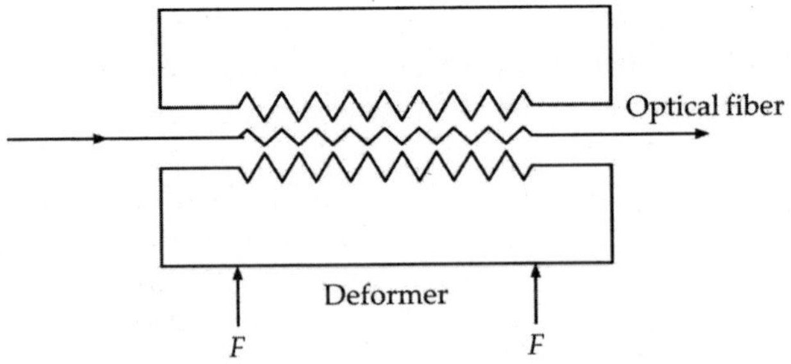

Fig. 11.8: Microbending in an optical fiber

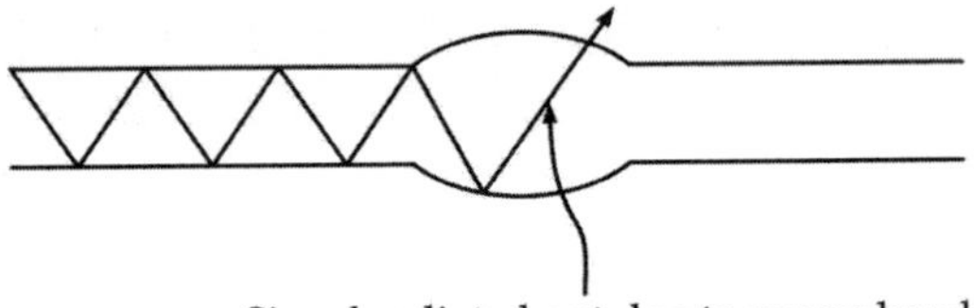

Fig. 11.9: Signal loss due to the presence of macro bend in the core of a fiber

This kind of microbending brings about a redistribution of guided power between modes of the fiber and also coupling of power from one mode to another. If the spatial wavelength Λ of the periodic deformation satisfies the phase matching condition $\beta_a - \beta_b = \frac{2\pi}{\Lambda}$, (where β_a and β_b represents the propagation constants for the two modes a and b) then power transfer takes place from the a – th mode to the b – th mode. For the case when b – th mode is the radiation mode, such a condition of power transfer results in a transmission loss of the guided modes. The critical spatial wavelength Λ_c of the deformer required to induce such a power transfer from the highest order mode to the radiation mode, for a step index fiber is given by

$$\Lambda_c = \frac{\pi a}{\sqrt{n_1 - n_2}} = \frac{\sqrt{2}\pi a n_1}{\Delta} \qquad ...(11.5)$$

where a is the core radius, n_1 being its refractive index, and Δ is the numerical aperture of the fiber. The microbend induced loss in a multimode fiber is given by

$$L_{MB} = K\frac{a^4}{(n_1 - n_2)^3 b^6} \qquad ...(11.6)$$

where b represents the cladding radius and K is a proportionality constant. A pressure or a displacement sensor is constructed by noting the reduced intensity of light when it passes through such microbends. In the case of graded index fiber, adjacent guided modes are separated from each other by a constant difference in wave number given as

$$\Delta\beta = \frac{\sqrt{2(n_1 - n_2)}}{a} \quad ...(11.7)$$

Hence, the spatial wavelength of the deformer, required to couple power between adjacent modes is given by

$$\Lambda = \frac{2\pi}{\Delta\beta} = \frac{2\pi a n_1}{NA} \quad ...(11.8)$$

The power received by a photodetector $W\Gamma$ is corresponding to an optical power of W watt at the fiber input and a transmission coefficient of Γ. Hence, a change in pressure ΔP applied to the deformer leads to an effective force on the microbend. As a result of this applied force, ΔF the amplitude of the micro bend changes from A to ΔA. The transmission coefficient of the fiber changes by

$$\Delta\Gamma = \left(\frac{\Delta\Gamma}{\Delta A}\right) D.\Delta P \quad ...(11.9)$$

Here, $\Delta A = D.\Delta P$ and D is a constant that depends on the pressure change. To realize a temperature sensor, the transmission coefficient Γ changes according as

$$\Delta\Gamma = \left(\frac{\Delta\Gamma}{\Delta A}\right) \alpha_s l_s.\Delta T \quad ...(11.10)$$

In the above expression, α_s and I_s represent the thermal expansion and the length of the spacers, ΔT denotes the change in temperature.

A procedure has been laid out in Fig. 11.10 to illustrate this. Microbending losses are produced in the fiber when the top block as shown in this diagram pushes the fiber due to an externally applied force. These losses are found to increase with the force applied to the top block. As a result of microbendings produced in the fiber, the intensity of light passing through it changes which is proportional to the external force. The changes in intensity due to the force are measured by comparing it with that of an unmodulated reference beam. This process eliminates errors in measurement produced because of source variations. Multimode intensity modulated sensors are classified into two categories depending upon the type of intensity modulation. These are (i) hybrid sensors and (ii) internal effect sensors. The hybrid sensors allow a broad range of modulation schemes and transmit light to a remote sensor which is usually a miniaturized device at the end of the fiber. It is a very simple sensor that simply acts as a switch for detecting the presence or absence of a stimulus at the sensor site.

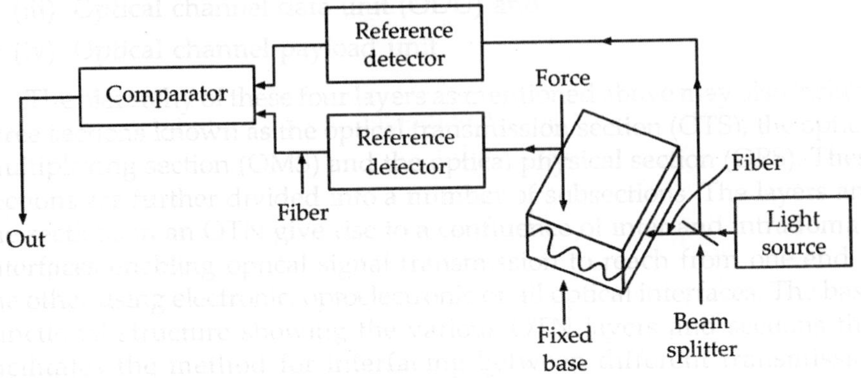

Fig. 11.10: Microbend train sensor

For a pressure sensor, the transmission coefficient of light propagating through the bend of the fiber is changed by an amount equal to the applied pressure is given as 11.10.

The inference from these discussions is that the mechanical parameter affects the sensitivity of the sensor. The applied force and the resulting displacement are related as

$$\delta F = k_f \, \delta x \qquad \text{...(11.11)}$$

Under the assumption of the bent fiber modelled as a bar loaded at the centre and clamped at its ends, we have

$$k_f = \frac{3\pi Y D \eta}{\Delta^3} \qquad \text{...(11.12)}$$

where *D* is the fiber diameter and η is the number of bent intervals, Y is young's modulus of the fiber.

11.3.3. Diffraction Grating Based IM Sensors

Another class of intensity modulated sensors is built on the diffraction gratings. Such a sensor for displacement measurement is illustrated in Fig. 11.11. In this sensor two diffraction gratings are butted together to form a single grating. As one of the gratings is displaced, the effective grating distance and the diffraction angle are changed as shown in Fig. 11.12. The detector senses this as a result of change in intensity. As the gratings move, the high and low intensity cycles are counted. This gives a direct measure of digital displacement measurement.

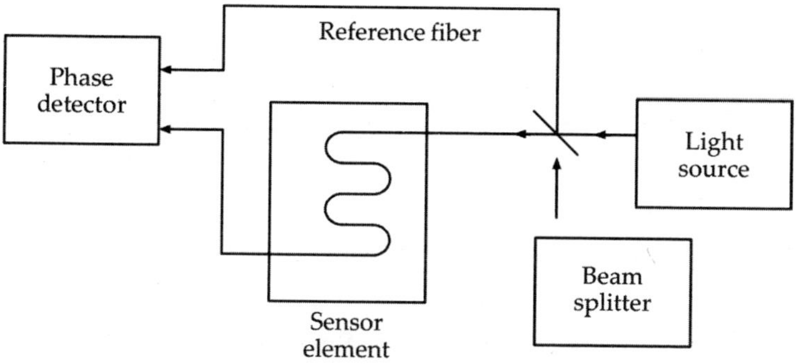

Fig. 11.11: Intensity modulation based on diffraction grating

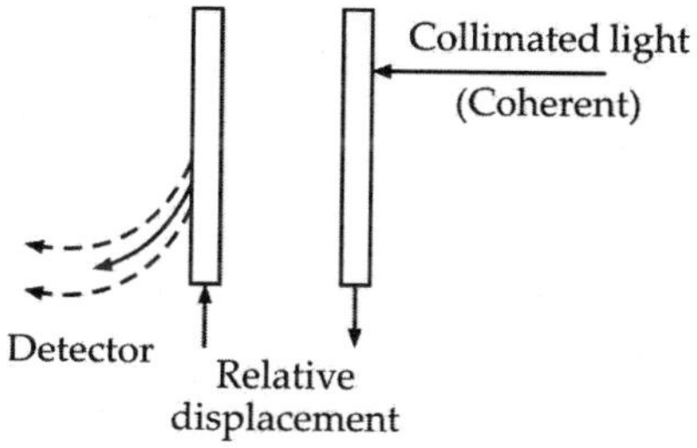

Fig. 11.12: Diffraction grating sensor

11.3.4. Reflective Displacement Sensors

Displacement measurement is carried out by extrinsic reflective or photonic optical sensor by intensity modulation of the transmitted light beam. The arrangement for doing this is shown in Fig. 11.13. Light reflected from the target is collected by a return fiber and subsequently detected by means of an optical detector. The detected power is a function of the distance between the fiber end and the target. Therefore, the position or the displacement of the target may be registered at the optical detector. The sensitivity of such a sensor may be improved considerably by placing the axes of the feed fiber and the return fiber at an angle to each other and also with respect to the target. However, a better sensitivity realized through this technique results in a poorer stability of the optical components which is a common feature of intensity modulated optical fiber based sensors. This reduced physical stability issue of such a sensor is taken care of by Moire fringe modulation sensors which is discussed next.

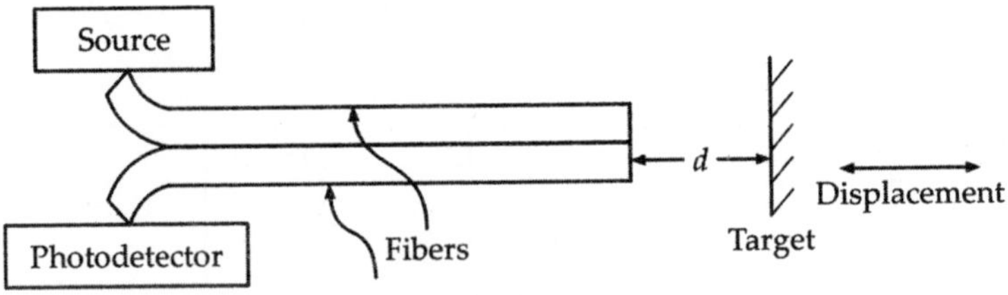

Fig. 11.13: Reflective or photonic displacement sensor

11.3.5. Moire Fringe Modulation Sensors for Displacement Measurement

A digital modulation scheme is used in this sensor. The arrangement in such a sensor is illustrated in Fig. 11.14. Displacement is measured by using a Moire fringe modulator. As shown in the diagram, the opaque line gratings produce dark Moire fringes. The transverse movement of one of the gratings with respect to the other results in a transverse (up and down) movement of the fringes. A count of such fringes as the grating moves serves as an indication of the displacement. As we may observe, the counting of fringes is independent of the instabilities of the sensor components that affect sensitivity. However, mechanical vibration may severely affect the accuracy of the measurement and may not be eliminated completely. An interrupted optical power to the sensors results in missed counts.

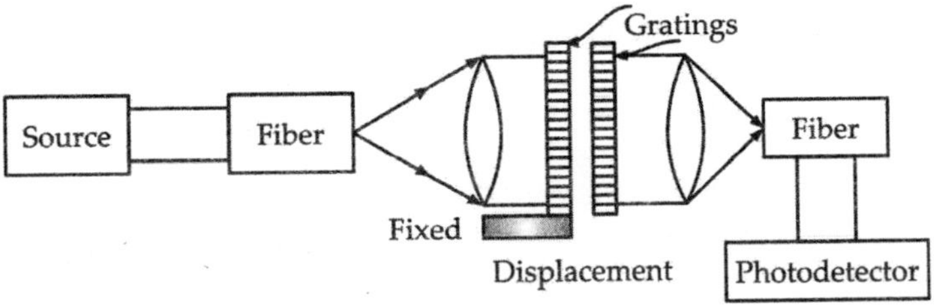

Fig. 11.14: Moire fringe modulation based displacement sensor

11.3.6. Liquid Level Type Hybrid Sensor

This sensor shown in Fig. 11.15 is used to detect the presence of a liquid in the gap between the two fibre ends. In this sensor, the fibre ends are tapered as a result of which the angle at the fibre end is greater than the critical angle required for total internal reflection of the central ray of the fibre transmitted beam. The total internal reflection occurring at the air-to-glass interface is eliminated when a liquid is present in the gap. The refractive index of the liquid, hence is approximately equal to that of glass. Under this condition, optical power is transmitted from one fiber end to the other end instead of being totally internally reflected.

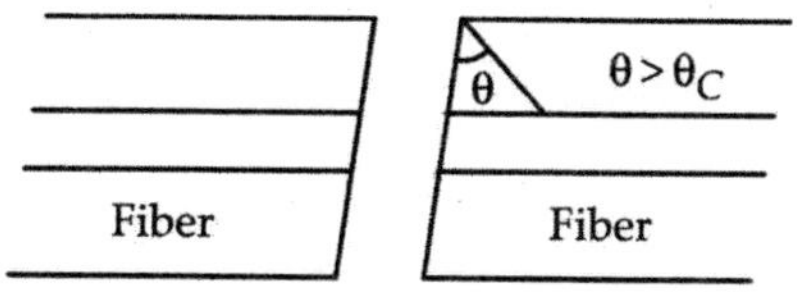

Fig. 11.15: Liquid level type hybrid sensor

11.3.7. Hybrid Pressure Sensor

A sensor of this type measures pressure in a glass transducer using photo elastic effect. This as shown in Fig. 11.16. When a pressure is applied to the glass, it causes stress induced birefringence, and thereby changing the polarization of the beam transmitted through the transducer. As shown in the diagram, there are two polarizers. The first polarizer polarizes light incident on the glass at an angle of 45° with respect to the applied stress P. The plane of polarization changes from linear to elliptical. The second polarizer which acts as an analyzer selects the component of the transmitted beam perpendicular to the original linear polarization.

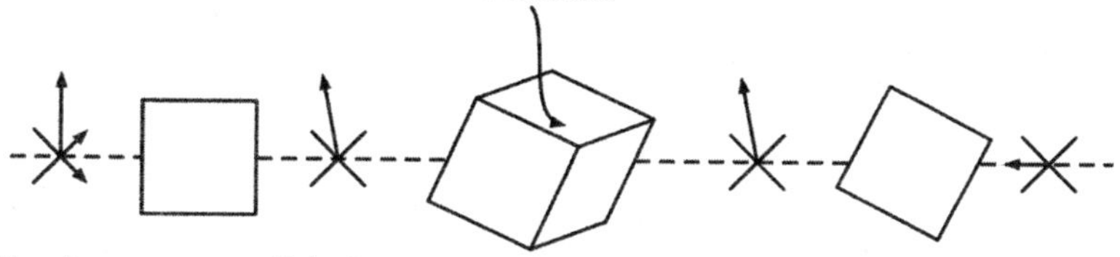

Fig. 11.16: Pressure sensor

The optical power transmitted by the sensor is mathematically expressed as

$$P = P_0 \sin^2\left(\frac{\pi P t}{f}\right) \qquad ...(11.13)$$

where P_0 is the stress required to make the transmitted power through the transducer go from maximum to minimum, f is the material fringe value, t is the thickness of the glass transducer and being the applied stress. The maximum measurement range is upto 31 Mpa. This sensor has a resolution better than 10^{-4}. These two types of sensors make use of high quality low loss multimode fibers in realizing the functionalities.

The third type of hybrid sensor consists of fibre bundles as shown in Fig. 11.17. It is a proximity sensor which uses light launched in one arm of the fibre bundle. This light is reflected from a surface as shown at the output end of the fiber and part of the reflected light is accepted by the second arm of the bundle which is transmitted to a detector. The fraction of the light returned to the detector depends on the distance between the end of the bundle and the surface being monitored.

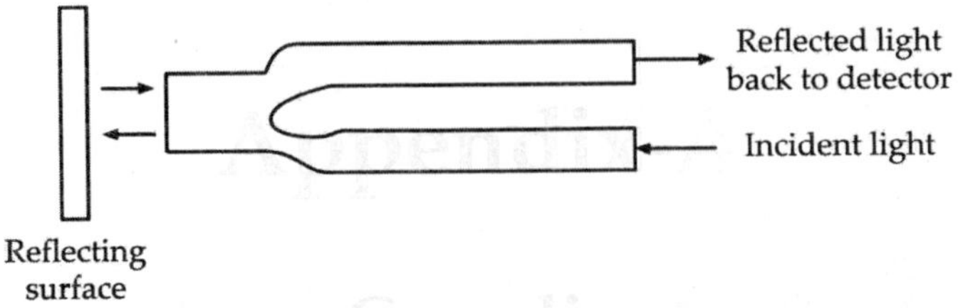

Fig. 11.17: Proximity Sensor

11.3.8. Liquid Level Detector

A sensor to measure a given liquid level is shown in Fig. 11.18. This is carried out by placing an optical dipstick between a laser and a photodetector. The liquid under consideration has a refractive index higher than that of the glass used to manufacture the dipstick. When the lower end of the dipstick does not touch the liquid, light from the laser undergoes a number of total internal reflections at the surface of the dipstick and is able to reach the photodetector. However, when the liquid touches the bottom end of the dipstick, light from the laser can not undergo total internal reflection and hence can not reach the detector. As the detector does not get any light beam it indicates the liquid level. It is a simple but crude method for determining the level of inflammable liquids.

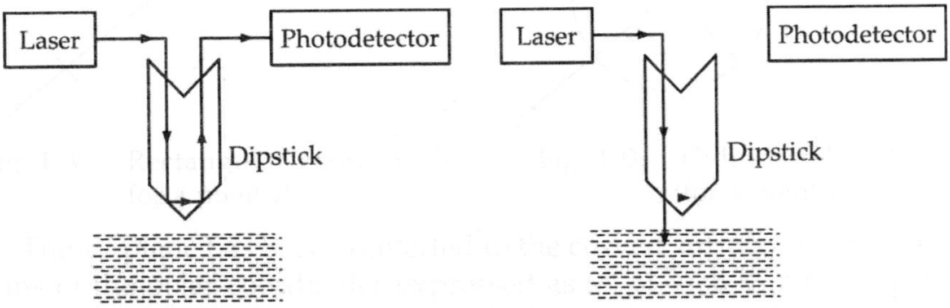

Fig. 11.18: Liquid level sensor through the use of a optical fiber based dipstick

11.3.9. Flow Sensor

An optical fiber based flow sensor as shown in Fig. 11.19 is based on vortex shedding principle observed in the downward flow of a liquid when the flow gets obstructed by placing an object in it. Every obstruction placed in the liquid has a critical flow speed for the formation of vortex shedding.

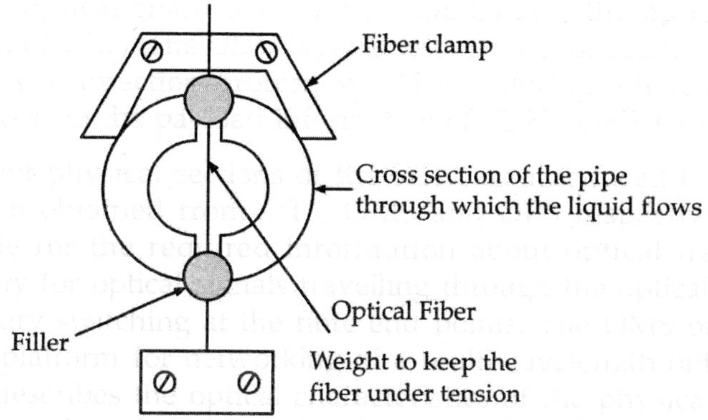

Fig. 11.19: Flow sensor for measuring liquid flow rate

Vortex shedding refers to a kind of oscillating flow when a liquid/fluid flows past a bluff body (a opposed to a streamlined body) at certain velocities that depends on the shape and size of the body. This refers to the formation of vortices of alternating low pressure zones in the downstream of the liquid. As a result of these alternating low pressure zone, the obstruction moves toward it. Sensing the vortices gives an indication of the strength of the liquid. The vibration of the vortices is related to the velocity of the liquid. The vortex shedding frequency f is given as

$$f = S_r \frac{v}{d} \qquad ...(11.14)$$

where S_r is known as Strouhal number which dimensionless (typically used to analyze unsteady oscillatory kind of fluid flows), v is the liquid velocity in m/s and d is the obstruction diameter.

To construct an optical fiber based flow sensor, a multimode optical fiber acts like the bluff body that needs to be placed in the liquid flow path. To do this, a circular pipe is bent in the form of a circle and a liquid is made to flow through it. The fiber is then stretched across the pipe such that when a liquid flows through this pipe, the flow occurs in a transverse manner to the fiber. As a result of the fiber insertion across the liquid, turbulences or vortices are formed around the fiber which makes it to oscillate as mentioned earlier. The resulting oscillation of the fiber which is proportional to the velocity of the liquid changes the mode power distribution within the fiber. Therefore, a similar intensity modulation profile is detected at the photodetector. This technique is however, limited to flow measurements in the range of 0.3 to 3 m/s. As we may expect, it does not work below the critical flow rate at which vortices are shed.

11.3.10. Acoustic Wave Sensor

This sensor shown in Fig. 11.20 commonly known as hydrophone is used for detecting sound waves. The principle is based on cantilever arrangement. Sound waves make the moving fiber vibrate. This leads to a relative transverse misalignment between the cores of the two fibers. This relative misalignment leads to a coupling loss that intensity modulates the transmitted light signal. A small core diameter results in a relatively higher sensitivity. For single mode fibers having typical core diameters of as small as 5μm, displacement as small as even 5μm would lead to 100% modulation. On the other hand, multimode fibers having core diameters to the tune of 100μm, a fairly linear response is found to occur over displacements of 50μm. This sensor has sufficient sensitivity to detect deep sea noise levels in the frequency range of 100 Hz to 1 KHz and to detect static displacements of a few angstroms.

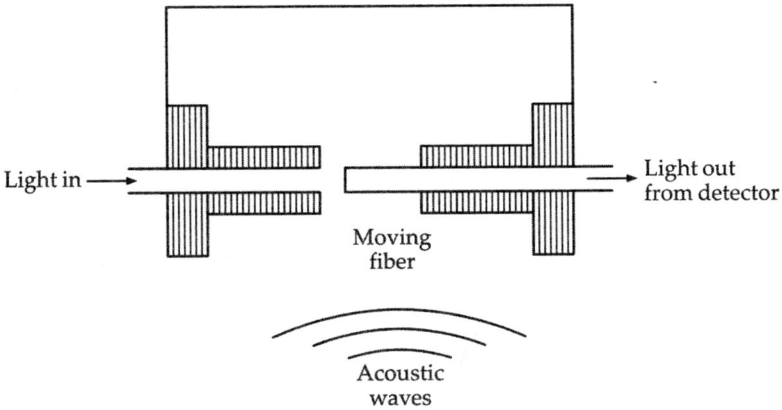

Fig. 11.20: Acoustic wave sensor (Hydrophone)

11.3.11. Temperature Sensor

The temperature sensor shown in Fig. 11.21 uses a multimode fiber. This is based on the changing absorption characteristics of silicon as a function of temperature near 1μm wavelength. This sensor is coated at one end with a thin layer of silicon which is then covered with a reflective coating. Light coupled into the fiber from one end passes first through the fiber, then through the silicon and after reflection from the reflection coating back to the same fiber end again passing through the silicon layer. The amount of light absorbed by silicon varies widely with temperature as a result of which the intensity of light received by the detector depends sensitively on temperature. With such a technique, differential temperature measurements of less than 0.001°C are possible with response times 1000 times smaller than those of a typical domestic thermostat.

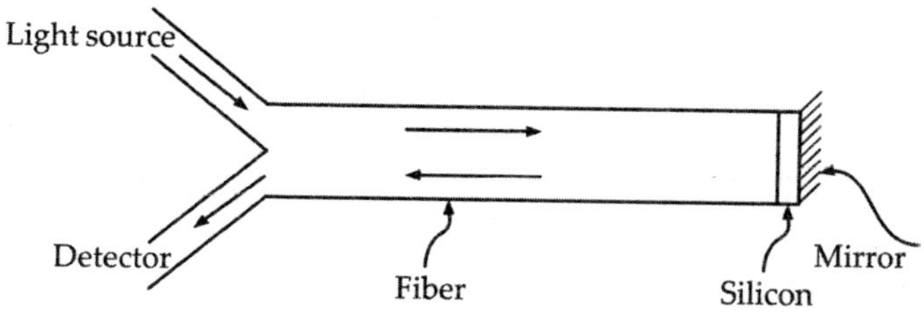

Fig. 11.21: A optical fiber based temperature sensor

11.4. PHASE MODULATED SENSORS

These kinds of optical fiber sensors utilize the phase modulation of light. The mesaurand brings about the phase modulation in a number of ways. Various interferometers best serve this purpose. We discuss a couple of such sensors next.

11.4.1. Interferometric Sensors

Fiber optic sensors built using interference phenomena of light waves essentially splits a given light beam into two light beams using a beam splitter and delay those by unequal distances. These two light beams are redirected using mirrors and recombined again using another beam splitter. The resultant intensity of the two light beams so combined becomes a function of the phase offset ϕ occurring between them as a result of unequal distances traversed. The phase offset is expressed as

$$\phi = \frac{2\pi nd}{\lambda} \qquad ...(11.15)$$

where n is the refractive index of the medium through the light waves of wavelength λ travel, d is their optical path difference

There are broadly four categories of interferometers. These are

(a) Mach-Zehnder

(b) Michelson

(c) Sagnac and

(d) Fabry-Perot type

These sensors are used to measure temperature and pressure. This has high sensitivity and high resolution. However this requires a excellent fiber technology available for its construction. Its principle is to detect the influence of physical perturbations on the phase of light propagating in a single mode fiber. An interferometer of type (a) is shown in Fig. 11.22. As shown in the diagram, there are two fibers right after the directional coupler. One fiber

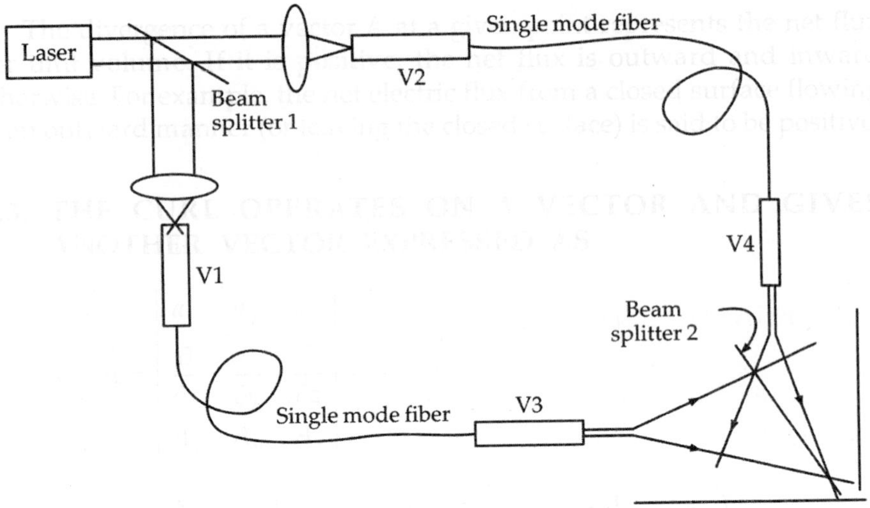

Fig. 11.22: Block schematic of an optical fiber based Mach-Zehnder Interferometer

The light beams emerging from the two single mode optical fibers are made to superpose with each other with the beam splitter 2. By adjusting the output ends of the two fibers, the centres of curvature of the two diverging light beams coming out from them are made collinear with the fiber axis for which case circular fringes are formed on a screen placed as shown. This arrangement produces an interference pattern similar to the one produced by two point sources on a screen placed perpendicular to the line joining the sources. When the output ends of the fiber are brought closer to each other, the circular fringes expand, and when they are moved away from each other, they shrink. As we have seen previously, the total phase difference between the interfering beams at a given point P is given as

$$\phi = \frac{2\pi}{\lambda_0}\left(\sqrt{S_2^2 + d^2} - \sqrt{S_1^2 + d^2}\right) + (\phi_2 - \phi_1) \qquad \text{...(11.16)}$$

where $d = O_2P$ is the distance of the observation point from the centre of the screen, S_1 and S_2 are the distances from the fiber ends of the screen, ϕ_1 and ϕ_2 and are the phase shifts introduced by the light beam in propagating through the two fiber arms which includes the constant phase difference in fiber lengths as well as any difference due to external disturbance. Here, $S_1 = D + d/2$ and $S_2 = D - d/2$

If the interferometer uses 50:50 beam splitting, then the time averaged intensity produced by a photo detector becomes $I = I_0 \cos^2\left(\frac{\phi}{2}\right)$ where I_0 is the input intensity. We observe from this expression that the intensity as produced at any point on the screens is a function of the phase offset of the two light beams. Thus, the intensity at any point on the screen is a function of the phase difference between the two light beams propagating in the two fibers and as we may note this phase difference is a constant as long as the experimental set up is not altered. However, any external perturbation affecting the two light beams in a differential manner will change $(\phi_2 - \phi_1)$ which would alter the interference pattern. This differential phase change by a external stimulus is the basis for optical fiber based sensors designed to sense physical parameters like pressure, strain, temperature, magnetic field etc.

11.4.2. Fiber Based Fabry-Perot (FFP) Sensors

This is an interferometric type of sensor that is realized from single mode fibers. Single mode fibers are not suitable for industrial applications due to the small size of their core. Hence multimode fiber based sensors having comparable sensitivity and versatility of single mode fibers are used in such applications. An FFP is basically a piece of fiber with reflectors at its ends. The reflectors may be external attachments or thin film depositions. In the core of FFP resonator, the phase of monochromatic light having wavelength becomes

$$\phi = \frac{2\square 2\pi}{(\lambda/n)} L = \frac{4\pi n}{\lambda} L \qquad \text{...(11.17)}$$

where the fiber piece has length L and the core has a refractive index of n.

A shift in the optical wavelength or a change in the optical path length leads to a corresponding change in the phase $d\phi$ given as

$$d\phi = \frac{4\pi}{\lambda}\left[d(nL) - \left(\frac{nL}{\lambda}\right) d\lambda\right] \qquad \text{...(11.18)}$$

The above two conditions lead to the operation of the FFP as a sensor and also as a spectrum analyzer. Any physical quantity x that affects the optical length produces a corresponding phase change.

$$d\phi = \frac{4\pi}{\lambda}\left[n\left(\frac{dL}{dx}\right) + L\left(\frac{dn}{dx}\right)\right] \qquad \text{...(11.19)}$$

To realize such a sensor, as shown in Fig. 11.23, light from a single mode laser is coupled into a single mode fiber which is butt-jointed to the FFP. Light waves coming out of the FFP are detected by passing it through a multimode fiber. Any change in temperature causes direct variations in the optical path length of the FFP. However, measurands like electric current, pressure, vibration etc. need to be converted into a suitable compatible form before they can cause change in the optical path length. The sensor is bonded to a cantilever and the FFP signal is measured during this bonding. The bonding amplitude is measured by an inductive transducer used in a thin film thickness measuring instrument. The FFP fringes thus counted are plotted against the bonding amplitude as shown in Fig. 11.24. As we may observe, the number of fringes increase with the bonding amplitude.

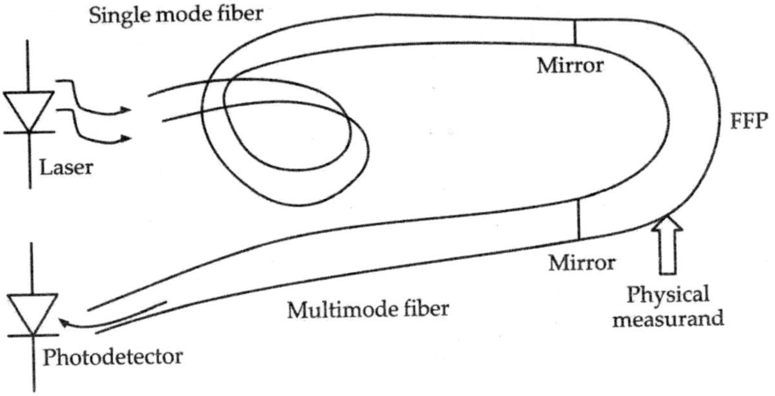

Fig. 11.23: FFP based Interferometer

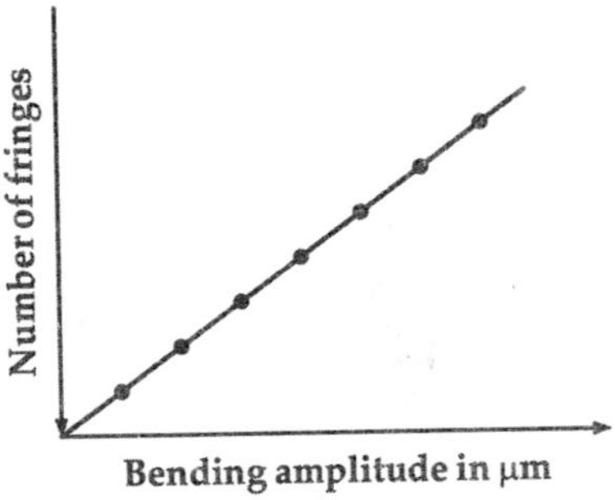

Fig. 11.24: Plot of no. of fringes as a function of bending amplitude as measured from the centre of the FFP

The FFP is a sensor that works exclusively on the principle of interference. Hence, it is necessary to count the number of fringes produced which are due to the measurands affecting the optical path length of light waves propagating through the multimode fiber as shown in Fig. 11.23. We further, note that, the number of fringes is independent of light intensity launched from the laser into the single mode fiber as well as that of the detected light. Moreover, we note that the aging of the optoelectronics components do not affect the number of fringes produced, thereby ensuring its accuracy of calibration over a prolonged use of the sensor.

11.4.3. Interferometric Temperature Sensor

This is a temperature sensor that utilizes the principle of phase variations. The fiber gets lengthened out when subjected to a rise in temperature which increases the path length of light waves propagating along it. This changes the phase of the light reaching the detector. The reference fiber provides a phase reference for the phase comparator. The working principle of this sensor is based on the change of a given fiber length L due to temperature. The fiber subject to a change of length is placed as the sensor arm of the interferometer. A change in temperature brings about a length change in the following two ways:

(i) Thermal expansion or contraction of length

(ii) Temperature dependence of the refractive index that results in a change of the propagation constant.

The resulting change in phase because of this propagation constant change is

$$\delta\phi = \beta\delta L + L\delta\beta$$

$$\Rightarrow \qquad \frac{\delta\varphi}{\delta T} = \left(\frac{2\pi}{\lambda}\right)\left[\left(\frac{n}{L}\right)\left(\frac{\delta L}{\delta T}\right) + \left(\frac{\delta n}{\delta T}\right)\right] \qquad \text{...(11.20)}$$

We write, further the change in length only as a function of temperature as

$$\frac{\delta L}{\delta T} = \frac{L(T_2) - L(T_1)}{T_2 - T_1} \qquad \text{...(11.21)}$$

where $L(T_2)$ and $L(T_1)$ are the fiber specimen length at temperature and respectively

and
$$\frac{\delta n}{\delta T} = \frac{n(T_2) - n(T_1)}{T_2 - T_1} \qquad \text{...(11.22)}$$

and $n(T_2)$ and $n(T_1)$ are the fiber core refractive indices at temperature and respectively.

11.4.4. Sagnac Effect Based Rotation Sensors (Optical Gyroscope)

A sensor to sense and consequently measure angular rotation constructed using optical fibers is based on the Sagnac effect. Such a sensor is shown in Fig. 11.25. This has a loop of polarization maintaining single mode fiber connected to a pair of directional couplers, a polarized source and a detector. Light from the source is split into two equal parts by the coupler close to the fiber loop. One part of light goes clockwise while the other part travels counter clockwise through the loop. After completing the loop travel, the two light beams are recombined at the same coupler. The part of light returning to the other coupler DC1 then has 50% of it detected and processed. We observe that when the entire arrangement is not rotating, the times of loop traversal by the two light beams are the same.

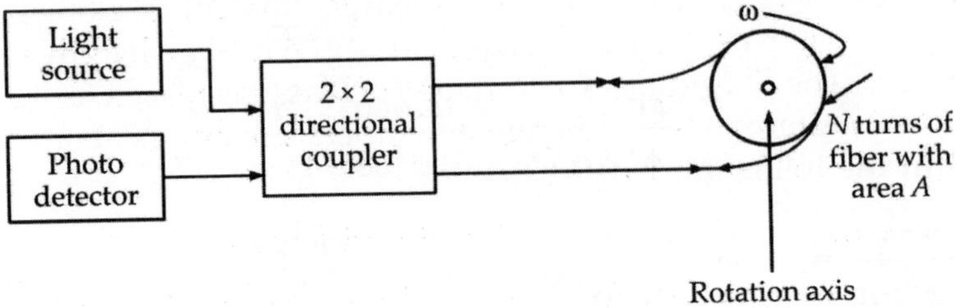

Fig. 11.25: Rotation Measurement by optical fiber based sensor or gyroscope

As a directional coupler introduces a phase shift of 90° between the two output arms, the coupler DC2 provides an additional phase shift of 90° if the loop itself does not introduce any phase shift. As a result, all of the original power returns to DC1. However, when the entire arrangement rotates with an angular velocity, the phase difference introduced between the two beams is given by

$$\Delta\phi = \frac{8\pi NA\omega}{c\lambda_0} \qquad ...(11.23)$$

Here N is the number of turns in the fiber loop, A is the area enclosed by one turn and of course, λ_0 is the free space wavelength. This is derived as follows.

As we have observed, when the fiber loop is placed on a nonrotating plane, the traversal time by the light beams is given by $t = \frac{2\pi r}{c}$ where is the radius of the loop. For non circular loops, the total length travelled by the beams would be equal to the perimeter of it irrespective of its shape. And as we may note, a loop having number of turns would correspondingly make the traverse time higher by a factor of N. When the fiber is placed on a plane rotating in a clockwise direction, the clockwise beam has a longer path to cover whereas the counter clockwise beam gets a shorter path for travel. Hence, the longer path length becomes

$$L_{long} = 2\pi r + r\omega \text{ and } L_{short} = 2\pi r - r\omega t \qquad ...(11.24)$$

Therefore, the difference between the times corresponding to the two paths become

$$\Delta_t = \frac{\Delta L}{c} = \frac{L_{long} - L_{short}}{c} = \frac{2r\omega}{c}t = \frac{2r\omega}{c}\cdot\frac{2\pi r}{c} = \frac{4\pi r^2}{c^2}\omega \qquad ...(11.25)$$

As there is a time difference between the two light beams in travelling a given length of the fiber loop, this would give rise to a corresponding phase difference given as

$$\Delta\phi = 2\pi\Delta t = \frac{8\pi^2 r^2}{c^2}\omega = \frac{8\pi A}{c^2}\omega \qquad ...(11.26)$$

In the above we have used the area of a circular loop as $A\pi r^2$. It is obvious from all these discussions that a loop with N turns would correspondingly produce a time difference times that of a single turn loop. Accordingly, the phase difference N between the two light beams is given by

$$\Delta\phi = N2\pi\Delta t \qquad ...(11.27)$$

11.5. FIBER SENSORS BASED ON PHASE AND POLARIZATION MODULATIONS

Interferometric technique is used in a very sensitive single mode fiber as phase sensors. The basic principle of such types of sensors is illustrated in Fig. 11.26(a). The sensor consists of a laser that emits a monochromatic light beam. This passes through a beam splitter realized as a glass plate inclined at an angle of 45° to the direction of the laser beam. The beam splitter is used to divide the beam in two directions at right angles to each other. The main beam passes through lens L_1 and then subsequently through a reference fiber that is made free from any external disturbance.

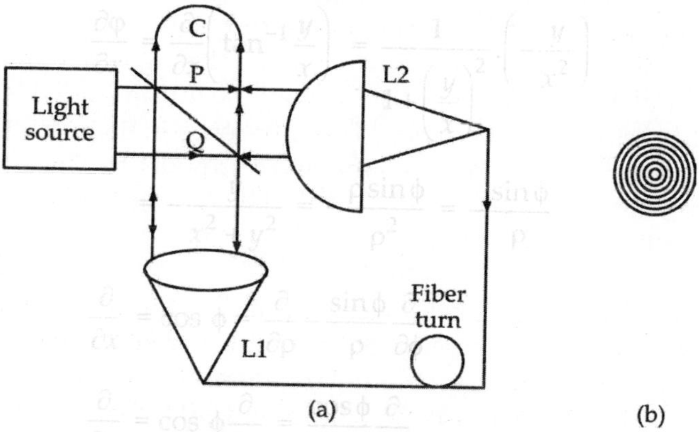

Fig. 11.26: (a) Optical fiber based Sagnac effect interferometer (b) Circular interference pattern

The beam strikes a second lens L_2 after emerging from the other end of the reference fiber. The split beam obtained in this manner (which is at right angles to the main beam) passes through a third lens L_3 from which it is launched into the test fiber. This test fiber is kept in an environment which is to be measured and finally this beam passes to the lens L_2 from the other end of the test fiber.

An interference of the main laser beam with the reference fiber guided laser beam is thus produced in the focal plane of L_2 in a fiber subject to strain or temperature change. The effect of a strained pressure or temperature change may give rise to a change in fiber length, core diameter or refractive index with respect to the reference fiber resulting in a differential path length. This produces a phase difference of the light beams emitted from the two fibers giving rise to circular interference pattern as shown in Fig. 11.26(b). A very accurate measurement of temperature or pressure is possible through the use of such interference pattern. Such an arrangement is known as a Mach Zehnder configuration. It is observed that the temperature sensitivity is about 10^7 rad/C/m for a typical two arm intefereometer that uses fused silica glass.

11.5.1. Stress Sensor

The photo elastic effect is an effect observed in certain transparent materials that exhibit a phenomenon known as birefringence. This is a phenomenon in which light rays passing through photo elastic materials experience two refractive indices. The photo elastic materials exhibit birefringence when subjected to stress. The magnitude of the refractive indices at each point in the material is directly related to the states of stress at that point. A polariscope provides information on maximum shear stress and its orientation by analyzing the birefringence obtained from such materials. The electromagnetic wave components are resolved into two principal stress directions when a ray of light passes through this kind of photo elastic material. Each component of the wave so thus resolved is associated with is a different refractive index. The difference in the refractive indices creates a phase difference between the two components. The magnitude of the relative phase difference ϕ is expressed as (for an isotropic material)

$$\phi = \beta C(S_1 - S_2)t \qquad \text{...(11.28)}$$

where C is the stress optic coefficient, S_1 and S_2 are the first and second principal stresses, t is the photoelastic material specimen. This phase difference obviously changes the polarization of light. The polarizations are combined by the polariscope before and after light is made to pass through the given material. We get interference between the two waves that gives rise to a fringe pattern. The number of fringes is expressed as

$$N = \frac{\phi}{2\pi} \qquad ...(11.29)$$

Hence by counting the number of fringes, it is possible to find out the state of stress at each point in the material.

11.5.2. Current Sensors

Measurement of current by means of optical fiber based sensors is an attractive proposition. This is because of the fact that electrical isolation is not an issue in these sensors as the fiber itself is a dielectric. Such sensors are also immune to electromagnetic interference, have a high bandwidth and potentially have a lower cost. The principle of Faraday effect, (a magneto optic effect) in glass in used to build such a sensor. This effect refers to the rotation of the plane of polarization of a linearly polarized wave travelling through a medium in which a strong magnetic field is applied along the direction of propagation. Application of a longitudinal magnetic field to the medium makes it circularly birefringent-that is the right circularly and left circularly polarized waves travel with different velocities. The amount of rotation is, it may be argued that is proportional to the applied magnetic field and the length of interaction. Mathematically, this becomes

$$\theta = V \int \mathbf{H}.d\mathbf{l} = VI \qquad ...(11.30)$$

where θ is the amount of rotation of the plane of polarization, **H** is the applied longitudinal magnetic field. The quantity $d\mathbf{l}$ is an elementary length of the total length over which the magnetic field and light interact. We note that, the quantity inside the integrand is nothing but the total current enclosed by a circular path of length dl as per Ampere's circuital law. Further, the direction of rotation can be reversed by reversing either the direction of the magnetic field or the direction of light propagation. Hence, the constant V known as Verdet constant must have the dimensions of degree/A. For example, silica has a V of 2.64×10^{-4} deg/A = 4.6×10^{-6} rad/A. This constant gives an idea about how strong the rotation is in a given material. An experimental setup as shown in Fig. 11.27 is used for current measurement. As illustrated, the set up has a single mode fiber which is wound around the current carrying conductor in the form of a helix. The magnetic field created due to this current carrying conductor brings about a rotation of the plane of polarization of the light passing through it. The current becomes, for N turns of the fiber,

$$\mathbf{H}.dl = NI \qquad ...(11.31)$$

Therefore, we have

$$\theta = VNI \qquad ...(11.32)$$

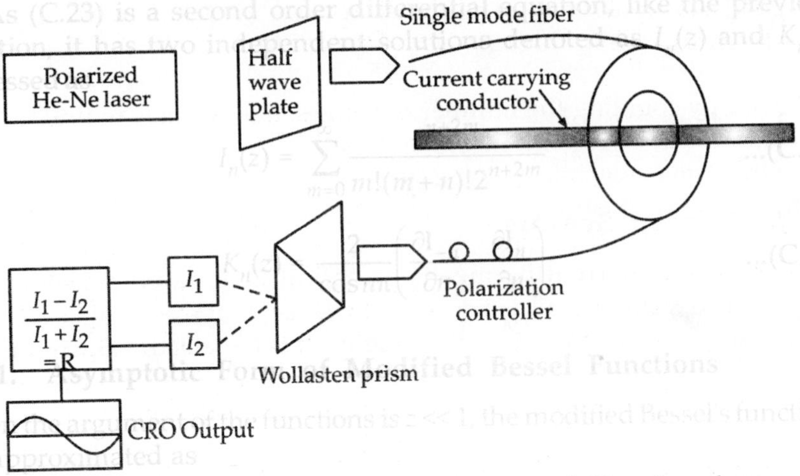

Fig. 11.27: Measurement of current by optical fiber based sensor based on the principle of Faraday rotation

We have assumed that the fiber maintains the state of polarization during the interaction of light and the magnetic field and the only rotation is due to the applied magnetic field. However, this is not true in practice. This is because real fibers do have some amount of random birefringence that leads to a change in state of polarization (SoP) even in the absence of an external field. In addition, bending the fiber also causes linear birefringence. Smaller the loop radius, larger is the birefringence. The signal R, in the presence of linear and circular birefringence is given as

$$R = 2\theta\frac{\sin\Delta}{\Delta} \quad ...(11.33)$$

where

$$\Delta^2 = 4\theta^2 + \delta^2 \quad ...(11.34)$$

and

$$\delta = \frac{2\pi}{\lambda_0}\Delta n_{eff}\, 2\pi RN \quad ...(11.35)$$

Here Δ represents the total phase shift due to the linear birefringence Δn_{off} and θ is the Faraday rotation in radians. This is because the phase offset between the two circularly polarized modes is 2θ. We may approximate the signal as

$$R \approx 2\theta\frac{\sin\delta}{\delta} \text{ for } \delta >> \theta \quad ...(11.36)$$

and

$$R \approx \sin 2\theta \text{ for } \delta << \theta \quad ...(11.37)$$

We may note that a dominant linear birefringence makes the first condition prevail giving a low sensitivity to the sensor. On the other hand, a dominant circular birefringence makes $\theta >> \delta$ and this makes a very good sensitive sensor. The signal in the fiber is independent of any linear birefringence in this case.

11.5.3. Magnetic Field Sensor

Using the same principle we described for a current sensor, we can also use the Faraday effect to sense the magnetic field. We further note that, the Verdet constant V depends on the refractive index dispersion given as $dn/d\lambda$. This has been shown to be

$$V = \frac{\theta}{t\mathbf{B}} = -\frac{1}{2}\frac{q\lambda}{mc}\frac{dn}{d\lambda} \qquad ...(11.38)$$

In the above expression, q/m represents the ratio of one electronic charge to its mass. We note that, by measuring the rotation angle, we do have a method for determining the magnetic field strength **B** when the thickness of the sample through which the light beam propagates t is known. For certain type of materials, the refractive index decreases with respect to increasing wavelength as a result of which the quantity $dn/d\lambda$ is negative. It means that a light beam travelling along the direction of the magnetic field has its plane of polarization a experiencing a counter clockwise rotation. With a good Nicole prism, rotation to the tune of one-half a minute can be measured. As we have an interaction of the light beam and the magnetic field spanning over a few centimetres of length in a typical experimental setup, it becomes quite easy to observe the Faraday effect with a magnetic field of the order of a few thousand Gauss.

As we may observe from the above, the current and magnetic field sensors are based on polarization modulation of a given light beam.

11.6. FIBER BRAGG GRATING (FBG) BASED SENSORS

11.6.1. Acceleration Sensor

The conventional sensors used to measure acceleration are based on piezoresistive, piezoelectric or capacitive principles. These measure the motion of the specimen under test through the current induced by the inertia forces acting on the specimen. As the output of these sensors is usually small, it needs signal amplification before being converted to a voltage that can be directly calibrated in terms of acceleration. However, an optical fiber based accelerometer is built around a Fiber Bragg Grating (FBG). The underlying principle of such an accelerometer is the photosensitive effect first observed by [10-11] as observed in optical fibers. A sensor employing such photosensitive effect is shown in Fig. 11.28.

This refers to a permanent change in the refractive index of the optical fiber when it is subjected to UV light. Photosensitivity results in a permanent change of the refractive index/opacity of the glass when it gets exposed to UV light. In case of the Bragg grating formation, the change in the refractive index of the optical fiber is brought about by exposure to intense UV light. An interference pattern of the UV light is usually created to be focussed onto the core of the optical fiber so that it experiences a periodic change in its refractive index. As we might guess, the refractive index of the core experiences the change where the intensity of the interference pattern is the maximum. The length of one single period of the grating structure is known as the grating pitch Λ which is controlled during the manufacturing process. This is typically of the order of typically ~0.5 μm and the range of refractive index change observed is of the order of 0.1 to 0.01 percent of the original refractive index. Each FBG has a unique wavelength that is related to the effective refractive index and the grating pitch as

$$\lambda_B = 2n_{off}\Lambda \qquad ...(11.39)$$

The average refractive index of the core is denoted as

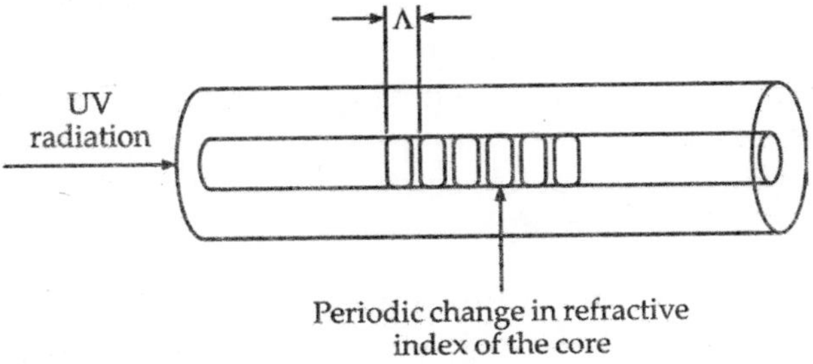

Fig. 11.28: Illustration of the photosensitive effect

Strain and temperature change the effective refractive index as well as the grating pitch. The Bragg wavelength shifts to lower or higher wavelengths in response to the applied thermal mechanical fields. This shift in the Bragg wavelength is considered to be a linear function of the applied thermal mechanical load for most of the applications. This dependence is indicated by the following expression

$$\frac{\Lambda\lambda_B}{\lambda} = \left[1 - \frac{n_{eff}^2}{2}(p_{12} - v(p_{11} + p_{22})\right]\varepsilon_z \qquad ...(11.40)$$

This is further simplified as

$$\frac{\Lambda\lambda_B}{\lambda} = P_e\varepsilon_z \qquad ...(11.41)$$

The FBG is coupled to a mechanical load in order to measure acceleration. The optical fiber is modelled as a mechanical rod with a spring constant of $K = Y/AL$ where Y is Young's modulus of elasticity which is 69 GPa for a typical monomode fiber, A is its area and the length is L. The FBG is optically inscribed into the optical fiber. The FBG sensor measures acceleration as the mass applies a resultant force to the optical fiber. The strain experienced by the fiber is expressed as

$$\varepsilon = \left(\frac{M}{YA}\right)a \qquad \text{...(11.42)}$$

where the acceleration is a assuming a linear model for such a sensor. A diagram illustrating this is shown in Fig. 11.29.

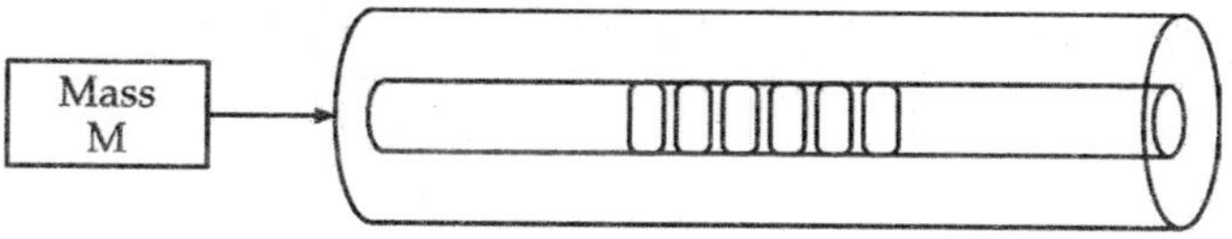

Fig. 11.29: Measurement of acceleration

This arrangement can also be used for measuring force when the mass of the object is known.

11.6.2. Humidity Sensor

Humidity can be detected by an FBG. The principle of such a sensor is the strain effect induced in the FBG due to the swelling of the moisture sensitive polymer coating given to the FBG. One such sensor is diagrammed in Fig. 11.30. We may anticipate that the performance of such a humidity sensor is largely determined by the type of polymer coating. The humidity response of the polymer coating in terms of the Bragg wavelength shift is monitored as a function of the swelling of the coating that stretches the fiber. The material should have high moisture swelling property and and the strength of the hydrated material should be quite good. A polymer coating like polyimide has a linear expansion of its volume with respect to moisture. Hence humidity sensors constructed with this kind of coating exhibit a linear and reversible response with humidity change. The sensor is fabricated by dip-coating the gratings in a polyimide solution. The Bragg wavelength shift increases with increase in humidity. At a constant temperature, a linear response is exhibited when the humidity content increases from 23 to 97%. A thicker polymer coating increases the sensitivity of this sensor. However, it results in a slow response time. The optimum polymer coating thickness has been reported to be 24 μm.

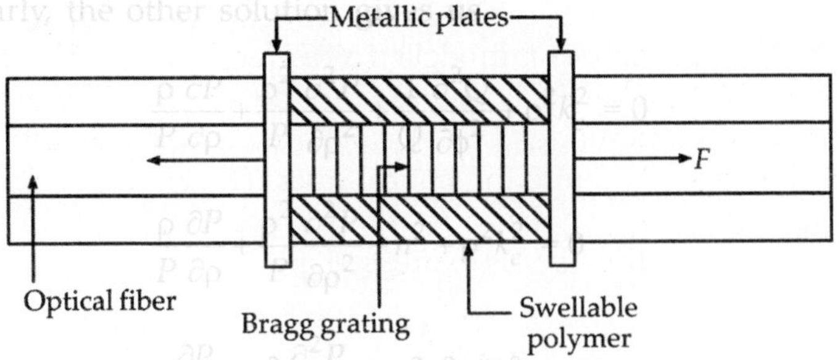

Fig. 11.30: Swell able polymer based FoS humidity sensor

11.7. Laser Doppler Velocimeter (LDV)

Measurement of point velocity in certain fluids is an important task. For example, in water, instantaneous point velocities are measured close to a solid boundary as in boundary value problems and random fluctuations of point velocity in turbulent flows. The two methods commonly available for carrying out this point velocity measurement are the hot film anemometer and laser Doppler velocimeter. The second category is a no probe method for measuring instantaneous point velocities in fluid flows. It measures point velocity at a given point in a fluid flowing in a glass walled conduit or channel by detecting the shift in frequency of a scattered light coming from suspended particles embedded in a given fluid flow that happen to cross the point of measurement defined by two intersecting laser beams.

Doppler effect as observed in light is used to measure velocity of particles in a turbulent flow. When light is reflected from a moving object, the frequency of the scattered light gets shifted, by an amount proportional to the velocity of the object. If this shift can be measured, then it is possible to determine the velocity of the object. The turbulent flow is impregnated with small, neutral buoyant particles that scatter light. The particles are illuminated by a known frequency that is readily provided by a laser. The scattered light is collected by a suitable photodetector (may be a photomultiplier tube). An instrument designed to measure particle velocity contained in a fluid has four units: a He-Ne laser, a beam splitter, a light pick-up unit and a signal processor.

We observe from Fig. 11.31 that the angle between the velocity vector and the perpendicular bisector of the angle PQR is denoted by θ which is also known as the direction of motion of the particle. The orientation of the photodetector is observed to be at an angle of ϕ with respect to the incident light beam. The Doppler shift in frequency is given by

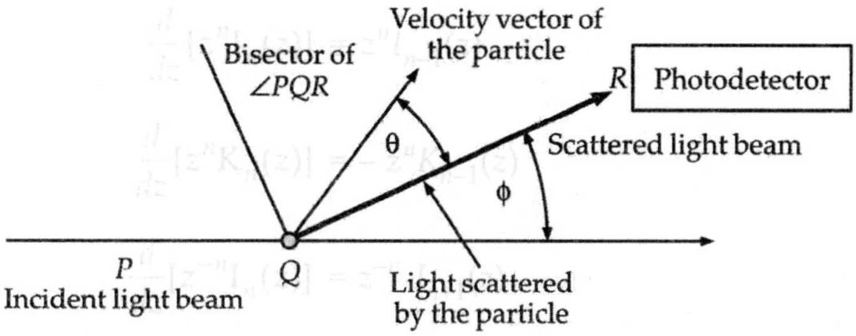

Fig. 11.31: Schematic of the principles of laser Doppler velocimetey

$$f_D = \frac{2V_x}{\lambda}\cos(\theta)\sin\left(\frac{\phi}{2}\right) \qquad ...(11.43)$$

The Doppler frequency shift can be measured by determining the frequency observed at the photodetector and that of the incident light. However, the shift in frequency being very small as compared to the incident frequency, a direct measurement is prone to gross errors. To circumvent this, two beams are utilized. In this method, the initial light beam is split into two halves of equal intensity. The beams are directed to intersect at a given point and this point serves as the measurement volume. This is illustrated in Fig. 11.32.

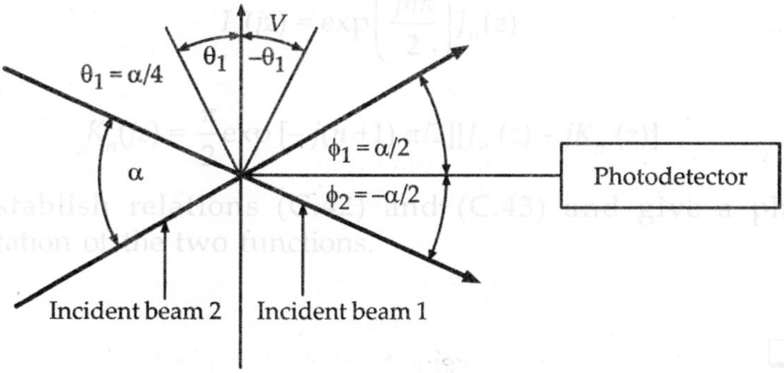

Fig. 11.32: Modified difference method to measure the Doppler frequency shift

We can observe from Fig. 11.32 that particles that pass through the measurement volume scatter both the light beams. The frequency of the two light beams thus scattered are in general different. The complete schematic of such a method is shown in Fig. 11.33.

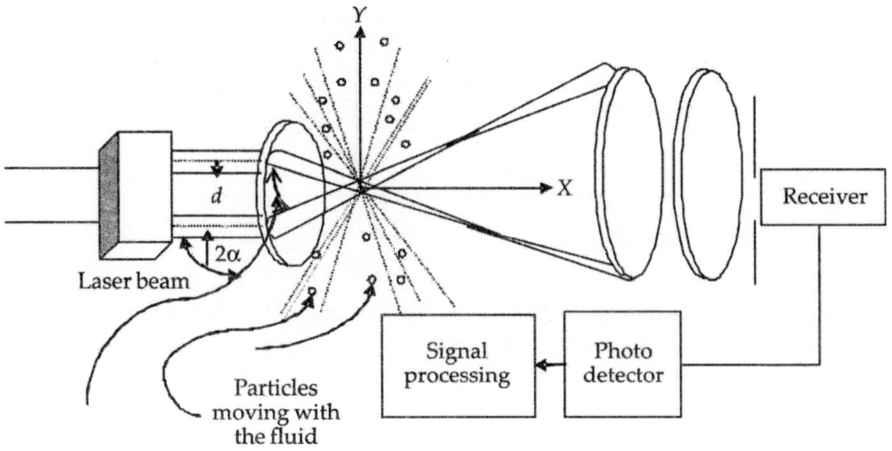

Fig. 11.33: Schematic of LDV

We observe from Fig. 11.33 that a laser beam is divided into two halves by the transmit optics that also focuses and crosses the two beams by means of a suitable lens arrangement. As we have seen earlier, the intersection region of the two beams define the measurement volume in which the fluid strewn with particles pass through. The particles, in this volume scatter the two beams that are collected by the receiving unit. The photodetector converts this light into an appropriate electrical signal. The frequency of this signal is proportional to the particle velocity. The signal processing unit extracts this frequency information from the signal developed by the photodetector and provides a digital output that indicates the instantaneous velocity of the particle. An optical heterodyning takes place at the photodetector. As a result, a difference frequency known as the Doppler frequency is produced. The Doppler frequency of the heterodyned signal at the output of the photodetector is given as

$$f_D = \frac{2V_x}{\lambda_0}\sin\alpha \qquad ...(11.44)$$

where V_x denotes the particle velocity along the x-direction.

We further define a quantity

$$d_f = \frac{\lambda_0}{2\sin\alpha} \qquad ...(11.45)$$

Hence,

$$f_D = \frac{V_x}{d_f} \qquad ...(11.46)$$

Therefore, measurement of f_D gives an indication of the points velocity v_x.

11.8. pH SENSOR

An ion selective fiber optic chemical sensor (FoCS) is constructed by immobilizing the distal end of a fiber optic probe with a fluorescent dye. For the detection of aluminium ions, morin is immobilized. When aluminium trace is to be tested, the chemical reaction between the two generates a highly fluorescent complex that can be monitored by fluorescent measurement. To detect the presence of oxygen molecules, fluorescence quenching of immobilized fluorophore trapped behind a gas permeable chamber is carried out. A gas permeable membrane separates the sample from a layer that contains the pH indicator dye. The acidic or basic gas is made to cross the membrane, enters the indicator layer and undergoes a proton transfer reaction with the dye. The extent of this reaction is monitored spectroscopically through a fiber optic probe. Gases like ammonia, carbon dioxide, hydrogen cyanide and nitrogen dioxide can be analyzed by this method. Non gaseous substances can be tested by direct spectroscopic methods. The fiber optic probes are used to guide light to a distant sampling site and then to transmit a portion of the resulting radiation from the sample back to the spectrometer. This method is known as remote spectroscopy. This is a reagent less process and relies upon the high quality spectra obtainable through optical fibers. Such pH sensors are shown in Fig. 11.34(a) and (b).

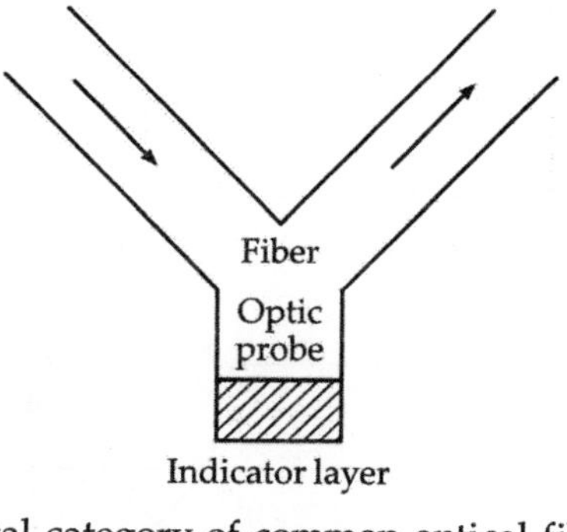

Fig. 11.34: (a) Distal category of common optical fiber based chemical sensor design

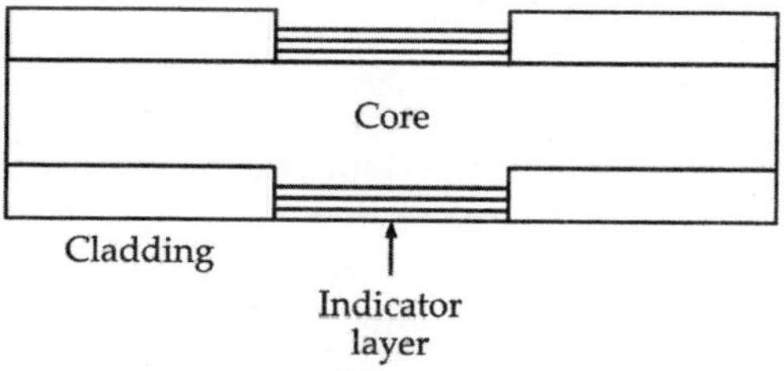

Fig. 11.34: (b) Evanscent wave type of common optical fiber based chemical sensor design

11.9. DISTRIBUTED OPTICAL FIBER SENSORS

Distributed optical fiber sensors make simultaneous measurement of both the spatial and temporal behaviour of measurands field by utilizing certain special properties of an optical fiber. The distributed nature of sensing makes them attractive especially for large structures with a fine monitoring of mesaurand behaviour. It is possible to monitor the behaviour of a mesaurand continuously as a function of position along a length of a suitably configured fiber. It is important and necessary to obtain accurate information of the spatial/temporal behaviour of strain and temperature from the point of view of both safety monitoring and of improved understanding of their behaviour under non conducive conditions in typically large structures such as dams, bridges, multi-storied buildings, air crafts, space crafts, boilers, electrical generators etc. The flexibility of fiber makes it relatively easy to install over the chosen measurement path and thus allows retrospective fitting unlike other sensor systems. Both intrinsic and extrinsic fiber sensors may be multiplexed onto a single fiber using a host of criteria to carry out sensing and measurement of a large number of measurands. The criteria used are time, frequency, wavelength, coherence, polarization and spatial multiplexing.

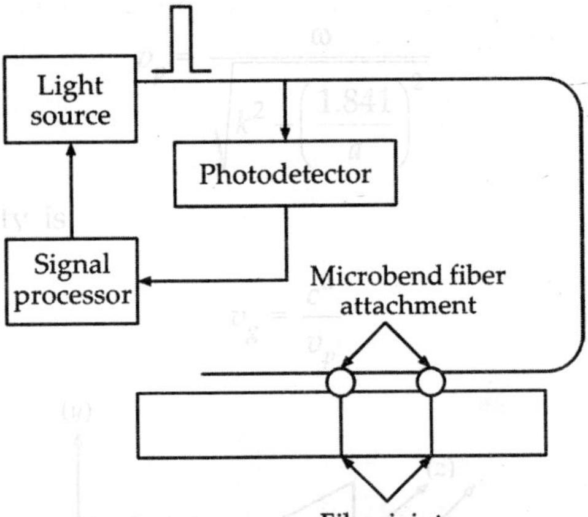

Fig. 11.35: A TDM system using microbend sensitive areas on fiber joints

A distributed sensing system using time division multiplexing is shown in Fig. 11.35. Here the light source launches a light pulse onto the fiber. The light pulse travelling down a given length of the fiber experiences different delays due to different sensors. Analysis of these delays gives an indication of the location of the sensors. The entire fiber can be made micro bend sensitive and Rayleigh scattering loss ex encountered on these distributed sensors predominantly used to measure strain. Other types of scattering from optical pulses propagating down optical fiber have been used to support distributed sensing, notably Raman scattering for temperature sensor. Brillouin scattering has been used in laboratory experiments to measure strain as well as temperature.

A distributed sensing built using the principle of frequency division multiplexing is illustrated in Fig. 11.36. Output from a laser is chirped by a saw tooth current waveform. Successive Mach-Zehnder type interferometers are offset with incremental lengths which differ from each other in such a manner that the carrier frequency used by each sensor is easily separable from the other sensors through suitable electrical signal processing.

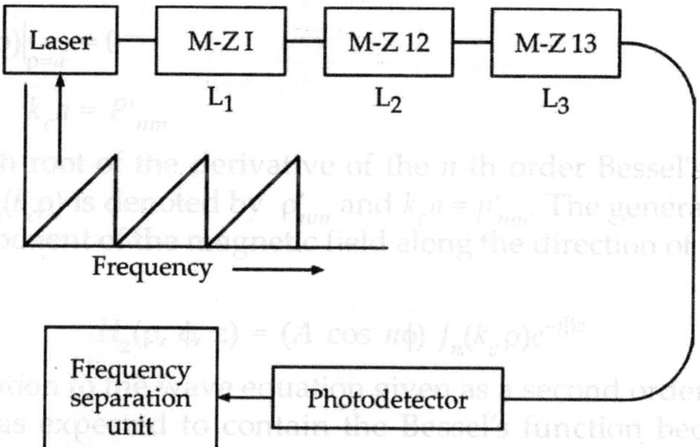

Fig. 11.36: FDM based optical fiber based distributed sensing

A very efficient method of utilization of input optical power for multiplexing applications is wavelength division multiplexing (WDM). A broad band light source such as an LED is coupled onto a series of fiber sensors that reflect signal over a wavelength band which is a subset of the source spectrum. A dispersive element such as a grating or a prism is used to separate out the signals from the sensors to the corresponding receivers. A method is illustrated in Fig. 11.37.

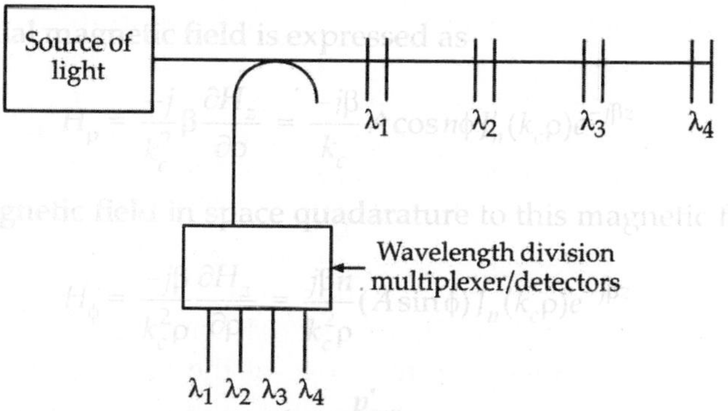

Fig. 11.37: WDM based optical fiber distributed sensing

11.9.1. Optical Time Domain Reflectometry (OTDR)

Optical reflectometry is a powerful technique for the characterization of optical fibers and components. Time domain techniques utilizing OTDR are used for the measurement of several kilometres with the resolution of a few meters. Intrinsic distributed sensors based on Rayleigh backscattering mechanism utilize either the mesaurand dependent loss coefficient or the backscattering coefficient in a single length of optical fiber that forms a kind of extended sensor. A typical experimental setup is illustrated in Fig. 11.38 to carry out loss measurement using OTDR.

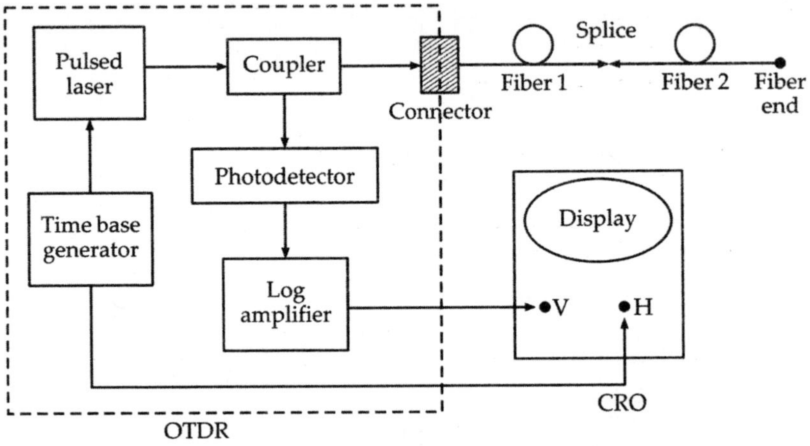

Fig. 11.38: Experimental setup for carrying out OTDR

Pulses coming from the injection laser are coupled into the fibers through the connector so that they arrive at fiber 2. The directional coupler changes the direction of the light pulse such that a part of it finds its way to the photodetector. This detector is blanked out to avoid saturation during the times the light pulses are moving on to the forward direction. Light pulses reflected from fiber 2 are collected and photodetected. For an ideal fiber, only Rayleigh scattering takes place uniformly along the whole length of both the fibers. For fibers with uniform loss, the intensity of the returned pulses decrease at a constant rate corresponding to the measured power expressed in dB. For ideal fibers, the returned pulses, hence maintain a constant amplitude with respect to time. The photodetected pulses pass through the logarithmic amplifier which in turn is connected to the vertical deflection plates of a CRO. The time base generator controlling the light pulses is fed to the horizontal deflection plates of the CRO. As a result, a fixed pattern is observed on the display. The loss per unit length of the fiber is obtained by measuring the slope of the pattern observed on the display.

However, OTDR measurement techniques are prone to dispersion effects (as a laser pulse is the driver) and hence are affected by a broadened spatial resolution at longer measurement distances. This further implies that the hot spot sensitivity of measurements relying upon such laser pulses deteriorate with distance.

The basic mechanisms as well as their applications in various sensing of OTDR utilizing Rayleigh backscattering is presented in Fig. 11.39.

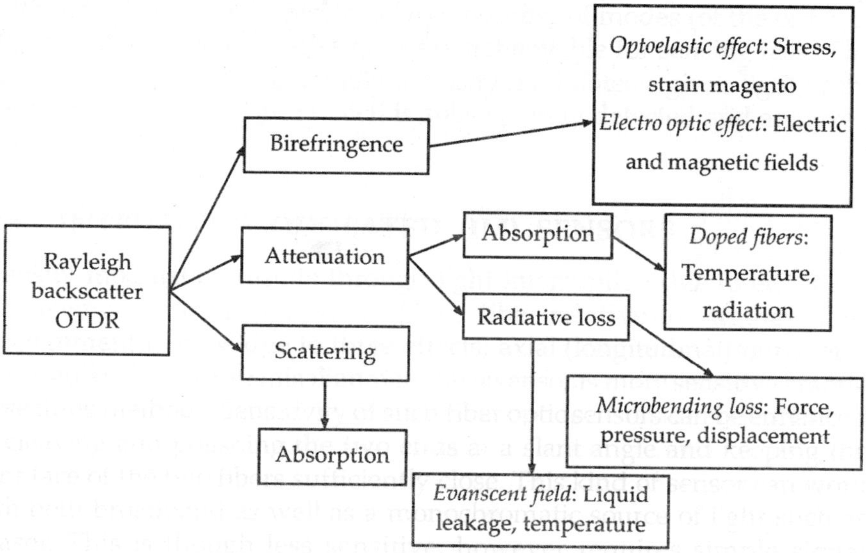

Fig. 11.39: Mechanisms and sensing applications utilizing OTDR

11.9.2. Optical Frequency Domain Reflectometry

OFDR techniques provide resolution in the range of millimetres. Measurands like pressure, temperature and tensile forces do affect glass fibers which locally change characteristics of light transmission through the fiber. Light waves propagating in doped quartz glass fibers are attenuated following backscattering. This can be exploited to determine the location of the external physical effect. Quartz glass is a form of silicon dioxide (SiO2) with amorphous solid structure. Thermal effects induce lattice oscillations within the solid. When light falls onto these thermally excited molecular oscillations, an interaction occurs between the light particles (photons) and the electrons of the molecule. Light scattering, also known as Raman scattering, occurs in the optical fiber. Unlike incident light, this scattered light undergoes a spectral shift by an amount equivalent to the resonance frequency of the lattice oscillation. The light scattered back from the fiber therefore contains three different spectral components:

(a) Rayleigh scattering with the wavelength of the laser being used
(b) Stokes component with a longer wavelength corresponding to photon generation
(c) And the anti Stokes component with smaller wavelength than Rayleigh scattering leading to destruction of photons

The Stokes and anti-Stokes light components are the basis for distributed temperature sensing by optical fibers. The intensity of the so-called anti-Stokes band is temperature-dependent, while the so-called Stokes band is practically independent of temperature. The local temperature of the optical fiber is determined from the ratio of these two components of light intensity.

An example of a typical distributed temperature sensing system utilizes OFDR. An OFDR system makes use of backscattered light similar to the OTDR described earlier, however with an important difference. As the name suggests, OFDR furnishes information about the mesaurand by measuring the backscatter light as a function of frequency unlike the time domain dependence as in OTDR. An OFDR system provides information on the local characteristics only when the backscatter light detected during the entire measurement time is measured as a function of frequency in a complex manner upon which subsequent Fourier transformation is applied. The advantages of such a method are the quasi continuous wave mode employed by the laser and the narrow band detection of the backscattered light. This results in an improved sensitivity in terms of a better signal to noise ratio. It also allows use of readily available components for signal averaging. The continuously back scattered Raman light is filtered in the optical processor that gets converted into an appropriate electrical signal by photodetectors. The resulting signal is then amplified and subjected to mixing in the low frequency range. Fourier transformation of the averaged low frequency signals produces two Raman backscatter curves.

The heart of an OFDR system is the Michelson interferometer. OFDR systems can be divided into two types: noncoherent and coherent. In the former, a continuous optical carrier is intensity modulated by a constant-amplitude RF signal whose frequency is changed periodically over a certain frequency range either in a stepwise manner or continuously known otherwise as the sweep frequency method. This carrier is then coupled into the system under consideration. Rayleigh backscattered and backreflected lights are detected as a function of the modulation frequency and processed suitably to obtain the frequency response of the fiber.

In the coherent OFDR method, the carrier frequency of the optical source is swept linearly and phase-modulated in time, and the frequency-modulated optical signal split into two paths, one of which goes to the system under consideration and the other is used as reference signal. Heterodyning the probe signal and the reference signal on the detector yields the difference frequencies that are related to the optical amplitude and phase response of the system under consideration. Fourier transform of the photocurrent provides the graph of reflections as a function of distance. This method is often called as coherent FMCW.

The OFDR as shown in Fig. 11.40 is based on coherent detection of the signal reflected from the system under consideration getting heterodyned with a fixed reflection coming from the local oscillator (LO). The laser frequency is swept linearly with respect to time so that the difference frequency produced between the two reflections is proportional to the distance between the LO and the reflection point. Fourier transformation of the photodetected signal allows visualization of multiple reflections. Let us consider a single reflection $R(l)$ coming from the fiber under test located at a distance of l from the LO. The time varying current generated at the photo detector as a result of the interference of the two light waves is

$$I(f) = I_0\sqrt{R_{LO}R(l)}\exp(-2l/L_c)\cos\left(\frac{4\pi fn}{c}l - \theta\right) \qquad ...(11.47)$$

where n is the effective refractive index of the fiber, R_{Lo} and $R(l)$ are the reflection coefficients of the local oscillator and the measured reflection respectively, I_0 is the laser amplitude and θ is a phase. We note that, at a given distance l, the amplitude of the current is a sinusoidal function of frequency f. The frequency of the sinusoidal component is directly proportional to the distance between the reflection and the LO. This forms the basis of an OFDR in which reflections from FuTs placed at different lengths or distances can be distinguished from one another by means of the Fourier transform of the photodetected current as given by (11.46).

A peak is obtained in the temporal Fourier transform of this current that corresponds to a reflection from a given FuT for a linear frequency such as the one shown in Fig. 11.40.

A comprehensive summary of the effects utilized in OFS s is made in Table 11.4.

Finally, a number of important parameters to consider while selecting a sensor for a given measurand are normally the following:

- Dynamic range (the minimum and the maximum value of the measurand)

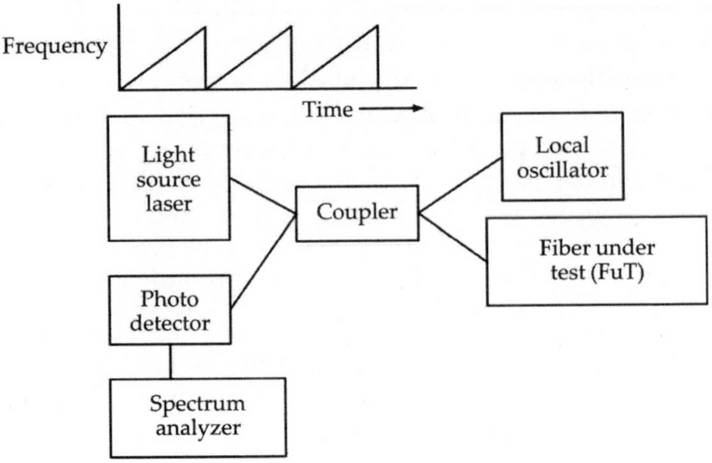

Fig. 11.40: A typical coherent OFDR set up

Table 11.4: Effects/Modulations based Measurand sensing

Sl. No.	*Measurand*	*Type of Sensor*
1	Strain	FBG, Phase modulation based
2	Angular rotation	Sagnac effect interferometer (phase modulation)
3	Pressure	Polarization based
4	Displacement, position	Intensity modulation
5	Liquid level	Intensity modulation
6	Biomedical	Spectroscopic methods, Laser Doppler velocimeter
7	Electric field, magnetic field	Polarization based

- Linearity (whether the output of the sensor follows a linear variation with the measurand value over its dynamic range)
- Accuracy (usually expressed as a percentage of the nominal value of the measurand). It tells us how close is the measured value is to the true value.
- Resolution: often known as the least count of the sensor. This gives an idea of the least value that can be measured by the sensor
- Precision: This is the ability to reproduce the measured value correct to an acceptable value over long times of sensor usage, aging or temperature variation
- Sensor size and weight: In what condition is the sensor being used? For example, a velocity of 1m/s may require a sensor that is significantly smaller in size and weight when a velocity of 1 km/s is being measured.

- Type of readout: analog or digital
- Frequency response: If the sensor is able to respond quickly to changes occurring in the value of the measurand
- Signal conditioning requirement: what kind of signal conditioning is required in order for the output from the OFS to produce a reliable measurement of the mesaurand

11.10. CHAPTER SUMMARY

The optical fibre based sensors provide an easy, rugged, low cost, light weight, electromagnetically immune solution to sensing a panorama of physical parameters in harsh and otherwise dangerous, hazardous environments. The basic principle of any such sensor is modulation of one or more one properties of light (light is viewed as a carrier). We have, therefore, intrinsic as well as extrinsic sensors that are based on intensity, phase, polarization or combinations thereof modulation of light carrier.

SUGGESTED READING

[1] Li, X., Yang, C., Yang, S. and Li, G. "Fiber-optical Sensors: Basics and Applications in Multiphase Recators", *Sensors*, vol. 12, no. 12, pp. 12519-12543, Sep 2012.

[2] Fraden, J. "*Handbook of Modern Sensors*", AIP Press, Woodbury, New York, 1997.

[3] http://www.howstuffworks.com

[4] Wang, W. "Intensity Modulation", University of Washington

[5] Hecht, E. "Optics", 2nd. Ed, 1987, Addison-Wesley

[6] Jenkins and White, "*Fundamentals of Optics*"

[7] Hill, K.O. *et al.*, "Photosensitivity in optical fiber waveguides: Application to reflection filter fabrication," *Applied Physics Letters*, Vol. 32, May 1978.

[8] Al-Bassyiouni, M., Yu, M., Balachandran B. and Oh, J. "Active acoustics control with fiber optic sensors," *Proceedings of SPIE 2002 Conference on Modelling, Signal Processing and Control*, Vol. 4693, Paper No. 4693-44, March 2002.

[9] Baldwin, C., Niemczuk, J., Kiddy, J., and Salter, T. "Review of Fiber Optic Accelerometers", Systems Planning and Analysis, Inc., Advanced Engineering Development Program, 4865 Walden Lane, Lanham, MD 20706.

[10] Kashyap, R. "Fiber Bragg Gratings", Academic Press, San Diego, 1999.

[11] Hill, K., Fuji, Y., Johnson, D. and Kawasaki, B. "Photosensitivity in optical waveguides: Application to reflection filter fabrication," *Applied Physics Letters*, Volume 32, No. 10, pp. 647, 1978.

[12] Systems Planning and Analysis, Inc., http://www.spa.com/aedp.htm

[13] Oberson, P., Huttner, B., Guinnard, O., Guinnard, L., Ribordy, G., Gisin, N. "OFDR with a narrow linewidth fiber laser", *IEEE Photon. Technol. Lett., June*, 2000.

[14] Weid von der, J.P., Passy, R., Mussy, G., Gisin, N. "On the characterization of optical fiber network components with OFDR", J. *Lightwave Technolgy.*, vol. 15, no. 7, pp. 1131-1 141, 1997.

[15] Yüksel, K., Wuilpart, M. and Mégret, P. "Optical-frequency domain reflectometry: roadmap for high-resolution distributed measurements", Proceedings Symposium IEEE/LEOS Benelux Chapter, Brussels, pp. 231-234, 2007.SS

12

Performance of Optical Receivers

12.1. INTRODUCTION

Data is transmitted in the form of electrical rectangular pulses (assumed) in either binary or non binary form. These data may come from various sources. Data may be generated by the keystrokes of a computer or may come from digitized voice or video signals. A LASER or an LED may be used as the optical transmitter that takes these digital pulses and produce a modulated output as determined by the application, cost and so on. The data rate of a typical system is considered to be R_b bits per second (bps) so that the duration of each bit is $T_b = 1/R_b$. For slow speed applications, R_b is in the range of a few bps or may be a few Kbps while Mbps of data rate is usually used for high speed applications. We can clearly see that it is just not enough to transmit data at a given rate. The data transmitted through an optical fibre are subject to losses and distortions of various kinds, as we saw in previous chapters. A pulse is attenuated due to the losses in the fibre and distorted simultaneously because of a number of distortions. Noise of several types also affect the data pulses in a number of ways. The receiver, as a result of all these impairments receives noisy attenuated and distorted pulses. The minimum signal that can be detected by a receiver depends on the noise(s) generated by the receiver itself whereas the maximum signal that can be handled is primarily determined by the overload or saturation conditions of the amplifiers used in a receiver. This range of minimum to maximum values of a signal that can be handled by a receiver is its dynamic range.

For recovery of the original message signal from the impaired received data pulses, the receiver needs to distinguish a given pulse from its adjacent pulses and satisfy a certain performance index. The performance index may be the average probability of bit error (BEP), the mean square error or signal to noise ratio at the output of the detector. Usually for systems communicating with digital data streams in binary or non binary format, the BEP is the most important performance metric. What is more important is the recovery of the original data from the noisy distorted data output coming from the optical fiber. The BEP is an important parameter for judging the suitability of a receiver for a given application. We next take a look at the ways a given signal or a data pulse undergoes the impairments while being transmitted from the optical transmitter to the receiver through the optical fibre.

We have shown the schematic of a typical optical transmitter/receiver pair including the optical fibre in Fig. 12.1. As we may observe, the optical fibre acts as the physical channel between the transmitter and the receiver. The transmitter is a LASER or a simple LED that acts the binary data bits as input. The light output of either of the transmitter gets modulated by these binary pulses. The modulated pulses are then launched onto the optical fibre. Unlike the channels used to carry digital information that are characterized mainly by additive white Gaussian noise with bandwidth limitations, the optical fibre has its own set of problems to be addressed to from an information transfer point of view. There are various formats to represent the ones and the zeros occurring in a digital data stream. For example, a one may be represented by a certain positive voltage whereas zero may be represented as simply zero volt. The formats known as line codes are specific to a given application and we do not make any attempt here to provide a detailed description of these line codes. We use the terms bits and symbols in the present context to mean data. A symbol is a group of bits obtained by some kind of serial to parallel conversion in the transmitter. However, again our interest here is not to focus on the symbols.

It may, therefore be summarized that the objective of an optical receiver is twofold; (a) to convert the incoming optical power into a corresponding electrical output (usually carried out by mean of an appropriate photodetector) and (b) to retrieve the original data embedded in the optical signal.

We can clearly see from Fig. 12.1(a) that the binary data pulses corresponding to ones and zeros have become rounded and look like kind of analog waveforms after passing through the optical fibre.

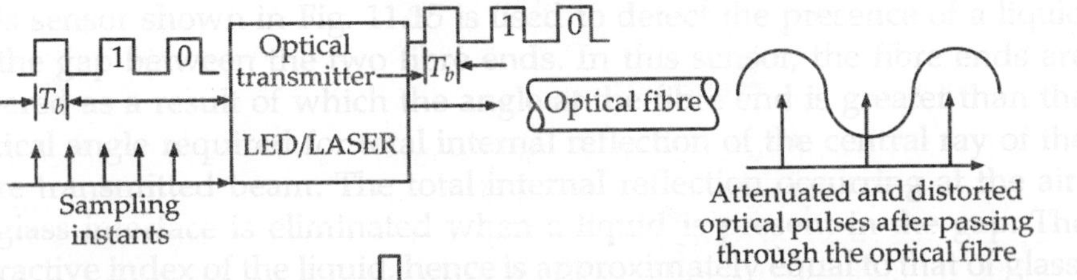

Fig. 12.1: (a) Input electrical pulses to an optical transmitter and effect of the optical fibre on these modulated pulses

Noise is also another impairment that alters the pulse shapes as we saw in the study of photodetectors. The effect of photodetector noise generated by these devices on transmitted data in the form of rectangular pulses is shown in Fig. 12.1(b). We may observe the zigzag portions of the current pulses at the output of the photodetectors. This is due to the various kinds of noises generated by the optoelectronic conversion process. We have shown a part of the optical receiver that accepts the attenuated distorted pulses coming from the fibre output. The photodetector acts as the demodulator that produces noisy electrical current pulses in response to these optical pulses. As we can see, the photodetector generates its own noise that gets added to the signal. The amplifier compensates the losses incurred by the signal while the filter attempts to provide a clean signal and corrects the distortions present in the signal. Therefore, the filter needs to be designed in such a manner than its frequency response corrects the rounding of the sharp edges of the pulses. We feed these clean amplified voltage pulses to a suitable decision making circuit that also is responsible for regenerating the pulses. This part is shown in Fig. 12.1(c). The decision making circuits work on a certain threshold that is predetermined by the data format and make a decision about a given data bit/symbol at the sampling instant. We have not shown in all the circuit diagrams, the method(s) by which the sampling instants are obtained. This is usually

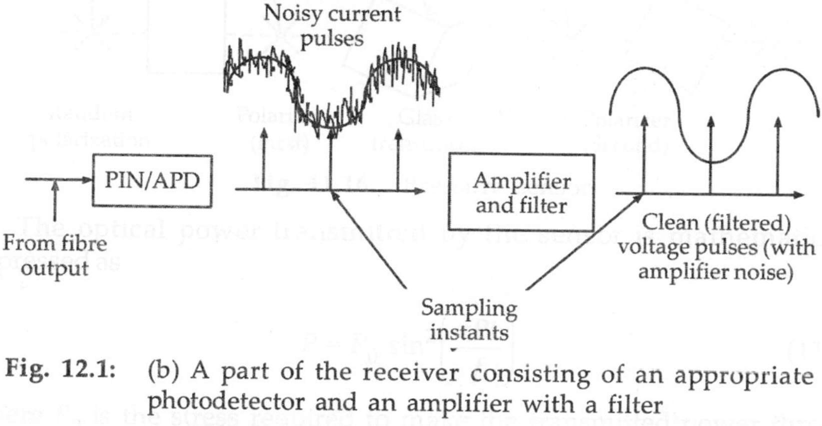

Fig. 12.1: (b) A part of the receiver consisting of an appropriate photodetector and an amplifier with a filter

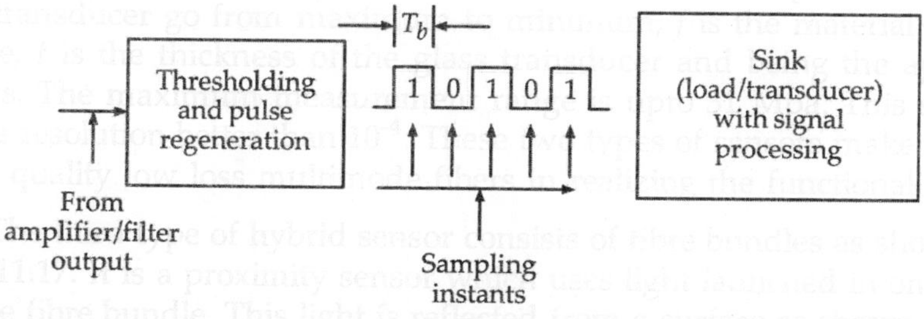

Fig. 12.1: (c) Binary data bits obtained after appropriate thresholding that are further fed to the data sink

carried out by a symbol timing recovery unit. The sampling instants are very crucial in all the digital parts of the receiver. However, for our purpose here we skip the details of bit/symbol timing recovery and assume the timings to be perfect.

A typical fibre optic receiver is illustrated in Fig. 12.2.

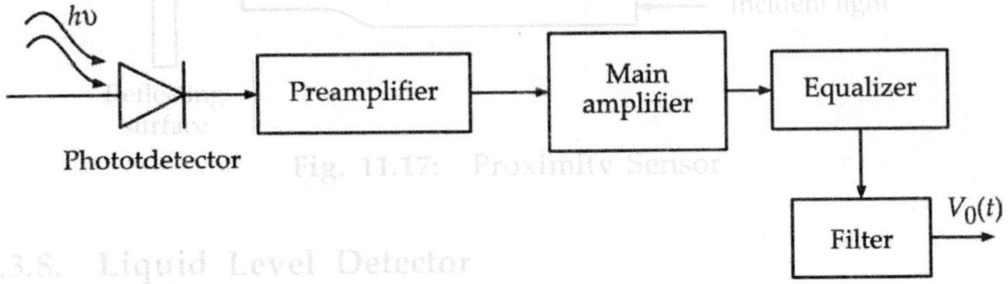

Fig. 12.2: An optical receiver block schematic

The first stage of an optical receiver as shown in Fig. 12.2 is a photodetector which may be a PIN photodiode or a avalanche photodiode that converts a given optical input into an appropriate electrical signal; usually a current. The electrical current has the same waveform as the envelope of the optical signal. This current is usually small in amplitude and cannot be directly given to an amplifier for increasing its level. It is, therefore, fed to a preamplifier having a low noise. The preamplifier boosts this current to a level that becomes suitable for amplification subsequently. The amplifiers provide linear functionality. However, as the signal has travelled down an optical fibre, it has become distorted due to a number of factors as we have seen previously. The input resistance and capacitance of the preamplifier may also accord some undesirable effects to the electrical current. The equalizer attempts to compensate for these effects. The filter ensures a certain signal to noise ratio while maintaining the essential features of the original signal. A detailed illustration of the photodetector along with the amplifier only is shown in Fig. 12.3.

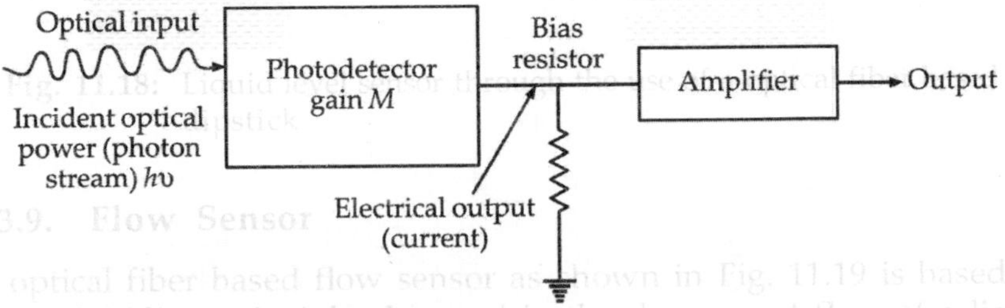

Fig. 12.3: Optical pulse detection in a receiver

As we have observed earlier, the performance of a typical photodetector is characterized by the following.

- Sensitivity
- Responsivity
- Noise generated during the process of opto- electric conversion
- Reliability and
- Compatibility with the incoming fiber for better power coupling

We may note that, it is the inherent noise in the detector that ultimately limits the performance of a receiver. Hence, let us take a look at the effect of noise sources and the way the performance of a receiver is affected by them. The performance is measured in terms of the average probability of bit error for a digital receiver and carrier (signal) to noise ratio for an analog receiver.

12.2. ERRORS IN THE DATA STREAM

The data bits/symbols are subject to errors due to a number of factors. As we have seen in Fig. 12.1, they get attenuated while propagating through the optical fibre due to a number of loss mechanisms occurring inside the fibre and distortions of various kinds. Moreover, the photodetector adds to these impaired signals, its own set of noises that make the signals become noisy. All of these impairments act upon the data stream to make look very different from the one that was transmitted. The decision making device is likely to make errors when the impairment level is such that the voltages originally associated to the data bits get reversed. The noises contributed by the photodetector are the shot noise and the thermal noise. Shot noise is due to the spontaneous fluctuations of current or voltage in electric circuits. This noise is generated in the receivers due to the discrete nature of current flow in the device. Thermal noise is manifested at all temperatures above absolute zero and is due to the random motion of electrons in a metal. It is generated by resistors and usually modelled as white; a voltage that contains theoretically all possible frequencies. As the generation of this type of noise is a random process, a certain probability distribution function is associated with it. Normally, the Gaussian distribution describes the amplitude of this noise.

12.2.1. Receiver Configuration

A typical optical receiver schematic is illustrated in Fig. 12.4. We observe three basic stages contained here; the photodetector (PD), the amplifier and the equalizer. The PD may be an avalanche PD type having a mean gain M or a PIN type with unity gain. The PD also is characterized by a quantum efficiency η and a capacitance C_d. The detector has been biased by the resistor R_b that is responsible for generating the thermal noise current $i_b(t)$. The amplifier has an input impedance which is the parallel combination of its input resistance R_{in} and input shunt capacitance C_{in}. The input voltage appearing across this impedance is $v_{in}(t)$ that causes the input current $i_{in}(t)$ to flow. The amplification factor is given by the transconductance parameter g_m as it is modelled as a voltage controlled current source. The two important noises limiting the amplifier usefulness are the input noise current source $i_a(t)$ due to the thermal noise produced by the amplifier input resistance R_{in} and the thermal noise $v_a(t)$ of the amplifier channel. As these are thermal noises, they follow a Gaussian distribution having a flat power spectral density (PSD). The PSDs associated with these two noises are given as S_I and S_E respectively.

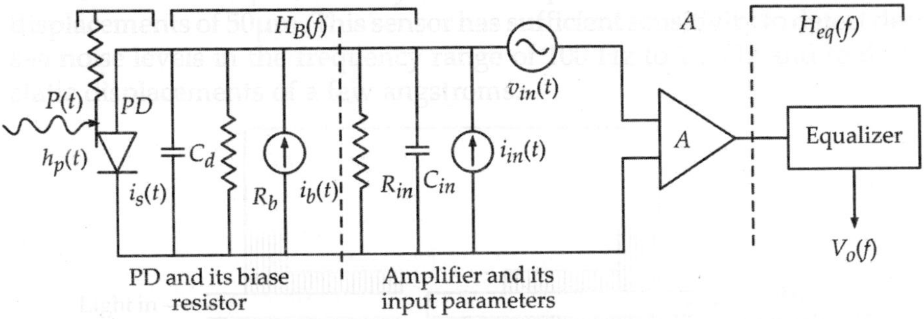

Fig. 12.4: Schematic diagram of a typical optical receiver

The equalizer is a signal processor that is a linear frequency shaping filter used to take care of signal distortion and intersymbol interference. Ideally, the input to the equalizer is the combined frequency response of the transmitter, the channel and the receiver. The equalizer performs a signal processing upon this combined response in such a manner that the output becomes suitable for further signal processing. In some cases, the equalizer may be used to correct the electrical frequency response of the photodetector and the amplifier.

The binary pulse train arriving at the photodetector is expressed as

$$P(t) = \sum_{k=-\infty}^{\infty} b_k h_p(t - kT_b) \qquad \text{...(12.1)}$$

where $P(t)$ is the received optical power, b_k is the k-th binary message bit generated at the transmitter, $h_p(t)$ is the received pulse shape and T_b is the bit duration. As we may see, for binary date transmission, b_k takes on just two values b_1 and b_0 corresponding to message bits 1 and 0 respectively. Usually, the received pulse shape is assumed to be of unit energy so that

$$\int_{-\infty}^{\infty} h_p(t)dt = 1 \qquad \text{...(12.2)}$$

The mean output current from the photodiode due to this pulse train at its input is expressed as

$$\langle i(t)\rangle = \frac{\eta q}{h\upsilon} MP(t) = \Re_0 M \sum_{k=-\infty}^{\infty} b_k h_p(t - kT_b) \qquad ...(12.3)$$

where $\frac{\eta q}{h\upsilon} = \Re_0$ is the photodiode sensitivity, and M is the mean avalanche gain for an APD.

In writing the output current of the photodetector, the dc current component has been ignored as it is due to the dark currents only. The current is amplified and filtered to produce a mean voltage at the equalizer output which is expressed as

$$\langle v_{out}(t)\rangle = A\Re_0 MP\,(t) * h_B\,(t) * h_{eq}\,(t) \qquad ...(12.4)$$

$$= \Re_0\, GP(t) * h_B(t) * h_{eq}(t)$$

We have taken $G = AM$ in the above expression, $h_B(t)$ and $h_{eq}(t)$ are the impulse responses of the bias circuit and the equalizer respectively. We may note that,

$$h_B(t) = \mathfrak{F}^{-1}[H_B(f)] = \int_{-\infty}^{\infty} H_B(f)\exp(j2\pi ft)df \qquad ...(12.5)$$

where $\mathfrak{F}$ denotes the Fourier transform operator. The bias current transfer function $H_B(F)$ is simply the impedance as offered by the parallel combination of R_b, R_{in}, C_d and C_{in}.

We have, therefore,

$$H_B(f) = \frac{R.1/j2\pi fc}{R + (1/J2\pi fC)} = \frac{R}{1 + j2\pi fRC} = \frac{1}{(1/R + j2\pi fC} \qquad ...(12.6)$$

where $$R = R_b \parallel R_{in} \qquad ...(12.7)$$

and $$C = C_d \parallel C_{in} \qquad ...(12.8)$$

Similarly, the mean equalizer output voltage is expressed as

$$\langle v_{out}(t)\rangle = \sum_{k=-\infty}^{\infty} b_k h_o(t - kT_b) \qquad ...(12.9)$$

where $$h_0(t) = \Re_0\, Gh_p(t) * h_B(t) * h_{eq}(t) \qquad ...(12.10)$$

represents the combined effect of the amplifier and the equalizer on the basic pulse shape. The transform of the equalizer output voltage, is therefore

$$H_0(f) = \Re_0\, GH_p\,(f)\, H_B\,(f)\, H_{eq}\,(f)$$

$$\Rightarrow \quad H_B(f)\, H_{eq}\,(f) = H_0\,(f)/\Re_0\, GH_p\,(f) \qquad ...(12.11)$$

12.3. NOISE SOURCES

The noise in optical receivers differs considerably from those in electrical receivers. For electrical receivers, the main noise that deteriorates receiver BER is thermal noise that is modelled to have a Gaussian distribution. However, in optical receivers, there is another noise component that must be addressed to. This is the shot noise current. The photocurrent, we have seen previously is proportional to the incident optical power. Hence, the same current does not flow in response to the powers received corresponding to data bit 1 and 0. The primary photocurrent generated by the photodiode is a time-varying Poisson process that is due to the random arrival of photons at the receiver. The average number of electron-hole pairs generated in a time in response to an optical signal $P(t)$ is

$$\bar{N} = \frac{\eta}{h\upsilon}\int_0^{\tau} P(t)dt = \frac{\eta E}{h\upsilon} \qquad ...(12.12)$$

where E is the energy received during the time interval τ, η is the quantum efficiency of the photodetector and hv is the energy of the photon. As the arrival of photons at the detector input is modelled by a Poisson random process, the actual number of electron-hole pairs generated fluctuates from the average value according to the Poisson distribution as

$$P_r(n) = \bar{N}^n \frac{\exp(-\bar{N})}{n!} \qquad ...(12.13)$$

where $P_r(n)$ is the probability that τ n electrons are emitted in an interval of τ. As we may observe, it is not possible to predict exactly how many electron-hole pairs are generated by a known optical power incident on the detector. This gives rise to the origin of quantum noise. The random nature of the avalanche multiplication process gives rise to another type of shot noise. The excess noise factor $F(M)$ for electron injection is

$$F(M) = kM + \left(2 - \frac{1}{M}\right)(1-k) \qquad ...(12.14)$$

which is often approximated by the empirical expression

$$F(M) = M^x \qquad ...(12.15)$$

where the factor x ranges between 0 and 1 depending on the photodiode material. In the above expression,M represents the mean avalanche gain and k is the effective ionization constant defined as

$$k = \frac{\text{hole ionization rate}}{\text{electron ionization rate}}$$

Intersymbol interference (ISI) is another source of error in digital communication systems that employ the photodetectors besides the white Gaussian noise, quantum noise and shot noise affecting the output from photodetectors. As we have seen earlier, pulses launched onto the optical fibres do spread out in the time domain while travelling down a given length of the fibre due to a number of factors. ISI causes some of the transmitted energy to spill into the neighbouring time slots. This spilling of energy becomes a source of interference for a given pulse. In other words, a pulse allotted to stay in its own time slot does not do so, however extends into its neighbouring pulses because of the temporal spreading.

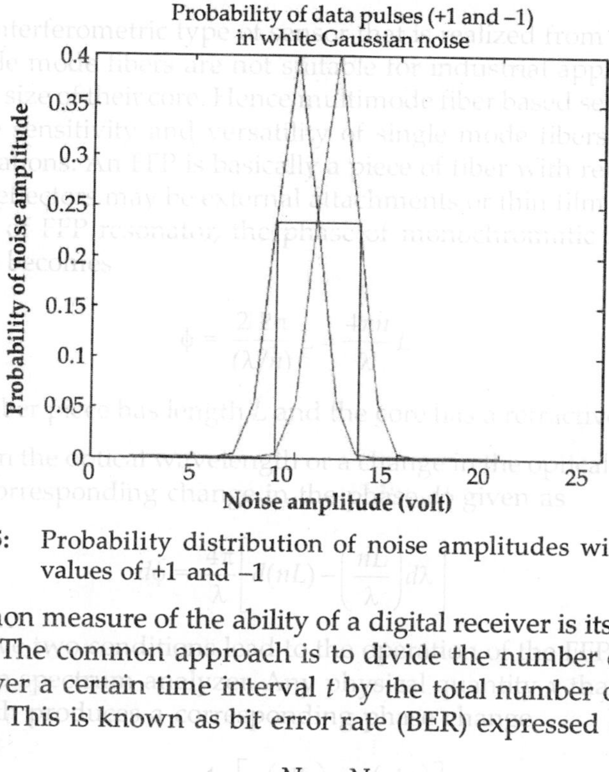

Fig. 12.5: Probability distribution of noise amplitudes with mean values of +1 and –1

A common measure of the ability of a digital receiver is its probability of bit error. The common approach is to divide the number of errors N_e occurring over a certain time interval t by the total number of pulses N_t transmitted. This is known as bit error rate (BER) expressed as

$$\text{BER} = \frac{N_e}{N_t} = \frac{N_e}{R_b t} \qquad \text{...(12.16)}$$

where $R_b = \dfrac{1}{T_b}$ is the bit rate or the number of pulses transmitted per second. Typical error rates for optical fibre communication systems vary from 10^{-6} to 10^{-10}. This error rate depends on the signal to noise ratio at the receiver input. A BER of 10^{-6} indicates that the receiver is likely to make an error every second in every 10^6 number of transmitted bits; one bit is likely to be misinterpreted in a transmitted data stream of 10^6 bits. As we can see quickly, a better BER calls for sophisticated signal processing at the receiver. The system error rate requirements and the receiver noise levels set a lower limit on the optical signal power that is required at the photodetector.

The BER computation is made by having the knowledge about the noise statistic. For Gaussian noise, the shape of the equalizer output becomes

$$p_1(v < v_{th}) = \int_{-\infty}^{v_{Th}} p(v|1)dv \quad ...(12.17)$$

This is the probability that a data pulse is misinterpreted as 0 while a 1 has been transmitted. Similarly, the probability of making an erroneous decision in favour of 1 while a minus 1 has been transmitted becomes

$$p_0\ (v > v_{Th}) = \int_{v_{Th}}^{\infty} p(v|0)dv \quad ...(12.18)$$

In these two integrals, we see the use of a variable v_{Th} which gives us a threshold. The receiver decision is compared with this threshold. A plus one is declared if the equalizer output exceeds this voltage whereas decision is made in favour of minus 1 for the equalizer output voltage falling less than this threshold. These two events are independent and hence the total probability of making an error is expressed as

$$p_e = a_1 p_1(v_{th}) + a_2 p_0(v_{th})$$

The weighting factors are decided by the *a priori* probabilities of the data bits/symbols. That is a_1 and a_2 are the probabilities that either a 1 or a 0 occurs respectively. For independent and identically distributed (i.i.d.) or unbiased data, we see that the bits 1 and 0 occur with equal probability. Hence, $a_1 = 0.5 = a_2$. We should select the threshold in such a manner that the probability of bit error will be minimized. For example, if we instead choose to send data with 1 and -1's, then the threshold would be set at zero and so on.

We note these conditional probabilities from Fig. 12.5. The blue curve represents the conditional probability for $p_1(v)$ whereas the left red curve represents $p_0(v)$. These have been obtained in MATLAB. We can clearly see that the rectangle drawn around the overlapping regions of the two curves represents the error probability p_e.

To calculate the probability of bit error, we need to have knowledge about the noise statistic. Here, as we have assumed white Gaussian noise, this is completely specified by its mean and variance. Thus, noise that affects the receiver performance has a probability distribution function (PDF) given as

$$p(n) = \frac{1}{\sqrt{2\pi\sigma^2}} \exp\left(-\frac{n^2}{2\sigma^2}\right) \qquad \text{...(12.19)}$$

Here the mean of this noise is zero and variance is σ^2. We use this to evaluate the probability of error for a data stream in which all the data pulses are of equal amplitude *A* volt. First we find out the probability that a 1 was transmitted, but noise voltage at the sampling instant is such that the receiver decides in favour of a zero because the equalizer output is less than the threshold. This probability of deciding in favour of a 0 when a 1 was transmitted is

$$p_1(v) = \int_{-\infty}^{A/2} \frac{1}{\sqrt{2\pi\sigma^2}} \exp\left[-\frac{(\upsilon - A)^2}{2\sigma^2}\right] d\upsilon \qquad \text{...(12.20)}$$

Similarly, the probability of deciding in favour of a 0 when a 1 was transmitted becomes

$$p_0(v) = \int_{A/2}^{\infty} \frac{1}{\sqrt{2\pi\sigma^2}} \exp\left(-\frac{\upsilon^2}{2\sigma^2}\right) d\upsilon \qquad \text{...(12.21)}$$

As we have seen earlier, for i.i.d. data bits having equal amplitude *A*, the threshold is $\frac{A}{2}$ and that explains the use of $\frac{A}{2}$ in the above two integrals.

The total probability of error, therefore becomes

$$= \frac{1}{2}[P_1(v) + P_0(v)]$$

$$p_e = \frac{1}{2}\left[1 - erf\left(\frac{A}{2\sqrt{2}\sigma}\right)\right] \qquad \text{...(12.22)}$$

where the function *erf*(*x*) is defined as

$$erf(x) = \frac{2}{\sqrt{\pi}} \int_0^x \exp(-t^2) dt \qquad \text{...(12.23)}$$

is known as the error function tabulated in various mathematical handbooks. One important observation made from this error probability is that it depends only on the signal amplitude to the standard deviation of noise; the factor $\frac{A}{\sigma}$. Since σ is usually known as the root mean square (RMS) noise, (standard deviation) the factor $\frac{A}{\sigma}$ is the peak signal to RMS noise ratio. Thus, the probability of bit error in a typical digital communication system depends on the signal-to-noise ratio, often designated as . *S/N* For a given signal amplitude, it is noise power or its variance that determines the overall probability of bit error. As the noise power increases, the receiver has to decide from very noisy samples, at the sampling instants, the data bits either in favour of a 1 or a 0 and the probability of making an error increases.

Expressed in dB, the SNR becomes

$$\left(\frac{S}{N}\right)_{dB} = 20\log\frac{A}{\sigma} \quad ...(12.24)$$

P-12.1 Verify (12.23).

12.3.1. Noise Equivalent Bandwidth (NEB) of the Receiver

To calculate the NEB of the optical receiver, we proceed by evaluating the equalizer output voltage. This is written as

$$V_0(t) = \overline{V_0(t)} + v_n(t) \quad ...(12.25)$$

where $\overline{V_0(t)}$ is the statistical average of the equalizer output voltage and $v_n(t)$ is the random noise component that makes the actual output voltage to differ from this average value.

This noise voltage may be represented by

$$v_n^2(t) = v_s^2(t) + v_R^2(t) + v_i^2(t) + v_v^2(t) \quad ...(12.26)$$

where $v_s(t)$ is the quantum or shot noise resulting from the random multiplied Poisson nature of the photocurrent $i_s(t)$ generated by the photodetector, $v_R(t)$ is the thermal noise generated by the bias resistor R_b, amplifier input noise currents and input noise voltages giving rise to $v_i(t)$ and $v_v(t)$ respectively. The amplifier noise sources are assumed to be independent of each other as they are generated by independent processes and also Gaussian in their statistics. The mean square noise voltage is expressed as

$$\overline{v_n^2(t)} = \left[V_0(t) - \overline{V_0(t)}\right]^2$$

$$= \overline{V_0(t)}^2 - \overline{V_0^2(t)} \qquad \text{...(12.27)}$$

$$= \overline{v_S^2(t)} + \overline{v_R^2(t)} + \overline{v_i^2(t)} + \overline{v_v^2(t)}$$

The last three components are thermal noise components at the equalizer output. The thermal noise power developed across a load resistor R_b is

$$\overline{v_R^2(t)} = \frac{4K_BT}{R_b} B_n R^2 A^2 \qquad \text{...(12.28)}$$

where A is the amplifier gain, B_n is the noise equivalent bandwidth of the combined effect of the amplifier including biasing and the equalizer.

The receiver mean square shot noise voltage has been simplified by [1-2]

$$\overline{v_S^2(t)} = 2q\overline{i_{unity}}\,\overline{m^2} B_n R^2 A^2 \qquad \text{...(12.29)}$$

where $\overline{i_{unity}}$ is the average unity gain photocurrent over the bit interval T_b and $\overline{m^2}$ is the mean square avalanche gain that usually takes the form M^{x+2} with $0 < x \le 1.0$. The factor 2 is present in the shot noise variance due to the fact that there is an equal contribution to the noise from generation as well as recombination of carriers crossing the photoconductive channel.

The average unity gain photocurrent is evaluated next. To do so, we take into account the not only shot noise contribution from a pulse within a given time slot but also the shot noises as generated by all neighbouring pulses that spill into the pulse under consideration at the sampling instant. The shot noise within a time slot, therefore depends on the received pulse shape and also on the data sequence. As we may guess, the worst case of shot noise in any particular time slot occurs when the bit under consideration is zero and all other neighbouring pulses are 1s as such a received data pattern causes the greatest amount of ISI. The average value of the unity gain photocurrent averaged over the bit interval T_b corresponding to a 1 is

$$\overline{i_{unity_1}} = \sum_{k=-\infty}^{\infty} \frac{\eta q}{h\upsilon} b_1 \frac{1}{T_b} \int_{-T_b/2}^{T_b/2} h_p(t - kT_b)\,dt$$

$$\frac{\eta q}{h\upsilon} \frac{b_1}{T_b} \int_{-\infty}^{\infty} h_p(t)\,dt = \frac{\eta q}{h\upsilon} \frac{b_1}{T_b} \qquad \text{...(12.30)}$$

The mean value of this current corresponding to a zero is,

$$\overline{i_{unity\,0}} = \sum_{k=-\infty, k\neq 0}^{\infty} \frac{\eta q}{h\upsilon} b_1 \frac{1}{T_b} \int_{-T_b/2}^{T_b/2} h_p(t-kT_b)dt$$

$$= \frac{\eta q}{h\upsilon}\frac{b_1}{T_b}\left[\sum_{k=-\infty}^{\infty} \int_{-T_b/2}^{T_b/2} h_p(t-kT_b)dt - \int_{-T_b/2}^{T_b/2} h_p(t)dt\right] \qquad ...(12.31)$$

$$= \frac{\eta q}{h\upsilon}\frac{b_1}{T_b}(1-\gamma)$$

where

$$\gamma = \int_{-T_b/2}^{T_b/2} h_p(t)dt \qquad ...(12.32)$$

is the fractional energy of a 1 pulse that is contained within its own bit period. The quantity 1 – X is the fractional energy of a pulse that has spread outside its bit period as it travels through an optical fibre.

The NEB is defined, however for positive frequencies only as

$$B_n = \frac{1}{\left|H_B(0)H_{eq}(0)\right|^2}\int_0^{\infty}\left|H_B(f)H_{eq}(f)\right|^2 df$$

$$= \frac{1}{\left|H_o(f)/\Re_0 G H_p(f)\right|^2}\int_0^{\infty}\left|H_o(f)/\Re_0 G H_p(f)\right|^2 df \qquad ...(12.33)$$

$$= \frac{1}{\left|H_o(0)/H_p(0)\right|^2}\int_0^{\infty}\left|H_o(f)/H_p(f)\right|^2 df$$

and R has been defined earlier.

We assume the thermal noises contributed by the amplifier input noise current source $i_{in}(t)$ and $v_{in}(t)$ to be statistically independent and Gaussian in nature. They are characterized by their power spectral densities. We have, therefore,

$$\overline{v_I^2(t)} = S_I B_n R^2 A^2 \qquad ...(12.34)$$

and

$$\overline{v_V^2(t)} = S_V B_{eq} A^2 \qquad ...(12.35)$$

for the PSD in the time domain due to the amplifier input noise voltage source $v_{in}(t)$. In these expressions, the spectral density of the amplifier input noise current source has been considered to be S_1 in A^2/Hz and S_V is the spectral density of the amplifier input noise voltage source in V^2/Hz. The quantity B_{eq} representing the NEB of the equalizer is defined as

$$B_{eq} = \frac{1}{\left|H_{eq}(0)\right|^2} \int_0^{\infty} \left|H_{eq}(f)\right|^2 df$$

$$= \frac{R^2}{\left|H_o(0)/H_p(0)\right|^2} \int_0^{\infty} \left|\frac{H_o(f)}{H_p(f)}\left(\frac{1}{R} + j2\pi fC\right)\right|^2 df \qquad ...(12.36)$$

12.3.2. Evaluation of Receiver Sensitivity

Receiver sensitivity is defined as the minimum optical power that ensures a BER of at least 10^{-9}. To derive an expression for this, we note that the electrons generated in the photodetector follow a Poisson distribution given as

$$p(k,\Delta\tau) = \frac{(\rho\Delta\tau)^k \exp\left[-(\rho\Delta\tau)\right]}{k!} \qquad ...(12.37)$$

which gives the probability that k electrons are generated in time t. In the above expression,

$$\rho = \left[\frac{\eta q}{h\upsilon} P_0 + \frac{I_d}{q}\right] \qquad ...(12.38)$$

The mean and variance of such a probability distribution function is $\rho\Delta\tau$. The current generated due to these electrons becomes

$$i(t, \tau) = Mqkh(t - \tau) \qquad ...(12.39)$$

The average value of this current is

$$\overline{i(t,\tau)} = qh(t-\tau)\overline{Mk} = q\overline{M}(\rho\Delta\tau)h(t-\tau)$$

$$= \overline{M}\left[\frac{\eta q}{h\upsilon} P_0(\tau) + I_d\right] h(t-\tau)\Delta\tau \qquad ...(12.40)$$

The total current, is therefore given as

$$i(t) = \int_{-\infty}^{\infty} \overline{M}\left[\frac{\eta q}{h\upsilon} P_0(\tau) + I_d\right] h(t-\tau)d\tau$$

$$= \overline{M} \int_{-\infty}^{\infty} \frac{\eta q}{h\upsilon} P_0(\tau) h(t-\tau) d\tau \quad ...(12.41)$$

Let the incident optical power at the photodetector be given as

$$P_0(t) = P_0[m_T + m(t)] \quad ...(12.42)$$

The mean square value of the photocurrent due to the Poisson distributed electron-hole pairs in response to this optical power is

$$\overline{i_t^2} = \left(\overline{M}\frac{\eta q}{h\upsilon}\right)^2 P_0^2 \int_{-\infty}^{\infty}\int_{-\infty}^{\infty} \left[m_T^2 + m_T \overline{m(\tau)m(\mu)} + \overline{m(\tau)m(\mu)}\right] h(t-\tau) d\tau h(t-\mu) d\mu$$

$$= \left(P_0 \overline{M} \frac{\eta q}{h\upsilon}\right)^2 \left\{ m_T^2 H^2(0) + \int_{-\infty}^{\infty} S_m(f) |H(f)|^2 df \right\} \quad ...(12.43)$$

where the squared signal current is expressed as

$$\overline{i_t^2} = \left(\frac{\overline{M}\eta q}{h\upsilon}\right)^2 \int_{-\infty}^{\infty} S_m(f) |H(f)|^2 df \quad ...(12.44)$$

In this, $S_m(f)$ denotes the message signal power spectral density.

Mean square noise current is

$$\overline{i(t,\tau)^2} - \overline{i(t,\tau)}^2 \quad ...(12.45)$$

The squared current is

$$\overline{i(t,\tau)^2} = \left[Mqkh(t-\tau)\right]^2 = \overline{M^2} q^2 \left[(\rho\Delta\tau)^2 h^2(t-\tau) + \rho\Delta\tau\right] \quad ...(12.46)$$

Hence, the mean square noise current becomes

$$\overline{M^2} q^2 \left[(\rho\Delta\tau)^2 h^2(t-\tau) + \rho\Delta\tau\right] - (\overline{M})^2 q (\rho\Delta\tau)^2 h^2(t-\tau)$$

$$= \left\{ q^2 \left[\overline{M^2} - (\overline{M})^2\right] (\rho\Delta\tau)^2 + \overline{M^2}\rho\Delta\tau \right\} h^2(t-\tau) \quad ...(12.47)$$

$$\Rightarrow \quad i_n^2(t,\tau) = q^2 \overline{M^2} \rho h^2(t-\tau) \Delta\tau$$

The total noise current is, therefore

$$i_n^2(t) = q^2 \overline{M^2} \rho \int_{-\infty}^{\infty} h^2(t-\tau) d\tau$$

$$= q^2\overline{M^2}\int_{-\infty}^{\infty}\left(\frac{\eta}{h\upsilon}P_0(\tau)+\frac{I_d}{q}\right)d\tau \qquad ...(12.48)$$

$$= q^2\overline{M^2}\left(\frac{\eta}{h\upsilon}P_0+\frac{I_d}{q}\right)B_n$$

Therefore, the figure of merit for the receiver defined in terms of carrier to noise ratio becomes

$$\frac{C}{N}=\frac{i_{sig}^2}{i_n^2+i_{th}^2}$$

$$=\frac{(\eta qP_0/h\upsilon)^2\int_{-\infty}^{\infty}S_m(f)\left|H(f)\right|^2 df}{2q\left[\left(\frac{\eta qP_0}{h\upsilon}(m_T+\overline{m})\right)+I_d\right]F(\overline{M})B_n+\frac{4K_BT}{R_e}B_n} \qquad ...(12.49)$$

$$=\frac{(\eta qP_0/h\upsilon)^2\int_{-\infty}^{\infty}S_m(f)df}{2q\left[\left(\frac{\eta qP_0}{h\upsilon}\right)+I_d\right]F(\overline{M})B_n+\frac{4K_BT}{R_e}B_n}$$

Errors occur in the decision process when the noise $n_1(t)$ either goes below $m_T\Re_0P_0$ and the noise $n_0(t)$ goes above this value of $m_T\Re_0P_0$. Here $n_1(t)$ is that noise affecting a data bit of 1 and similarly $n_0(t)$ is the noise that affects a 0. This has been illustrated in Fig.12.6 that shows a noise like $n_0(t)$ takes the threshold above $m_T\Re_0P_0$ and similarly $n_1(t)$ affects the data pulse for an one such that the it falls below the threshold.

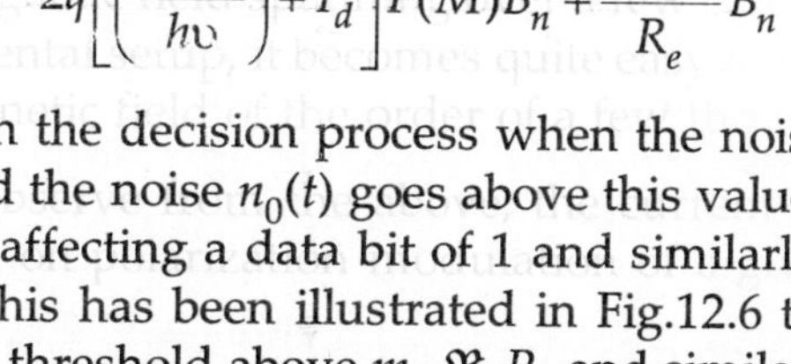

Fig. 12.6: Noises affecting data bits 1 and 0 and threshold

We note that, if

$$n_1(t) + 2\Re_0P_0 < m_T\Re_0P_0 \qquad ...(12.50)$$

then an error occurs for the case when a 1 is transmitted. The effect of noise is such that the above condition is satisfied and decision is made in favour of a 0 as the combined noise and data falls below the threshold $m_T \Re_0 P_0$. The combined event probability for 1 and 0 becomes

$$p(n_1 < -2\Re_0 P_0 + m_T \Re_0 P_0 | 1) + p(n_0 > m_T \Re_0 P_0 | 0)$$

$$= \frac{1}{2}\left[\int_{-\infty}^{\Re_0 P_0(m_T-2)} \frac{1}{\sqrt{2\pi\sigma_1^2}}\exp\left[-\frac{(x-1)^2}{2\sigma_1^2}\right]dx + \int_{m_T \Re_0 P_0}^{\infty} \frac{1}{\sqrt{2\pi\sigma_0^2}}\exp\left(-\frac{x^2}{2\sigma_0^2}\right)dx\right] \quad ...(12.51)$$

For optimum detection, we set

$$\frac{\Re_0 P_0(2-m_T)}{\sigma_1} = \frac{\Re_0 P_0 m_T}{\sigma_0}$$

$$\Rightarrow \qquad m_T = \frac{2\sigma_0}{\sigma_0 + \sigma_1} \quad ...(12.52)$$

For a modulated message expressed as $P_0 = P_i[m_T + m(t)]$, the total power contained in the message is

$$\int_{-\infty}^{\infty} S_m(f)df = \overline{m^2(t)} \quad ...(12.53)$$

For message signal corresponding to 1, we get $m_1 = 2 - m_T$ and the photodetector current is $2\Re_0 P_0$ and similarly for 0, we have $m_0 = -m_T$ and the photodetector signal current is zero.

The total power of the message containing equiprobable 1's and 0's become (this is the case for a very long message)

$$\overline{m^2(t)} = \frac{(2-m_T)^2 + (-m_T)^2}{2} = \frac{2(\sigma_0^2 + \sigma_1^2)}{(\sigma_0 + \sigma_1)^2} \quad ...(12.54)$$

This gives us a mean squared noise current as

$$i_n^2(t) = q^2 \overline{M^2} \int_{-\infty}^{\infty}\left[\frac{\eta P_0(\tau)}{h\upsilon} + \frac{I_d}{q}\right] h^2(t-\tau)d\tau \quad ...(12.55)$$

$$= q^2$$

The variance of noise affecting a 0 is

$$\sigma_0^2 = \left[2qI_dF(M) + \frac{4K_BT}{M^2R_e}\right]B_n \qquad ...(12.56)$$

and similarly,

$$\sigma_1^2 = \left[2q(2\Re_0P_0 + I_d)F(M) + \frac{4K_BT}{M^2R_e}\right]B_n \qquad ...(12.57)$$

$$= \sigma_0^2 + 4q\Re_0P_0F(M)B_n$$

Combining the two variances expressed in (12.56) and (12.57, we have

$$= \sigma_1^2 - \sigma_0^2 = 4q\Re_0P_0F(M)B_n \qquad ...(12.58)$$

Hence, the figure of merit or performance index of the receiver in terms of peak signal power to RMS noise power is expressed as

$$\frac{S}{N} = \frac{4(\Re_0P_0)^2}{(\sigma_0 + \sigma_1)^2}$$

$$\Rightarrow \qquad \sigma_0 + \sigma_1 = \frac{2\Re_0P_0}{\sqrt{S/N}} \qquad ...(12.59)$$

This is because the photocurrent, as we have seen earlier is $2\Re_0P_0$ in response to a data bit of 1 and 0 corresponding to a data bit of 0. Thus, the power due to signal component only is

$$(2\Re_0P_0)^2 = 4R_0^2P_0^2$$

From this figure of merit, we now have an expression for the receiver sensitivity. At baseband, this is expressed as

$$P_0 = \frac{\sigma_0 + \sigma_1}{2\Re_0}\sqrt{S/N}$$

$$= \frac{\sigma_0 + \sigma_1}{2(\eta q/h\upsilon)}\sqrt{S/N} \qquad ...(12.60)$$

$$= \frac{h\upsilon}{2\eta q}(\sigma_0 + \sigma_1)\sqrt{S/N}$$

Substituting this in the expression for the difference of the two variances,

$$\sigma_1 - \sigma_0 = 2qF(M)B_n\sqrt{S/N}$$

$$\Rightarrow \quad \sigma_1 = \sigma_0 + 2qF(M)B_n\sqrt{S/N} \qquad ...(12.61)$$

The receiver sensitivity at baseband may also be expressed in another way as follows.

$$\sigma_0 + \sigma_1 = \frac{2\Re_0 P_0}{\sqrt{S/N}}$$

$$\Rightarrow \quad \sigma_0 + \sigma_0 + 2qF(M)B_n\sqrt{S/N} = \frac{2\Re_0 P_0}{\sqrt{S/N}}$$

$$\Rightarrow \quad q\left[\frac{\sigma_0}{q} + F(M)B_n\sqrt{S/N}\right] = \frac{\Re_0 P_0}{\sqrt{S/N}} = \frac{\eta q}{h\upsilon} P_0 \frac{1}{\sqrt{S/N}} \qquad ...(12.62)$$

$$\Rightarrow \quad \frac{h\upsilon}{\eta}\sqrt{S/N}\left[\frac{\sigma_0}{q} + F(M)B_n\sqrt{S/N}\right]$$

The bit rate supported by the transmitter receiver combo is related to the noise equivalent bandwidth as

$$R_b = 2B_n \qquad ...(12.63)$$

Substituting the bit rate in the expression for receiver sensitivity in (12.62), we also get

$$P_0 = \frac{h\upsilon}{\eta}\sqrt{S/N}\left[\frac{\sigma_0}{q} + F(M)\frac{R_b}{2}\sqrt{S/N}\right]$$

$$= \frac{h\upsilon}{2\eta}\sqrt{S/N}\left[\frac{2\sigma_0}{q} + F(M)R_b\sqrt{S/N}\right]$$

$$= \frac{h\upsilon}{2\eta}\left[F(M)R_b\frac{S}{N} + \frac{2\sigma_0}{q}\sqrt{S/N}\right]$$

$$= \frac{h\upsilon}{2\eta}\frac{S}{N}\left[F(M)R_b + \frac{2\sigma_0}{q\sqrt{S/N}}\right] \qquad ...(12.64)$$

Let us substitute the expression for σ_0 as obtained in (12.56) in (12.64) to obtain a different expression for the receiver sensitivity. Doing so gives us

$$P_0 = \frac{h\upsilon}{2\eta}\frac{S}{N}\left[F(M)R_b + \frac{2\sigma_0}{q\sqrt{S/N}}\right]$$

$$= \frac{h\upsilon}{2\eta}\frac{S}{N}F(M)R_b\left[1+\frac{2\sigma_0}{qF(M)R_b\sqrt{S/N}}\right] \quad ...(12.65)$$

$$= \frac{h\upsilon}{2\eta}\frac{S}{N}F(M)R_b\left[1+\frac{2}{F(M))R_b}\sqrt{\frac{\left[2qI_dF(M)+\frac{4K_BT}{M^2R_e}\right]B_n}{q^2(S/N)}}\right]$$

To find out the optimum value of M, we differentiate the SNR with respect to it and set the result to zero.

$$\frac{\partial}{\partial M}\left(\frac{S}{N}\right)=0$$

$$\rightarrow M_{opt}=\left[\frac{1}{k}\left(\frac{2}{q}\sqrt{\frac{2\pi K_BTC_d}{(S/N)}}\right)+k-1\right]^{\frac{1}{2}} \quad ...(12.66)$$

We also have, from the foregoing,

$$\frac{S}{N}=\frac{\frac{2(\sigma_0^2+\sigma_1^2)}{(\sigma_0+\sigma_1)^2}}{(\sigma_0^2+\sigma_1^2)}$$

$$=\frac{2}{(\sigma_0+\sigma_1)^2} \quad ...(12.67)$$

We note that, the maximum frequency that can be handled by the photodetector is expressed as

$$f_m=\frac{1}{2\pi R_bC_d} \quad ...(12.68)$$

12.3.3. Derivation of the two Variances Affecting 1 and 0

We are interested to derive expressions for the two noise variances that affect transmission of 1' and 0's. The thermal noise power generated in the receiver is independent of the incident input optical power. However, this is not true for shot noise. This is because, the photocurrent generated in response to an optical pulse corresponding to 1 is different from that corresponding to zero. In the ideal case, the optical power representing an electrical zero does not produce any photocurrent as there is simply no power in the bit interval corresponding to a zero. In this case then, the receiver noise is due to the thermal noise only. In case of an optical pulse representing an electrical one, there is a photocurrent proportional to the responsivity of the photodetector and the receiver noise now contains the shot noise plus the thermal noise.

The contribution of ISI distorting the pulse shape is ignored. The assumption made here is that perfect equalization is carried out so that the equalized pulse stream is ISI free at the sampling instant KT_b and the maximum value of a given pulse attained at $t = 0$ is unity. Therefore, we write

$$h_0(t = 0) = 1 \text{ and}$$
$$h_0(t = kT_B) = 0 \text{ for } k \neq 0 \qquad \text{...(12.69)}$$

Substitution of this in the expression for the actual equalizer output voltage gives us

$$v_{out} = b_l h_0(0) + v_n(kT_b) \qquad \text{...(12.70)}$$

Two dimensionless quantities are introduced; $\tau = t/T_b$ and $f = fT_b$. The numerical value of the integrals defined earlier become independent of the bit period T_b and depends only on the received and equalized pulse shapes.

$$H_p(f) = \int_{-\infty}^{\infty} h_p(t)\exp(-j2\pi ft)dt$$

$$\int_{-\infty}^{\infty} h_p(\tau T_b)\exp(-j2\pi f\tau T_b)d\tau T_b \qquad \text{...(12.71)}$$

$$= \frac{1}{T_b}.T_b H_p(fT_b)$$

$$= H'_p(\phi)$$

$$h_p(\tau) \Leftrightarrow H'_p(\phi)$$

where $x(t) \Leftrightarrow X(f)$ is a Fourier transform pair

Similarly we can show that

$$H'_o(\phi) = \frac{1}{T_b} H_o(f) \qquad \text{...(12.72)}$$

P-12.2: Prove (12.72).

We have, normalized bandwidth integrals defined as,

$$I_Z = \frac{1}{T_b}\int_0^{\infty}\left|\frac{H_o(f)}{H_p(f)}\right|^2 df$$

$$= \frac{1}{T_b}\int_0^{\infty}\left|\frac{T_b H_0'(\phi)}{H_p'(\phi)}\right|^2 d\phi \qquad \text{...(12.73)}$$

$$= \int_0^{\infty}\left|\frac{H_o'(\phi)}{H_p'(\phi)}\right|^2 d\phi$$

The first line of (12.73) gives us

$$I_Z T_b = \int_0^{\infty}\left|\frac{H_o(f)}{H_p(f)}\right|^2 df \qquad \text{...(12.74)}$$

Similarly, $$I_2 = \frac{1}{T_b}\int_0^{\infty}\left|\frac{H_o(f)}{H_p(f)}\right|^2 f^2 df$$

$$= \int_0^{\infty}\left|\frac{T_b H_o'(\phi)}{H_p'(\phi)}\right|^2 \frac{f^2 T_b^2}{T_b^2} d\phi . \frac{1}{T_b} \qquad \text{...(12.75)}$$

$$= \frac{1}{T_b^3}\int_0^{\infty}\left|\frac{H_0'(f)}{H_L'(f)}\right|^2 \phi^2 d\phi$$

and $$I_2 T_b = \int_0^{\infty}\left|\frac{H_o(f)}{H_p(f)}\right|^2 f^2 df$$

$$\frac{I_2 T_b}{T_b^2} = \frac{1}{T_b^2}\int_0^{\infty}\left|\frac{H_o(f)}{H_p(f)}\right|^2 f^2 df \qquad \text{...(12.76)}$$

$$\frac{I_2}{T_b} = \frac{1}{\left|H_o(v)/H_p(v)\right|^2}\int_0^{\infty}\left|\frac{H_o(f)}{H_p(f)}\right|^2 f^2 df$$

The NEB of the combined effect of biasing circuit, the amplifier and the equalizer becomes

$$B_n = \frac{1}{\left|H_o(0)/H_p(0)\right|^2}\int_0^{\infty}\left|H_o(f)/H_p(f)\right|^2 df$$

$$= \frac{I_Z T_b}{\left|H_o(0)/H_p(0)\right|^2} \quad \text{...(12.77)}$$

$$= \frac{I_Z T_b}{T_b^2}$$

$$= \frac{I_Z}{T_b} = I_Z B$$

where $B = 1/T_b$

This result follows from the fact that represents the areas bounded by the curves corresponding to the equalizer output $H_0(0)$ and $H_p(0)$ and the basic pulse shapes respectively. The area $H_0(0)$ is equal to T_b and that for $H_p(0)$ is unity. Therefore,

$$B_n = \frac{I_Z}{T_b} = I_Z B \quad \text{...(12.78)}$$

and that of the equalizer alone is

$$B_e = \frac{I_Z}{T_b} + \frac{(2\pi RC)^2}{T_b^3} I_2 = I_Z B + (2\pi RC)^2 I_2 B^3 \quad \text{...(12.79)}$$

Now, we can write down the receiver noise variances in terms of these two normalized integrals as by making use of (12.27)

$$\overline{v_n^2(t)} = R^2 A^2\left(2q\overline{i_{unity}}M^{2+x} + \frac{4K_B T}{R_b} + S_I + \frac{S_V}{R^2}\right) I_Z B + (2\pi RC)^2 A^2 S_V B^3$$

$$= (qRAB)^2\left(\frac{2}{q}\overline{i_{unity}}M^{2+x} T_b I_Z + \mathrm{N}\right) \quad \text{...(12.80)}$$

Here, $$N = \frac{1}{q^2 B}\left(S_I + \frac{4K_B T}{R_b} + \frac{S_V}{R^2}\right) I_Z + \frac{(2\pi C)^2}{q^2} S_V I_2 B \quad \text{...(12.81)}$$

represents a dimensionless parameter that characterizes the thermal noise of the receiver. This is called the thermal noise characteristic associated with the receiver amplifier. We observe that the current generated by a photodetector is proportional to the input optical power. In the worst case, the variance of the total noise affecting the receiver would serve as the input power. Our objective, then is to evaluate this power.

$$i = \frac{\eta q}{h\upsilon} P$$

$$P = \frac{h\upsilon}{\eta q} t \qquad \text{...(12.82)}$$

$$\sigma_1^2 \leftrightarrow \frac{h\upsilon}{\eta q}\overline{i_{unity,1}}$$

We make use of these results now to derive appropriate expressions for the two variances affecting 1 and 0. For 1, we have

$$\sigma^2_1 = (qRAB)^2 \left(\frac{2}{q}\overline{i_{unity_1}} M^{2+x} T_b I_Z + \text{N}\right)$$

$$= (qRAB)^2 \left[2\frac{\eta q}{h\upsilon}\frac{b_{on}}{T_b} M^{x+2} T_b I_Z + \text{N}\right] \qquad \text{...(12.83)}$$

The effect of photomultiplication, amplifier gain can be clubbed together and the noise power corresponding to a unit bandwidth then becomes

$$\sigma_1^2 = \left[2q\frac{\eta q}{h\upsilon}\frac{b_{on}}{T_b} M^x T_b I_Z + \frac{\text{N}}{M^2}\right] \qquad \text{...(12.84)}$$

$$= \left[2\frac{q^2\eta}{h\upsilon} M^x b_{on} I_Z + \frac{\text{N}}{M^2}\right]$$

and similarly, the variance affecting a zero is expressed as

$$\sigma^2_1 = (qRAB)^2 \left(\frac{2}{q}\overline{i_{unity_0}} M^{2+x} T_b I_Z + \text{N}\right)$$

$$(qRAB)^2 \left[2\frac{\eta q}{h\upsilon}\frac{b_{on}(1-\gamma)}{T_b} M^{x+2} T_b I_Z + \text{N}\right] \qquad \text{...(12.85)}$$

The decision threshold voltage v_{Th} is set so that there is an equal error probability for 0 and 1 pulses and under the assumption of equiprobable 1's and 0's, we have

$$p_0(v_{Th}) = p_1(v_{Th}) = 0.5p_e \qquad \text{...(12.86)}$$

Under the assumption of a Gaussian random variable for the equalizer output, the error probability is given as

$$p(1/0) = \frac{1}{\sqrt{2\pi}\sigma_0} \int_{v_{Th}}^{\infty} \exp\left[-\frac{(v-b_0)^2}{2\sigma_0^2}\right] dv \qquad \text{...(12.87)}$$

corresponding to the case of deciding in favour of 1 when a 0 was transmitted because the total noise affecting the decision statistic at the sampling instant exceeds the threshold voltage. This gives the conditional probability of deciding a 1 when the actual transmitted bit was a 0. Similarly, the conditional probability that a decision is made in favour of a 0 when a 1 was transmitted becomes

$$p(0/1) = \frac{1}{\sqrt{2\pi}\sigma_1} \int_{-\infty}^{v_{Th}} \exp\left[-\frac{(-v-b_1)^2}{2\sigma_1^2}\right] dv \qquad \text{...(12.88)}$$

These two conditional probabilities are equal. Let us define a parameter known as Q.

$$Q = \frac{v_{Th} - b_0}{\sigma_0} = \frac{b_1 - v_{Th}}{\sigma_1} \qquad \text{...(12.89)}$$

This gives us an expression for the threshold which now becomes

$$v_{Th} = \frac{b_{on}\sigma_0 + b_{off}\sigma_1}{\sigma_1 + \sigma_0} \qquad \text{...(12.90)}$$

Then the probability of error becomes

$$p_e(Q) = \frac{1}{\sqrt{\pi}} \int_{Q/\sqrt{2}}^{\infty} \exp(-x^2)dx \qquad \text{...(12.91)}$$

$$= \frac{1}{2}\left[1 - erf\left(\frac{Q}{\sqrt{2}}\right)\right]$$

As may note the parameter Q is related to the signal to noise ratio required to achieve the desired BEP. The threshold voltage must be at least Q standard deviations above b_0 relative to the noise at b_0. We also have

$$b_0 = v_{Th} - \sigma_0 Q \qquad \text{...(12.92a)}$$

and

$$b_1 = v_{Th} + \sigma_0 Q \qquad \text{...(12.92b)}$$

The receiver sensitivity also is defined as

$$b_1 - b_0 = v_{Th} + \sigma_0 Q - v_{Th} + \sigma_1 Q = Q(\sigma_0 + \sigma_1)$$

$$\Rightarrow \qquad Q = \frac{b_1 - b_0}{\sigma_0 + \sigma_1} \qquad \text{...(12.93)}$$

The required energy per pulse needed to have a certain desired BER as characterized by the Q parameter becomes, corresponding to $b_0 = 0$ (substituting the values of s_1 and s_2)

$$b_1 = \frac{Q}{M}\left\{ \left[2\frac{q^2\eta}{h\upsilon} M^{2+x} b_{on} I_2 + \text{N} \right]^{1/2} + \left[2\frac{q^2\eta}{h\upsilon} M^{2+x} b_{on}(1-\gamma) I_2 + \text{N} \right]^{1/2} \right\} \qquad \text{...(12.94)}$$

12.4. PREAMPLIFIER TYPES

An optical preamplifier is used to increase the sensitivity of an optical receiver. The sensitivity and the bandwidth of a receiver are primarily determined by the noise sources at the receiver front end or the preamplifier stages. Therefore, the performance of a given receiver can be improved by placing low noise preamplifiers. The objective, hence is to minimize noise or maximize sensitivity while maintaining a suitable bandwidth. Preamplifiers used in optical fibre communication receivers are broadly of three categories: low impedance, high impedance and transimedance types.

12.4.1. Low Impedance Type Preamplifiers

A photodiode operates into a low impedance amplifier for example, a 50 Ω in the low impedance type as shown in Fig. 12.7. A bias or load resistor R_b (not shown in this figure) is used to match the amplifier impedance in order to suppress the formation of standing wave patterns due to impedance mismatch. The current developed in the photodetector in response to a light is converted into a suitable voltage across the load impedance. The resulting signal is buffered by the main amplifier. C_T The capacitance represents the total capacitance of the photodetector and the amplifier. The value of this biasing resistor is such that in association with the amplifier input capacitance, the amplifier's bandwidth is equal to or greater than the signal bandwidth. The intrinsic bandwidth of this preamplifier is $1/(2\pi R_L C_T)$. They satisfy bandwidth requirement at the cost of increased noise. Hence, these amplifiers can operate wideband. However, they do not have a high receiver sensitivity as only a small signal voltage can be developed across the amplifier input terminals. This limits their use to special short distance applications where high receiver sensitivity is not a major issue.

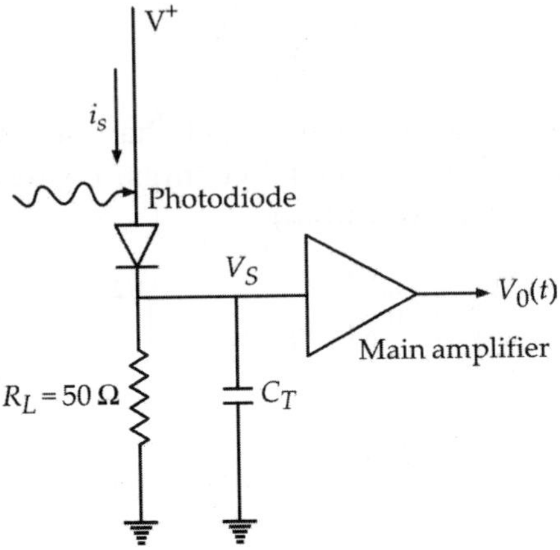

Fig. 12.7: Low Impedance preamplifier

In the high impedance type, the load resistance R_L is made large that allows a substantial amount of voltage to be developed in response to a given optical input. Hence, these amplifiers have better sensitivity as compared to their low impedance types, however, precisely because of a large R_L, the bandwidth of this amplifier becomes less than that of the signal it handles. Hence, additional filtering or equalization in the form of differentiation is needed to achieve the desired bandwidth. Although they have the highest sensitivity among the three types, however, their dynamic range is quite limited owing to the use of a large R_L that accords a kind of integrating effect to the input signal to the main amplifier. They are, hence characterized by low noise performance.

12.4.2. Transimpedance Type

The transimedance kind of amplifiers provide a good compromise between the wide bandwidth of the provided by the low impedance type of amplifier and the greater sensitivity of the high impedance type. Here the output is a voltage while input is a current source. Hence the transfer function takes the dimensions of impedance. A simple realization of such an amplifier is shown in Fig. 12.8.

As we may observe in Fig. 12.8, a feedback resistor R_f has been placed in the feedback path of the operational amplifier. The current developed by the photodetector in response to a given optical input directly flows through this feedback resistor as the amplifier's input impedance is very high. This current is responsible for developing an output voltage $V_0(t)$ across the output terminals of the amplifier. Here the feedback resistor can be made very large as the photodiode, in effect sees an impedance reduced by a factor of 1 + A where A is the open loop gain of the amplifier that is responsible for providing a larger bandwidth of the amplifier. Hence, the bandwidth of such an amplifier can be matched to that of the input signal without requiring additional equalization. In addition to this, the thermal noise contribution of the feedback resistor is also minimized which makes sensitivity better. The closed loop gain of this amplifier is, then

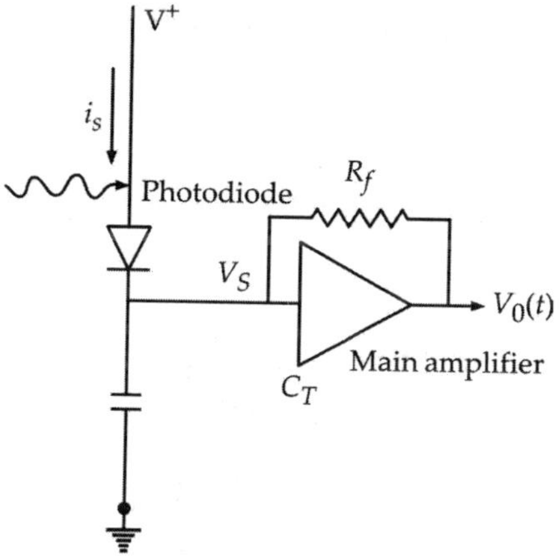

Fig. 12.8: A transimedance type of optical preamplifier

$$A_f = -\frac{V_0}{i_S} = -R_f \qquad ...(12.95)$$

The equivalent circuit model to analyze a transimpedance type of amplifier from the SNR point of view is drawn in Fig. 12.9. As we may observe from this figure, all noise sources are reduced to a minimum in case of high impedance preamplifiers. This is accomplished by reducing the input capacitance by going for low capacitance high frequency devices, minimizing thermal noise generated by biasing resistors and also by selecting detectors with low dark currents. The thermal noise can further be reduced by using high impedance amplifiers (like emitter followers, FET based preamplifiers or MOSFETs in source follower configurations) together with a large photodetector bias resistor R_b. The high impedance at the input gives rise to a longer time constant at the input terminals resulting in a receiver front end bandwidth being less than the signal bandwidth. As a result of this, the input signals tend to experience an integrating effect at output which needs appropriate equalization techniques to compensate this effect.

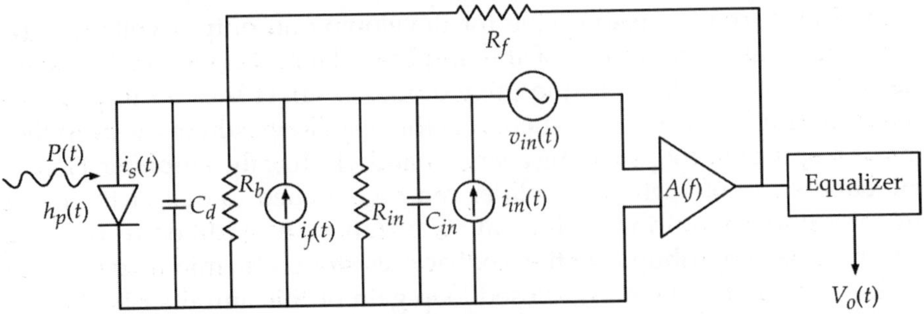

Fig. 12.9: An equivalent of a transimpedance receiver

The thermal noise characteristic N of such amplifiers represented by an equivalent circuit of Fig. 12.9 is expressed as

$$N_{TZ} = \frac{1}{q^2 B}\left[\frac{4K_BT}{R'_b} + \frac{2K_BT}{g_m(R_b^{'})^2} + \frac{S_E}{(R')^2}\right]I_Z + \left(\frac{2\pi C}{q}\right)^2 S_E I_2 B \quad ...(12.96)$$

In (12.96), $R_b = R_b \,||\, R_f$ and $R' = R_{in} \,||\, R_b \,||\, R_f$

Usually $R_f >> R_{in}$ then $R' \approx R$ so that,

$$NTZ = \mathrm{N}_{HZ} + \frac{4K_BT}{q^2 R_f} I_Z B \quad ...(12.97)$$

where N_{HZ} is the thermal noise characteristic of the high impedance amplifier depending on whether an FET or a BJT is used as the main amplifying device. We may observe that the thermal noise characteristic of a transimpedance amplifier may thus be modelled as the sum of the output noise of an amplifier without feedback plus the thermal noise associated with a feedback resistor as given by the second term in RHS of (12.97). The transfer function of the amplifier without feedback is

$$H(f) = \frac{A}{(1/R) + j2\pi fC} \quad ...(12.98)$$

where A is the midband gain of the amplifier, R and C are defined by (12.7) and (12.8) respectively. This gives a NEB equal to. For the transimedance amplifier, however, the transfer function is

$$H_{TZ}(f) = \frac{A}{(1/R) + j2\pi fC}$$

To derive this, we note that

$$i_s + i_f = i_r + i_c$$

$$= Mi_p(f) + \frac{V_o - v_{in}}{R_f} = \frac{v_{in}}{R} + j2\pi fCv_{in}$$

$$= v_{in}\left(\frac{1}{R} + j2\pi fC\right) \quad ...(12.99)$$

$$\Rightarrow \quad Mi_p(f) = v_{in}\left(\frac{1}{R} + j2\pi fC\right) - \frac{V_o - v_{in}}{R_f}$$

We know that,

$$V_0(t) = Av_{in} \quad ...(12.100)$$

Substituting (12.100) in (12.99), we get

$$Mi_p(f) = -\frac{V_o}{A}\left(\frac{1}{R} + j2\pi fC\right) - \frac{V_o}{R_f} + \frac{V_o}{AR_f}$$

$$\Rightarrow \quad Mi_p(f) = -\frac{V_o}{A}\left(\frac{1}{R} + j2\pi fC\right) - \frac{V_o}{R_f} + \frac{V_o}{AR_f} \quad ...(12.101)$$

$$\Rightarrow \quad Mi_p(f) = \left(\frac{R - AR - R_f - j2\pi fCRR_f}{ARR_f}\right)V_o$$

$$\Rightarrow \quad H_{TZ}(f) = \frac{V_o}{Mi_p(f)} = \frac{ARR_f}{R - AR - R_f - j2\pi fCRR_f}$$

$$= \frac{1}{\left(\frac{1}{AR_f} - \frac{1}{R_f} - \frac{1}{AR} - \frac{j2\pi fC}{A}\right)}$$

We have used an APD of average gain M. This factor is unity for a PIN photodiode. In any case, the expression for the transfer function would be in the same. The gain of the amplifier is very large. Thus, the first and third terms in the denominator in (12.101) are neglected in comparison to the two other terms. Doing so gives us the approximate simplified transfer function as

$$H_{TZ}(f) = \frac{1}{\left(-\frac{1}{R_f} - \frac{j2\pi fC}{A}\right)}$$

$$= -\frac{R_f}{1+\frac{j2\pi f R_f C}{A}} \quad \text{...(12.102)}$$

The negative sign here simply indicates that we are making use of the inverting terminal of the amplifier. The NEB of such a transimpedance amplifier can be evaluated to be equal to $B_{TZ} = \frac{A}{4RC}$ which is A times that of the high impedance amplifier. This also makes equalization simpler in such amplifiers with feedback.

12.4.3. High Impedance Type

In the third category of preamplifier types suggested in [3-5], the drawbacks of the high impedance amplifier are overcome by utilizing a low noise, high impedance amplifier with a negative feedback resistor R_f with an equivalent thermal noise current $i_f(t)$ that provides a shunting effect. The amplifier has an input equivalent series voltage noise source $v_{in}(t)$, an equivalent shunt current noise $i_{in}(t)$ and an input impedance given by the parallel combination of s and C_{in}.

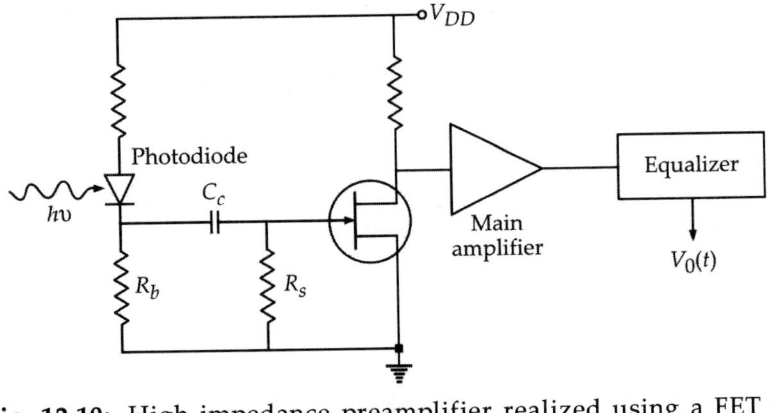

Fig. 12.10: High impedance preamplifier realized using a FET in common source configuration

The lowest-noise receivers are realized using GaAs MESFET preamplifiers. At lower frequencies silicon MOSFETs or JFETs are usually used. The circuit of a simple high impedance preamplifier implemented with a common source configuration is shown in Fig. 12.10. Typical FETs have very large input impedance. Therefore, the total input impedance at the input terminals reduces to R_b. Here the main noise sources are the thermal noise associated with the FET's channel transconductance, load or feedback resistor and shot noise due to gate leakage current. Another noise source is the FET $1/f$ noise which becomes significant at low bit rates. The input current noise spectral density S_I is

$$S_{I,FET} = \frac{4K_BT}{R_{in}} + 2qI_g \cong 2qI_g \qquad ...(12.103)$$

as the input impedance is very large. In this, I_g represents the gate leakage current of the FET. The thermal noise associated with the channel's transconductance is

$$S_E = \frac{\Gamma 4K_BT}{g_m} \qquad ...(12.104)$$

where Γ is the FET channel noise factor that accounts for thermal noise and gate induced noise plus the correlation between these two noises. The thermal noise characteristic N at the equalizer output is

$$N_{HZ,FET} = \frac{1}{q^2B}\left(2qI_g + \frac{4K_BT}{R_b} + \frac{4\Gamma K_BT}{g_mR_b^2}\right)I_Z + \left(\frac{2\pi C}{q}\right)^2 \frac{4\Gamma K_BT}{g_m} I_2B \qquad ...(12.105)$$

The bias resistor should be very large in order to minimize noise in a high impedance design. The effect of this is that the detector output signal is integrated by the amplifier input resistance. This can be compensated by carrying out differentiation in the equalizer filter. The integration-differentiation is known as the high impedance amplifier design technique. This yields a low noise, however at the cost of a lowered dynamic range. The gain of a silicon FET approaches unity for signal frequencies in the vicinity of 25 to 50 MHz. Hence, for frequencies in excess of these values, either a GaAs MESFET or a silicon bipolar transistor can be used as the preamplifiers.

12.4.4. High Impedance Bipolar Transistor Amplifiers

A simple common emitter amplifier using voltage divider biasing has been used as a high impedance preamplifier in Fig. 12.11. The biasing is carried out by the resistors R_1 and R_2. The input resistance of such an amplifier is given as

$$R_i = \frac{K_BT}{qI_B} \qquad ...(12.106)$$

where I_B is the base bias current. For this amplifier, the input resistance R_{in} is given as the following:

$$R_i = R_i \parallel R_1 \parallel R_2 \qquad ...(12.107)$$

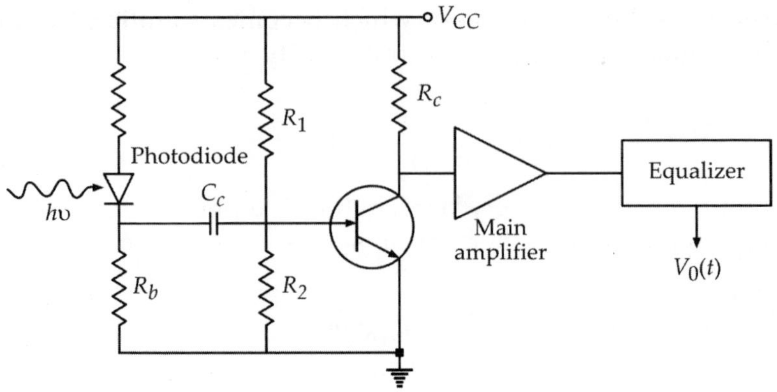

Fig. 12.11: High impedance preamplifier realized with a bipolar transistor

For low noise design, the values of the biasing resistors R_1 and R_2 are chosen to be much greater than the amplifier's original input resistance R_i; that $R_1, R_2 >> R_i$ is so that the actual input resistance becomes

$$R_{in} \cong R_i \qquad ...(12.108)$$

The input noise current power spectral density resulting from the shot noise of base current is

$$S_I = 2qI_B = \frac{2K_BT}{R_{in}} \qquad ...(12.109)$$

The corresponding input noise voltage power spectral density is

$$S_V = \frac{2K_BT}{g_m} \qquad ...(12.110)$$

As we may observe the transconductance of the BJT is related to the shot noise by the following:

$$g_m = \frac{qI_c}{K_BT} = \frac{\beta}{R_{in}} \qquad ...(12.111)$$

where I_c is the dc collector current and β is the dc current gain of the transistor. The corresponding thermal noise characteristic N at the equalizer output, for such an amplifier is

$$N_{HZ,BJT} = \frac{T_b}{q^2} 2K_BT\left[\left(\frac{1}{R_{in}} + \frac{2}{R_b} + \frac{R_{in}}{\beta R^2}\right)I_Z + \left(\frac{2\pi C}{T_b}\right)^2 \frac{R_{in}}{\beta} I_2\right] \qquad ...(12.112)$$

Unlike the FET based preamplifier, the contribution of the transistor's input capacitance to the overall capacitance is negligible so that we usually have $R \approx R_{in} \approx R_i$ because of the use of large valued bias resistors. Therefore,

$$N_{HZBJT} = \frac{2K_BT}{q^2}\left[\left(\frac{T_b}{R_{in}}\frac{\beta+1}{\beta}\right)I_Z + \frac{(2\pi C)^2}{\beta T_b}R_{in}I_2\right] \quad ...(12.113)$$

The impedance loading the photodetector also integrates the photodetector's output signal. Hence, the amplified signal is differentiated before it is fed to the equalizing filter to undo the undesirable effect of integration.

12.5. ANALOG RECEIVERS

The baseband signal or the message signal (audio, video, image, sensor measurements etc.) modulates certain parameters of an optical carrier. This modulation may take the form of AM, FM or PM very similar to analog modulation techniques. A number of such subcarriers are then frequency division multiplexed which then modulates the optical source. The intensity of the optical source is modulated in accordance with the subcarriers. This gives us modulation techniques like AM-IM (Intensity modulation), FM-IM or PM-IM.

Direct Intensity Modulation (D-IM): An analog receiver that employs D-IM is drawn in Fig.12.12 where the baseband signal changes the intensity of the light of the optical source directly.

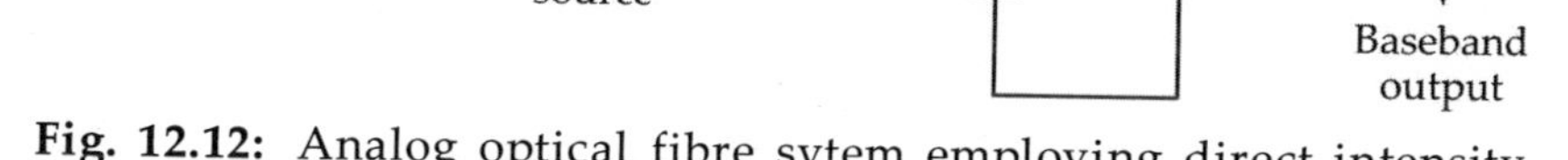

Fig. 12.12: Analog optical fibre sytem employing direct intensity modulation

For a baseband signal (t), the total optical power transmitted is

$$P_0(opt) = P_i[1 + m(t)] \quad ...(12.114)$$

where P_i is the power of the unmodulated optical source. For example, if the message is a pure tone, then we have

$$P_0(opt) = P_i[1 + m_a \cos \omega_m t] \quad ...(12.115)$$

In (12.115), m_a is the modulation index as defined for an AM signal. In response to such a sinusoidally modulated signal, the current generated by the photodetector is

$$I = MI_p(1 + m_a \cos \omega_m t) = M\mathfrak{R}_0 P_0(1 + m_a \cos \omega_m t) \quad ...(12.116)$$

We know that $$I = \frac{\eta q}{h\upsilon} P_o$$

The mean square value of the signal current given in (12.116) is, therefore,

$$\overline{i_{sig}^2} = \frac{1}{2}(m_a^2 M^2 I_p^2) = \frac{1}{2}(m_a M I_p)^2 p \quad ...(12.117)$$

ignoring the dc term as it does not contribute to the signal power.The noise introduced by the photodetector is the combined effect of the quantum noise and dark current noise which is expressed as

$$2p(I_p + I_d)\, M^2 F(M) B$$

where B is the effective post detection bandwidth. The thermal noise component has a power given as

$$N_o = \frac{4K_B T B F_n}{R_L} \quad ...(12.118)$$

where T is the temperature, F_n is the preamplifier noise figure and R_L is the load resistance across which this noise voltage appears.

Therefore, the figure of merit in terms of peak signal power to RMS noise power is defined as

$$\left(\frac{S}{N}\right)_{rms} = \frac{(1/2)(m_a M I_p)^2}{2qB(I_p + I_d)M^2 F(M) + \dfrac{4K_B T B F_n}{R_L}}$$

$$= \frac{(m_a I_p)^2}{4qB(I_p + I_d)F(M) + \dfrac{8K_B T B F_n}{R_L M^2}} \quad ...(12.119)$$

dividing the numerator and the denominator of (12.119) by M^2.

For a PIN photodetector, the RMS SNR reduces to

$$\left(\frac{S}{N}\right)_{rms,\ PIN} = \frac{(m_a I_p)^2}{4qB(I_p + I_d) + \dfrac{8K_B T B F_n}{R_L}} \quad ...(12.120)$$

Moreover, the SNR for video transmission is often defined in terms of the peak-to-peak picture signal power to the RMS noise power and may include the ration of luminance to composite video *b*. This definition of SNR gives us

$$\left(\frac{S}{N}\right)_{peak-to-peak,\ PIN} = \frac{(2m_a I_p b)^2}{2qB(I_p + I_d) + \dfrac{4K_B TBF_n}{R_L}} \qquad ...(12.121)$$

For the receiver being shot noise limited, the contribution of the second term in the denominator of (12.120) and (12.121) is negligible in comparison with the first term. Hence, the receiver figure of merit becomes

$$\left(\frac{S}{N}\right)_{RMS,shot-noise} = \frac{(m_a I_p)^2}{4qB(I_p + I_d)F(M)} \approx \frac{m_a^2 I_p}{4qBF(M)} \quad \because I_d << I_p \quad ...(12.122)$$

As we may observe, (12.122) applies to receivers built using APDs. The low noise preamplification may not be always obtained and higher thermal noise levels affect the receiver sensitivity adversely for a given SNR. The performance of a system that is considered to be thermal noise current limited is

$$\left(\frac{S}{N}\right)_{RMS,\ thermal} = \frac{(m_a I_p)^2}{8K_B TBF_n / M^2 R_L} \qquad ...(12.123)$$

For a PIN photodetector, $M = 1 = F(M)$. Hence the two expressions simplify to

$$\left(\frac{S}{N}\right)_{RMS,shot-noise} = \frac{m_a^2 I_p}{4qB}$$

$$= \frac{m_a^2 \left(\dfrac{\eta q}{h\upsilon}\right) P_o}{4qB} \qquad ...(12.124)$$

Hence, the relationship between the average received optical power and P_0 the primary photocurrent is expressed as, for a PIN type photodetector

$$\left(\frac{S}{N}\right)_{rms,shot-noise} = \frac{m_a^2 I_p}{4qB} \qquad ...(12.125)$$

The required input optical power in order to result in an SNR as expressed in (12.125), therefore becomes

$$P_0 = \frac{4h\upsilon}{m_a^2\eta} B \left(\frac{S}{N}\right)_{rms,shot-noise} \qquad \text{...(12.126)}$$

For a quantum noise limited analog receiver, the optical input power is directly proportional to the effective noise or post detection bandwidth B. A similar result can be derived for a thermal noise limited receiver.

and

$$\left(\frac{S}{N}\right)_{rms} = \frac{(m_a I_p)^2}{8K_B TBF_n/R_L} \qquad \text{...(12.127)}$$

P-3: Derive (12.127).

12.5.1. Subcarrier Intensity Modulation

Direct intensity modulation of a given optical source is suitable for the transmission of a baseband analog signal. However, the wideband nature of the optical fibre medium is best exploited by multiplexing a number of baseband signals and then launching this multiplexed message signals onto the optical fibre. This may be achieved with analog transmission through frequency division multiplexing of the individual baseband channels. Initially, the baseband signals must be translated onto carriers of different frequency by amplitude modulation (AM), frequency modulation (FM) or phase modulation (PM) prior to being simultaneously transmitted as an frequency division multiplexing signal. The frequency translation may be performed in the electrical regime where the baseband analog signals modulate electrical subcarriers and are then frequency division multiplexed to form a composite electrical signal prior to intensity modulation of the optical source.

A block schematic of an analog system employing this technique known as subcarrier intensity modulation is illustrated in Fig. 12.12. The baseband signals are modulated onto radio frequency (RF) subcarriers by either AM, FM or PM and multiplexed before being applied to the optical source drive circuit. Hence an intensity modulated (IM) optical signal is obtained which may be AM-IM, FM-IM or PM-IM. In practice, however, system output SNR considerations dictate that generally only the latter two modulation formats are used. Systems may incorporate two levels of electrical modulation whereby the baseband channels are initially amplitude modulated prior to FM or PM. The FM or PM signal so obtained is then used to intensity modulate the optical source. At the receiver, the transmitted optical signal is detected prior to electrical demodulation and demultiplexing to obtain the originally transmitted baseband signals.

A further major advantage of subcarrier intensity modulation is the possible SNR improvement. We first, hence derive general SNR expressions for an IM optical carrier that may be applied to the subcarrier intensity modulation formats. An electrical signal intensity modulates the source.

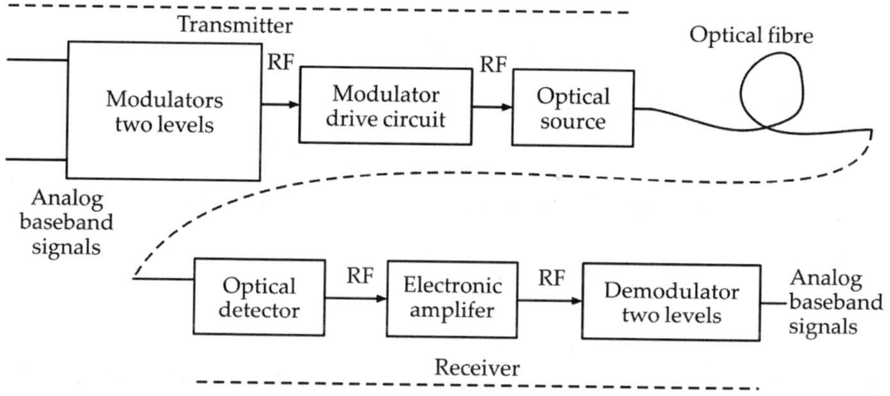

Fig. 12.13: Subcarrier intensity modulation system for analog optical fibre transmission

Let us derive the performance index for the case of an AM-IM receiver. A block schematic for such a transmitter-receiver combination is shown in Fig. 12.13. As we have seen before,

$$P_0(tot) = P_i[1 + m(t)] \quad \text{...(12.128)}$$

The current in response to (12.128) at the photodetetor's input is given as

$$i(t) = I_p M[1 + m(t)]$$

$$\Rightarrow \qquad \overline{i_{sig}^2} = (I_p M)^2 P_M \quad \text{...(12.129)}$$

We note that, in (12.129),

$$P_M = \frac{1}{2\pi} \int_{-2\pi B_m}^{2\pi B_m} S_m(\omega) d\omega = \int_{-B_m}^{B_m} S_m(f) df \quad \text{...(12.130)}$$

The RMS SNR is expressed as

$$\left(\frac{S}{N}\right)_{rms} = \frac{(MI_p)^2 P_M}{2qB_m(I_p + I_d)M^2 F(M) + \dfrac{4K_B T B_m F_n}{R_L}}$$

$$= \frac{(\Re P_o)^2 P_M}{2qB_m(I_p + I_d)F(M) + \dfrac{4K_BTB_mF_n}{R_LM^2}} \qquad ...(12.131)$$

$$= \frac{(\Re P_o)^2 P_M}{2B_mN_0}$$

In (12.131), we have made use of

$$N_0 = q(I_p + I_d)F(M) + (4k_BTF_n/R_L) \qquad ...(12.132)$$

When the message signal is given as $m(t) = A_c a(t) \cos\omega_c t$, we get

$$P_M = \frac{P_a}{2} \qquad ...(12.133)$$

12.5.2. Subcarrier double sideband modulation (DSB-IM)

A simple way to translate the spectrum of a baseband message signal is by direct multiplication with a subcarrier waveform $A_c \cos \omega_c t$ giving a double sideband suppressed carrier (DSBSC) type of amplitude modulation expressed as

$$v_{AM}(t) = A_c a(t) \cos \omega_c t \qquad ...(12.134)$$

where A_c is the amplitude and F_c is the frequency of the subcarrier. For such a waveform, the subcarrier electric field becomes

$$E_m(t) = \frac{A_c}{2}\left[\cos(\omega_c - \omega_m)t + \cos(\omega_c + \omega_m)t\right] \qquad ...(12.135)$$

giving the upper and the lower sidebands. This DSBSC signal intensity modulates an optical source. The transmitted power, therefore is given by

$$P_{opt}(t) = P_i[1 + a(t)]$$
$$= P_i[1 + a(t) \cos \omega_c t] \qquad ...(12.136)$$

where $A_c = 1$.

In order to prevent overmodulation, the value of the message signal is normalized such that $|a(t)| \le 1$ with power $P_a \le 1$. The DSB modulated electrical subcarrier occupies a bandwidth of $B_m = 2B_a$, with a carrier amplitude of unity, $P_m = \frac{P_a}{2}$. Therefore, the ratio of the RMS signal power to RMS noise power at the input to the DSB demodulator is given by

$$\left(\frac{S}{N}\right)_{rms} \text{input DSB} = \frac{(\Re P_o)^2}{2\times 2B_a N_o}\frac{P_a}{2} = \frac{(\Re P_o)^2 P_a}{8B_a N_o} \quad ...(12.137)$$

An ideal DSB modulator gives a detection gain of 2 or 3 dB improvement in SNR [6]. This yields an output SNR of

$$\left(\frac{S}{N}\right)_{rms} \text{output DSB} = 2\left(\frac{S}{N}\right)_{rms} \text{input DSB}$$

$$= \frac{(\Re P_o)^2 P_a}{4B_a N_o} \quad ...(12.138)$$

A comparison of (12.131) and (12.138) shows that a 3 dB SNR degradation results from the use of DSB-IM as compared to the use of D-IM under the same conditions of bandwidth ($B_m = B_a$), modulating signal power ($P_m = P_a$), detector photocurrent and noise. For this reason DSB-IM systems are usually not considered efficient for fibre optic communications. Hence attention is given to both FM-IM and PM-IM systems.

12.5.3. Subcarrier Frequency Modulation

Here the subcarrier is frequency modulated by the message signal. An FM signal has the following form:

$$v_{FM}(t) = A_c \cos\left[\omega_c t + k_f \int_0^t a(\lambda)d\lambda\right] \quad ...(12.139)$$

where k_f is the proportionality constant in radians/sec per unit of message signal. The carrier amplitude is taken such that $A_c \le 1$ to prevent overmodulation. The bandwidth of the modulated signal, from Carson's rule is

$$B_m = 2(\Delta_f + f_a) = 2(\beta + 1)f_a \quad ...(12.140)$$

where

$$\beta = \frac{\Delta f}{f_a} = \frac{k_f |a(t)|_{\max}}{f_a} \quad ...(12.141)$$

The quantity Δ_f is known as the peak frequency deviation and it is related to the maximum amplitude of the message signal. The SNR at the input to the subcarrier FM demodulator is

$$\left(\frac{S}{N}\right)_{rms} \text{input FM} = \frac{(\Re P_o)^2 (A_c^2/2)}{2B_m N_o} \quad ...(12.143)$$

The subcarrier demodulator operating above threshold yields an output SNR [7]

$$\left(\frac{S}{N}\right)_{rms} \text{output FM} = 6\beta^2(\beta^2+1)\frac{P_a(\Re P_o)^2(A_c^2/2)}{2B_m N_o} \qquad ...(12.144)$$

This however, assumes that we are using wideband subcarrier frequency modulation. Substitution of the modulated signal bandwidth from (12.139) in (12.142) gives us

$$\left(\frac{S}{N}\right)_{rms} \text{output FM} = 3\beta^2\frac{P_a(\Re P_o)^2(A_c^2/2)}{2f_a N_o} \qquad ...(12.145)$$

This result indicates that a significant improvement in the post detection SNR may be obtained by using wideband FM-IM.

12.5.4. Subcarrier Phase Modulation

In this modulation technique, the instantaneous phase of the subcarrier is made proportional to the amplitude of the modulating signal. The resulting PM-IM signal may be written as

$$v_{PM}(t) = A_c \cos A_c \cos\left[\omega_c t + k_p a(t)\right] \qquad ...(12.146)$$

where k_p is the proportionality constant expressed in radians per unit of $a(t)$. Here also we make the carrier amplitude $A_c \le 1$ in order to prevent overmodulation. The bandwidth of the modulated signal is expressed as

$$B_m = 2(\beta + 1)f_a \qquad ...(12.147)$$

where β_p is the frequency deviation ratio for the PM-IM system. We define β_p as

$$\beta_p = k_p\left|\frac{da(t)}{dt}\right|_{\max} \qquad ...(12.148)$$

The SNR at the input to the subcarrier PM modulator is

$$\left(\frac{S}{N}\right)_{rms} \text{input PM} = \frac{(\Re P_o)^2(A_c^2/2)}{2B_a N_o} \qquad ...(12.149)$$

The output SNR from an ideal subcarrier PM demodulator operating above threshold is [7]

$$\left(\frac{S}{N}\right)_{rms} \text{output PM} = \beta_p^2\frac{(\Re P_o)^2(A_c^2/2)}{2B_a N_o}P_a \qquad ...(12.150)$$

This result suggests that an improvement in SNR over D-IM may be obtained from the use of PM-IM especially for large SNR conditions with $A_c = 1$. However, between FM-IM and PM-IM the former yields the highest SNR.

12.6. CHAPTER SUMMARY

The performance of optical receivers has been defined in terms of probability of bit error. This is determined by signal to noise ratio available at the output of the amplifier. The factor affecting the output SNR is indicated by the NEB of the receiver. The performance is affected by shot noise and thermal noise for both p-i-n as well as avalanche type of photodetectors. In addition, the avalanche noise multiplication process contributes to an additional source of noise in receivers built using APD type of detectors. The mechanisms of generation of these predominant noises are different. However, all noises require a statistical description. While the thermal noise mostly follows a Gaussian distribution, the shot noise is characterized by a Poisson distribution. Depending on the magnitudes of these noises, a receiver is said to be thermal noise limited if its magnitude is higher than the shot noise or shot noise limited otherwise. Three types of preamplifiers suitable for implementation in such optical receivers at different frequencies have been discussed. The thermal noise characteristic for each has been elaborated. For analog receivers, performance is evaluated in terms of SNR as available at the amplifier output. This SNR is different for different subcarrier modulation systems. A comparison of four different subcarrier modulation types has been done and wideband FM has been found to yield the highest improvement in SNR.

APPENDIX-12.1

Derivation of noise equivalent bandwidth of the equalizer only

$$H_0(f) = \Re_0 \, GH_p(f) \, H_B(f) H_{eq}(f)$$

$$\Rightarrow \quad H_{eq}(f) = H_0(f)/[\Re_0 GH_p(f) \, H_B(f)]$$

$$B_v = \frac{1}{|H_{eq}(0)|^2} \int_0^{\infty} |H_{eq}(f)|^2 \, df$$

$$= \frac{1}{|H_0(0)/[\Re_0 GH_p(0) H_B(0)]|^2} \int_0^{\infty} |H_o(f)/[\Re_0 GH_p(f) H_B(f)]|^2 \, df$$

$$= \frac{1}{|H_o(0)/[H_p(0) H_B(0)]|^2} \int_0^{\infty} |H_o(f)/[H_p(f) H_B(f)]|^2 \, df$$

$$= \frac{1}{\left|H_o(0)/[H_p(0)R]\right|^2} \int_0^\infty \left| H_o(f) \Big/ \left[H_p(f) . \left(\frac{1}{(1/R) + j2\pi fC} \right) \right] \right|^2 df$$

APPENDIX-12.2

Derivation of (12.79)

$$B_{eq} = \frac{R^2}{|H_o(0)/H_p(0)|^2} \int_0^\infty \left| \frac{H_o(f)}{H_p(f)} \left(\frac{1}{R} + j2\pi fC \right) \right|^2 df$$

$$= \frac{R^2}{|H_o(0)/H_p(0)|^2} \int_0^\infty \left| \frac{H_o(f)}{H_p(f)} . \frac{1}{R} + j2\pi fC . \frac{H_o(f)}{H_p(f)} \right|^2 df$$

$$= \frac{R^2}{|H_o(0)/H_p(0)|^2} \left\{ \int_0^\infty \left[\frac{H_o(f)}{H_p(f)} \right]^2 . \frac{1}{R^2} df + (2\pi fRC)^2 \int_0^\infty \left[\frac{H_o(f)}{H_p(f)} \right]^2 df \right\}$$

$$= \frac{I_Z}{T_b} + \frac{(2\pi RC)^2}{T_b^2} . \int_0^\infty \left[\frac{H_o(f)}{H_p(f)} \right]^2 df$$

$$= \frac{I_Z}{T_b} + \frac{(2\pi RC)^2}{T_b^2} \frac{I_2}{T_b} = \frac{I_Z}{T_b} + \frac{(2\pi RC)^2}{T_b^3} I_2$$

$$= I_Z B + (2\pi RC)^2 I_2 B^3$$

APPENDIX-12.3

A different expression for the receiver sensitivity is derived here.

Let the average receive power is $P_{rx} = \frac{P_1 + P_0}{2}$. We assume 0 bits to carry no optical power. Hence the current that flows in the photodetector in response to a data bit of 0 becomes $I_0 = 0$ and the current flowing in response to a data bit of 1 is

$$I_I = M\Re P_1 = 2M\Re P_{rx}$$

The noise currents σ_1 and σ_0 include the effects of shot noise and thermal noise. Hence, we have $\sigma_1 = (\sigma_S + \sigma_T)^{1/2}$ and $\sigma_0 = \sigma_T$ and now the shot noise current in response to a data bit of 1 is

$$\sigma_s^2 = 2qI_1B_n$$

$$= 2q.2MMF\ (M)\Re P_{rx}\ B_n$$

$$= 4qM^{x+2}\ \Re P_{rx}B_n$$

Thus, $$\sigma_1 = (\sigma_S + \sigma_T)^{1/2}$$

$$= \left[4qM^{x+2}\Re P_{rx}B_n + \frac{4K_BT}{R_L}B_n\right]^{1/2}$$

$$= 2M\left(\Re P_{rx} + \frac{4K_BT}{R_LM^2}\right)^{1/2}\sqrt{B_n}$$

The Q parameter has been defined as

$$Q = \frac{I_1}{\sigma_1 + \sigma_0} = \frac{2M\Re P_{rx}}{\sigma_1 + \sigma_0}$$

$$\Rightarrow \qquad P_{rx} = \frac{Q(\sigma_1 + \sigma_0)}{2M\Re}$$

SUGGESTED READING

[1] Smith, D.R. and Garrett, I. "A Simplified Approach to digital optical receiver design", Opt. *Quantum Electronics,* vol. 10, pp. 211-221 2, 1978.

[2] Personick, S.D. "Receiver Design for optical fiber systems", *Proc. of IEEE,* vol. 65, pp. 1670-1678, Dec. 1977.

[3] Personick, S.D. "Receiver design", in S.E. Miller and A.G. Chynoweth, eds., *Optical Fiber Communications, Academic,* New York, 1979.

[4] Kasper, B.L. "Receiver design", in S.E. Miller and I.P. Kapinow, eds., *Optical Fiber Telecommunications, Academic,* New York, 1988.

[5] Smith, R.G. and Personick, S.D. "Receiver design for optical fiber communication systems", in H. Kressel, ed., Semiconductor Devices for Optical Communications", 2nd, Springer-Verlag, New York, 1982.

[6] Lathi, B.P. "*Communication Systems*", McGraw Hill, 1965.

[7] Shanmugan, K. Sam. "Digital and Analog Communication Systems", Wiley, 1979.

❑❑❑

13

Coherent Optical Communications

13.1. INTRODUCTION

The two transmission regions of an optical fibre that can be used for coherent systems are from 1270 to 1350 nm and from 1480 to 1600 nm, in the wavelength window from 1270 to 1350 nm A system designer can accommodate, theoretically 1000 channels with a channel spacing of 10-GHz while about 1500 channels can be accommodated in the window from 1480 to 1600 nm. Coherent optical detection is analgous to superheterodyne detection in modern radios.

The possibility of actually realizing these optical coherent systems lies in the fact of the ability to develop truly coherent semiconductor lasers that can be tuned over a significant portion of the 30,000 GHz bandwidth of a single mode optical fibre, implementation of appropriate line codes in order to superimpose data on optical carrier and the development of a practical polarization control method for the receiver.

The main advantages of a coherent optical communication system are nearly ideal receiver sensitivity and a high degree of frequency stability. The selectivity of coherent systems is due to the fact that narrowband electronic filters rather than broad optical filters pick out the channels. A nominal 2-nm wavelength passband of an optical filter is equivalent to a frequency passband of 240 GHz. Comparing this with a 2-GHz wide electrical filter we observe that electrical filtering provides several orders of magnitude more frequency selectivity. Therefore, the channel density of a multichannel coherent system can be 100 times that for direct detection.

13.2. FUNDAMENTAL PRINCIPLES

The invention of the laser diode in 1960 paved the way for coherent optical communications by exploiting the coherent properties of the laser light.

Gas lasers provided improved SNRs over direct detection of intensity modulated signals. The effort to provide a truly coherent optical communication took shape with the possibility of having a single mode transmission from a narrow-linewidth AlGaAS kind of laser. Polarization stability of the transmission medium is crucial towards achieving a coherent detection. Ideally, the fibre is required to maintain a single linear polarization state throughout its length for achieving a coherent communication.

A block schematic of a generalized coherent optical communication system is illustrated in Fig. 13.1. The transmitter is a CW narrow-linewidth laser that acts as the optical frequency oscillator. An external optical modulator provides modulation of this laser output in terms of amplitude, frequency or phase in accordance with the message signal which is digital in the present case. The external modulators are fabricated from lithium niobonate or group-III-V compound semiconductors. Internal modulation of the injection laser drive current may be utilized to produce either ASK or FSK. The three basic modulation formats that are typically used in a coherent optical communication system are described briefly as follows.

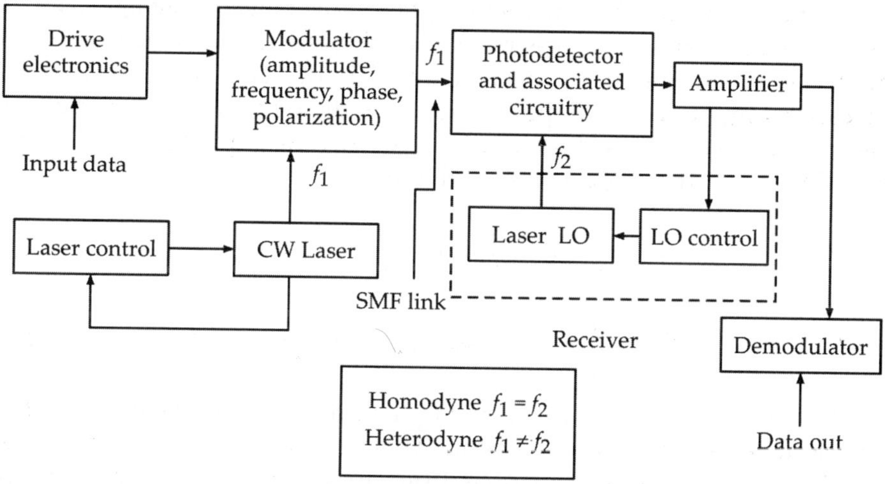

Fig. 13.1: Block schematic of a generalized coherent optical communication system

The fundamental principle of coherent optical detection is illustrated by considering the electric field of a transmitted signal to be a plane wave of the following form:

$$E_s = A_s \cos[\omega_s t + \phi_s(t)] \qquad \text{...(13.1)}$$

where A_s is the amplitude of the optical signal, $f_s = \omega_s/2\pi$ is its frequency and $f_s(t)$ is its phase. Information is sent over this carrier by modulating its amplitude, frequency or phase in accordance with the message signal.

We, can have, therefore the following systems.

(a) Amplitude shift keying (ASK) or on-off keying (OOK): As the name suggests, the amplitude of the carrier is shifted between two values (one corresponding to 'on' and the other for 'off' condition of the message signal) depending on whether a 1 or a 0 is being transmitted at a given time instant. As making the amplitude 'on' or 'off' is analgous to electrical keying (corresponding to an electrical circuit being closed by pressing the key or the switch or opened by releasing the key), it is named so.

(b) Frequency shift keying (FSK): Here also we can understand that the frequency of the optical carrier is made to take only two values corresponding to data bit of 1 or 0. The amplitude A_s and phase $\phi_s(t)$ are kept unchanged in this system.

(c) Phase shift keying (PSK): The phase $\phi_s(t)$ is varied in accordance with a sine wave $\phi_s(t) = \beta \sin \omega_m t$ where β is known as the modulation index and f_m is the modulation frequency.

In a direct detection system, the electrical signal coming into the transmitter amplitude modulates the optical power level of the light source. Therefore, the optical power is proportional to the signal current level. At the receiver, the incoming optical signal is directly converted into a corresponding electrical output (to serve as the demodulated output) by the photodetector. This directly detected current is proportional to the intensity I_{DD} of the optical signal, giving us

$$I_{DD} = E_s E_s^* = \frac{1}{2} A_s^2 \cos^2\left[\omega_s t + \phi_s(t)\right]$$

$$= \frac{1}{2} A_s^2 \left[1 + \cos\left(2\omega_s t + \phi_s\left[t\right]\right)\right] \quad \text{...(13.2)}$$

We observe that, there is a dc term plus a double frequency term in the directly detected current expression. This double frequency term, being very high in frequency is rejected from appearing at the output of the receiver as this is beyond the response capability of the detector. Hence, we get, for the direct detection receiver

$$I_{DD} = \frac{1}{2} A_s^2 \quad \text{...(13.3)}$$

At the coherent optical receiver as shown in Fig. 13.2, the receiver adds a locally generated optical carrier to the incoming modulated signal and this combination is detected forthwith. There are four demodulation formats depending on how the optical signal is mixed with the local oscillator (that gives heterodyning or homodyning) and how the electrical signal is detected. For a given modulation format, homodyne receivers are more sensitive than heterodyne receivers, and synchronous detection is more sensitive than asynchronous detection.

The mixing of the information bearing and local oscillator signals is carried out on the surface of the photodetector. If the local oscillator has a waveform that corresponds to the following

$$E_{LO} = A_{LO} \cos[\omega_{LO} + \phi_{LO}(t)] \quad ...(13.4)$$

where A_{LO} is the amplitude of the local oscillator's waveform, ω_{LO} is its angular frequency ϕ_{LO} being the phase. The detected photocurrent is proportional to the input power which, in turn is proportional to the square of the amplitude of the waveform falling on the photodetector. Hence,

$$I_P(t) \; \alpha \; (E_s + E_{LO})^2$$

$$= \frac{1}{2}A_s^2 + \frac{1}{2}A_{LO}^2 + 2A_s A_{LO} \cos[\omega_s t + \phi_s(t)]\cos[\omega_{LO} t + \phi_{LO}(t)]\cos\theta(t)$$

$$+ \frac{1}{2}A_s^2 \cos 2[\omega_s t + \phi(t)] + \frac{1}{2}A_{LO}^2 \cos 2[\omega_{LO} t + \phi(t)]\cos\theta(t)$$

$$= \frac{1}{2}A_s^2 + \frac{1}{2}A_{LO}^2 + \cos[(\omega_s - \omega_{LO})t + \phi_s(t) - \phi_{LO}(t)]\cos\theta(t); \quad ...(13.5)$$

neglecting higher frequency terms

$$= \frac{1}{2}A_s^2 + \frac{1}{2}A_{LO}^2 + \cos[(\omega_s - \omega_{LO})t + \phi(t)]\cos\theta(t)$$

Here in the above expression

$$\theta(t) = \frac{E_s . E_{LO}}{|E_s||E_{LO}|} \quad ...(13.6)$$

represents the polarization misalignment between the incoming modulated signal wave and the local oscillator's wave. The quantity $\phi_s(t) - \phi_{LO}(t)$ is the relative phase difference between the incoming signal and the local oscillator wave. We have here assumed that the photodetector does not respond to the double frequency terms or the sum frequency term. As the optical power is proportional to intensity, at the photodetector, we have

$$P(t) = P_s + P_{LO} + 2\sqrt{P_s P_{LO}} \cos[(\omega_s - \omega_{LO})t + \phi(t)]\cos\theta(t) \quad ...(13.7)$$

In this expression, we have made use of the fact that

$$P_s = \frac{1}{2} A_s^2 \qquad \text{...(13.8)}$$

is the power contained in the incoming signal,

$$P_{LO} = \frac{1}{2} A_{LO}^2 \qquad \text{...(13.9)}$$

represents the power of the local oscillator sinewave and $P_{LO} >> P_s$. The frequency difference $\omega_{IF} = \omega_s - \omega_{LO}$ is known as the intermediate frequency. The phase angle $\phi(t)$ gives the phase difference between the incoming signal and the local oscillator waveform which is time-varying. The frequency ω_{IF} is normally in the radio frequency range of a few tens or hundreds of MHz.

The electrical demodulator block shown in Fig. 13.1 is required in particular for an optical heterodyne system that can utilize either synchronous or asynchronous electrical detection. Synchronous or coherent demodulation means an estimation of phase of the IF signal in transferring it to the baseband. Such an approach requires phase lock loops to track phase changes occurring in the incoming and local oscillator signals. The basic block structure for an optically coherent receiver according to this principle is shown in Fig. 13.2. Asynchronous or noncoherent IF detection does not require phase estimations or stringent phase locked loops. This is usually achieved by recovering information contained in the envelope of the IF signal. Hence, the envelope detection techniques are less demanding from a signal processing view point. Optical homodyne detection, by definition is a synchronous or coherent demodulation technique as it requires the local oscillator frequency and phase to be in synchronism with that of the transmitter laser parameters.

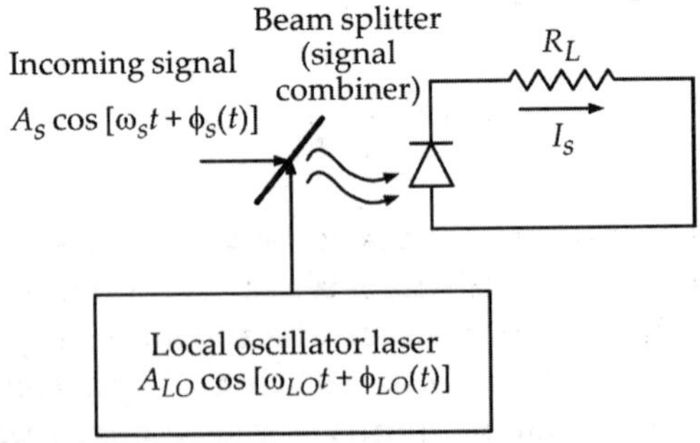

Fig. 13.2: Basic coherent receiver model

13.2.1. Homodyne Detection

Homodyne detection shown in Fig. 13.3 refers to the particular case when $\omega_{IF} = \omega_s - \omega_{LO} = 0$. In other words, the local oscillator frequency is made equal to that of the incoming wave. Therefore, the power expression becomes

$$P(t) = P_s + P_{LO} + 2\sqrt{P_s P_{LO}}\cos\phi(t)\cos\theta(t) \qquad ...(13.10)$$

We may use either OOK [varying signal level P_s while keeping $\phi(t)$ fixed] or PSK [varying the phase $\phi_s(t)$ of the signal and keeping P_s constant] modulation schemes to transmit information. We may observe that the cross product term in (13.10) contains the transmitted information under the assumption of $P_{LO} >> P_S$. This term increases with increasing laser power, the local oscillator effectively acts as a signal amplifier, thereby giving greater receiver sensitivity than direct detection.

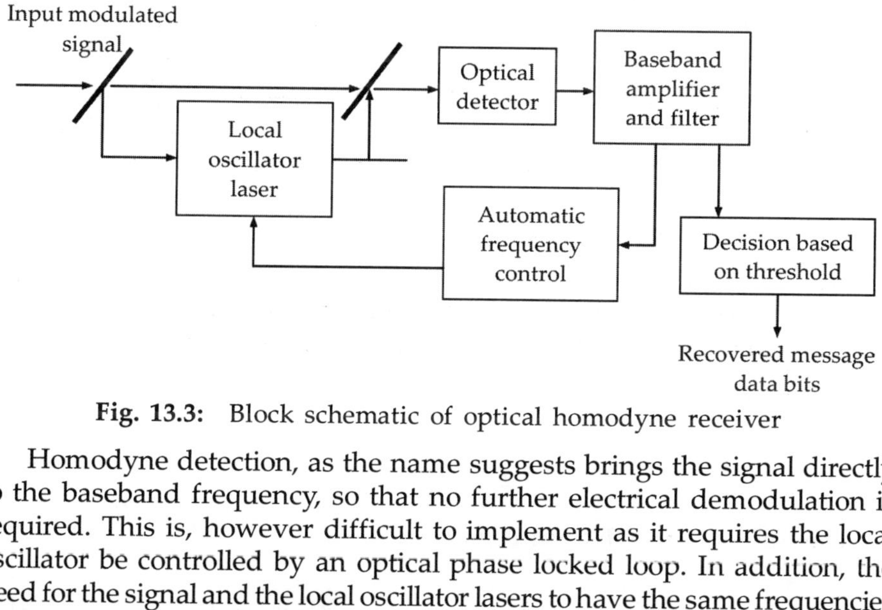

Fig. 13.3: Block schematic of optical homodyne receiver

Homodyne detection, as the name suggests brings the signal directly to the baseband frequency, so that no further electrical demodulation is required. This is, however difficult to implement as it requires the local oscillator be controlled by an optical phase locked loop. In addition, the need for the signal and the local oscillator lasers to have the same frequencies puts very stringent requirements on these two optical sources. These criteria call for an extremely narrow spectral width and a high degree of wavelength tunability.

13.2.2. Heterodyne Detection

The block schematic of an optical heterodyne receiver is shown in Fig. 13.4. The demodulator may be synchronous or nonsynchronus type. We may observe from this diagram that the frequency of the input modulated signal and the local oscillator's (or laser here) frequency are different. That is why the name heterodyne and it works with a certain value of intermediate frequency (IF). The IF here has a finite value and there is no need to provide an optical phase locked loop at the receiver. Hence, these kind of receivers are very easy to implement. Any of the three modulation techniques mentioned above can be utilized for heterodyne detection. In order to calculate the output current of the receiver, we note that under the assumption of a large local oscillator power as compared to the incoming signal power, the time-varying power becomes

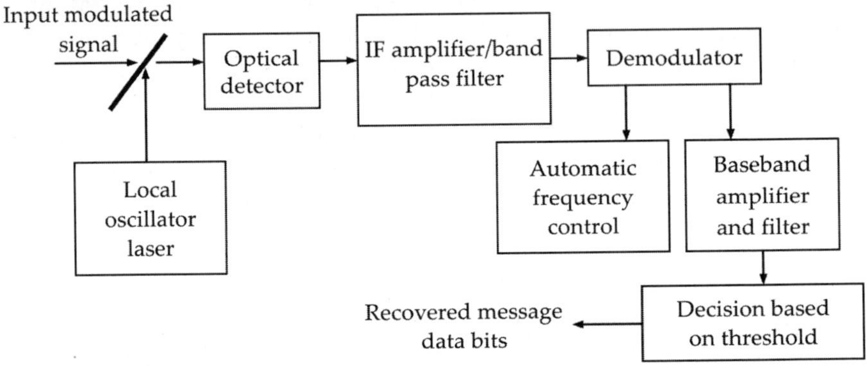

Fig. 13.4: Block schematic of optical heterodyne receiver

$$P(t) = P_{LO} + 2\sqrt{P_{LO}P_s}\cos[\omega_{IF} + \phi(t)]\cos\theta(t) \quad \text{...(13.11)}$$

This power contains a constant power equal to that of the local oscillator power and a time-varying cross term. Therefore, the detector current would contain a dc term and a time-varying term.

The dc term of the receiver output current is

$$i_{dc} = \frac{\eta q}{h\upsilon} P_{LO} \quad \text{...(13.12)}$$

The time-varying IF term is given as

$$i_{IF}(t) = 2\frac{\eta q}{h\upsilon}\sqrt{P_{LO}P_s}\cos[\omega_{IF} + \phi(t)]\cos\theta(t) \quad \text{...(13.13)}$$

Hence, the signal power as due to (13.13) becomes

$$P_s = \left[\frac{i_{IF}(t)}{\sqrt{2}}\right]^2 = \left(\frac{\eta q}{h\upsilon}\right)^2\left(\frac{4P_{LO}P_s}{2}\right)$$

$$= 2\left(\frac{\eta q}{h\upsilon}\right)^2 P_{LO}P_s \quad \text{...(13.14)}$$

We know that the shot noise current has a variance given as

$$\overline{i_{sh}^2} = 2qI_pB$$

$$= 2q.\frac{\eta q}{h\upsilon}P_{LO}B \text{ in photodetector bandwidth of } B$$

$$= 2\frac{\eta q^2}{h\upsilon}P_{LO}B \qquad \text{...(13.15)}$$

The SNR becomes

$$\frac{P_S}{\sigma^2} = \frac{(\eta q/h\upsilon)^2\ P_{LO}P_s}{2(\eta q^2/h\upsilon)\ P_{LO}B} \qquad \text{...(13.16)}$$

$$= \frac{\eta}{h\upsilon}\frac{P_s}{2B} = \frac{\eta}{h\upsilon}\frac{P_s}{B_{IF}}$$

In the SNR expression given in (13.16), we have considered the average signal energy which is half of the value given by (13.14). This is due to the fact that we have an optical pulse for a '1' while it is off for a binary data bit of '0'. This SNR expression is true for the heterodyne receivers where the IF amplifier has a bandwidth equal to $B_{IF} = 2B$. For homodyne receivers, the SNR expression is modified to

$$\frac{P_S}{\sigma^2} = \frac{\eta}{h\upsilon}\frac{P_s}{B} \qquad \text{...(13.17)}$$

A comparison of (13.16) and (13.17) tells us that a homodyne receiver has a 3 dB (10log 2) higher SNR than its heterodyne counterpart for the same detector bandwidth and same value of incoming signal power.

The dc current is normally filterer out in the receiver and the IF current gets amplified. The original information is recovered from the time-varying component as given in (13.13) by conventional RF demodulation methods.

13.3. MODULATION TECHNIQUES

We have observed from the foregoing that information can be transmitted by three fundamental ways. The techniques are described as ASK, FSK and PSK corresponding to the changes in the amplitude, frequency and phase of the carrier respectively. In binary digital systems a common ASK technique is on-off keying (OOK). Information embedded in such an OOK waveform may be retrieved by either homodyne or heterodyne optical detection methods together with either synchronous or asynchronous electrical detection. The particular choice of modulation and demodulation determines the fundamental receiver sensitivity.

13.3.1. OOK Homodyne System

OOK modulation can use either homodyne or heterodyne modulation techniques. Here, the modulated waveform remains at either high corresponding to a data bit of 1 or low for a data bit of 0. When the 'low' of the modulated waveform is received, the average number $\overline{N_0}$ of electron-hole pairs generated is simply the number generated by the local oscillator. This gives us

$$\overline{N_0} = A_{LO}^2 T \quad \text{...(13.18)}$$

where T is the duration of the '0' pulse. Similarly, the average number $\overline{N_0}$ of electron-hole pairs generated in response to a pulse of '1' is

$$\overline{N_1} = (A_{LO} + A_s)^2 T \approx (A_{LO}^2 + 2A_{LO}A_s)T \quad \text{...(13.19)}$$

under the assumption of a large local oscillator power. The voltage V as seen by the receiver during a '1' pulse is

$$V = \overline{N_1} - \overline{N_0} = 2A_{LO}A_s T \quad \text{...(13.20)}$$

The associated noise has an RMS value given as

$$\sigma = \sqrt{\overline{N_1}} = \sqrt{\overline{N_0}} \quad \text{...(13.21)}$$

The SNR of this modulation format, hence is

$$\frac{S}{N} = \left(\frac{V}{\sigma}\right)^2 = \frac{(2A_{LO}A_s T)^2}{A_{Lo}^2 T} = 4A_s^2 T \quad \text{...(13.22)}$$

The corresponding BER, from chapter 12 is

$$P_e = \frac{1}{2}\left[1 - erf\left(\frac{1}{2}\sqrt{\frac{S}{2N}}\right)\right]$$

$$= \frac{1}{2}\left[1 - erf\left(\frac{V}{2\sqrt{2}\sigma}\right)\right] = \frac{1}{2} erfc\left(\frac{V}{2\sqrt{2}\sigma}\right) \quad \text{...(13.23)}$$

$$= \frac{1}{2} erfc\left(\frac{2A_{LO}A_s T}{A_{LO}\sqrt{2T}}\right)$$

$$= \frac{1}{2} erfc\left(\frac{A_s\sqrt{T}}{\sqrt{2}}\right)$$

In the above, we have made use of the fact that erfc(u) = 1 – *erf*(*u*) is the complementary error function. We have seen previously that we need a ratio (V/σ) = 12 in order to achieve a BER of 10^{-9}. Using this we get

$$\left(\frac{V}{\sigma}\right)^2 = 144$$

$$\Rightarrow \qquad \frac{4A_{LO}^2 A_s^2 T^2}{A_{LO}^2 T} = 144 \qquad ...(13.24)$$

$$\Rightarrow \qquad 4A_s^2 T = 144$$

$$\Rightarrow \qquad A_s^2 T = 36$$

This is the required energy per optical pulse which must be given to it in order to have a BER of 10^{-9} for a receiver that employs homodyne OOK type of demodulation. In other words, an optical pulse needs to generate 36 photons for this BER. From the photodetector viewpoint, this amounts to generating 36 electron-hole pairs. In the ideal case of the receiver's quantum efficiency being equal to unity, a BER of 10^{-9} is achieved with an average receive optical energy of 36 photons per pulse.

The BER of a given modulation format is a function of the signal to noise ratio as we observe from the previous discussions. We wish to translate this SNR into a term containing the average number of photons per bit. To do so, we note that, the signal power is due to the photocurrent generated by the incoming optical power. Hence,

$$\frac{P_S}{\sigma^2} = \frac{\eta}{h\upsilon}\frac{P_s}{B} = \eta\overline{N_p} \qquad ...(13.25)$$

This is because, the total energy carried by the average number of photons in one bit interval is

$$P_s T = \frac{P_s}{B} \qquad ...(13.26)$$

Hence, the average number of photons in one bit interval is

$$\overline{N}_p = \frac{\frac{P_s}{B}}{h\upsilon} = \frac{P_s}{Bh\upsilon} \qquad ...(13.27)$$

The BER of such an OOK system, in terms of the average number of photons per bit interval is expressed as

$$P_e = \frac{1}{2} erfc\left(\sqrt{\eta \overline{N_p}}\right) \quad ...(13.28)$$

13.3.2. PSK Homodyne System

The incoming optical signal is first combined with a strong optical wave being generated by the local oscillator in this method. This is carried out using either a fibre directional coupler or a partially reflecting plate called a beam splitter as shown in Fig. 13.5. When a beam splitter is used, it is made almost completely transparent, since the incoming signal is much weaker than the local oscillator output. In this modulation format, information is transmitted by modulating the phase of the carrier in accordance with the message data bits. The phase of the carrier remains unaltered when a data bit of '1' is present and the phase of the signal and local oscillator are made 180° apart corresponding to a '0'. For a '1' the resultant number of electron-hole pairs generated is

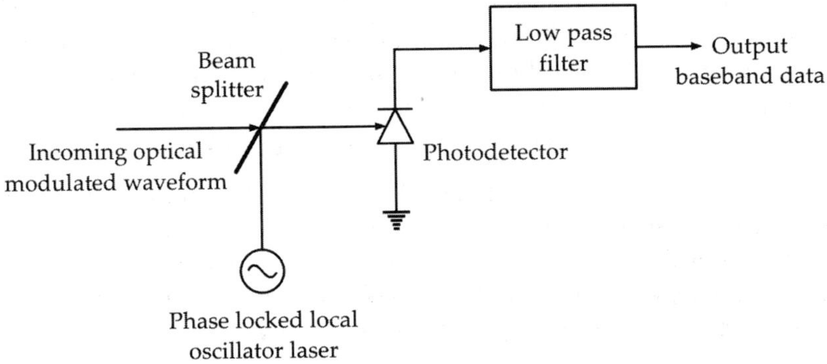

Fig. 13.5: PSK homodyne receiver

$$\overline{N_1} = (A_{LO} + A_s)^2 T \quad ...(13.29)$$

Similarly, for a '0', the corresponding figure is

$$\overline{N_0} = (A_{LO} - A_s)^2 T \quad ...(13.30)$$

The voltage as observed by the photodetector, as before becomes

$$V = \overline{N_1} - \overline{N_0} = 4A_{LO}A_s T \quad ...(13.31)$$

The associated noise variance is

$$\sigma^2 = A^2_{Lo} T \quad ...(13.32)$$

For the PSK homodyne receiver, the SNR per bit, therefore is

$$\frac{S}{N} = \left(\frac{V}{\sigma}\right)^2 = \frac{(4A_{LO}A_sT)^2}{A_{LO}^2 T} = 16A_s^2 T \qquad ...(13.33)$$

It may noted here that (13.32) is true, however, under the assumption of a strong local oscillator power. Therefore, the figure of merit for such a scheme that decides the BER is

$$\begin{aligned} P_e &= \frac{1}{2} erfc\left(\frac{1}{2\sqrt{2}}\sqrt{\frac{S}{N}}\right) \\ &= \frac{1}{2} erfc\left(\frac{4A_s\sqrt{T}}{2\sqrt{2}}\right) \qquad ...(13.34) \\ &= \frac{1}{2} erfc(A_s\sqrt{2T}) \end{aligned}$$

A comparison of (13.23) and (13.34) reveals that for the same SNR condition, the BER of a PSK homodyne system is better than its OOK counterpart. This is because the function *erfc*(x) is a monotonically decreasing function of its argument. The SNR provided per bit is two times that provided by the OOK modulation scheme for the same operating conditions and bit rate. As the probability of bit error is a function of SNR, a higher SNR yields a smaller value of the BER. We also note that, the PSK modulation scheme is 3 dB better in SNR per bit than the OOK system for identical bit rate for the two systems. However, the PSK homodyne system requires a coherent detection while the OOK modulated signal can be detected with simple envelope detection. As the SNR per bit of a PSK modulation format is two times that of the corresponding OOK system, the BER is translated to the number of average photons in one bit interval as follows:

$$\left(\frac{S}{N}\right)_{PSK} = 2\left(\frac{S}{N}\right)_{ASK} \qquad ...(13.35)$$

$$P_e = \frac{1}{2} erfc\left(\sqrt{2\eta \overline{N}_p}\right) \qquad ...(13.36)$$

The significance of (13.36) lies in the fact that for ideal PSK homodyne detection, an average of 9 photons per bit is required to achieve a BER of 10^{-9}.

We may note from Fig. 13.2 that the requirement for having a phase locked laser to act as the local oscillator for demodulating a PSK modulated optical signal is the greatest drawback of this format although it provides a lower BER than ASK.

13.3.3. Heterodyne ASK Demodulation

The IF output current from the photodetector $i_{IF}(t)$ is expressed as

$$i_{IF}(t) = \begin{cases} I_s \cos\left[\omega_{IF} + \phi\right] & for \quad 1 \\ 0 \quad for \quad 0 \end{cases} \qquad ...(13.37)$$

The current amplitude is $I_s = 2\frac{\eta q}{h\upsilon}\sqrt{P_{LO}P_s}$ from (13.13). The phase terms are not of any interest to us for the present case as information is contained in the amplitude of the current appearing at the IF amplifier output. The block schematic of an optical receiver employing the heterodyne method is shown in Fig. 13.6 that is observed to use an envelope detector for recovering the message signal. Hence it is a nonsynchronous type of detector.

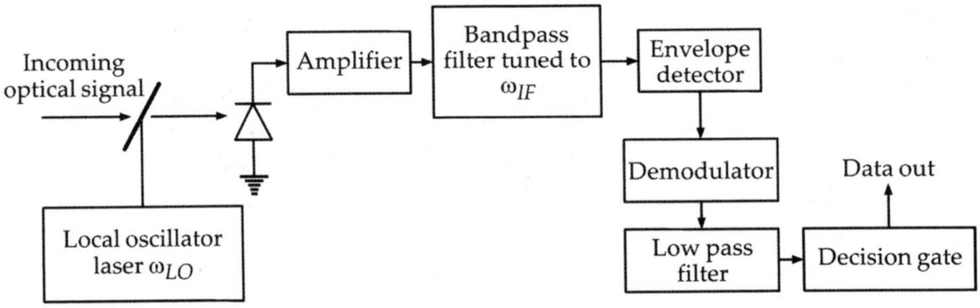

Fig. 13.6: Heterodyne nonsynchronous ASK receiver

Next we proceed to obtain a suitable expression for the noise current appearing at the output of the IF amplifier so that we can evaluate the receiver sensitivity. To do so, we make two assumptions: the local oscillator power is much larger than the incoming signal power so that the total noise current is approximately equal to i_{sh}^2 given by (13.15) and this is applicable to both the 1 and 0 pulses. The IF noise current $N(t)$ is modelled as a typical narrowband noise that has both an inphase and quadrature phase component both components being centred around the intermediate frequency. Therefore, we write

$$N(t) = x(t) \cos \omega_{IF}t + y(t) \sin \omega_{IF}t \qquad ...(13.38)$$

where $x(t)$ is the inphase component and $y(t)$ is the quadrature phase component, both the components varying much slower than the IF signal. The mean square values of both the components are equal and given as

$$\overline{x^2(t)} = \overline{y^2(t)} = \overline{i_{sh}^2} \qquad \text{...(13.39)}$$

The IF amplifier is typically a combination of an amplifier and a bandpass filter with a centre frequency equal to ω_{IF}. Most of the gain and bandwidth (hence the receiver sensitivity) of the heterodyne receiver is primarily decided by this stage. This amplifier is followed by a demodulation circuit that has a phase synchronized reference signal proportional to cos $\omega_{IF}t$. The envelope detector produces at its output, a voltage that is proportional to the envelope of the noisy signal appearing at the IF amplifier output. This envelope is

$$V(t) = \{[i_{IF}(t) + x(t)]^2 \cos^2 \omega_{IF}t + y^2(t) \sin^2 \omega_{IF}t\}^{1/2}$$

$$= \left[i_{IF}(t) + x(t)\right]\cos\omega_{IF}t\left\{1 + \frac{y^2(t)}{\left[i_{IF}(t) + x(t)\right]^2}\tan^2\omega_{IF}t\right\}^{1/2} \qquad \text{...(13.39)}$$

We note from (13.39) that the second term is the bracket is usually small as compared to unity as it is a noise term. The signal plus noise appears in the denominator of this second term which is higher than the quadrature noise component only under high SNR conditions.

Hence, the demodulator output is expressed as

$$V(t) = k[i_{IF}(t) + x(t)] \qquad \text{...(13.41)}$$

The quadrature phase component is typically ignored in this expression due to the reasons mentioned earlier and the cosine term is also ignored as we are discussing an envelope detector. The optimum decision threshold is set at midway between the zero current and the peak signal current corresponding to 1. Thus,

$$I_D \approx \frac{i_{IF}}{2} = \frac{\eta q}{h\upsilon}\sqrt{P_{LO}P_s} \qquad \text{...(13.42)}$$

The receiver makes an error when the envelope detector output falls below this threshold defined by (13.42) corresponding to a 1 being transmitted. Similarly, when a 0 has been transmitted but the demodulator output becomes higher than the threshold, an error is made. These two events are independent and are decided by the conditional probability density function of the demodulator output. For Gaussian noise assumptions, the conditional PDFs become, corresponding to the error events

$$P(1/0) = \int_{I_D}^{\infty} \frac{1}{\sqrt{2\pi \overline{i_{sh}^2}}} \exp\left[-\frac{(i_{IF} - I_D)^2}{2\overline{i_{sh}^2}}\right] di_{IF}$$

$$= \frac{1}{2} erfc\left[\frac{|i_{IF} - I_D|}{\sqrt{2}\,i_{sh}}\right] \qquad ...(13.43)$$

Similarly, the conditional probability of mistaking a 0 for a 1 is

$$P(0/1) = \int_{-\infty}^{I_D} \frac{1}{\sqrt{2\pi i_{sh}^2}} \exp\left[-\frac{(-I_D)^2}{2i_{sh}^2}\right] di_{IF}$$

$$= \frac{1}{2} erfc\left[\frac{|-I_D|}{\sqrt{2}\,i_{sh}}\right] \qquad ...(13.44)$$

The argument of the complementary error function expressed in (13.43) and (13.44) is simplified to, upon the substitution of the threshold current from (13.42)

$$i_{IF} - I_D = i_{IF} - \frac{i_{IF}}{2} = \frac{i_{IF}}{2} \qquad ...(13.45)$$

Assuming equiprobable bits, we have, for the total probability of bit error

$$P_e = P(0)P(1/0) + P(1)P(0/1)$$

$$= \frac{1}{2}\left\{\frac{1}{2} erfc\left[\frac{|i_{IF}/2|}{\sqrt{2i_{sh}^2}}\right] + \frac{1}{2} erfc\left[\frac{|-i_{IF}/2|}{\sqrt{2i_{sh}^2}}\right]\right\}$$

$$= \frac{1}{2} erfc\left(\frac{i_{IF}}{2\sqrt{2i_{sh}^2}}\right) \qquad ...(13.46)$$

We can see from (13.46) that BER of such an ASK modulation demodulation combo is primarily decided by the square root of the SNR appearing at the demodulator output and nothing else under the assumption of small noise components. To simplify (13.36) further, we make the necessary substitutions from (13.14) and (13.15) and obtain

$$P_e = \frac{1}{2} erfc\left(\frac{i_{IF}}{2\sqrt{2i_{sh}^2}}\right)$$

$$= \frac{1}{2} erfc\left[\frac{(2\eta q/h\upsilon)\sqrt{P_{LO}P_s}}{2\sqrt{2(2\eta q^2/h\upsilon)P_{LO}B}}\right] \qquad ...(13.47)$$

$$= \frac{1}{2} erfc\left(\sqrt{\frac{\eta}{2h\upsilon}\frac{P_s}{B_{IF}}}\right) \quad \because B_{IF} = 2B$$

It is more appropriate to specify the BER in terms of the bit rate of transmission B_T rather than the receiver bandwidth. To do so, we assume the transmission at a rate equivalent to twice the baseband bandwidth so that $B_T = 2B \approx B_{IF}$. The use of this in (13.47) gives us

$$P_e = \frac{1}{2} erfc\left(\sqrt{\frac{\eta}{2h\upsilon}\frac{P_s}{B_{IF}}}\right) \quad \because B_{IF} = 2B$$

$$= \frac{1}{2} erfc\left(\sqrt{\frac{\eta}{4h\upsilon}\frac{P_s}{B_T}}\right) \qquad ...(13.48)$$

We have derived an expression for the BER of an ASK heterodyne receiver under the shot noise limited conditions (in (13.38) where the effect of thermal noise has been ignored. A close look at this expression tells us that, the probability of bit error is dependent on the transmission rate. An increase in this rate is likely to produce more errors keeping the SNR the same before and after the increase of data rate. Hence, an increase in transmission data rate must be commensurate with a corresponding increase in SNR to maintain the BER at the same value.

We can have a simplified expression for (13.48) by making use of the following:

$$erfc(u) \approx \exp[-(u)^2]$$

This reduces (13.47) to, under high SNR conditions and shot noise limited receiver operation

$$P_e \approx \frac{1}{2}\exp\left[-\frac{i_{IF}^2}{8i_{sh}^2}\right] \qquad ...(13.49)$$

Similarly, we have for (13.48)

$$P_e \approx \frac{1}{2}\exp\left(-\frac{\eta}{4h\upsilon}\frac{P_s}{B_T}\right) \qquad ...(13.50)$$

This is true for a large argument of the *erfc*(.) function and low BER conditions.

13.3.4. Heterodyne Binary Frequency Shift Keying (BFSK)

In this modulation format, information is transmitted by switching the frequency of the optical oscillator (the laser) between two different values corresponding to message bits of 1 and 0. Thus, the receiver as shown in Fig. 13.7 has to decide, in a bit interval of which T of the frequencies is present. This requires the use of two bandpass filters tuned to the two frequencies ω_i, $i = 1, 2$. Hence, the binary decision is made to ascertain if one of the following is present at the sampling instant:

$$A_s \cos[\omega_1 t + \phi_s(t)] \text{ or } A_s \cos[\omega_2 t + \phi_s(t)]$$

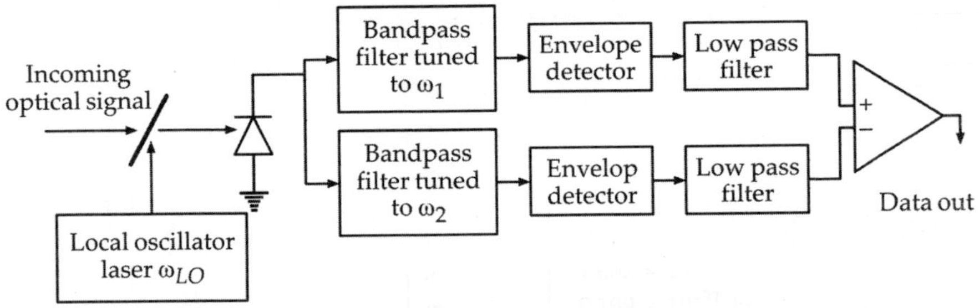

Fig. 13.7: Heterodyne non synchronous FSK receiver with two filters

We may observe that, the received power is proportional to A_s^2. The super heterodyne receiver mixes the incoming signal with a locally generated signal and feeds this sum signal to the photodetector. The sum signal, therefore is

$$R(t) = A_s \cos[\omega_i t + \phi_1(t)] + B_s \cos[\omega_i t + \phi_2(t)] \; i = 1,2 \qquad \text{...(13.51)}$$

The IF frequency, as we have noted earlier, is the intermediate frequency $\omega_{IF} = \omega_i - \omega_{LO}$ and $\phi_i(t)$ is a message independent phase. The output noise of the photodetector is modelled as a shot noise process. As message is recovered by envelope detection, we are interested in calculating the envelope of the received signal at the sampling instant. This is

$$V^2(t) = A^2 + B^2 + 2AB \cos[\omega_i t + \theta(t)] \qquad \text{...(13.52)}$$

where $\theta(t) = \phi_1(t) - \phi_2(t)$

The first two square terms in (13.52) are constants as far as detection of the two frequencies embedded in noise is concerned. The component of our interest from detection viewpoint is the cross term appearing as the third term of (13.52). Therefore, at the sampling instant, the electrical demodulator has at its input, a noisy signal given as

$$E_0(t) = 2AB\cos[\omega_i t + \theta(t),\ i = 1,2 \quad ...(13.53)$$

The noise $N(t)$ affecting the detection process is modelled as a white Gaussian noise of power spectral density $N_0/2$ W/Hz. We assume no cross talk between the two frequencies and any additional voltage generated is due to shot noise process only. We consider an interval when a 1 is transmitted. However, the receiver wrongly decides in favour of a '0'. The probability of this event, under the assumption of a Gaussian PDF from the envelope detector 1, is

$$P_1(V) = \frac{1}{\sqrt{2\pi \overline{i_{sh}}}}\exp\left[-\frac{(i_{IF} - V_1)^2}{2\overline{i_{sh}^2}}\right] \quad ...(13.54)$$

V_1 is the voltage corresponding to ω_1.

The local oscillator power is assumed to be much higher than the incoming signal power. Therefore, the total noise current is given by (13.15). The noise output from envelope detector 2 is given as

$$p_2(V) = \frac{1}{\sqrt{2\pi \overline{i_{sh}}}}\exp\left(-\frac{V_2^2}{2\overline{i_{sh}^2}}\right) \quad ...(13.55)$$

V_2 is the voltage corresponding to ω_2.

An error occurs when the output from receiver 2 exceeds that from receiver 1 that is supposed to favour the decision of a '1'. Thus, the probability of bit error is equivalent to making $V_1 - V_2 < 0$. We note that, the PDF of voltage produced by envelope detector 1 has a Ricean form whereas that from receiver 2 has a Rayleigh distribution. The probability of error is expressed as

$$P_e = \int_0^\infty p(v_1 \mid 1)\left[\int_{v_1}^\infty p(v_0 \mid 1)\,dv_0\right]dv_1 \quad ...(13.56)$$

The inner integral has a Rayleigh distribution and it evaluates to $\exp\left(-\frac{v_1^2}{2\overline{i_{sh}^2}}\right)$. Substituting this in (13.56) and the Rician PDF for the outer integral, we have the following result

$$P_e = \int_0^\infty \frac{v_1}{\overline{i_{sh}^2}}\exp\left\{-(v_1^2 + i_{IF}^2)/2\overline{i_{sh}^2}\right\} I_0\left(\frac{i_{IF}v_1}{\overline{i_{sh}^2}}\right)\exp\left(-\frac{v_1^2}{2\overline{i_{sh}^2}}\right)dv_1$$

$$= \int_0^\infty \frac{v_1}{i_{sh}^2} \exp\left\{-\left(2v_1^2 + i_{IF}^2\right) / 2i_{sh}^2\right\} I_0\left(\frac{i_{IF} v_1}{i_{sh}^2}\right) dv_1 \qquad \text{...(13.57)}$$

To evaluate (13.57), we make the following substitution. Let $\lambda = \sqrt{2}v$ and $\alpha = \frac{i_{IF}}{\sqrt{2}} \Rightarrow \alpha^2 = \frac{i_{IF}^2}{2}$. Hence, $i_{IF}v_1 = \alpha\lambda, dv = \frac{d\lambda}{\sqrt{2}}$,

$$P_e = \int_0^\infty \frac{v_1}{i_{sh}^2} \exp\left\{-(2v_1^2 + i_{IF}^2) / 2i_{sh}^2\right\} I_0\left(\frac{i_{IF} v_1}{i_{sh}^2}\right) dv_1$$

$$= \frac{1}{\sqrt{2}} \cdot \frac{1}{\sqrt{2}} \int_0^\infty \frac{\lambda}{\sqrt{2} i_{sh}^2} \exp\left\{-\frac{\lambda^2 + 2\alpha^2}{2i_{sh}^2}\right\} I_0\left(\frac{\alpha\lambda}{i_{sh}^2}\right) d\lambda]$$

$$= \frac{1}{2} \exp\left(-\frac{\alpha^2}{2i_{sh}^2}\right) \int_0^\infty \frac{\lambda}{\sqrt{2} i_{sh}^2} \exp\left\{-\frac{\lambda^2 + \alpha^2}{2i_{sh}^2}\right\} I_0\left(\frac{\alpha\lambda}{i_{sh}^2}\right) d\lambda \qquad \text{...(13.58)}$$

$$= \frac{1}{2} \exp\left(-\frac{i_{IF}^2}{4i_{sh}^2}\right) \int_0^\infty \frac{\lambda}{\sqrt{2} i_{sh}^2} \exp\left\{-\frac{\lambda^2 + \alpha^2}{2i_{sh}^2}\right\} I_0\left(\frac{\alpha\lambda}{i_{sh}^2}\right) d\lambda$$

The integrand in (13.58) reduces to the Rician PDF whose total area equals unity. The signal power is given by (13.14) and the shot noise variance is given by (13.15). Substituting these two into (13.58), we get

$$P_e = \frac{1}{2} \exp\left(-\frac{i_{IF}^2}{4i_{sh}^2}\right) = \frac{1}{2} \exp\left[-\left(\frac{\eta P_s}{2h\upsilon B_{IF}}\right)\right] \qquad \text{...(13.59)}$$

$$= \frac{1}{2} \exp\left[-\left(\frac{\eta P_s}{2h\upsilon B_T}\right)\right]$$

This is because the noise power is evaluated over the bandwidth of the IF amplifier for the heterodyne receiver. Hence, the shot noise variance given in (13.15) is evaluated over a bandwidth of B_{IF}.A comparison of (13.50) and (13.59) tells us that for identical values of transmission bit rate, the SNR at the envelope detector output of an FSK receiver is better than its ASK counterpart by 3 dB.

13.3.5. Heterodyne PSK Demodulation

In this modulation format information is transmitted by a carrier of one phase for a binary data bit of 1 and a different phase for binary 0. The phase shift normally used is π radians. Thus, the IF amplifier output current is expressed as

$$i_{IF}(t) = \begin{cases} I_s \cos\left[\omega_{IF}t + \varphi\right] & \text{for } 1 \\ I_s \cos\left[\omega_{IF}t + \varphi + \pi\right] & \text{for } 0 \\ \text{or} & \\ -I_s \cos\left[\omega_{IF}t + \varphi\right] & \text{for } 0 \end{cases} \quad \text{...(13.60)}$$

The detection has to be synchronous as information about the message signal resides in the phase of the carrier. We cannot use envelope detection for demodulating a BPSK modulated signal simply because the amplitude and frequency of the modulated signal are constants. Therefore, the synchronously detected signal $i_{IF}(t)$ is positive for a 1 and is genitive for a binary 0. The optimum decision threshold, in this case, therefore is set $I_D = 0$ as unlike (13.45) where it is set at half of $i_{IF}(t)$. The probability of bit error derivation will follow a very similar approach as that outlined for the ASK case with the difference that the limits of integration would change from $I_D/2$ to zero. The conditional probabilities are:

$$P(1/0) = \int_0^{\infty} \frac{1}{\sqrt{2\pi \overline{i_{sh}^2}}} \exp\left[-\frac{(i_{IF}+1)^2}{2\overline{i_{sh}^2}}\right] di_{IF}$$

$$= \frac{1}{2} erfc\left[\frac{i_{IF}}{\sqrt{2\overline{i_{sh}^2}}}\right] \quad \text{...(13.61)}$$

Similarly, the conditional probability of mistaking a 0 for a 1 is

$$P(0/1) = \int_{-\infty}^{0} \frac{1}{\sqrt{2\pi \overline{i_{sh}^2}}} \exp\left[-\frac{(i_{IF}-1)^2}{2\overline{i_{sh}^2}}\right] di_{IF}$$

$$= \frac{1}{2} erfc\left[\frac{i_{IF}}{\sqrt{2\overline{i_{sh}^2}}}\right] \quad \text{...(13.62)}$$

The total probability of bit error is, hence

$$P_e = P(0)P(1/0) + P(1)\,P(0/1)$$

$$= \frac{1}{2}\left\{\frac{1}{2}erfc\left(\frac{i_{IF}}{\sqrt{2\overline{i_{sh}^2}}}\right) + \frac{1}{2}erfc\left(\frac{i_{IF}}{\sqrt{2\overline{i_{sh}^2}}}\right)\right\}$$

$$= \frac{1}{2}erfc\left(\frac{i_{IF}}{\sqrt{2\overline{i_{sh}^2}}}\right) \qquad ...(13.63)$$

Substitution of the current and shot noise variance from (13.14) and (13.15), we note that the argument of the *erfc*() function takes on the following form:

$$\frac{i_{IF}}{\sqrt{2\overline{i_{sh}^2}}} = \frac{2(\eta q/h\upsilon)\sqrt{P_{LO}P_s}}{\sqrt{2.2(\eta q^2/h\upsilon)P_{LO}B_{IF}}}$$

$$= \sqrt{\frac{\eta P_s}{h\upsilon B_{IF}}} = \sqrt{\frac{\eta P_s}{h\upsilon B_T}} \qquad ...(13.64)$$

Therefore, the BER for PSK heterodyne detection becomes, with the help of (13.62) and (13.63)

$$P_e = \frac{1}{2}erfc\left(\sqrt{\frac{\eta P_s}{h\upsilon B_T}}\right) \qquad ...(13.65)$$

We may observe from the foregoing that in the shot noise limit PSK heterodyne synchronous detection exhibits 3 dB and 6 dB more sensitivity than the FSK and ASK heterodyne synchronous detection schemes respectively. In practice, however, very small levels of phase fluctuation at the transmitter can significantly degrade the potential for low BER attainments by a PSK receiver.

13.6. CHAPTER SUMMARY

The term coherent in optical fibre communication systems refer to any technique of nonlinear mixing of two optical beams. This is in contrast to the electrical coherent systems wherein the absolute phase of the incoming carrier is to be accurately tracked by the receiver using some suitable technique such as a phase locked loop. The two main advantages of such coherent optical communication systems lie in excellent receiver sensitivity and high degree of frequency selectivity. The coherent optical receivers make use of homodyne or heterodyne detection methods. In homodyne detection no intermediate frequency is used. Here the local oscillator is controlled by means of an optical phase locked loop which is difficult to build and maintain in practice. However, such type of detection brings the signal directly to the base band so that no further electrical demodulation is required. As can be seen, such a method yields the most sensitive coherent receiver. In the heterodyne type, an intermediate frequency of a suitable value is used in the receiver. This is easy to build and gives a performance 3 dB inferior to a corresponding homodyne receiver. Coherent optical fiber systems require single mode semiconductor laser diodes with narrow linewidths, stable frequencies and wave length tuning capabilities. Here modulation formats commonly used in electrical radio systems are used. These are amplitude modulation or on off keying, frequency and phase modulation of an optical carrier. Unlike its electrical counterpart, an optical carrier cannot have negative amplitudes. It can be either present or absent depending on the base band data. The performance of these receivers is also governed by the signal to noise ratio available at the receiver output as the receiver output follows a Gaussian distribution for high local oscillator laser powers. The receiver sensitivity is described in terms of the average number of photons required to achieve a BER of 10^{-9}. Of all the modulation formats, the PSK modulation when used in conjunction with a homodyne detection yields the most sensitive receiver of all; requiring just nine photons per optical pulse to result in the above BER.

❑❑❑

14

Optical Networks

14.1. INTRODUCTION

A network is a physical and logical way of interconnecting homogeneous or heterogeneous systems designed to perform a group of specific tasks aimed at member subscribers. For example, voice service is provided to subscribers by the conventional wired public switched telephone network (PSTN). These days we see a plethora of networks all aiming to provide a number of services to subscribers. We have the cellular networks that provide voice, video and data services in a wireless manner. The conventional computer networks provide data services in excess of mega bits per second in both wired and wireless configuration as is evidenced by a local area network. Optical fibers have played a significant role in replacing the old copper wires used to connect a subscriber to a PSTN or in connecting computers as in a typical LAN scenario due to the inherent advantages they offer as mentioned in chapter one. The most important and significant functionalities that must be performed by a network are

(a) Multiplexing
(b) Switching

Multiplexing is the task of providing the network resources to all the subscribers based on certain criteria. We have seen multiplexing in time domain and frequency domain as applied to electrical signals. In time division multiplexing, subscribers are allotted a certain time slot in a periodic manner and bidirectional communication takes place in this time slot only. In frequency domain, a chunk of the designated frequency band is given to a subscriber and usually a guard band is provided between two consecutive subscribers to prevent interference. We can thus see that the network resources are made available to a subscriber when he demands for it either on a time division or a frequency division basis. This also brings us to the fact that different subscribers are kept separate based on their allotted time slot or a frequency band of a fixed width. Switching is the task of providing a path to two subscribers so that they can communicate. We have basically two types of switching; circuit switching and packet switching that will be discussed later.

For optical transreceivers connected to each other in a networked configuration, an option of multiplexing subscribers based on wavelength is also exploited. Let us, hence take a look at multiplexing subscribers based on wavelength division next.

14.2. Wavelength Division Multiplexing (WDM)

Point to point communication links a single optical fibre has one optical source at its transmitting end and one photodetector at the receiving end. Signals from different light sources require separate and uniquely assigned optical fibres. We know that an optical source has a narrow spectral width. This implies that these kind of point to point links making use of a single optical fibre and just one source and one receiver utilize a very small portion of the tremendous bandwidth available by a typical fibre. This obviously is a gross underutilization of the transmission capacity of a fibre.

A dramatic increase in the information carrying capacity of an optical fibre can be realized by simultaneous transmission of optical signals over the same fibre from many different light sources that have properly spaced peak emission wavelengths. By operating each source at a different peak wavelength, the integrity of the independent messages from each source is maintained for subsequent conversion to electric signals at the receiving end. This is the basis of wavelength division multiplexing. As we may observe, the WDM scheme is conceptually similar to the frequency division multiplexing scheme used in terrestrial microwave links and satellite systems. A number of WDM techniques have been investigated from various angles for their suitability in optical fibre networks in literature.

We have illustrated two different WDM layouts in the following diagrams.

A unidirectional WDM device is used to combine different signal carrier wavelengths onto a single optical fibre at one end and to separate them into their respective detectors at the receiving end in Fig. 14.1. A bidirectional scheme has been illustrated in Fig. 14.2.

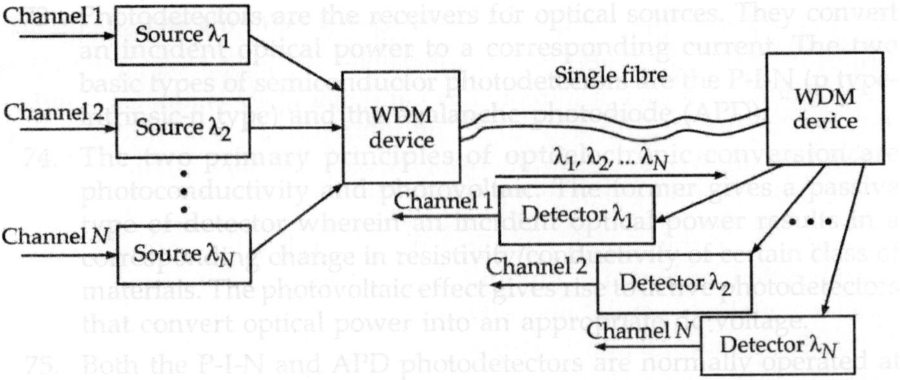

Fig. 14.1: A unidirectional WDM system that multiplexes N number of independent input signals for transmission over a single optical fibre

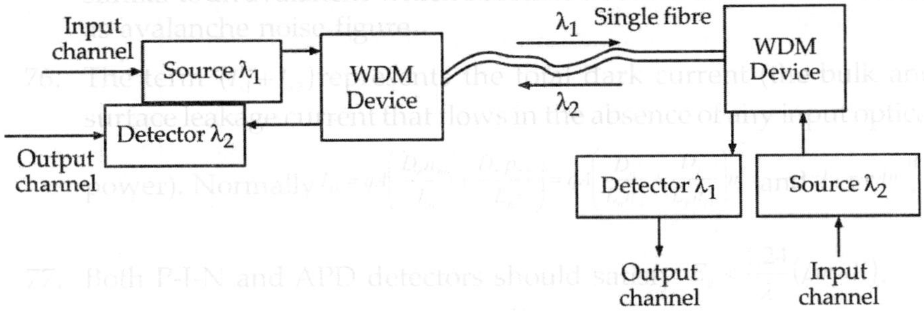

Fig. 14.2: Block schematic of a bidirectional WDM in which two or more wavelengths are transmitted simultaneously in opposite directions over the same optical fibre

A bidirectional WDM scheme, as Fig. 14.2 indicates makes use of two independent sources operating at wavelengths λ_1 and λ_2, however located at the opposite ends of the same fibre. We have two detectors corresponding to these two wavelengths similarly placed at the opposite ends of the same fibre.

The three performance metrics that dictate the system requirements in implementing a typical WDM scheme are insertion loss, channel width and cross talk. The amount of power loss that arises in the optical fibre from the addition of a WDM coupling device is known as insertion loss. This also includes losses occurring at the connection points of the WDM element to the fibre and any intrinsic losses within the multiplexing element itself. Insertion losses of a few decibels at each end can be, however tolerated.

The wavelength range allocated to a particular optical source is known as the channel width. Channel widths of several tens of nanometers are required to ensure zero crosstalk resulting from the source instability when lasers are used as the sources. However, when LEDS are used as the sources, channel widths that are 10 to 20 times larger are required because of the wider spectral output width of these sources.

The amount of signal coupling from one channel to another is referred to as cross talk. The tolerable interchannel cross talk levels very widely depending on the application. In general, a –10 dB level is not sufficient whereas a –30 dB level is.

A multiplexer is needed at the transmitting end to combine optical signals from several light sources onto a single fibre in case of a unidirectional system. This necessitates the use of a demultiplexer at the receiving end to separate out into appropriate detection channels. In such a scheme, as the optical signals being combined generally do not emit a significant amount of optical power outside of their designated channel spectral width, interchannel cross talk is hardly an issue. However, the basic design problem is the ability of the multiplexer to provide a low loss path from each optical source to the multiplexer output. A different requirement exists for the demultiplexer due to the fact that the photodetectors are usually sensitive over a broad range of wavelengths that could include all the WDM channels. Therefore, to prevent significant amounts of the wrong signal from entering each receiving channel, either the demultiplexer should be designed carefully or very stable optical filters with sharp wavelength cut offs must be used.

WDM is implemented by multiplexers at the transmitting end and demultiplexers at the receiving end. In principle, an optical demultiplexer can also be used as a multiplexer. The multiplexers or the demultiplexers are implemented using two broad category of devices; namely the angularly dispersive type such as prisms or gratings and filter based devices such as multilayer thin film interference type or single mode optical devices. Henceforth, we shall refer to the multiplexer/demultiplexer as mux/demux for the sake of convenience. The block schematic of an angularly dispersive type of mux/demux is shown in Fig. 14.3 for a three wavelength system where $\frac{d\theta}{d\lambda}$ is the angular dispersion of the device. As we may observe from this figure, the angularly dispersive device is used as a demux as it is observed to split the incoming collimated light beam into three constituent components λ_1, λ_2 and λ_3. The lens L_1 is used to collimate the light beam coming from the left hand fibre containing the mixture of the three wavelengths. The right hand lens L_2 is known as the focusing lens that focuses the output beams into the appropriate receiving fibres or photodetectors. The linear dispersion $\frac{dx}{d\lambda}$ at the output fibres shown on the right hand side has been shown[1] to be

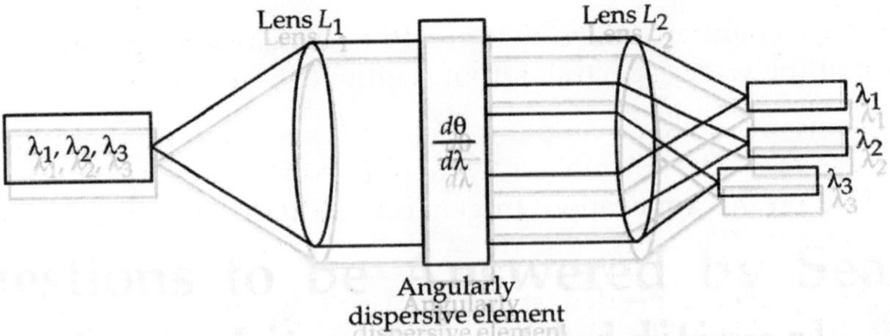

Fig. 14.3: Schematic representation of an angularly dispersive WDM element shown for three wavelengths.

$$\frac{dx}{d\lambda} = f\frac{d\theta}{d\lambda} \quad ...(14.1)$$

In (14.1), the focal length of lens L_2 has been considered to be f. In the ideal case of aberationless optics and zero source spectral width, the intrinsic insertion loss and cross talk is zero when the output signals are separated by more than the diameter d which implies that

$$\frac{dx}{d\lambda}\Delta\lambda \geq d \quad ...(14.2)$$

The spectral separation between the channels has been taken to be $\Delta\lambda$ in (14.2). It is assumed here that all the input and output end fibres have the same diameter d and numerical aperture NA.

The collimating lens L_1 must have a diameter that satisfies the following condition in order to collect all the light from the input fibre;

$$b \geq 2f\frac{NA}{n'} \quad ...(14.3)$$

The refractive index of the medium between the dispersive device and the lens L_1 has been taken to be n' in 14.3. From (14.1), we get

$$f = \frac{(dx/d\lambda)}{(d\theta/d\lambda)} \quad ...(14.4)$$

Similarly, we have, from (14.2),

$$\frac{dx}{d\lambda} \geq \frac{d}{\Delta\lambda} \quad ...(14.5)$$

Use of (14.4) and (14.5) in (14.3) gives us

$$b \geq 2\frac{(dx/d\lambda)}{(d\theta/d\lambda)}\frac{NA}{n} = \frac{2d}{\Delta\lambda}\frac{NA}{n'(d\theta/d\lambda)}$$

$$\Rightarrow \qquad b \geq \frac{2(NA/n')d}{\Delta\lambda(d\theta/d\lambda)} \qquad ...(14.6)$$

The output beam tends to spread out in any real system because of the finite size of the light sources and the angular dispersion resulting from the wavelength spread of the source spectral width. The fractional increase in the beam diameter is approximately given by

$$S = \frac{b' - b}{b} \cong (1 + p)\frac{Wd(NA)}{b^2 n'} \qquad ...(14.7)$$

where p is the number of wavelength channels, b' is the diameter of lens L_2 and W is the total path length from the output of lens L_2 to the input of lens L_1. The total beam spread must be a small fraction of the collimating lens diameter in order to avoid overfilling the numerical aperture of the output fibre which means that $S \ll 1$.

A large number of channels can be combined and separated with these kind of angularly dispersive multiplexing elements. Most of these devices use a grating with lens combination. Sometimes a prism is used for the angularly dispersive element. Insertion losses vary typically from 1 to 3-dB range while cross talk levels vary between –20 and –30 dB.

A filter type of multiplexing element corresponding to a two wavelength operation is illustrated in Fig. 14.4. The filters are designed to transmit light in a specific wavelength channel and either to absorb or to reflect all other wavelengths. Reflection type filters are used commonly as the insertion loss of the absorption type of filters tend to be rather high sometimes in excess of 1 dB. The reflection filter consists of a flat glass substrate upon which multiple layers of different dielectric films are deposited for wavelength selectivity. These filters can be used in series to separate additional wavelength channels. The complexity involved in stacking the filters in series and the corresponding increase in signal loss that occurs with the addition of successive multiplexers generally limit their operation to two or three filters which means that the substrate has three or four channels.

A WDM system designer must take into account to minimize the factors responsible for link margin degradation. The insertion loss of the coupling devices should be kept low. Care must be exercised to minimize reflections of the optical power that occur at the coupling devices or at connectors and splices, since they can give rise to signal cross talk. These effects are of particular importance in bidirectional links.

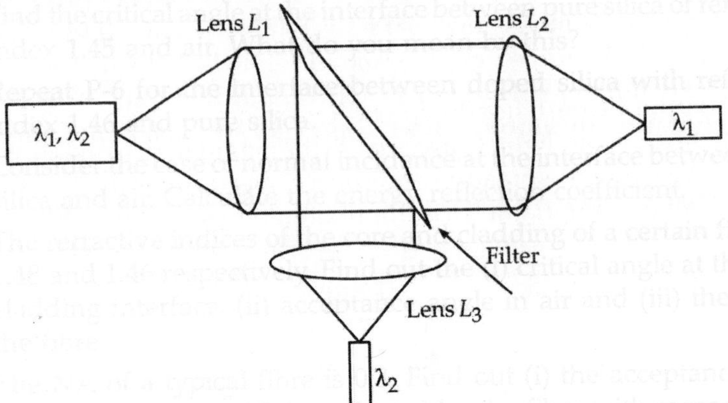

Fig. 14.4: Filter type multiplexer that uses a multilayer thin film reflector type structure. This device is observed to be transparent at the wavelength λ_1 and λ_2 reflects the wavelength.

The mux/demux strictures discussed above are passive types in which the wavelength selectivity is rather fixed. We also come across active mux/demux devices. In these WDM devices, the wavelength to be transmitted further can be actively switched or tuned. The multiwavelength source and detector integrated optic arrays, wavelength tuneable lasers and wavelength tuneable filters have been reported in literature[2-8]. The concept of tuneable filtering is illustrated in Fig. 14.5.

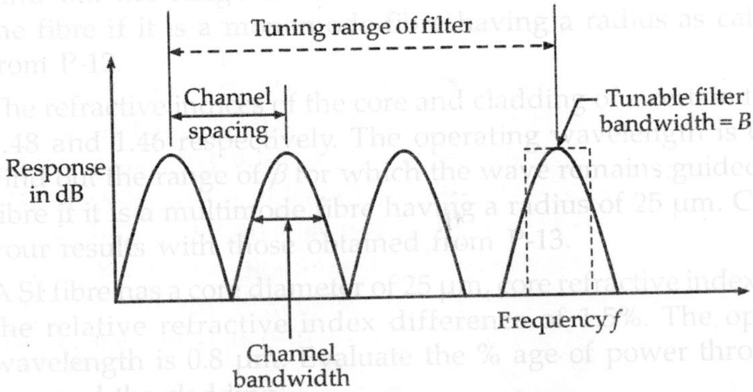

Fig. 14.5: Different signals being sent in individual frequency channels of bandwidth in a tunable filtering method. A tunable filter of passband selects the desired channel.

In the tunable filtering schemes such as one shown in Fig. 14.5, different information signals are transmitted in the individual frequency channels each of bandwidth B. By using a filter of passband width B that is tunable over the frequency range of these channels, a user can select the desired channel.

14.3. LOCAL AREA NETWORKS

The increased demand for bandwidth in local environment has led to the deployment of fibre optics in local area networks although any type of transmission medium can be used to support data transfer within the information carrying capacity of that medium. The traditional media such as parallel wires, coaxial cables, waveguides can support data transfer upto a few hundreds/thousands of megahertz. Beyond this GHz range and the need to multiplex a large number of channels augurs well in favour of an optical fibre as the transmission medium. The manner in which the nodes are geometrically arranged and connected is known as the topology of the network[9]. Two major concerns for LAN in terms of the fibre optic transmission medium are the deployment of suitable topologies and fail-proof architectures for these topologies. The three basic topologies widely used in a typical LAN setup are illustrated in Fig. 14.6. Each of the topologies has its own merit and demerit in terms of reliability, scalability and performance measures.

14.3.1. Network Topology

Data is input and removed from the optical fibre bus via four-port couplers located at the nodes. Data is circulated in a kind of unidirectional manner being looped through the nodes at each coupling point in the ring topology. Therefore, data is made to be regenerated in amplitude and phase in such a configuration. However, in a star topology, a central hub is created by an optical star coupler that may be passive or active. In the passive operation, the star coupler at the hub splits the data in terms of power so that the signal is transmitted to all nodes. Overall losses are therefore primarily decided by the number of splits. With an active hub, the data is split in an optoelectronics star coupler that carries out both opto-electronic and electronic-opto conversions and therefore this type of active coupler does not contribute to the overall network losses.

The ring and star topologies are combined in the mesh configuration to provide multiple interconnections among the network nodes. It is referred to as full mesh when each network node is interconnected with all nodes in the network. Therefore, the full mesh configuration provides more interconnectivity than the other three topologies as shown. Optical networks can be implemented using either a single or a combination of these topologies. The full mesh topology, though complex is often referred for provision of either a logical or virtual topology due to its high flexibility and interconnectivity features.

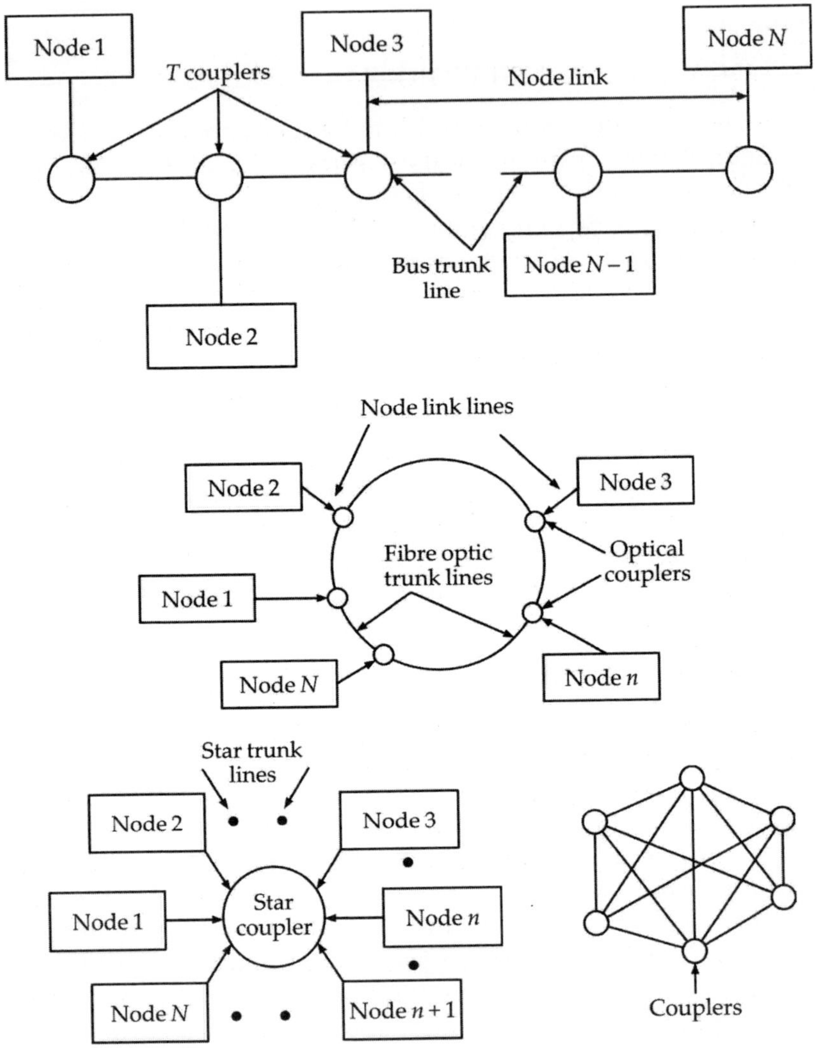

Fig. 14.6: (a) Bus type of topology, (b) the ring topology, (c) star topology, (d) mesh topology

14.3.2. Network Architectures

The architecture provides for the implementation of networking functionalities in different layers of a reference model[9]. The partitioning of functions following a logical methodology defines communication tasks into a set of hierarchical layers within a standardized network architecture.

Each layer in the network architecture performs a related set of functions in order that a user system will be able to communicate with another user system. The ultimate goal of this approach is to produce a network that can be physically and operationally interconnected. and that offers large network capacity while ensuring network protection/survivability. Therefore, a major functionality of the optical network physical layer is to provide appropriate routing and switching strategies within optical multiplexed networks.

14.4. NETWORKING MODES

There are two networking modes in which a transmission path is established and these are referred to as either connection oriented or connectionless. In the first type, an end-to-end connection setup, generally referred to as handshaking is established first before any real data transfer is initiated. The standard public switched telephone networks employ a scheme almost similar to this for conversation to be established between two users. These type of networks employ a bidirectional communication environment to initiate a connection so that both source and destination can communicate with each other. Once the path is determined, all the subsequent information and data are sent to and from the destination using the established connection. This is also referred to as circuit switching that is discussed next.

However, in networks operating with connectionless modes, no dedicated end-to-end connection is available and no explicit connection setup is performed before the actual data transmission. The transmission is launched onto a common fibre channel such as the bus or ring that interconnects the various nodes. The connectionless networking mode is most often achieved by transmitting information in one direction without confirming the presence or readiness of the destination to receive the information. In these networks, both the addressing information and the data are organized in small blocks called packets wherein each packet is assigned a unique identification number and sent in a sequence such that the destination node can reassemble them to construct the complete message.

Network Switching Modes

Switching in a network may be realized by different means. Two primary modes are circuit switching and packet or cell switching. An end-to-end circuit is required to be set up before establishing a connection in the circuit switched mode. All the specific network resources necessary for the switching are dedicated for the particular transmission during which no other transmission can access these resources. Furthermore, such transmissions are continuous and hence arrive in real time. Once the transmission is complete, the circuit is terminated and the network resources are available to other users. Circuit switching is therefore, characterized by continuous, selective and temporary kind of switching. As we may see, it is used only in connection oriented class of networks.

Packet or cell switching does not require a dedicated circuit to be established prior to actual data transfer between two nodes. It is rather a store and forward kind of switching in which the message to be sent is broken down into smaller units known as packets that are numbered and forwarded to a neighbour node with the recipient as well as the sender addresses. Sending or receiving a complete message in a single block requires a buffering stage where information can be gathered before it is sent to the destination. Alternatively, messages may be sent in small blocks which do not require significant buffering stages. The message is divided into packets that comprise variable length blocks of data with a tight upper bound limit or a cell which incorporates a fixed length block of data. Packets or cells from many different sources are statistically multiplexed and sent on to their destinations. In statistical multiplexing, a fixed bandwidth communication channel is divided into several variable channels and the desired or requested bandwidth is allocated to each user. This contrasts with simple time or wavelength division multiplexing where a fixed bandwidth is assigned. Finally, both packet and cell switching are applied in connection oriented networks. Packets, however are often employed in connectionless networks.

Virtual Circuits

Unlike a physical circuit, which terminates the physical medium on specific physical ports, a virtual circuit is a series of logical connections between the sending and the receiving devices. A virtual circuit is a connection between source and destination nodes which may consist of several different routes. These routes can change at any time, and the incoming return route does not have to mirror the outgoing route. When the connections are set up in advance, based on expected traffic patterns, these are termed provisional virtual circuits. In switched networks, however, where connections can be set up on demand and released after the data exchange is complete, such connections are known as switched virtual circuits.

Network Routing

Routing refers to the process whereby a node finds one or more paths to possible destinations in a network. In this process control and data processing functions are performed to identify the route and to handle the data during the journey from source to destination. A simple routing process known as the control plane comprises three stages which are neighbour discovery, topology discovery and path selection. Neighbour discovery is the first step whereby the nodes in a network discover the identity of their immediate neighbours and their connections to the neighbouring nodes. Topology discovery is the process by which a node discovers all other nodes in the network and how they are connected to each other.

Once the network topology is known, paths from a source to a destination can be determined using path computation algorithms or routing protocols. A protocol is a set of standard rules for data representation, signalling, authentication and error detection required to send information over a communication channel. The control plane therefore acquires and then provides the reliability information throughout the network. This reachability or routing information disseminated by the routing protocol is consolidated in a forwarding table at each node which contains the identity of the next node and the outgoing interface to forward packets addressed to each destination. In order to handle the data more efficiently it requires an additional procedure known as the data plane which allows an optical node to forward the incoming packets towards their destinations by using information contained in a forwarding table. This operation involves examining the destination address of the packets, searching the forwarding table to determine the next hop and the corresponding outgoing interface, and then forwarding the packets on this next hop.

Modularity and Scalability

Modularity defines the characteristic of a network which allows the addition or reduction of networking nodes in a modular fashion. Such modular networks therefore permit the use of wavelengths (wavelength reuse) in different sections of a large optical network without causing wavelength conflict.

Scalability is the property of a network which enables it to progressively accommodate a large number of nodes and end user systems without incurring excessive overheads. It is one of the most important design aspects of any network or protocol. For a network with a high number of nodes, scalability becomes difficult to realize as a large number of alternative routes are available. Scalability can be achieved in several different ways such as by employing a routing hierarchy, or by performing route aggregation[10]. Both these techniques help contain the size of the routing databases. Another common technique for achieving scalability is to reduce the number of routing updates using threshold schemes[11].

14.5. Optical Network Node and Switching Element

The optical node as we have shown in previous diagrams marked as a circle is an essential component of an optical network. It is a multifunctional network element that is required to perform a host of tasks depending on its type and network requirement. It is required top send, receive and resend or redirect optical signals to its neighbouring nodes. The resending or redirecting of an optical signal to the desired networking nodes requires the node to perform either a routing or a switching function. It should be noted that when several optical signals travel in a multiplexed form, the networking node becomes transmission system dependent since different optical multiplexing techniques can be used such as time division or wavelength division multiplexing. An optical node can also function as a router directing am input signal wavelength to a specified port. A router essentially consists of optical couplers where the desired signal from an incoming multiplexed signal can be isolated or demultiplexed and then directed or routed to the output port. Therefore it is commonly referred to as a wavelength router. It may be possible to send the wavelength of the signal before sending it to the specified output port. The node then is known as a wavelength converting router and since the device is used to switch the wavelength by changing it to the compliant signal wavelength it is also known as a wavelength routing switch[12]. An optical switch performs wavelength operational functions like multiplexing, demultiplexing and switching unlike its semiconductor counterparts. The physical medium remains fixed with only the information contained in the messages being changed to send it to the corresponding destination node through the assigned route. Such wavelength switches may be constructed using integrated optical or optoelectronics devices.

The four different functions of an optical router are illustrated in Fig. 14.7. A 2 × 1 multiplexer is shown in Fig. 14.7(a). This is observed to be a three port device that combines two incoming optical signals operating at two different wavelengths and outputs them as a single signal.

The four different functions of an optical router are shown in Fig. 14.7. A 2 × 1 wavelength multiplexer is shown in Fig. 14.7(a) where two different wavelengths λ_2 and λ_2 appearing at input ports 1 and 2 are combined onto a single output port 3. The demultiplexer shown in Fig. 14.7(b) splits the optical signal appearing at input port 3 to two signal wavelengths that appear at the output ports 1 and 2. An optical add/drop multiplexer (OADM) which also comprises a wavelength add/drop device is shown in Fig. 14.7(c) where an incoming multiplexed signal consisting of three wavelengths is partially demultiplexed by dropping the λ_2 signal at an intermediate port 2. The OADM further adds another wavelength signal λ_4 at intermediate port 3 which is then multiplexed with the transmitted signal wavelengths so that the combined signal leaving port 4 contains a multiplexed signal stream having $\lambda_1\,\lambda_3\,\lambda_4$. Devices which are configured to drop a particular wavelength (here λ_2) and add another specific wavelength (here λ_4) are also known as fixed OADM.

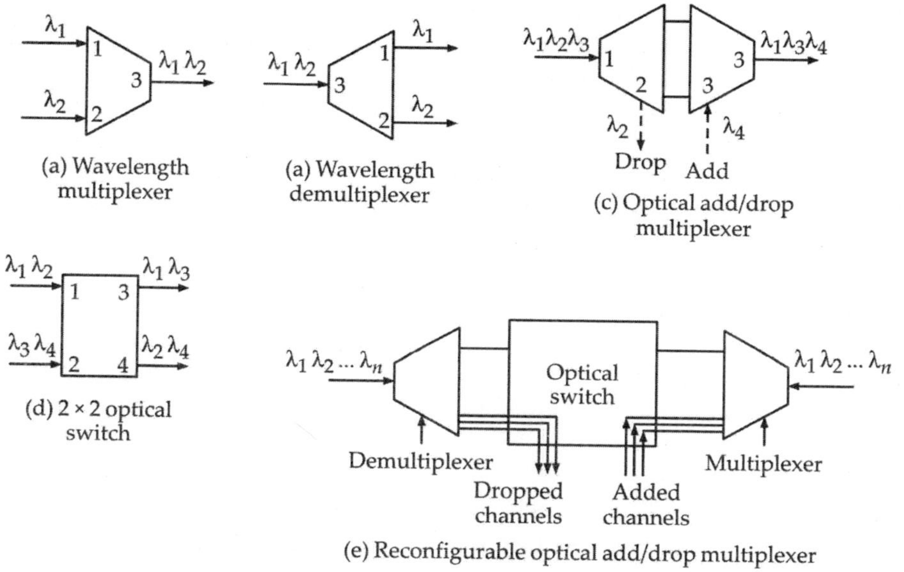

Fig. 14.7: Optical networking node elements

A simple 2 × 2 optical switch is shown in Fig.14.7(d) that has two input and two output ports. The two optically multiplexed signals comprising wavelengths $\lambda_1\,\lambda_2$ and $\lambda_3\,\lambda_4$ are present at input ports 1 and at 2, respectively. At the output ports the wavelengths are required to be switched and multiplexed as $\lambda_1\,\lambda_3$ and $\lambda_2\,\lambda_4$ emerging at the output ports 3 and 4, respectively. It is possible to produce an optical switch with a greater number of ports ($N \times N$), (where N is usually a large number) using this approach. A combination of an OADM and an optical switch producing a reconfigurable optical add/drop multiplexer (ROADM) is shown in Fig. 14.7(e). This device can drop one or a desired number of wavelength channels after demultiplexing a wavelength multiplexed signal and, similarly, it can also add a new single or more wavelength channels through an optical switch. The multiplexer in Fig. 14.7(e) brings together all the wavelength channels to produce a combined wavelength multiplexed output signal. A passive ROADM with this functionality can be constructed using a fibre Bragg grating and an optical circulator.

The ROADMs are, however not suitable for facilitating complex mesh topologies that incorporate a large number of nodes. Such networks therefore require an additional element to provide a means of cross network connection with the added features of an ROADM. This additional element is an optical cross connect (OXC) which has the capability of switching the connection between two interfaced points.

A block schematic of different OXC structures is shown in Fig. 14.7. A large number of wavelength signals ($\lambda_1\ \lambda_2 \ldots \lambda_n$) can be demultiplexed at the input ports of an OXC as shown in Fig. 14.8(a) and then these are internally connected to the desired output ports where the different wavelength signals are multiplexed for onward transmission. Furthermore, as indicated the OXC makes it possible to provide an interface between different transmission methodologies. It may be noted here that the OXC basically consists of an ($N \times N$) optical switch that provides the requisite interconnections between network nodes. Cross network interconnections between optical and electronic network nodes, however require optoelectronic conversion. Moreover, wavelength conversion may also be needed to establish a connection that is otherwise not possible using a single lightpath. An OXC having such wavelength conversion features can switch different wavelength signals and therefore known as a wavelength interchangeable cross connect(WIXC)[13-14]. In addition, an OXC facilitating optoelectronic conversion is often referred to as a opaque OXC whereas one incorporating all optical/optical wavelength converters is known as a transparent OXC. The ($N \times N$) optical switching fabric in an OXC facilitates either physical or logical interconnections. Physical interconnection is provided by the fibre switching capability within the

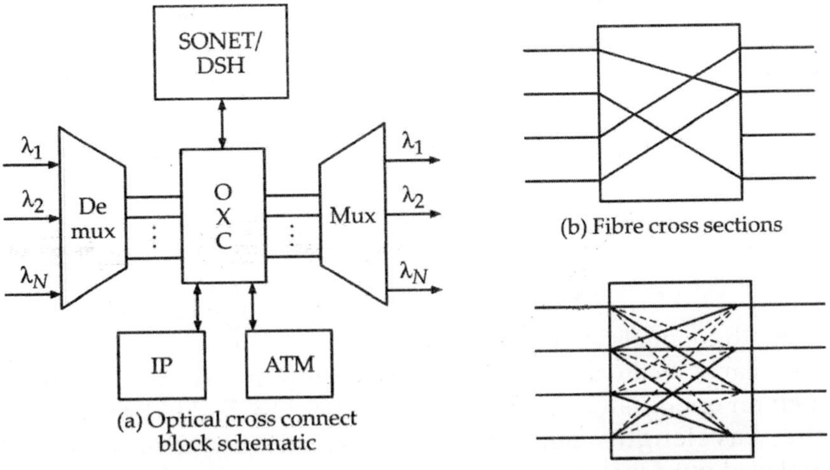

(a) Optical cross connect block schematic

(b) Fibre cross sections

(c) Logical cross sections

Fig. 14.8: Optical cross connect switches

OXC as illustrated in Fig. 14.8(b). In this case optical fibres are switched and connected to the desired network node. The OXC can also facilitate the logical wavelength interconnection features as shown in Fig. 14.8(c) for multiple wavelength signals which can be routed to their destination either individually or in a group of different wavelength signals. In the latter case, several desired wavelength signals can be multiplexed and routed together to their destination nodes.

The various types of logical interconnection (single wavelength, multiple wavelength, wavebands) and physical interconnections can also be combined together in a multilayer structure. In such multilayered OXCs each layer, in addition to providing the specific switching capability, can be connected to another layer, thus offering increased flexibility. A multilayered OXC is sometimes referred to as a multigranular OXC[15].

14.5.1. Wavelength Division Multiplexed Networks

Optical fibre networks that use WDM can be categorized as either broadcast and select networks or wavelength routing networks. A broadcast- and- select network shown in Fig. 14.9 is based on a star coupler. The optical transmission is broadcast to all other nodes using fixed transmitters and a tunable receiver at the destination node retrieves the desired signal from the entire group of wavelength multiplexed transmitted signals. All transmissions are broadcast to all network nodes and hence most of the transmitted power is depleted on the receivers which do not use it. As a result, each station receives only a small amount of the overall transmitted power as the number of nodes increase. A wavelength routing network, on the other hand, avoids this power wastage due to a limited number of connections given to the nodes. Instead of distributing the message over the entire network, the signal is routed to the specific destination through either a single node or using multiple nodes.

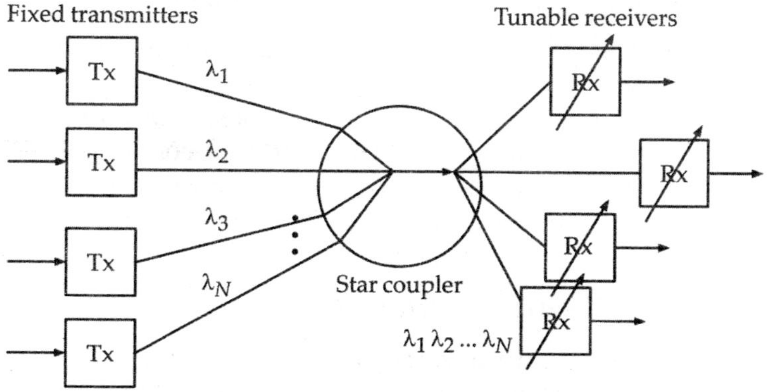

Fig. 14.9: Broadcast and select network

The concept of wavelength routing is illustrated in Fig. 14.10. Physical bidirectional interconnections between five nodes (for example) (*A, B, C, D* and *E*) are shown in Fig. 14.10(a). We can see from this diagram that any node can transmit/receive data from any other node through single/multiple hops using just three wavelengths λ_1 λ_2 and λ_3. The strategy of wavelength implementation or path selection as shown in Fig. 14.10(b) is known as routing and wavelength assignment (RWA). For example, node *C* can transmit to node *B* using λ_3 while it can receive data from node *B* on wavelength λ_2 only when the signal is routed through node *A*. The routing and wavelength assignment strategy is simplified by using virtual topologies that describe only the enabled path lengths. The virtual topology is indicated in Fig. 14.10(c) where only wavelength signals identify the possible interconnections between nodes as given by the RWA. We may observe that the nodes *B* and *C* are not connected physically, however there exists a virtual connection between them using the wavelengths λ_2 and λ_3. Both the broadcast and select and wavelength routing networks can be further divided into single hop and multi hop networks. In single hop networks, data communication between two nodes is accomplished using just one direct communication that allows complete optical domain data transfer.

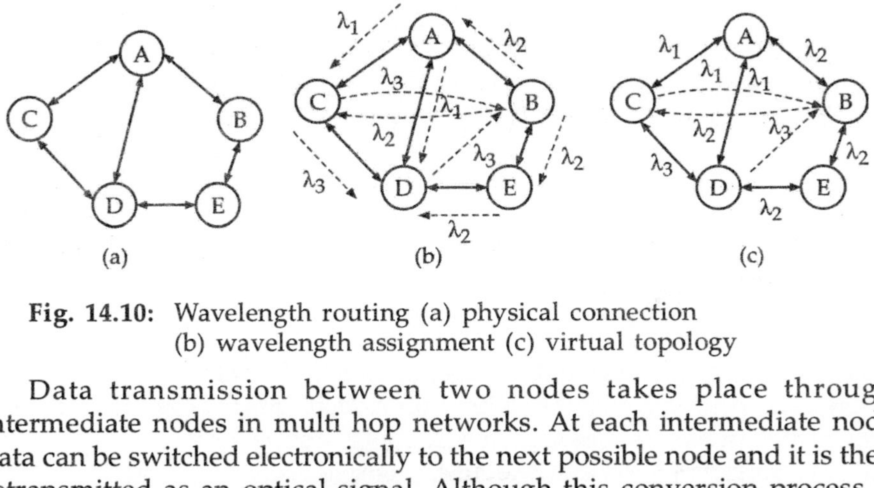

Fig. 14.10: Wavelength routing (a) physical connection (b) wavelength assignment (c) virtual topology

Data transmission between two nodes takes place through intermediate nodes in multi hop networks. At each intermediate node data can be switched electronically to the next possible node and it is then retransmitted as an optical signal. Although this conversion process is inefficient, it is necessary if there is no common wavelength path between two nodes in order to establish a direct connection.

14.5.2. Optical Circuit Switched Networks

In circuit switching, a path is established between two communicating nodes prior to actual data transfer and the dedicated path or circuit remains fixed till the data transfer process is complete. Following the full data transfer, the process is terminated and the path or circuit so established is released for use by other nodes of the network. As we may see, this type of switching requires an end-to-end connection before the actual data transfer which implies that a fixed share of network resources is then reserved for the specific transmission that no other transmission can use. A request signal must, however travel from the source to the destination and it should also be acknowledged before the transmission begins. Voice and data are usually transmitted using this type of switching over the conventional wired PSTN.

A block schematic illustrating an optical circuit switched network (OCS) is shown in Fig. 14.11.

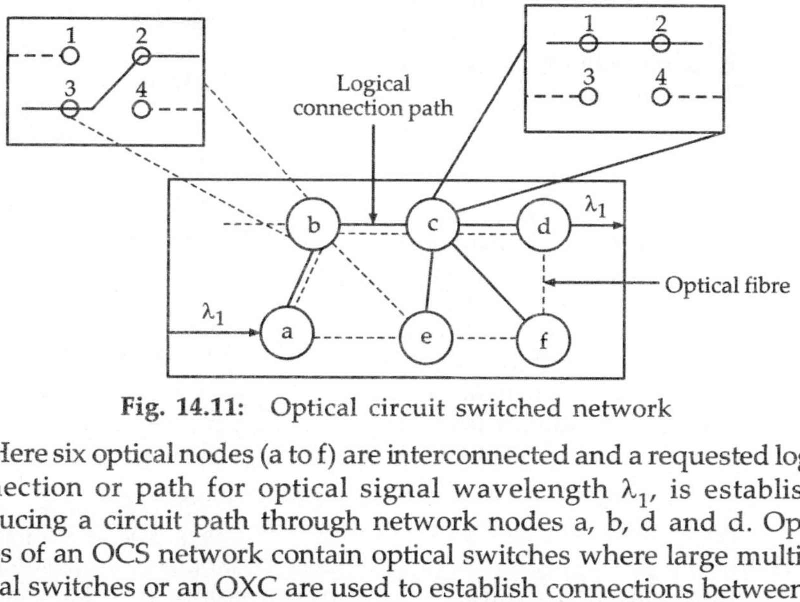

Fig. 14.11: Optical circuit switched network

Here six optical nodes (a to f) are interconnected and a requested logical connection or path for optical signal wavelength λ_1, is established producing a circuit path through network nodes a, b, d and d. Optical nodes of an OCS network contain optical switches where large multiport optical switches or an OXC are used to establish connections between the desired input and output ports. Although different component enabling technologies can be used to construct such optical switches, the basic optical switching function remains the same. This functionality has been illustrated in Fig. 14 at optical nodes b and c. For example, a optical switch at node b enables cross connection using ports 2 and 3 whereas node c employs a direct connection for ports 1 and 2. Hence a logical path or an optical path from node a to d is created for the lightpath using signal wavelength λ_1.

It should be noted that an optical switch can handle traffic from both SDH/SONET and IP networks. However, it has certain limitations in the form that only inservice signals can use the available network resources. Furthermore, the wavelength continuity constraint also limits the flexibility of OCS networks as lightpaths using the same wavelength can only carry the traffic from the source to the destination. However, this constraint can be overcome if the wavelength conversion function is used at the output ports of the optical nodes to change the wavelength of the signal as required. In general OCS networks are suitable for implementation in public switched telephone networks where a large volume of traffic is required to be switched in real time.

The OCS networks have the disadvantage that they cannot handle bursty traffic as given by computer networks. In such traffic conditions the data is sent in optical bursts of different lengths and therefore the resources cannot be readily assigned. For example, an active user needs large bandwidth when sending a burst during the peak hours of network activity as compared with relatively little bandwidth for a nonactive user. Bursty traffic is therefore characterized by both peak and average bandwidth and much of the bandwidth remains unused with the transmission of bursty traffic in as OCS network[16]. This has triggered the search for another type of switching such that the network resources can be utilized in an optimized manner.

14.5.3. Optical Packet Switched Network

Data is transported entirely in the optical domain without intermediate optelectronic conversions in an optical packet switched network (OPS). The four fundamental functions that are typically performed by an optical switch in such networks are routing, forwarding, switching and buffering. The network connectivity information is provided by the routing function often through pre assigned routing tables. Forwarding defines the output for each incoming packet based again on a routing table. The switch directs each packet to the correct output as defined by the forwarding process while buffering provides data storage for packets to resolve any contention problems that may occur during packet transmission.

The overall structure of an optical packet is shown in Fig. 14.12. It contains a header or label, the actual data known as payload and a guard band to ensure the data is not overwritten. The label points to an entry in a lookup table that specifies the forwarding address. Such a labelling

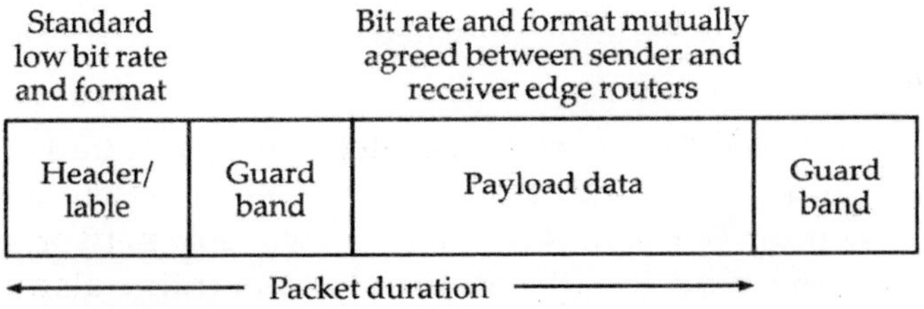

Fig. 14.12: Typical structure of a data packet

technique is much faster than the traditional routing method where each packet is examined before a forwarding decision is made. At the receiving node these labels are required to be recognized from the look up table and then the data is reassembled sequentially. Since labels define the routing criteria it is therefore sometimes referred to as label switching.

The packets in an OPS network can carry different types of traffic such as voice, data and video. Therefore, an optical packet switching scheme can integrate existing SDH/SONET and IP based optical networks.

The clients of the network operator receive service in the access nodes (AN) as shown in Fig. 14.13. The ANs may take the form of a SONET or an IP client network. The incoming IP packets are grouped into optical packets for better optical core utilization. Optical packets thus generated are sent to the CNs and they hop between these CNs. The core nodes switch these optical packets. In the destination EN, the packets undergo optoelectronic conversion before these can be sent to the ANs. The ENs acting as routers perform the functions of attaching and detaching a label as identified for labels 1 and 2 attached to two different optical packets being transmitted to specific networks (to the SONET and IP client networks). The CN router processes only the labels keeping the payload content intact. The routing function involves four jobs which are extraction of the label from the packet, processing of the label to find out the routing information, routing of payload and contention resolution if necessary and rewriting of the label and combining it with the payload.

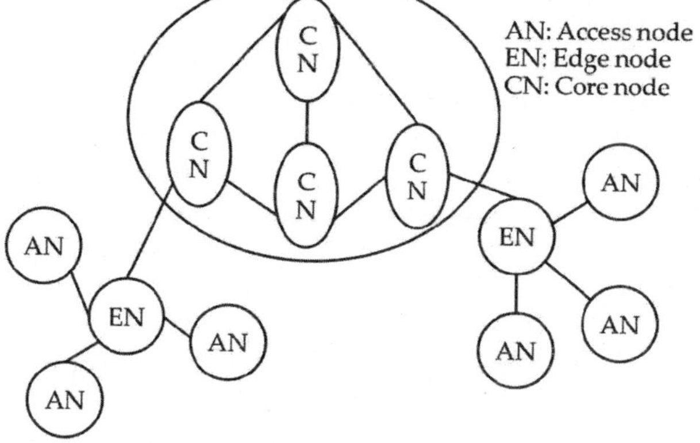

Fig. 14.13: Client Server Model in an optical packet switched network

The OPSs can be synchronous or asynchronous. In the former category, time is divided into slots during which packets are transmitted and received with the constraint that the packet length be no longer than the time slot. In the latter category, packets can arrive at or move out at any time slot.

The general model of an optical packet switch is illustrated in Fig. 14.14.

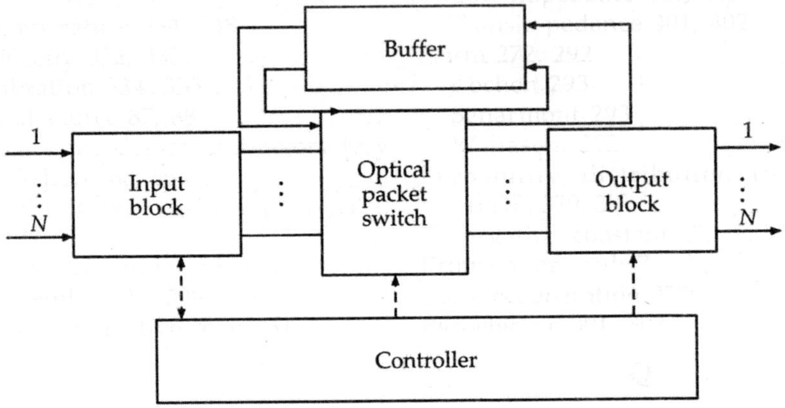

Fig. 14.14: A typical optical switch

14.6. THE SYNCHRONOUS OPTICAL NETWORK (SONET)

The very high data rates possible with optical fibres require certain standards for digital data transmission. The familiar DS signals and T-series carriers, starting with T-1 at basic 1.544 Mbps are developed for use within the telephone networks using coaxial cables and microwave transmission. These standards do not exploit the tremendous information carrying capacity of a fibre. The SONET is a standard that has been developed specifically cater to the high data rate capacity of a fibre optic transmission. This is an American standard while its European counterpart is known as synchronous digital hierarchy (SDH). Hence the abbreviation SONET/SDH is typically used to indicate both the standards and as we may observe are international standards. The SONET standard has four functional layers following the ISO-OSI paradigm. As shown in Fig. 14.15, these are (i) photonic layer, (ii) section layer, (iii) line layer and the (iv) path layer. We may observe that the photonic layer is responsible for all opto electronic conversions. They correspond to both the physical and the data link layers of the conventional computer networks following a hierarchical configuration.

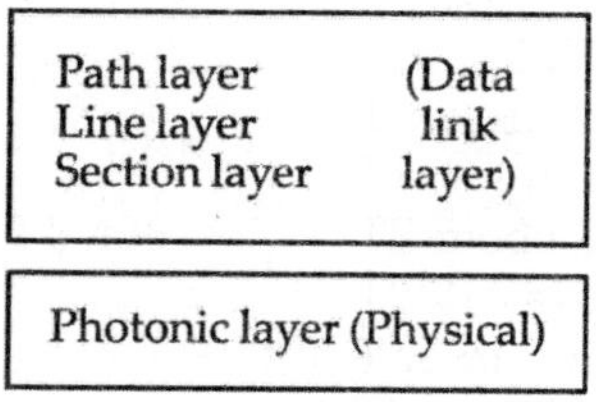

Fig. 14.15: SONET layers compared with OSI or Internet layers

The data link layer (DLL) performs two functions in the conventional electrical computer networks; logical link control (LLC) and the medium access control (MAC). For the SONET, the equivalent of the DLL is observed to comprise of three sections. The path layer is responsible for the signal movement from an optical source to an optical destination. At the optical source, the electrical to optical conversion is carried out, multiplexing of this signal with others is performed and the resulting multiplexed data stream is encapsulated in a frame. At the optical destination, the inverse operation is carried out, namely, the received frame is demultiplexed and the individual optical signals are converted into their corresponding electrical forms. Path layer overhead is added at this layer. Synchronous transport signal (STS) multiplexers provide the path layer functions. The line layer is responsible for the signal movement across a physical line. The line layer overhead is added to the frame at this layer. STS multiplexers and the OADMs carry out this line layer function. Movement of the signal across a physical section is taken care of by the section layer. This layer also provides framing, scrambling and error control. Section layer overhead is added to the frame at this layer. An STS multiplexer is responsible for providing the functionalities of photonic layer and the DLL.

The photonic layer is the equivalent of the physical layer of the standard computer networks. It provides physical specifications for the optical fibre channel, sensitivity of the receiver, multiplexing functions and aspects related to the physical characteristics of the signal. SONET uses NRZ encoding with the presence of light indicating 1 and the absence of light representing 0. A typical network using the STS mux/demux is illustrated in Fig. 14.16.

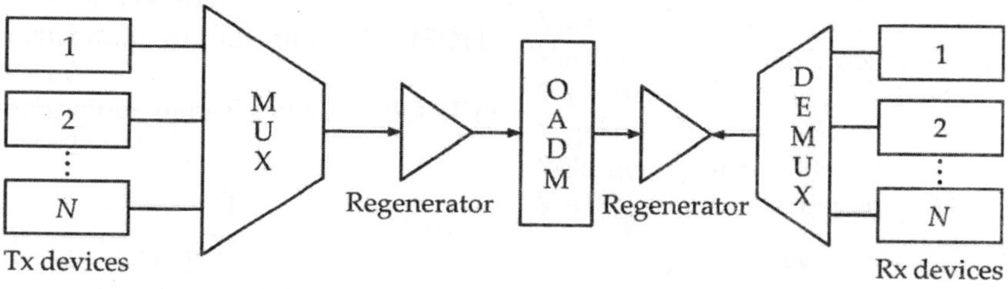

Fig. 14.16: Device-layer relationship in a SONET

Each synchronous transfer signal STS-n is made of 8000 frames per second. Hence, the duration of each frame is (1/8000)s = 125μs. Each frame is a two dimensional matrix of bytes with 9 rows and 90×n columns. Hence, STS-1 frame has 9 rows and (90 × 1 = 90) columns resulting in 810 bytes. Similarly, an STS-3 signal has 9 rows and 270 columns. Standard PCM speech samples are digitized with 8 bits (1 byte) per sample. Hence, we may say that each byte in a SONET frame can carry information from a digitized voice channel. In other words, an STS-1 signal can carry 774 voice channels corresponding to 774 bytes in its frame. The remaining (810-774 = 36) bytes are required for the various overhead as we saw earlier.

To compute the data rate of the basic STS-1 signal, first of all we note that it has 810 bytes per frame and there are 8000 frames. Hence, the basic data rate for SONET as characterized by an STS-1 signal is (810 × 8000 = 648000 bytes) which is equal to 648000 × 8 = 5184000 bits per second or 51.840 Mbps.

The basic SONET rate as calculated above, therefore, is 51.840 Mbps and multiples of this basic data rate are possible. A signal at the basic rate is called a synchronous transport signal level one or STS-1. The same signal is referred to as optical carrier (OC-1) when transmitted over an optical fibre. The STS-1 is the basic building block of the SONET hierarchy that corresponds to OC-1 being the lowest level optical signal used at the SONET equipment and network interface. Higher level signals in the hierarchy are obtained by byte interleaving an appropriate number of STS-1 signals in a manner very similar to the one used in standard electrical PCM-TDM (pulse code modulation time division multiplexing) that results in the well known US T1 standard or the European CEPT standard.

Similarly, the data rate for an STS-3 signal is computed to be 51.840 × 3 = 15.52 Mbps.

Frames of lower rate can be synchronously time division multiplexed into a higher rate frame. For example, three STS-1 signals or channels can be combined into one STS-3 signal, four STS-3s can be multiplexed into one STS-12 and so on as shown in Fig. 14.17.

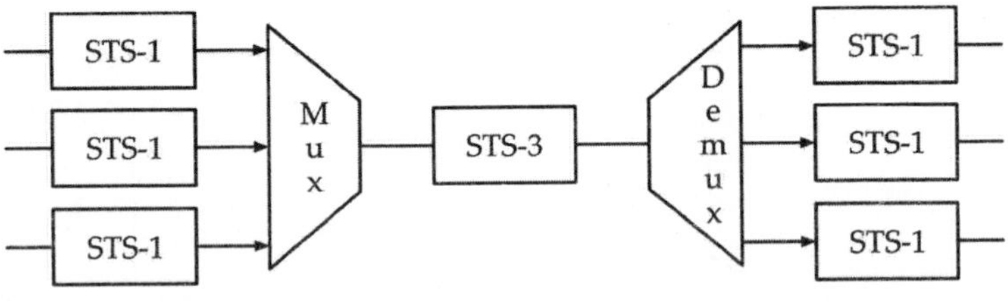

Fig. 14.17: STS multiplexing/demultiplexing (synchronous transmission)

As we may observe from the foregoing, the multiplexing carried out in SONET/SDH wherein lower order frames are multiplexed onto one larger frame, is synchronous TDM and all clocks in the network are locked to a master to achieve synchronization. Multiplexing takes place at higher data rates. Four STS-3 signals can be multiplexed into one STS-12 signal. However, the STS-3 signals need to be first demultiplexed into 12 STS-1 signals and then these 12 lower order STS-1 signals are multiplexed into one STS-12 signal. Some popular SONET data rates are shown in Table 14.1. STM here stands for synchronous transport module and the STM rates are those used for synchronous transmission. The lowest speed in the hierarchy is designed to accommodate the North American DS-3 data rate of 44.736 Mbps and the European E-3 rate of 34 Mbps. The last entry in the table gives a data rate of 2.5Gbps which is well within the present capability of optical links using single-mode fibre and laser diodes. Additional multiples of the SONET basic rate may be accommodated.

The SONET, as a standard differs from the older digital transmission schemes in certain ways. Like the DS-1 signal, the STS-1 signal starts with 8000 frames/sec corresponding to the sampling rate of digital telephony. One frame of DS-1 contains 193 bits; 8 bits for one sample for each of 24 channels plus one framing bit. Every sixth frame makes use of the least significant bit from each channel for signalling. The SONET-STS-1 signal, however has 6480 bits (810 bytes) per frame. This frame may be viewed as being arranged in the rows and columns as shown in Fig. 14.7. Transmission of a frame begins in the upper left hand corner and ends in the lower right hand corner 125μs later.

Table 14.1: SONET line rates

STS-level	*OC level*	*STM level*	*Line rate (Mbps)*
STS-1	OC-1		51.84
STS-3	OC-3	STM-1	155.52
STS-9	OC-9		466.56
STS-12	OC-12	STM-4	622.08
STS-18	OC-18		933.12
STS-24	OC-24		1244.16
STS-36	OC-36		1866.24
STS-48	OC-48	STM-16	2488.32

The SONET network can be used as a high speed backbone carrying loads from other networks such as ATM (asynchronous transfer mode) to be described later or IP. The SONET networks topologies may belong to linear, ring or mesh configurations.

The optical packet switched networks employ WDM mainly. The WDM networks are built upon OXCs and OADMs with support from optical amplifiers, regenerators and wavelength converters. WDM networks are able to establish lightpaths between network nodes that are sent to the networks at higher layers like PDH, SONET/SDH or IP.

14.7. ASYNCHRONOUS TRANSFER MODE (ATM)

Data communication at the data link layer is based on frame switching and frame networks. Different protocols use frames of varying size and intricacy. As networks become more complex, the header information becomes more extensive. As a result of this, headers become longer compared to the data packet. Some protocols have enlarged the size of data packets to make the header more efficient. However, a large data packet requires a large network time or delay and higher bandwidth. If there is not much information to transmit, much of the data field remains unutilized. This may be taken care of by using variable frame sizes to the users.

The synchronous time division multiplexing as discussed above for SONET has a serious drawback. If a particular time slot is empty as there is no data to be transmitted during this slot, then this represents a waste as it cannot be used by another user who must wait for his assigned (synchronous) time slot to turn up. This empty time slot, if can be allotted to a user having data to be sent on an immediate basis, then network resource utilization can be made in an optimum manner. Moreover, the time slots in a SONET can carry a fixed amount or constant data. These cannot handle variable data rates. These drawbacks of the conventional synchronous time multiplexing are taken care of in ATM. Thus, ATM is also a multiplexing and switching technique. Multiplexing of different data packets (known as cells in ATM literature) from different channels is carried out in ATM. ATM multiplexers fill a slot with a cell from any input channel that has a cell to send; the slot is empty of the channels have no

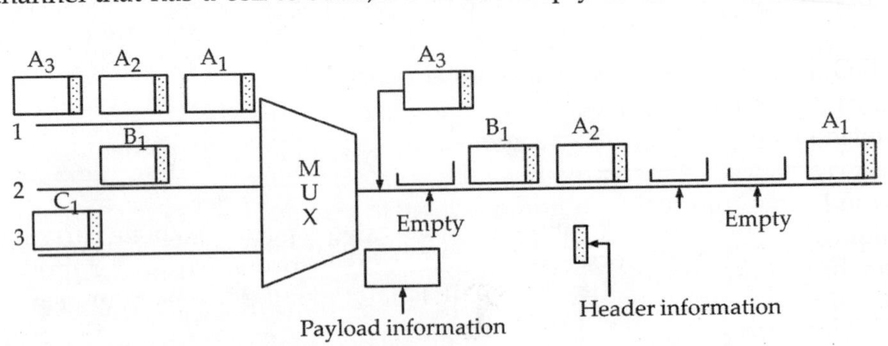

Fig. 14.18: Multiplexing in STM (SONET)

data to send. This is very unlike the synchronous transfer mode wherein a time slot must be filled with data from the assigned channel only and not "any" channel. This kind of asynchronous multiplexing has been illustrated in Fig. 14.18.

We can see the significant wastage of network resources through the use of synchronous time multiplexing as in SONET. Channels B and C do not have any packets to send in the slots assigned for them during the first cycle of multiplexing and these two slots go as a waste (assuming the multiplexing cycle to contain three slot corresponding to the three channels as in this diagram). During the second cycle similarly, slot 3 remains unoccupied as it belongs to channel C that does not have any packet to send. Similar situations take place during third cycle of multiplexing. This issue is addressed in ATM as shown in Fig. 14.19.

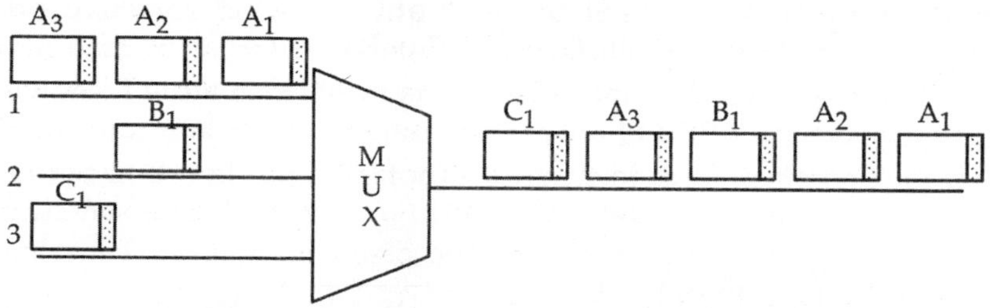

Fig. 14.19: Multiplexing in ATM

At the first tick of the clock, channel B does not have any packet to send. Unlike STM, it is not allowed to remain empty; rather it is assigned to channel A that has the packet A_2 to send. The third slot that would have catered to channel C, is now given to channel B that sends out its packet B_1. As we can see from this diagram, no time slot is allowed to remain unoccupied and it is given to a channel having a packet to send at this slot. The multiplexer output caters to a continuous data traffic unlike STM. It is quite obvious that, when all the packets from all the channels are multiplexed, the output slots become free.

The basic data unit in an ATM network is called a cell. In fact, all the packets which we have shown in Fig. 14.18 and Fig. 14. 19 are small data packets or cells having fixed size. A cell is only 53 bytes long with 5 bytes allocated to the header and 48 bytes meant for the payload data (user data may be less than 48 bytes). The header contains what are known as virtual packet identifier (VPI) and virtual circuit identifier (VCI). These two entities define a virtual connection through which the cell travels from an endpoint to a switch or from a switch to another switch. The typical cell structure is shown in Fig. 14.20.

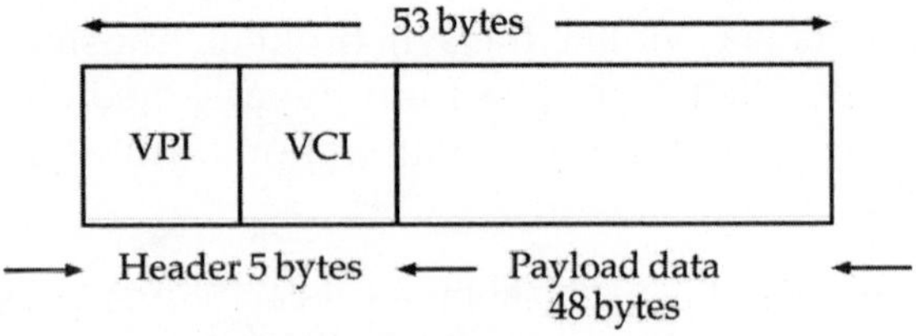

Fig. 14.20: An ATM cell

14.7.1. Routing with a Switch

ATM, as said earlier is also a switching technique. It provides switching of the cells using the VPI and the VCI information contained in the header part of its cell so that this cell can be routed from a source endpoint to a destination endpoint. Routing requires the complete identifier. Cell routing accomplished through a virtual permanent circuit (VPC) is illustrated in Fig. 14.21. A cell with a VPI of 95 and VCI of 33 arrives at switch interface port 1 (in this example). The switch checks its switching table which stores six pieces of information per row: arrival interface number (1 in this case), incoming VCI (95), incoming VPI (33), the corresponding outgoing port number, the new VCI and the new VPI. The switch finds entry with the port 1, VCI 95, VPI 33 and discovers that the combination corresponds to output port 3, VPI 112 and VCI 68. It changes the VPI and the VCI in the header to 112 and VPI to 68 respectively, and sends the cell out through the port 3.

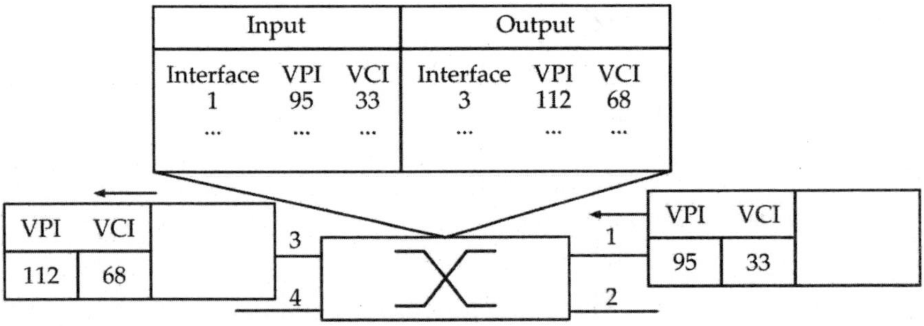

Fig. 14.21: Routing in an ATM switch using VPI and VCI

ATM functions using three layers. From top to bottom these are the application adaptation layer (AAL), the ATM layer and the physical layer just like other networks. The endpoints use all three layers while the switches use only the two bottom layers. The ATM cells can be carried by any physical layer just like the Ethernet (wired LAN) and the wireless LANs. The original design of the ATM was based on SONET as the physical layer. The ATM layer is responsible for routing, traffic management, switching and multiplexing functionalities. It processes the outgoing traffic by accepting 48 byte segments from the AAL sublayers and transforming them into 53-byte cells by the addition of a 5-byte header.

14.7.2. Architecture of an ATM Network

As we have understood now, ATM provides switching and multiplexing of cells, thereby making it a cell switched network. The user access devices known as endpoints are connected to the ATM switch through user-to-network interface (UNI) whereas the switches themselves are connected through network-to-network interface (NNI)s. This has been illustrated in Fig. 14.22. Connection between two endpoints is carried out through transmission paths comprising of VPIs and virtual circuits (VC).

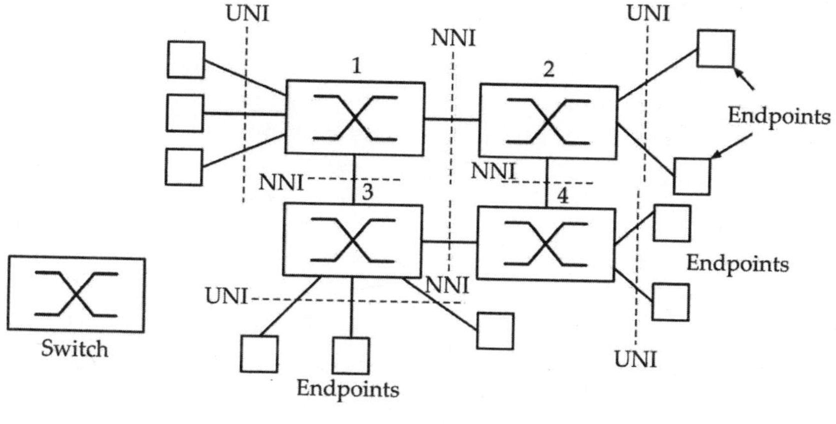

Fig. 14.22: A typical ATM network architecture

14.7.3. Open Systems Interconnection Reference Model

The OSI describes a standard architecture to be used for designing networks. The OSI model was specified jointly by the International Organization for Standardization (ISO) and ITU-T. The hierarchical structure of the OSI network reference model is shown in Table 14.1. That also specifies the functionalities performed by each of the seven layers. As we see, the application layer at the top of this structure provides a means for a user to access information on or utilize the network by receiving a service (database management, network management). This layer is the main interface for the users to interact through applications with the network.

Table 14.1: The hierarchical seven layer structure of the ISO-OSI model of a network

Layers	*Function*	*Data unit*		
7	Application	Network process to application	Data	Host layers
6	Presentation	Data presentation and encryption		
5	Session	Session management of machine communication		
4	Transport	End-to-end connections and reliability	Segments	
3	Network	Path determination an logical addressing	packets	Media layers
2	Data link	Physical addressing, medium access protocol	Frames	
1	Physical	Media, signal and binary transmission	Bits	

The sixth layer is the presentation layer which transforms data to provide a standard interface for the application layer. The session layer controls the sessions between the intelligent machines. Further, it establishes, manages and terminates the connections between the local and the remote application. It also provides for either duplex or half-duplex operation and establishes check pointing, adjournment, termination and restart operations. The top three layers are, therefore data driven layers. The fourth layer is the transport layer which provides transparent transfer of data between end users, thus relieving the upper layers from any transport concern while providing reliable data transfer. The transport layer controls the reliability of a specific link through flow control, segmentation/desegmentation and error control. Some protocols are both state and connection oriented. This means that the transport layer can keep track of the packets and retransmit those without fail. The best known example of a transport protocol is the transmission control protocol (TCP) which is a connection oriented protocol that provides basic data transfer between nodes offering interconnection, flow control, multiplexing with reliable priority and security.

The third layer of this structure known as the network layer provides the functional and procedural method for transferring variable length data sequences from a source to a destination *via* one or more networks while maintaining the quality of service requested by the transport layer. The network layer provides routing, might go for segmentation/desegmentation and is responsible for providing a report on delivery errors. Routers operate at this layer for sending data throughout the entire network that makes Internet possible.

The second layer known as data link layer (DLL) provides framing of data bits being sent by the physical layer. It is also responsible for giving medium access (MA) to competing users on a predetermined protocol that is designed to optimize network throughput at a reasonable delay. Best known protocols are carrier sense multiple access with collision detection

(CSMA/CD) or the Ethernet protocol. It provides physical addressing so that the frames can be appended with source and destination addresses. Some error correction based on parity check is also carried out by this layer. Network bridges and switches operate at this layer. Connectivity is provided among locally attached network nodes forming layer 2 domains for unicast or broadcast forwarding. High level data link control (HDLC) and the advanced data communications control procedure (ADCCP) are other data link layer protocols.

The physical layer located at the bottom of this OSI reference model defines all electrical, optical and physical media specifications for devices (cable specifications, repeaters, hubs, network adaptors, fibre type etc.). It both establishes and terminates a connection to the communications medium. Furthermore, it also ensures that communication resources are effectively shared among multiple users. For example, it implements contention resolution and flow control while carrying out data conversion between the equipment and the signal requirements of the communications channel.

14.7.4. Optical Transport Network

An optical transport network (OTN) is considered to be a set of facilities that use optical fibre interconnections to carry data between network elements that switch or route data from different customers into the network[17]. Such a transport network is responsible for the transport, multiplexing, routing, supervision and survivability for client signals as specified by the ITU-T recommendations. The functionalities of an OTN are indicated in Fig. 14.23. The OTN is viewed as a successor to SDH/SONET which is the current standard for voice and data traffic in the public switched telephone network. Therefore it combines the benefits of SDH/SONET with dense WDM. The four layers basic to the functionality of such an OTN are

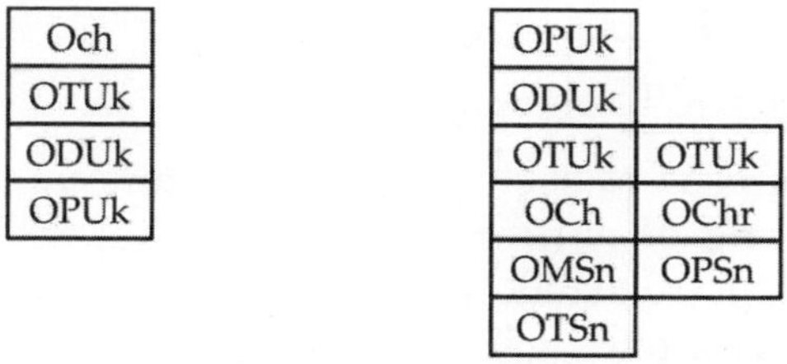

(Full functionality) (Reduced functionality)

Fig. 14.23: (a) Typical hierarchy of optical transport network layer
(b) hierarchy as specified in ITU-T recommendation

(i) The optical channel (Och)
(ii) Optical channel transport unit (OTU)
(iii) Optical channel data unit (ODU) and
(iv) Optical channel payload unit

The hierarchy of these four layers as mentioned above may also include three sections known as the optical transmission section (OTS), the optical multiplexing section (OMS) and the optical physical section (OPS). These sections are further divided into a number of subsections. The layers and the sections in an OTN give rise to a confluence of inter and intradomain interfaces enabling optical signal transmission to reach from one end to the other using electronic, optoelectronic or all optical interfaces. The basic functional structure showing the various OTN layers and sections that facilitates the method for interfacing between different transmission domains is illustrated in Fig. 14.24.

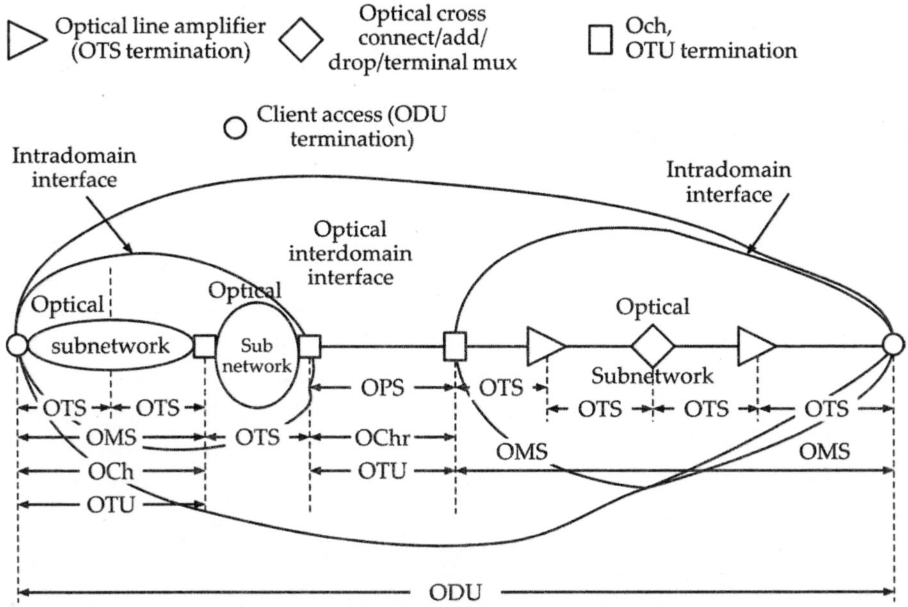

Fig. 14.24: Functional block diagram showing typical OTN layers

The OCh layer is used in such an OTN structure in order to access end-to-end networking of optical channels so that transparent user information transfer is realized. This layer also ensures connection rearrangement for flexible routing. Either full or partial channel availability may be employed as required. The OTU layer provides additional functionality by adding forward error correction to the network elements and by allowing carrier operators to reduce the number of optical devices and switches used in the network. It also encapsulates two additional layers, the optical channel data unit (ODU) and the optical channel payload unit (OPU). The ODU layer is responsible for client-independent connectivity, connection protection and monitoring whereas the OPU enables access to the payload information of SDH/SONET signals.

Different physical sections of the OTN are interfaced based on the information obtained from OTS, OMS and OPS respectively. OTS is responsible for the required information about optical transmission functionality for optical signals travelling through the optical fibres and the necessary switching at the fibre end points. The OMS provides the necessary platform for networking of a multiwavelength optical signal. The OPS describes the optical characteristics of the physical section in order to provide reduced functionality excluding those sections which are not required to establish an interface between two domains. A reduced functionality can be achieved by direct adaptation of the OPS into OChr functions since both OMS and OTS are not necessary in the interdomain interfaces.

The control plane of the OTN refers to a set of protocols related to signalling and routing for distributed routing control, connection management and resource discovery when a connection is required to be established between two network nodes. The OTN can be made reconfigurable depending on the provision of fast, efficient and reliable network connection. Such a reconfigurable OTN is known as an automatically switched optical network (ASON). Such anarchitecture is capable of switching the optical channels automatically when requested. The OTN is a transport layer that can carry 10Gbps Ethernet transmission from IP/Ethernet switches and routers at full bandwidth[18-19] unlike SONET/SDH. It also provides the requisite platform for the optical fibre networks to be combined onto a common network infrastructure.

14.8. CHAPTER SUMMARY

The two main functionalities that must be performed by a network are switching and multiplexing. Simply connecting two nodes (devices with signal processing and related computing facilities) does not give us a network. Networks have evolved from all voice using circuit switching and statistical multiplexing to voice and data of very fast rates that are implemented with packet switching combined with statistical/ synchronous/asynchronous kind of multiplexing. The networks have to comply with the ISO-OSI model of architecture. For an optical network, the equivalent of physical layer is provided by the photonic layer and other functionalities remain almost the same as found on a typical computer network. Higher data rate transfer in a network is achieved by multiplexing a number of lower data rates with a suitable architecture.

Optical networks aim to provide ultra fast data rates in 10 Gbps for which they rely upon IP with optical transport networks.

SUGGESTED READING

[1] Tomlinson, W.J. "Wavelength multiplexing in multimode optical fibers", *Applied Optics*, vol. 16, pp. 2180-2194, Aug 1977.

[2] Kanada, T., Okano, Y., Aoyama, K.I. and Kitami, T. "Design and performance of WDM transmission systems at 6.3Mb/s", IEEE Trans. on Comm., vol COM-31, pp. 1095-1102, Sep. 1983.

[3] Winzer, G. "Wavelength multiplexing components-A review of single mode devices and their applications", *J. Lightwave Tech.*, vol. LT-2, pp. 448-463, Aug 1984.

[4] Melman, P., Carlsen, W.J. and Foley, B. "Tunable birefringent wavelength-division multiplexers/demultiplexers", *Electronics Letters*, vol. 21, pp. 634-635, July 1985.

[5] Carlsen, W.J. and Buhrer, C.F. "Flat passband birefringent wavelength-division multiplexers", *Electronics Letters*, vol. 23, pp. 106-107, 1987.

[6] Frenkel, A. and Lin, C. "Angle-tuned etalon filters for optical channel selection in high density wavelength division multiplexed systems", *J. Lightwave Tech.*, vol. 7, pp. 615-624, Apr 1989.

[7] Henry, C.H., Kazarinov, R.F., Shani, Y., Kistler, R.C., Pol, V. and Orlowsky, K.J. "Four channel wavelength division multiplexers and bandpass filters based on elliptical Bragg reflectors", *J. Lightwave Tech.*, vol. 8, pp. 748-755, May 1990.

[8] Fow-Sen, C., Zhao, X., Xiuqin, Y., Lin, J., Zhang, P., Gu, Y., Ru, G., Guansong, Z., Longjun, L., Huiping, X., Hadimioglu, H. and Chao, H.J. "An optical packet switch based on WDM technologies", *J. Lightwave Tech.*, vol. 23, pp. 994-1014, March 2005.

[9] Tanenbaum, A. "Computer Networks", Upper Saddle River, NJ: Prentice Hall, 2003.

[10] Benjamin, A.S., Assaf, S. and Keren, B. "A modular, scalable, extensible and transparent optical packet buffer", *J. Lightwave Tech.*, vol. 25, pp. 978-985, Apr 2007.

[11] Lee, S., Sriram, K., Kim, H. and J. Song, "Contention-based limited deflection routing protocol in optical burst-switched networks", *IEEE J. Sel. Areas in Comm.*, vol. 23, pp. 1596-1611, Aug 2005.

[12] Fukushima, Y., Harai, H., Arakawa, S. and Murata, M. "Design of wavelength-convertible edge nodes in wavelength-routed networks", J. of Optical Networks", vol. 5, pp. 196-209, Mar 2006.

[13] Lei, Z., Ji, P., Ting, W., Matsuda, O. and Cvijetic, M. "Study on wavelength cross-connect realized with wavelength selective switches", Proc. of Optical Fiber Comm. and National Fibre Optic Engg. Conf., Anaheim, CA, p. 7, Mar 2006.

[14] Hamza, H.S. and Deogun, J.S. "Wavelength-exchanging cross connects (WEX)- a new class of photonic cross connect architectures", *J. Lightwave Tech.*, vol. 24, pp. 1101-1111, March 2006.

[15] Cao, X. Vishal, A., Li, J. and Xin, C. "Waveband switching networks with limited wavelength conversion", IEEE Comm. Lett., vol. 9, pp. 646-648, July 2005.

[16] Luo, H., Hu, G. and Li, L. "Burstification queue management in optical burst switching networks", Photonics Network Comm., vol. 11, pp. 87-97, Jan 2006.

[17] ITU-T, "Optical Transport Network:tutorial, *http://www.itu.int/ITU-T/studygroups/com15/otn/OTNtutorial.pdf*, 17 Oct 2007.

[18] Bigos, W., Cousin, B., Gosselin, S., Foll, M. Le and Nakajima, H. "Optimized design of survivable MPLS over optical transport networks", J. of Optical Switched Networks, vol. 3, pp. 202-218, Mar-Apr 2006.

[19] Welch, D.F. *et. al*, "Large-scale in P photonic integrated circuits: Enabling efficient scaling of optical transport networks" *IEEE J. on Selected Topics on Quantum Electronic*, vol. 13, pp. 21-31, Jan 2007.

❑❑❑

Appendix-A

Elementary Coordinate Systems

Relationship between rectangular (Cartesian) and cylindrical coordinates.

We know that the rectangular system of coordinates is given as (x, y, z) while the cylindrical system of coordinates is expressed as (ρ, ϕ, z).

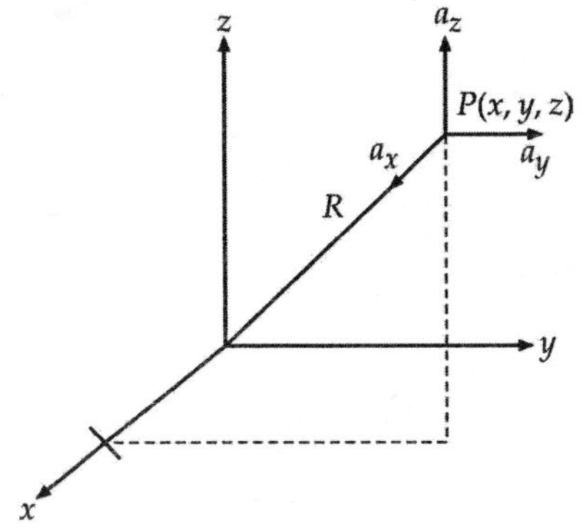

Fig. I-A: Rectangular coordinates for a point P

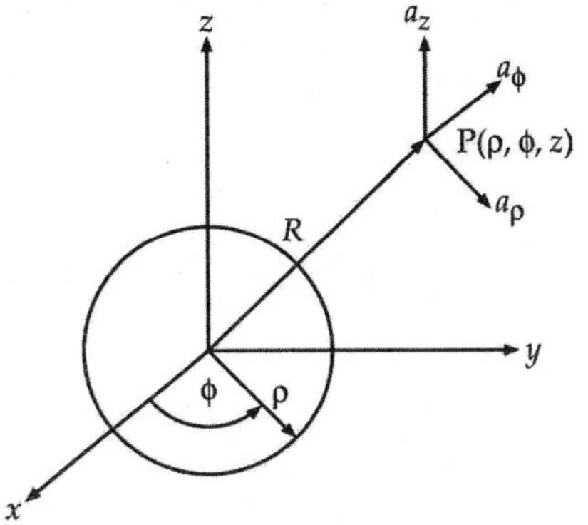

Fig. I-B: Cylindrical system for a point P

The xy-plane has been converted to the curved surface of a cylinder in terms of the polar coordinates expressed as (ρ, ϕ) such that the angle ϕ is rotated from the x-axis in a counterclockwise direction. Therefore,

$$x = \rho \cos \phi \quad \text{...(A.1)}$$

$$y = \rho \sin \phi \quad \text{...(A.2)}$$

(Rectangular to cylindrical/polar)

$$\rho = \sqrt{x^2 + y^2} \quad \text{...(A.3)}$$

and

$$\tan \phi = \frac{y}{x} \Rightarrow \phi = \tan^{-1}\left(\frac{y}{x}\right) \quad \text{...(A.4)}$$

(Polar/rectangular)

We also observe that the range of the axes are as follows

$$-\infty \leq x \leq \infty, -\infty \leq y \leq \infty \text{ and } -\infty \leq z \leq \infty$$

However, $\quad 0 \leq \rho \leq \infty, 0 \leq \phi \leq 2\pi \text{ and } -\infty \leq \infty$

The elementary lengths in the rectangular system are dx, dy and (ρ, ϕ). Hence, the elementary plane areas are $dxdy$, $dydz$ and $dzdx$ with the unit normals to the planes being a_z, a_x and a_y respectively. The elementary volume is $dxdydz$.

Similarly, the elementary lengths in the cylindrical system are $d\rho$, $\rho d\phi$ and dz. Hence, the elementary plane areas are $\rho d\rho d\phi$, $d\rho dz$ and $\rho d\phi dz$ with the unit normals to the planes being d_z, a_ϕ and a_ρ respectively. The elementary volume is $\rho d\rho d\phi dz$.

The unit normals at the point P are drawn in the directions of increasing magnitude.

Example A-1: Evaluate the area of the curved surface of a cylinder of radius a.

Solution: The elementary area is $\rho d\rho d\phi$. The total area is

$$\int_0^a \int_0^{2\pi} \rho d\rho d\phi = \left.\frac{\rho^2}{2}\right|_0^a 2\pi = \pi a^2$$

Example A-2: Establish the relationship between the unit vectors of both the coordinates.

Solution: From the figures we observe that, projections of a_x, s_y, a_z onto a_ρ, a_ϕ, a_z give us the following

$$a_x, a_\rho = \cos\phi, a_y.a_\rho = \sin\phi \text{ and } a_z, a_\rho = 0, \text{ and}$$

Similarly, $\quad a_x, a_\phi = \cos(90 + \phi) = -\sin\phi, a_y, a_\phi = \sin(90 + \phi)$

$$= \cos\phi \text{ and } a_z, a_\phi = 0$$

$$a_x, a_z = 0, a_y, a_z = 0 \text{ and } a_z, a_z = 1$$

Therefore, $\quad a_x = a_\rho \cos\phi - a_\phi \sin\phi$

$$a_y = a_\rho \sin\phi + a_\phi \cos\phi$$

Example A-3: Express a vector A given in rectangular coordinates in cylindrical coordinates.

Solution: We understand that a vector A is expressed as $A = A_x a_x + A_y a_y + A_z a_z$ where A_z, A_y and A_z are the constituent components along x, y and z respectively.

Using the relationships derived in I-2, we write

$$\begin{aligned} A &= A_x (a_\rho \cos\phi - a_\phi \sin\phi) + A_y (a_\rho \sin\phi + a_\phi \cos\phi) + A_z a_z \\ &= (A_x \cos\phi + A_y \sin\phi)\, a_\rho + (-A_x \sin\phi + A_y \cos\phi)\, a_\phi + A_z a_z \\ &= A_\rho a_\rho + A_\phi a_\phi + A_z a_z \end{aligned}$$

It is obvious that

$$A_\rho = A_x \cos\phi + A_y \sin\phi,\ A_\phi = A_y \cos\phi - A_x \sin\phi \text{ and } A_z = A_z$$

In a similar manner, it can be shown that

$$A_x = A_\rho \cos\phi - A_\phi \sin\phi$$

$$A_y = A\rho \sin\phi + A_\phi \cos\phi$$

Example A-4: Relate the differentials in cylindrical coordinates to the rectangular coordinates.

Solution: $x = \rho \cos\phi\ y = \rho \sin\phi\ \rho = \sqrt{x^2 + y^2}$, $\tan\phi = \dfrac{y}{x}$

$$\frac{\partial}{\partial x} = \cos\phi \frac{\partial}{\partial \rho} - \frac{\sin\phi}{\rho}\frac{\partial}{\partial \varphi}$$

$$\frac{\partial}{\partial y} = \sin\phi \frac{\partial}{\partial \rho} + \frac{\cos\phi}{\rho}\frac{\partial}{\partial \varphi}$$

❑❑❑

Appendix-B

Some Elementary Useful Vector Algebra

The three basic operations performed upon a vector are gradient, divergence and curl. The gradient operator acts on a scalar and gives a vector. The operator ∇ is known as the "del" operator and is defined as

$$\nabla = \frac{\partial}{\partial x}a_x + \frac{\partial}{\partial y}a_y + \frac{\partial}{\partial z}a_z$$

B.1. THEREFORE, FOR THE RECTANGULAR SYSTEM OF COORDINATES

$$\nabla\Psi = \frac{\partial\psi}{\partial x}a_x + \frac{\partial\psi}{\partial y}a_y + \frac{\partial\psi}{\partial z}a_z \quad \text{...(B.1)}$$

that gives the gradient of the scalar ψ. Hence, gradient is always measured with respect to a certain direction in space and represents the spatial variation of the physical parameter under consideration. We have potential gradient, temperature gradient, pressure gradient and so on. For example, the electrical potential is a scalar quantity. However, its gradient at a given point gives a measure of the electric field at that point. Electric field is a vector.

B.2. THE DIVERGENCE RELATES TO THE DIFFERENTIAL OF A VECTOR AND IS EXPRESSED AS A DOT PRODUCT OF THE ∇ OPERATOR AND THE VECTOR. THE RESULT IS A VECTOR EXPRESSED AS

$$\nabla.\,\mathbf{A} = \frac{\partial A_x}{\partial x}a_x + \frac{\partial A_y}{\partial y}a_y + \frac{\partial A_z}{\partial z}a_z \quad \text{...(B.2)}$$

The divergence of a vector **A** at a given point represents the net flux per unit volume. If it is positive, the net flux is outward and inward otherwise. For example, the net electric flux from a closed surface flowing in an outward manner (or leaving the closed surface) is said to be positive.

B.3. THE CURL OPERATES ON A VECTOR AND GIVES ANOTHER VECTOR EXPRESSED AS

$$\nabla \times \mathbf{A} = \begin{bmatrix} a_x & a_y & a_z \\ \dfrac{\partial}{\partial x} & \dfrac{\partial}{\partial y} & \dfrac{\partial}{\partial z} \\ A_x & A_y & A_z \end{bmatrix}$$

$$= \left(\frac{\partial A_z}{\partial y} - \frac{\partial A_y}{\partial z}\right) a_x - \left(\frac{\partial A_z}{\partial x} - \frac{\partial A_x}{\partial z}\right) a_y + \left(\frac{\partial A_y}{\partial x} - \frac{\partial A_x}{\partial y}\right) a_z \qquad \text{...(B.3)}$$

The Laplacian for a vector is defined as

$$\nabla^2 \mathbf{A} = \nabla (\nabla.\, A) - \nabla \times \nabla \times \mathbf{A} \qquad \text{...(B.4)}$$

Some useful vector identities

$$\nabla.\, (b\mathrm{A}) = \mathrm{A}.\, \nabla b + b\nabla.\mathbf{A} \text{ (Here } b \text{ is a scalar)} \qquad \text{...(B.5)}$$

$$\nabla \times (b\mathbf{A}) = \nabla b \times \mathbf{A} + b\nabla \times \mathbf{A} \qquad \text{...(B.6)}$$

$$\nabla.(\mathbf{A} \times \mathbf{B}) = \mathbf{B}.\nabla \times A - A.\nabla \times B \qquad \text{...(B.7)}$$

$$\nabla.(\mathbf{A}.\mathbf{B}) = (\mathbf{A}.\nabla)\, \mathbf{B} + (\mathbf{B}.\nabla)\, \mathbf{A} + \mathbf{A} \times (\nabla \times \mathbf{B}) + \mathbf{B} \times (\nabla \times \mathbf{A}) \qquad \text{...(B.8)}$$

$$\nabla.(\nabla \times A) = 0 \qquad \text{...(B.9)}$$

$$\nabla.\nabla = \nabla^2 \qquad \text{...(B.10)}$$

Example B.1: Compute $\nabla.\mathbf{A} = \dfrac{\partial A_x}{\partial x} + \dfrac{\partial A_y}{\partial y} + \dfrac{\partial A_z}{\partial z}$ in cylindrical coordinates

Solution: Substitution of the constituent vectors in rectangular coordinates from I- in the above expression gives us

$$\nabla.\mathbf{A} = \left(\cos\phi \frac{\partial}{\partial \rho} - \frac{\sin\phi}{\rho}\frac{\partial}{\partial \phi}\right)(A_\rho \cos\phi - A_\phi \sin\phi)$$

$$+ \left(\sin\phi \frac{\partial}{\partial \rho} + \frac{\cos\phi}{\rho}\frac{\partial}{\partial \phi}\right)(A_\rho \sin\phi + A_\phi \cos\phi) + \frac{\partial A_z}{\partial z}$$

This is

$$\nabla.\mathbf{A} = \cos^2\phi\frac{\partial A_\rho}{\partial\rho} - \frac{\sin\phi(-\sin\phi)}{\rho}A_\rho + \frac{\sin^2\phi}{\rho}\frac{\partial A_\phi}{\partial\phi}$$

$$+\sin^2\phi\frac{\partial A_\rho}{\partial\rho} + \frac{\cos\phi}{\rho}.A_\rho\cos\phi g + \frac{\cos^2\phi}{\rho}\frac{\partial A_\phi}{\partial\phi} + \frac{\partial A_z}{\partial z}$$

Therefore,

$$\nabla.\mathbf{A} = (\cos^2\phi + \sin^2\phi)\frac{\partial A_\rho}{\partial\rho} + \left(\frac{\sin^2\phi}{\rho} + \frac{\cos^2\phi}{\rho}\right)A_\rho$$

$$+\left(\frac{\cos^2\phi}{\rho} + \frac{\sin^2\phi}{\rho}\right)\frac{\partial A\phi}{\partial\phi} + \frac{\partial A_z}{\partial z}$$

$$= \frac{\partial A_\rho}{\partial\rho} + \frac{A_\rho}{\rho} + \frac{\partial A_\phi}{\rho\partial\varphi} + \frac{\partial A_z}{\partial z}$$

$$= \frac{1}{\rho}\frac{\partial}{\partial\rho}(\rho A_\rho) + \frac{\partial A_\phi}{\rho\partial\phi} + \frac{\partial A_z}{\partial z}$$

Example B.2: Transform the equation $\frac{\partial^2\Psi}{\partial x^2} + \frac{\partial^2\Psi}{\partial y^2} = 0$ into polar coordinates.

Solution: We know $x = \rho\cos\phi$ $y = \rho\sin\phi$ $\rho = \sqrt{x^2 + y^2}$ $\phi = \tan^{-1}\frac{y}{x}$,

Hence, $$\frac{\partial\rho}{\partial x} = (x^2 + y^2)^{-1/2}.2x = \frac{x}{\sqrt{x^2 + y^2}} = \frac{\rho\cos\phi}{\rho} = \cos\phi$$

and $$\frac{\partial\varphi}{\partial x} = \frac{\partial}{\partial x}\left(\tan^{-1}\frac{y}{x}\right) = \frac{1}{1 + \left(\frac{y}{x}\right)^2}.\left(-\frac{y}{x^2}\right)$$

$$= -\frac{y}{x^2 + y^2} = -\frac{\rho\sin\phi}{\rho^2} = -\frac{\sin\phi}{\rho}$$

Thus, $$\frac{\partial}{\partial x} = \cos\phi = \frac{\partial}{\partial\rho} - \frac{\sin\phi}{\rho}\frac{\partial}{\partial\phi}$$

$$\frac{\partial}{\partial y} = \cos\phi\frac{\partial}{\partial\rho} = \frac{\cos\phi}{\rho}\frac{\partial}{\partial\phi}$$

Thus $$\frac{\partial \Psi}{\partial x} = \frac{\partial \Psi}{\partial \rho}.\frac{\partial \rho}{\partial x} + \frac{\partial \Psi}{\partial \varphi}.\frac{\partial \phi}{\partial x} = \cos\phi\,\frac{\partial \Psi}{\partial \rho} - \frac{\sin\phi}{\rho}\frac{\partial \Psi}{\partial \varphi}$$

Therefore,

$$\frac{\partial^2 \Psi}{\partial x^2} = \frac{\partial}{\partial x}\left(\frac{\partial \Psi}{\partial x}\right) = \left(\cos\phi\frac{\partial}{\partial \rho} - \frac{\sin\varphi}{\rho}\frac{\partial}{\partial \phi}\right)\left(\cos\phi\frac{\partial \Psi}{\partial \rho} - \frac{\sin\phi}{\rho}\frac{\partial \Psi}{\partial \phi}\right)$$

$$= \cos^2\phi\frac{\partial^2 \Psi}{\partial \rho^2} - \frac{\cos\phi\sin\phi}{\rho}\frac{\partial^2 \Psi}{\partial \rho \partial \phi} - \frac{\sin\cos\phi}{\rho}\frac{\partial^2 \Psi}{\partial \rho \partial \phi} + \frac{\sin^2\phi}{\rho^2}\frac{\partial^2 \Psi}{\partial \phi^2}$$

$$- \cos\phi\sin\phi\frac{\partial \Psi}{\partial \phi}\frac{\partial}{\partial \rho}\left(\frac{1}{\rho}\right) - \sin\phi\cos\phi\frac{\partial \Psi}{\partial \varphi}\frac{\partial}{\partial \rho}\left(\frac{1}{\rho}\right) - \frac{\sin\phi}{\rho}\frac{\partial}{\partial \phi}(\cos\phi)\frac{\partial \Psi}{\partial \rho}$$

$$= \cos^2\phi\frac{\partial^2 \Psi}{\partial \rho^2} - 2\frac{\cos\phi\sin\phi}{\rho}\frac{\partial^2 \Psi}{\partial \rho \partial \phi} + \frac{\sin^2\phi}{\rho^2}\frac{\partial^2 \Psi}{\partial \phi^2}$$

$$- 2\cos\phi\sin\phi\left(-\frac{1}{\rho^2}\right) - \frac{\sin\phi}{\rho}(-\sin\phi)\frac{\partial \Psi}{\partial \rho}$$

$$= \cos^2\phi\frac{\partial^2 \Psi}{\partial \rho^2} - 2\frac{\cos\phi\sin\phi}{\rho}\frac{\partial^2 \Psi}{\partial \rho \partial \phi} + \frac{\sin^2\phi}{\rho^2}\frac{\partial^2 \Psi}{\partial \phi^2}$$

$$+ \frac{2\cos\phi\sin\phi}{\rho^2}\frac{\partial \Psi}{\partial \phi} + \frac{\sin^2\phi}{\rho}\frac{\partial \Psi}{\partial \rho}$$

Similarly,

$$\frac{\partial^2 \Psi}{\partial y^2} = \frac{\partial}{\partial y}\left(\frac{\partial \Psi}{\partial y}\right) = \left(\sin\phi\frac{\partial}{\partial \rho} + \frac{\cos\phi}{\rho}\frac{\partial}{\partial \phi}\right)\left(\sin\phi\frac{\partial \Psi}{\partial \rho} + \frac{\cos\phi}{\rho}\frac{\partial \Psi}{\partial \phi}\right)$$

$$= \sin^2\phi\frac{\partial^2 \Psi}{\partial \rho^2} + \frac{\cos\phi\sin\phi}{\rho}\frac{\partial^2 \Psi}{\partial \rho \partial \phi} + \frac{\cos\phi}{\rho}\frac{\partial}{\partial \phi}$$

$$(\sin\phi)\frac{\partial \Psi}{\partial \rho} + \frac{\sin\cos\phi}{\rho}\frac{\partial^2 \Psi}{\partial \rho \partial \phi} + \frac{\cos^2\phi}{\rho^2}\frac{\partial^2 \Psi}{\partial \phi^2} + \sin\phi\cos\phi\frac{\partial}{\partial \rho}\left(\frac{1}{\rho}\right)\frac{\partial \Psi}{\partial \phi}$$

$$+ \cos\phi\sin\phi\,\frac{\partial}{\partial \rho}\left(\frac{1}{\rho}\right)\frac{\partial \Psi}{\partial \phi}$$

$$= \sin 2\,\phi \frac{\partial^2 \Psi}{\partial \rho^2} + 2\frac{\cos\phi \sin\phi}{\rho}\frac{\partial^2 \Psi}{\partial \rho \partial \phi} + \frac{\cos^2 \phi}{\rho}\frac{\partial \Psi}{\partial \rho}$$

$$-2\frac{\sin\phi\cos\phi}{\rho^2}\frac{\partial \Psi}{\partial \phi} + \frac{\cos^2 \phi}{\rho^2}\frac{\partial^2 \Psi}{\partial \phi^2}$$

Adding the two terms we get

$$= \cos^2 \phi \frac{\partial^2 \Psi}{\partial \rho^2} - 2\frac{\cos\phi \sin\phi}{\rho}\frac{\partial^2 \Psi}{\partial \rho \partial \phi} + \frac{\sin^2 \phi}{\rho^2}\frac{\partial^2 \Psi}{\partial \phi^2}$$

$$+\frac{2\cos\phi\sin\phi}{\rho^2}\frac{\partial \Psi}{\partial \phi} + \frac{\sin^2 \phi}{\rho}\frac{\partial \Psi}{\partial \rho} +$$

$$\sin^2 \phi \frac{\partial^2 \Psi}{\partial \rho^2} + 2\frac{\cos\phi \sin\phi r}{\rho}\frac{\partial^2 \Psi}{\partial \rho \partial \phi} + \frac{\cos^2 \phi}{\rho}\frac{\partial \Psi}{\partial \rho}$$

$$-2\frac{\sin\phi\cos\phi}{\rho^2}\frac{\partial \Psi}{\partial \phi} + \frac{\cos^2 \phi}{\rho^2}\frac{\partial^2 \Psi}{\partial \phi^2}$$

$$= (\cos^2 \phi + \sin^2 \phi)\frac{\partial^2 \Psi}{\partial \rho^2} + \left(\frac{\sin^2 \phi}{\rho} + \frac{\cos^2 \phi}{\rho}\right)\frac{\partial \Psi}{\partial \rho}$$

$$+\frac{\sin^2 \phi}{\rho^2}\frac{\partial^2 \Psi}{\partial \phi^2} + \frac{\cos^2 \phi}{\rho^2}\frac{\partial^2 \Psi}{\partial \phi^2}$$

$$= \frac{\partial^2 \Psi}{\partial \rho^2} + \frac{1}{\rho}\frac{\partial \Psi}{\partial \rho} + \frac{\partial^2 \Psi}{\rho^2 \partial \phi^2} = \frac{1}{\rho}\frac{\partial}{\partial \rho}\left(\rho \frac{\partial \Psi}{\partial \rho}\right) + \frac{\partial^2 \Psi}{\rho^2 \partial \phi^2}$$

Example B.3: Convert $\frac{\partial \Psi}{\partial x}$ and $\frac{\partial \Psi}{\partial y}$ to the cylindrical coordinates

Solution: We know that

$$x = \rho \cos \phi \text{ and } y = \rho \sin \phi$$

Thus, $\rho^2 = x^2 + y^2$ and $\phi = \tan^1 (y/x)$

$$\tan \phi = \frac{y}{x}$$

$$\Rightarrow \qquad \cos \phi = \frac{x}{\rho}$$

$$\Rightarrow \qquad \cos^2\phi = \frac{x^2}{\rho^2}$$

We have

$$\Rightarrow \qquad \rho^2 = \frac{x^2}{\cos^2\phi}$$

$$\Rightarrow \qquad 2\rho\frac{\partial\rho}{\partial\phi} = \frac{x^2(2\cos\phi\sin\phi)}{\cos^4\phi} = \frac{x^2\sin\phi}{\cos^3\phi}$$

$$\Rightarrow \qquad \frac{\partial\rho}{\partial\phi} = \frac{x^2\sin\phi}{\rho\cos^3\phi} = (x/\cos\phi)\frac{x^2\sin\phi}{(x/\cos\phi)\cos^3\phi} = \frac{x\sin\phi}{\cos^2\phi}$$

We may write,

$$\frac{\partial\Psi}{\partial x} = \frac{\partial\Psi}{\partial\rho}\frac{\partial\rho}{\partial x}$$

$$\Rightarrow \qquad 2\rho\frac{\partial\rho}{\partial x} = 2x$$

$$\Rightarrow \qquad \frac{\partial\rho}{\partial x} = \frac{x}{\rho} = \frac{x\cos\phi}{x} = \cos\phi$$

Therefore,

$$\frac{\partial\Psi}{\partial x} = \cos\phi\frac{\partial\Psi}{\partial\rho}$$

$$\frac{\partial^2\Psi}{\partial x^2} = \frac{\partial}{\partial x}\left(\frac{\partial\Psi}{\partial x}\right) = \frac{\partial}{\partial\rho}\frac{\partial\rho}{\partial x}\left(\frac{\partial\Psi}{\partial x}\right)$$

Now,

$$= \frac{\partial}{\partial\rho}\cos\phi.\left(\cos\phi\frac{\partial\Psi}{\partial\rho}\right)$$

$$= \cos^2\phi\frac{\partial^2\Psi}{\partial\rho^2} + \frac{\partial\Psi}{\partial\rho}\frac{\partial}{\partial\rho}\left(\cos^2\phi\right)$$

and similarly,

$$\frac{\partial\Psi}{\partial y} = \sin\phi\frac{\partial\Psi}{\partial\rho}$$

❑❑❑

Appendix-C

Bessel Functions

We have made extensive use of Bessel functions in the study of modes in step index fibers in chapter 5 and the diffraction patterns obtained for circular apertures in chapter 8. Bessel functions are obtained as generalized series solutions to the wave equations propagating through waveguides having circular cross sections. The series solutions come as functions having an order and argument. We also get a number of recurrence relations from the functions of different orders and their derivatives. Therefore, we describe the Bessel functions here.

C.1. BESSEL FUNCTIONS

The following differential equation has a solution given as Bessel function. This differential equation is known as Bessel's equation.

$$\frac{d^2\psi}{dz^2} + \frac{1}{z}\frac{d\psi}{dz} + \left(1 - \frac{n^2}{z^2}\right)\psi = 0 \qquad \text{...(C.1)}$$

This is a second order differential equation which has two independent solutions. The first solution is expressed as a series as the following.

$$J_n(z) = \sum_{m=0}^{\infty} \frac{(-1)^m z^{n+2m}}{m!(m+n)!2^{n+2m}} = 0 \qquad \text{...(C.2)}$$

In (C.2), the function $J_n(z)$ is known as Bessel function of first kind and order n. The order of the equation is given by the value of n. This can be an integer or a non integer. In our case, we have considered only integer values for n.

The second independent solution to (C.1) is expressed as

$$K_n(z) = \frac{J_n(z)\cos n\pi - J_{-n}(z)}{\sin n\pi} \qquad ...(C.3)$$

In (C.3) K_n (z) is known as a Neumann function or Bessel function of second kind of order n. Hence, the complete solution to (C.1) is,

$$\Psi = AJ_n(z) + BK_n\,(z) \qquad ...(C.4)$$

Bessel functions of the second kind assume infinite values at z = 0 and hence cannot be used to represent physical fields except in those cases for which the region z = 0 is excluded.

C.1.1. Differentiation and Integration of Bessel functions

We may note from (C.2) that $J_0(z)$ is expressed as

$$J_0(z) = 1 - \frac{1}{1!}\left(\frac{z}{2}\right)^2 + \frac{1}{(2!)^2}\left(\frac{z}{2}\right)^4 - \frac{1}{(3!)^2}\left(\frac{z}{2}\right)^6 + ... \qquad ...(C.5)$$

and

$$J_1(z) = \frac{z}{2}\left[1 - \frac{1}{1!2!}\left(\frac{z}{2}\right)^2 + \frac{1}{2!3!}\left(\frac{z}{2}\right)^4 - \frac{1}{3!4!}\left(\frac{z}{2}\right)^6 + ...\right] \qquad ...(C.6)$$

We note that, if we differentiate (C.5) w.r.t. z once, then we have

$$\frac{d}{dz}J_0(z) = -\left(\frac{z}{2}\right) + \frac{4}{(2!)^2.2}\left(\frac{z}{2}\right)^3 - \frac{6}{(3!)^2.2}\left(\frac{z}{2}\right)^5 + \frac{8}{(4!)^2.2}\left(\frac{z}{2}\right)^7 - ...$$

$$= -\left(\frac{z}{2}\right)\left[1 - \frac{1}{1!2!}\left(\frac{z}{2}\right)^2 + \left(\frac{1}{2!3!}\right)\left(\frac{z}{2}\right)^4 - \frac{8}{4!3!}\left(\frac{z}{2}\right)^6 + ...\right]$$

$$= -J_1(z) \qquad ...(C.7)$$

We can similarly show that the following relations are true in general.

Let us denote the differentiation w.r.t. z as $J_n(z) = \frac{d}{dz}J_n(z)$. Hence,

$$J_0'(z) = -J_2(z) \qquad ...(C.8)$$

$$J_1'(z) = J_0(z) - \frac{1}{z}J_1(z) \qquad ...(C.9)$$

$$zJ_n'(z) = nJ_n(z) - zJ_{n+1}(z) = zJ_{n-1}(z) - nJ_n(z) \qquad ...(C.10)$$

$$\frac{d}{dz}[z^n J_n(z)] = z_n\, J_{n-1}(z) \qquad ...(C.11)$$

$$\frac{d}{dz}[z^{-n}J_n(z)] = -z^{-n}J_{n+1}(z) \qquad \text{...(C.12)}$$

$$2nJ_n(z) = z[J_{n-1}(z) + J_{n+1}(z)] \qquad \text{...(C.13)}$$

Similarly, the integrals of the Bessel functions are expressed as

From (C.14), $$\int J_1(z)dz = -J_0(z) \qquad \text{...(C.8)}$$

From (C.11), $$\int z^n J_{n-1}(z)dz = z^n J_n(z) \qquad \text{...(C.15)}$$

From (C.12), $$\int z^{-n} J_{n+1}(z)dz = -z^{-n}J_n(z) \qquad \text{...(C.16)}$$

C.1.2. Asymptotic Form of Bessel Functions

We observe that for $z \ll 1$, the J and N functions are approximated as

$$J_n(z) \approx \frac{z^n}{2^n n!} \qquad \text{...(C.17)}$$

$$N_0(z) \approx \frac{2^n(n-1)!}{\pi z^n} \qquad \text{...(C.18)}$$

In particular,

$$J_0(z) \approx 1 \qquad \text{...(C.19)}$$

$$N_0(z) \approx \frac{2}{\pi}(\ln z + C - \ln 2) \qquad \text{...(C.20)}$$

for a constant C.

However, for the argument $z \gg 1$, the asymptotic expressions become

$$J_n(z) \approx \sqrt{\frac{2}{\pi z}}\cos\left(z - \frac{n\pi}{2} - \frac{\pi}{4}\right) \qquad \text{...(C.21)}$$

$$K_n(z) \approx \sqrt{\frac{2}{\pi z}}\sin\left(z - \frac{n\pi}{2} - \frac{\pi}{4}\right) \qquad \text{...(C.22)}$$

C.2. MODIFIED BESSEL FUNCTIONS

The modified Bessel's equation of order n is

$$\frac{d^2\psi}{dz^2} + \frac{1}{z}\frac{d\psi}{dz} - \left(1 + \frac{n^2}{z^2}\right)\psi = 0 \qquad \text{...(C.23)}$$

As (C.23) is a second order differential equation, like the previous equation, it has two independent solutions denoted as $I_n(z)$ and $K_n(z)$ expressed as

$$I_n(z) = \sum_{m=0}^{\infty} \frac{z^{n+2m}}{m!(m+n)!2^{n+2m}} \qquad \text{...(C.24)}$$

$$K_n(z) = \frac{2}{\cos n\pi}\left(\frac{\partial I_{-n}}{\partial n} - \frac{\partial I_n}{\partial n}\right) \qquad \text{...(C.25)}$$

C.2.1. Asymptotic Form of Modified Bessel Functions

When the argument of the functions is $z \ll 1$, the modified Bessel's functions are approximated as

$$I_n(z) \approx \frac{z^n}{2^n n!} \qquad \text{...(C.26)}$$

$$K_n(z) \approx \frac{2^{n-1}(n-1)!}{z^n} \qquad \text{...(C.27)}$$

We may note that, for $z \ll 1$,

$$J_n(z) = I_n(z) \approx \frac{z^n}{2^n n!} \qquad \text{...(C.28)}$$

and

$$K_n(z) = \frac{\pi}{2}K_n(z) \approx \frac{2^{n-1}(n-1)!}{z^n} \qquad \text{...(C.29)}$$

In particular,

$$I_0(z) \approx 1 \qquad \text{...(C.30)}$$

$$K_0(z) \approx -(In\ z + C - ln\ 2) \qquad \text{...(C.31)}$$

For $z \gg 1$, $I_n(z) \approx \infty$ and $K_n(z) \approx 0$

C.2.2. Recurrence Relations for Modified Bessel Functions

$$zI'_n(z) = nI_n(z) + zI_{n+1}(z) \qquad \text{...(C.32)}$$

$$zK'_n(z) = nK_n(z) - zk_{n+1}(z) \qquad \text{...(C.33)}$$

$$zI'_n(z) = zI_{n-1}(z) - n\ In_n(z) \qquad \text{...(C.34)}$$

$$zK'_n(z) = -\ zK_{n-1}(z) - nK_n(z) \qquad \text{...(C.35)}$$

$$\frac{d}{dz}[z^n I_n(z)] = z^n I_{n-1}(z) \quad ...(C.36)$$

$$\frac{d}{dz}[z^n K_n(z)] = -z^n K_{n-1}(z) \quad ...(C.37)$$

$$\frac{d}{dz}[z^{-n} I_n(z)] = z^{-n} I_{n+1}(z) \quad ...(C.38)$$

$$\frac{d}{dz}[z^{-n} K_n(z)] = -z^{-n} K_{n+1}(z) \quad ...(C.39)$$

$$2nI_n(z) = z[I_{n-1}(z) - I_{n+1}(z)] \quad ...(C.40)$$

$$2nK_n(z) = -z[K_{n-1}(z) - K_{n+1}(z)] \quad ...(C.41)$$

From the asymptotic form of the Bessel and modified Bessel functions, it is obvious that the J and K functions correspond to cosine and sine functions that represent waves when used with the time harmonic factor exp ($j\omega t$). On the other hand, the modified Bessel functions represented as I and K are more appropriate for dissipative media. For lossless dielectrics, the arguments of these functions assume purely imaginary values, thereby reducing them to normal Bessel functions. For imaginary arguments, the relations between the normal and modified Bessel functions are

$$I_n(jz) = \exp\left(\frac{jn\pi}{2}\right) J_n(z) \quad ...(C.42)$$

$$K_n(jz) = \frac{\pi}{2} \exp\,[-j(n+1)\,\pi/2][J_n(z) - jK_n(z)] \quad ...(C.43)$$

Task: Establish relations (C.42) and (C.43) and give a physical interpretation of the two functions.

□□□

Appendix-D

TE Modes in Metallic Waveguides

For the TE modes propagating along the positive z-axis, there is no component of the electric field along this axis. Mathematically speaking, $E_z = 0$. Physically it means that the electric field is entirely transverse to the direction of propagation. The magnetic field has all three components; namely H_ρ, H_ϕ, H_z for a cylindrical waveguide. The TE modes are characterized by the presence of a transverse component of the electric field. Hence we solve the various field expressions starting from the longitudinal magnetic field. We write the generalized expression for the wave along the z direction as

$$\mathbf{H} = H_0 e^{-j\beta z}\,\hat{\mathbf{z}} \qquad \text{...(D.1)}$$

For a wave propagating along the z-axis, h_z is a function of (ρ, ϕ) only. Thus, to include the effects of time variation and the electric field has components along both the ρ and ϕ direction, we proceed as follows.

$$H_z = h_z(\rho, \phi)e^{-j\beta z}e^{j\omega t} \qquad \text{...(D.2)}$$

D.1. WAVE EQUATION AND ITS SOLUTION

The Laplacian in the cylindrical coordinates is derived in Appendix-B. We consider the cylindrical coordinates as the cross section of the fibre is assumed to be circular. he Laplacian of this field given in (D.2) gives us

$$\nabla^2 H_z + k^2 H_z = 0 \qquad \text{...(D.3)}$$

where $k = \sqrt{\mu\varepsilon}$

Substitution of this in the Laplacian function, we get

$$\nabla^2\Phi = \frac{1}{\rho}\left(\frac{\partial}{\partial\rho}\rho\frac{\partial H_z}{\partial\rho} + \frac{1}{\rho}\frac{\partial^2 H_z}{\partial\phi^2} + \rho\frac{\partial^2 H_z}{\partial z^2}\right) + k^2 H_z = 0$$

$$\Rightarrow \qquad \frac{1}{\rho}\frac{\partial}{\partial\rho}\rho\frac{\partial H_z}{\partial\rho} + \frac{1}{\rho^2}\frac{\partial^2 H_z}{\partial\phi^2} - \beta^2 H_z + k^2 H_z = 0$$

$$\Rightarrow \qquad \left(\frac{1}{\rho}\frac{\partial}{\partial\rho}\rho\frac{\partial}{\partial\rho} + \frac{1}{\rho^2}\frac{\partial^2}{\partial\phi^2} - \beta^2 + k^2\right)h_z = 0$$

$$\Rightarrow \qquad \left(\frac{1}{\rho}\frac{\partial}{\partial\rho} + \frac{\partial^2}{\partial\rho^2} + \frac{1}{\rho^2}\frac{\partial^2}{\partial\phi^2} + k_c^2\right)h_z = 0 \qquad \text{...(D.4)}$$

where $\quad k_c^2 = k^2 - \beta^2 \qquad$ ...(D.5)

Applying the method of separation of variables

$$h_z = P(\rho)\ Q(\phi)\ e^{-j\beta z} \qquad \text{...(D.6)}$$

where $P(\rho)$ is a function of alone and $Q(\phi)$ is a function of ϕ alone.

The above equation reduces to

$$\frac{Q}{\rho}\frac{\partial P}{\partial\rho} + Q\frac{\partial^2 P}{\partial\rho^2} + \frac{P}{\rho^2}\frac{\partial^2 Q}{\partial\phi^2} + k_c^2 PQ = 0 \qquad \text{...(D.7)}$$

Multiplying (D.7) by $\frac{\rho^2}{PQ}$, we obtain

$$\frac{\rho}{P}\frac{\partial P}{\partial\rho} + \frac{\rho^2}{P}\frac{\partial^2 P}{\partial\rho^2} + \frac{1}{Q}\frac{\partial^2 Q}{\partial\phi d^2} + \rho^2 k_c^2 = 0 \qquad \text{...(D.8)}$$

We break up (D.8) into two parts, one part being a function of ϕ only. The other is a function of ρ.

$$\frac{1}{Q}\frac{\partial^2 Q}{\partial\phi^2} = -k_\phi^2$$

$$Q(\phi) = A\cos k_\phi\phi + B\sin k_\phi\phi \qquad \text{...(D.9)}$$

Here k_ϕ must be an integer otherwise the function Q is multivalued. Thus,

$$Q(\phi) = A\cos n\,\phi + B\sin n\,\phi \qquad \text{...(D.10)}$$

Similarly, the other solution gives us

$$\frac{\rho}{P}\frac{\partial P}{\partial \rho}+\frac{\rho^2}{P}\frac{\partial^2 P}{\partial \rho^2}+\frac{1}{Q}\frac{\partial^2 Q}{\partial \phi^2}+\rho^2 k_c^2=0$$

$$\Rightarrow \qquad \frac{\rho}{P}\frac{\partial P}{\partial \rho}+\frac{\rho^2}{P}\frac{\partial^2 P}{\partial \rho^2}-n^2+\rho^2 k_c^2=0$$

$$\Rightarrow \qquad \rho\frac{\partial P}{\partial \rho}+\rho^2\frac{\partial^2 P}{\partial \rho^2}+(\rho^2 k_c^2-n^2)P=0 \qquad \text{...(D.11)}$$

Solution of the above Bessel's differential equation is

$$P(\rho) = CJ_n(k_c\rho) + DK_n(k_c\rho) \qquad \text{...(D.12)}$$

where $J_n(j_c\rho) = \sum_{m=0}^{\infty} \frac{(-1)^m \left[k_c(\frac{\rho}{2})\right]^{2m+n}}{m!(m+n)!}$ is Bessel's function of first kind of th

order and $K_n(k_c\rho) = \frac{J_n(k_c\rho)\cos n\pi - J_{-n}(k_c\rho)}{\sin n\pi}$ is modified Bessel's function of second kind of n th order. Also,

$$J_{-n}(J_c\rho) = \sum_{m=0}^{\infty} \frac{(-1)^m \left[k_c(\rho/2)\right]^{2m-n}}{m!(m-n)!} \qquad \text{...(D.13)}$$

The longitudinal component of the magnetic field, therefore becomes

$$H_z(\rho,\phi,z) = (A \cos n\phi + B \sin n\,\phi)[CJ_n(k_c\rho) + DK_n(k_c\rho)]e^{-j\beta z} \qquad \text{...(D.14)}$$

D.2. ELECTRIC AND MAGNETIC FIELD COMPONENTS IN THE CORE REGION ($\rho < a$)

The modified Bessel functions of second kind $K_n(k_c\rho)$are all infinite at the origin where $\rho = 0$. The value of the arbitrary constant D must be equal to zero if the field H_z is to remain finite at the origin. Thus, the electric field in the core region corresponding to ($\rho < a$) is represented by the Bessel function of first kind of order n. We note that,

$$k^2_{\text{c}} = k^2 - \beta^2 = \omega^2 \mu\varepsilon - \beta^2 \qquad \text{...(D.15)}$$

One of the two terms of $Q(\phi)$ may be eliminated. Both have the same field patterns except for a rotation of 90° in the ϕ direction. The excitation also can be chosen in such a way that only one field is present at any time. Thus, the magnetic field expression is further simplified as

$$H_z(\rho, \phi, z) = A \cos n\phi J_n(k_c\rho)e^{-j\beta z} \quad \text{...(D.16)}$$

Since the tangential component of the electric fields are continuous across the boundary of the waveguide (the electric field is zero inside the metallic walls of the waveguide), hence

$$E_\phi\Big|_{\rho=a} = \frac{j\omega\mu}{k_c^2}\frac{\partial H_z}{\partial \rho}\Bigg|_{\rho=a} = \frac{j\omega\mu k_c}{k_c^2} A\cos n\phi\, J'_n(k_c\rho)e^{-j\beta z}\Bigg|_{\rho=a}$$

$$J'_n(k_c\rho)\Big|_{\rho=a} = 0$$

$$k_c a = P'_{nm} \quad \text{...(D.17)}$$

The m th root of the derivative of the n th order Bessel's function of first kind $J'_n(k_c\rho)$ is denoted by ρ'_{nm} and $k_c a = p'_{nm}$. The general expression for the component of the magnetic field along the direction of propagation is given as

$$H_z(\rho, \phi, z) = (A \cos n\phi)\, J_n(k_c \rho)e^{-j\beta z} \quad \text{...(D.18)}$$

The solution to the wave equation given as a second order differential equation was expected to contain the Bessel's function because of the circular symmetry. This is verified in (D.18).

The electric field has a component along the radius of the optical fiber given as

$$E_\rho = \frac{-j\omega\mu}{k_c^2\rho}\frac{\partial H_z}{\partial \phi} = \frac{j}{k_c^2}\frac{\omega\mu n}{\rho} J_n(k_c\rho)(A\sin\phi)e^{-j\beta z} \quad \text{...(D.19)}$$

$$E_\phi = \frac{j\omega\mu}{k_c^2}\frac{\partial H_z}{\partial \rho} = \frac{j\omega\mu}{k_c} A\cos n\phi\, J'_n(k_c\rho)e^{-j\beta z} \quad \text{...(D.20)}$$

The radial magnetic field is expressed as

$$H_\rho = \frac{-j}{k_c^2}\beta\frac{\partial H_z}{\partial \rho} = \frac{-j\beta}{k_c} A\cos n\phi\, J'_n(k_c\rho)e^{-j\beta z} \quad \text{...(D.21)}$$

The magnetic field in space quadarature to this magnetic field is

$$H_\phi = \frac{-j\beta}{k_c^2\rho}\frac{\partial H_z}{\partial \rho} = \frac{j\beta n}{k_c^2\rho}(A\sin\phi)\, J_n(k_c\rho)e^{-j\beta z} \quad \text{...(D.22)}$$

where $$k_c = \frac{p'_{nm}}{a} \quad \text{...(D.23)}$$

For the TE_{nm} mode,

$$\beta_{nm} = \sqrt{k^2 - k_c^2} = \sqrt{k^2 - \left(\frac{p'_{mn}}{a}\right)^2} \quad \text{...(D.24)}$$

For cut-off frequency, $\beta_{nm} = 0$ which means that

$$f_c = \frac{1}{2\pi\sqrt{\mu\varepsilon}} \frac{p'_{nm}}{a} \quad \text{...(D.25)}$$

The mode with the lowest cut-off frequency is known as the dominant mode. Dominant mode is the TE_{11} mode for p_{11} as it has the smallest value of 1.841 with a corresponding cut off frequency of

$$f_{c11} = \frac{1}{2\pi\sqrt{\mu\varepsilon}}\left(\frac{1.841}{a}\right) \quad \text{...(D.26)}$$

The field expressions for E and H do not exist for $m = n = 0$. Thus, there is no TE_{00} mode propagating inside the circular waveguide. The wave impedance is

$$Z_{TE} = \frac{E_\rho}{H_\phi} = \frac{\omega\mu}{\beta} \quad \text{...(D.27)}$$

The guided wavelength is defined as

$$\lambda_g = \frac{2\pi}{\beta} \quad \text{...(D.28)}$$

and the phase velocity is defined as

$$v_p = \frac{\omega}{\beta} = \frac{\omega}{\sqrt{k^2 - \left(\frac{p'_{nm}}{a}\right)^2}} \quad \text{...(D.29)}$$

We note that $c = \frac{\omega}{k}$. But for the phase velocity inside the waveguide, the denominator is a smaller number than k. Hence, the phase velocity is greater than the velocity of light. This does not violate Einstein's law, since energy and information flow is dependent on the group velocity. Group velocity is defined as

$$E_z = J_n(k_c\rho)(C_n \cos n\phi + D_n \sin n\phi)\, e^{-j\beta z} \qquad \text{...(E.15)}$$

All the fields are expressed as

$$E_\rho = -\frac{j\beta}{k_c^2}\frac{\partial E_z}{\partial \rho} = -\frac{j\beta}{k_c^2} J'_n(k_c\rho) A \cos n\phi e^{-j\beta z} \qquad \text{...(E.16)}$$

$$E_\phi = -\frac{j\beta}{k_c^2\rho}\frac{\partial E_z}{\partial \phi} = -\frac{j\beta n}{k_c^2\rho} J_n(k_c\rho) A \sin n\phi e^{-j\beta z} \qquad \text{...(E.17)}$$

$$H_\rho = -\frac{j\omega\varepsilon}{k_c^2\rho}\frac{\partial E_z}{\partial \phi} = -\frac{j\omega\varepsilon n}{k_c^2\rho} J_n(k_c\rho) A \sin n\phi e^{-j\beta z} \qquad \text{...(E.18)}$$

$$H_\phi = -\frac{j\omega\varepsilon}{k_c^2\rho}\frac{\partial E_z}{\partial \rho} = \frac{j\beta n}{k_c} J'_n(k_c\rho) A \cos n\phi e^{-j\beta z} \qquad \text{...(E.19)}$$

The dominant mode corresponding to the TM waves is the TM_{01} mode for which the cut off frequency can be shown to be

$$f_c = \frac{1}{2\pi\sqrt{\mu\varepsilon}}\left(\frac{2.405}{a}\right) \qquad \text{...(E.20)}$$

where is the radius of the circular cross section.

Task 1: Establish the relations given in (E.16) to (E.19).

Task 2: Prove that the cut off frequency for the lowest order TM mode is satisfied by (E.20).

❑❑❑

Appendix-F

Some Useful Physical Constants

Parameter	*Symbol*	*Value*
Angstrom	λ	$1Å = 10^{-8}$ cm $= 10^{-10}$ m $= 0.1$ nm
Permittivity of free space	ε_0	8.850×10^{-12} F/m
Permeability of free space	μ_0	1.257×10^{-6} H/m
Velocity of light in vacuum	c	2.998×10^{8} m/s
Intrinsic impedance of free space	η_0	$120\pi = 377\ \Omega$
Electronic charge	q	1.602×10^{-19} C
Electronic mass	m	9.109×10^{-31} kg
Boltzman constant	K	1.381×10^{-23} J/K $= 8.620 \times 10^{-5}$ eV
Plank's constant	$h/2\pi = h$	1.054×10^{-34} J-S
Volt equivalent of temperature	$\frac{KT}{q} = V_T$	0.02586 V $= 26$ mV

100 Important Points

1. An optoelectronic system comprises of an optical source, an optical fibre and a photodetector. The source is an electro optic converter that generates light, the fibre carries the light over a certain distance and the photodetector is an opto electronic converter. The carrier of information is light.
2. As light occurs in the 10^{14} Hz of the electromagnetic spectrum, as a carrier, it offers enormous amount of transmission bandwidth to carry information in the form of very fast occurring digital pulses.
3. The three windows suitable for building an optical fibre based optoelectronic system are centred around 850, 1300 and 1550 nm due to the fact that signal loss is minimal at these wavelengths.
4. Very high data rate information transmission/reception is carried out at 1550 nm range of optical wavelength whereas lower bit rate data is supported by 800 nm.
5. An optical fibre is constructed from glass or plastic and has two basic units: the core and the cladding that provides a sheath to the core and lies outside it. The core has a higher refractive index than that of the cladding. Glass fibres are made from either silica (SiO_2) or silicate.
6. Light propagating in free space or vacuum is very different from light travelling in bounded media. The former is known as unguided wave whereas in the latter it becomes guided waves or some other type.
7. The constituent components of light wave are the electric field and the magnetic field. Their relative orientation in space with respect to the direction of wave propagation determines the type of wave.
8. Light waves travel as uniform plane waves in free space meaning the constituent components form a plane that travels transverse to the direction of wave propagation. This is also known as transverse electromagnetic waves. The electric field and the magnetic field aretransverse to each other also.

9. The term "uniform" in uniform plane waves signifies constant phase for all points lying on the plane.

10. A wave travelling in an arbitrary direction $\mathbf{k}$ is expressed as $\exp[j(\omega t - \mathbf{k.r})]$ where $\mathbf{r}$ is the position vector from the origin of reference.

11. $\mathbf{k.r} = k_x x + k_y y + k_z z = \beta(x\cos\phi_x + y\cos\phi_x + z\cos\phi_z)$. $\cos\phi_x$, $\cos\phi_y$ and $\cos\phi_z$ are known as the direction cosines. Therefore, $k_x = \beta_x \cos\phi_x = \beta\cos\phi_x$ and $\beta = 2\pi/\lambda$ is known as the propagation constant with λ being the free space wavelength.

12. A simple TEM wave may be described as $\mathbf{E} = E_0\cos(\omega t - \beta z)\hat{x}$, $\mathbf{H} = \frac{E_0}{\eta}\cos(\omega t - \beta z)\hat{y}$ where the direction of wave propagation is in the positive z – axis. The three entities namely, the electric field, the magnetic field and the direction of wave travel are assumed to follow the right handed corkscrew system.

13. The average power flow density in a wave is obtained by writing $\frac{1}{2}\text{Re}\{\mathbf{E}\times\mathbf{H}^*\}$.

14. When a plane wave meets the boundary of two different media, a part of it is reflected back into the medium of incidence and the rest part is transmitted or refracted into the second medium following Snell's law. A medium is characterized by its relative permittivity ε_r and relative permeability μ_r.

15. The velocity of a wave in a given medium is expressed as $v = \frac{1}{\sqrt{\mu\varepsilon}} = \frac{1}{\sqrt{\mu_0\mu_r\varepsilon_0\varepsilon_r}}$. Hence, the velocity of light in free space is 3×10^8 m/s.

16. The intrinsic impedance of a medium that it offers to a wave propagating through it is given as $\eta = \sqrt{\frac{\mu}{\varepsilon}}$. Thus, the intrinsic impedance of free space is 120π or 377Ω.

17. The velocity of light is reduced to $v = c/n$ when it travels in a medium having a refractive index n.

18. The basic wave equation is governed by a second order differential equation in space and time given as $\nabla^2\Psi = \frac{1}{v^2}\frac{\partial^2\Psi}{\partial t^2}$. This is also known as the scalar wave equation.
19. The amount of light reflected and transmitted at the boundary of two different media is determined by the coefficient of reflection and coefficient of transmission respectively. These coefficients are also known as Fresnel's coefficients.
20. A given wave incident at the boundary of two different media can do so either normal or oblique to the plane of incidence. For the latter, if the electric field is entirely transverse to the plane of incidence, the wave is known as transverse electric (TE). This is also known as perpendicular polarization.
21. For parallel polarization, in which the electric field is aligned along the plane of incidence, a transverse magnetic (TM) wave results. This is another type of oblique incidence.
22. Brewster's angle is that angle of incidence for which there is no reflection to the first medium and all of the incident wave is transmitted into the second medium. This angle does not exist if the two media happen to be two magnetic media with different permeabilities for the case of TM waves.
23. Information transfer through optical fibres by light waves is due to total internal reflection. This gives rise to surface waves. A noteworthy component of this is an evanescent wave which is characterized by a fast exponential decay with distance in the second medium.
24. The numerical aperture of a fibre determines the light gathering capability of it. It is proportional to the refractive index difference existing in it. The angle of incidence at the air-core interface decides which components of light would get through the fibre.
25. There are basically two types of optical sources; the light emitting diode (LED)s and the lasers. The LEDs produce non coherent light over a broad spectral width spanning several tens of optical wavelengths. The laser emits coherent light over a narrow spectral width.
26. Optical sources are basically square law devices meaning a linear variation in drive current results in a corresponding linear change in optical output power.
27. A lambertian source emits isotropically within the source medium. The term isotropic here refers to equal optical radiance intensity across its surface when viewed perpendicular to the source surface.

This is different from a collimated source that emits light only into a very narrow range of angles about the normal to its surface.

28. LEDs are the preferred choice of optical sources for systems operating at 100 to 200 Mbps when optical power in tens of microwatts needs to be coupled to multimode fibres. Lasers serve as the sources for optical communication systems working at bandwidths higher than 200 MHz.
29. Semiconductor materials are used to manufacture both types optical sources. An important material is AlGaAs and its compounds. For operation in the 800- to 900-nm wavelength region, the principal LED material is $Ga_{1-x}Al_xAs$. The quaternary alloy $In_{1-x}Ga_xAs_yP_{1-y}$ is the primary material for LEDs emitting at longer wavelengths.
30. The emission is LEDs is spontaneous whereas it is stimulated in lasers.
31. High radiance and high quantum efficiency are the two primary requirements of an LED that further dictate a high radiative type of recombination to take place in the active region of the PN junction.
32. Internal quantum efficiency decides the amount of optical power emitted from a given source and is directly proportional to the mean carrier lifetime.
33. LED configurations fall into two categories such as homojunction and single and double heterojunctions in order to achieve the two requirements as stated in 31.
34. LEDs are broadly divided into surface emitting (SLED) and edge emitting (ELED) types emitting different power patterns.
35. The frequency response or the modulation capability of an LED is determined by its diffusion capacitance. The 3-dB modulation bandwidth is inversely proportional to the effective carrier lifetime.
36. Lasers work on population inversion wherein a greater carrier concentration is available at excited higher energy states than those at lower energy levels. This requires pumping of external power. Stimulated emission in semiconductor lasers is due to optical transitions between valence and conduction band energy distributions.
37. Lasers are capable of coupling milliwatts of useful luminescent power into optical fibres with small physical sizes.
38. Semiconductor lasers typically utilize the double heterojunction structures. The materials used are compounds of group III-IV materials of the periodic table such GaAlAs, GaAs, InP and InGaAsP.

39. Modulation capability of a laser is determined by the carrier and photon lifetime parameters associated with its operation.
40. The angular distribution of source brightness is expressed as $B = B_0(\cos\theta)^m, \theta < \theta_{max}$, where θ_{max} is the maximum angle from the normal at which light is emitted and is determined by source geometry. For lambertian sources, $m = 1$ and it is higher for collimated sources.
41. An optical fibre is modelled as either a planer or a cylindrical waveguide. Propagation of light through the core by total internal reflection is understood by solving Maxwell's equations in rectangular/cylindrical coordinate system of coordinates by invoking appropriate boundary conditions. Solution of the wave equation gives rise to axial or longitudinal components for one/both the electric and magnetic fields.
42. Two basic types of fibres are manufactured: the step index and the graded index. The refractive indices of the core and the cladding change in a step like manner in the former. The change in refractive indices is gradual as governed by some exponent law in a graded index type.
43. Solution of wave equation in step index fibres with circular cross section gives rise to a combination of Bessel's function and modified Bessel's function. They are evaluated differently for the core and the cladding.
44. Wave propagation in a fibre takes place in the form of modes which are different from uniform plane waves. The modes correspond to particular electromagnetic field patterns usually sinusoidal waves having finite zeros in the core.
45. Each mode is discrete in the sense that it is characterized by integer mode numbers (m,n) and a characteristic propagation constant β_{mn}.
46. The four basic modes are the TE_{mn}, TM_{mn}, HE_{mn} and EH_{mn} modes. For the TE/TM modes, the electric/magnetic field is entirely transverse to the direction of wave travel. The HE and EH modes have axial components of both the fields along the direction of wave travel. The integer 'n' is associated with the roots of the Bessel function for a given integer 'm'.
47. Each mode is characterized by a cutoff frequency except the HE mode. The cutoff frequency determines the dominant mode of operation.
48. Fibres are also classified as single mode and multi mode depending on the number of modes that are possible in the core. Usually, a

step index fibre with a core diameter of 8-10μm supports just one mode and larger diameters support many modes. The graded index fibre inherently is multimoded.

49. The single mode that is supported by the single mode fibre is the HE_{11} mode which does not have any cutoff frequency.
50. The permissible value of β must satisfy $n_2k \le \beta \le n_1k$ in order for a mode to be guided by the core only.
51. The principle of a single mode fibre is dictated by $V = \frac{2\pi a}{\lambda}\sqrt{n_1^2 - n_2^2} \le 2.405$ which is satisfied by the HE_{11} mode only.
52. The constant 2.405 is the first root of order zero Bessel's function of first kind; $J_0(x)$.
53. The propagation constant β varies from mode to mode and each mode propagates with its own characteristic phase and group velocity.
54. For a given mode, phase velocity is given as $v_p = \omega/\beta = \frac{c}{n}$ whereas group velocity is expressed as $v_g = \frac{d\omega}{d\beta}$.
55. Wave propagation in graded index fibres is carried out by the WKB method. It is found that only certain discrete values of propagation constant satisfy the boundary conditions and give rise to allowed modes of propagation.
56. The scalar fields in the step index fibre are proportional to $J_n(ua)\cos n\phi$ in the core, $K_n(wa)\cos n\phi$ in the cladding. For a graded index fibre with a parabolic refractive index profile, the scalar field is proportional to $\exp\left(j\int K d\rho\right)$ where K is related to fibre constants and the frequency of the mode.
57. Light waves travel as axial rays in single mode fibres, multiple rays in multimode step index fibres and sinusoidal waves in graded index multimode fibres having a square law refractive index profile.
58. The optical pulse launched onto a fibre suffers from attenuation and dispersion. A pulse undergoes spreading in time domain which makes its duration longer than the original transmitted pulse. A significant pulse spreading causes intersymbol interference.
59. The two most common pulse dispersion types are the intermodal and intramodal. Various modes interact with each other in a

multimode fibre and different rays/waves take different times to traverse a given length of the fibre resulting in pulse widening at the fibre output end.

60. The intermodal dispersion also known as multipath dispersion is due to the reason cited in 53.
61. The intramodal dispersion occurs for a given mode and is further divided into types: material and waveguide dispersion. The refractive index becomes a function of wavelength giving rise to material dispersion.
62. Waveguide dispersion is determined by the fibre design. This is due to the fact that the propagation constant β becomes a function of core radius to operating wavelength.
63. An optical pulse undergoes absorption and scattering resulting in signal loss while travelling down a given length of fibre.
64. The amount of power coupled from a source onto a fibre is subject basically to three types of losses which are geometrical, Fresnel and angular.
65. Directional couplers are used for combining and dividing optical power between a number of fibres.
66. In butt coupling the fibre is fixed directly on the emitter with a matching index gel and glue in order to minimize Fresnel reflection losses at the fibre input end.
67. Butt coupling provides near optimal coupling of power when the source radius is higher than the core size. For emitting areas smaller than the fibre core radius, lensing arrangement provides more coupled power from source to fibre.
68. A splicer provides a permanent joint whereas a connector provides a demountable joining between fibres.
69. Interference between two waves due to their superposition occurs by division of wavefront as is the case with point sources and narrow slits or by division of amplitude due to reflected waves from upper and lower surfaces of thin films.
70. The intensity in an interference/diffraction follows a cosine square pattern known as 'Malu's law'.
71. Diffraction by a circular aperture results in alternate dark and bright fringes that gives rise to an intensity pattern proportional to Bessel's function first kind, first order.
72. GaAlAs-GaAS sources and Si detectors are used at 850 nm, InGaAsP-InP sources and either Ge or InGaAs-InP detectors are used at 1300 and 1550 nm.

73. Photodetectors are the receivers for optical sources. They convert an incident optical power to a corresponding current. The two basic types of semiconductor photodetectors are the P-I-N (p type-intrinsic-n type) and the avalanche photodiode (APD).
74. The two primary principles of optoelectronic conversion are photoconductivity and photovoltaic. The former gives a passive type of detector wherein an incident optical power results in a corresponding change in resistivity/conductivity of certain class of materials. The photovoltaic effect gives rise to active photodetectors that convert optical power into an appropriate dc voltage.
75. Both the P-I-N and APD photodetectors are normally operated at reverse bias condition. The high electric field present across the depletion layer of an APD results in carrier regeneration process similar to an avalanche which becomes a source of noise also known as avalanche noise figure.
76. The term $(I_{s1}+I_{s2})$ represents the total dark current (the bulk and surface leakage current that flows in the absence of any input optical power). Normally $I_{s1}=qA\left(\frac{D_n n_{po}}{L_n}+\frac{D_p p_{no}}{L_p}\right)=qA\left(\frac{D_n}{L_n n_A}+\frac{D_p}{L_p n_D}\right)n_i^2$ and $I_{s2}=qAn_i\frac{W}{\tau_0}$.
77. Both P-I-N and APD detectors should satisfy $E_g<\frac{1.24}{\lambda}(\mu m.eV)$.
78. Silicon based photodetectors are normally used to manufacture P-I-N detectors for wavelengths less than 1.0 μm, for wavelengths up to 1800 nm Ge based detectors are the choice, InGaAs-InP based P-I-N detectors are preferred at 1300 and 1550 nm.
79. The performance metrics for a photodetector are (i) quantum efficiency that indicates the efficiency of optoelectronic conversion, (ii) speed of response and (iii) the signal to noise ratio (SNR) available at the detector output.
80. High quantum efficiency calls for a low surface reflectivity, a depletion layer matched in thickness and position to the absorption region and minimal carrier recombination. This efficiency varies from 50% to 95% in practical detectors.
81. APDs offer higher sensitivity to low optical powers.
82. The main noise sources in a photodetector are (i) quantum noise, (ii) avalanche noise due to dark current resulting from bulk and surface currents , (iii) surface dark current and (iv) thermal noise.
83. The noise sources other than the thermal noise are noises generated due to the very process of optoelectronic conversion that ultimately limit the sensitivity of a photodetector.

84. The speed of response is determined by the width of the depletion layer, junction area and the RC time constant of associated electrical circuits.
85. PMT type of photodetectors working on photoelectric effect provide very large internal gains (of the order of 10^6) and bandwidths in excess of 1 GHz.
86. Optical sensors utilize electrooptic, optical birefringence, photoelectric, magneto optic effects of light.
87. Optical sensors typically utilize intensity, phase, polarization and wavelength modulation of a given light beam in intrinsic or extrinsic configurations to build a sensor.
88. Optical sensors can be used in point, multiplexed or distributed configurations.
89. Optical sensors can be measure virtually all mechanical, electrical and chemical measure and using principles cited in 86 and 87.
90. The shot noise is a characteristic of all optical receivers and it follows a Poisson distribution.
91. Thermal noise is generated in a receiver because of resistors, a finite temperature of working and it follows a Gaussian distribution.
92. Shot noise usually dominates thermal noise in optical receivers.
93. Optical receivers perform a basic opto electronic conversion with signal recovery while minimizing the probability of bit error.
94. Optical receivers are broadly divided as homodyne and heterodyne types. In the homodyne, the intermediate frequency is zero whereas the heterodyne type utilizes a finite value of the intermediate frequency.
95. Optical receivers employ OOK,FSK and PSK as the optoelectronic modulation with the homodyne PSK receiver requiring just 9 photons/optical pulse to result in a BER of 10^{-9} under unity quantum efficiency for an ideal receiver.
96. The two principal objectives that a given network must perform are switching and multiplexing.
97. Circuit switching is normally used for voice transmission whereas packet switching is used for data traffic.
98. The photonic layer responsible for all optoelectronic conversion sits at the bottom of the optical network hierarchy as it is a class of physical layer.
99. The main topologies that a network follows to connect a large number of subscribers are (i) bus, (ii) star, (iii) ring and (iv) mesh. Physical topology is usually different from logical topology.
100. ATM through the use of cells results in a very improved spectrum utilization than STM.

Questions to be Answered by Search from Literature (Additional and Optional)

1. Find out different refractive index profiles corresponding to a step index fibre and discuss their importance.
2. Discuss material dispersion in pure and doped silica.
3. Rewrite the wave equation corresponding to the TE mode as an eigenvalue equation.
4. Discuss the organic LED with specific reference to its applications.
5. Discuss an optical mouse with a block diagram and principle of working.
6. Find information about LED and LCD television and explain how are they different from their cathode ray tube (CRT) based counterparts.

Numerical Problems

1. Assume that the fields of an EM wave are expressed as:

   $\mathbf{E}(\mathbf{y},t) = E_0 \exp[j(\omega t - \mathbf{k.y})]\mathbf{u}_1$ and $\mathbf{B}(\mathbf{y},t) = B_0 \exp[j(\omega t - \mathbf{k.y})]\mathbf{u}_2$

   (i) Identify the time varying and space varying components of this wave.

   (ii) What condition must be satisfied in order for the wave to become transverse?

   (iii) Find out the direction of propagation for the $\mathbf{u}_1$ and $\mathbf{u}_2$ being unit vectors along two unspecified directions.

   (iv) Prove that both of the fields are transverse to the direction of propagation.

   (v) Show that $\mathbf{u}_1 = \dfrac{\mathbf{k} \times \mathbf{u}_2}{k}$.

2. Show that the time-averaged power flow density for a TEM wave is given as $P = \dfrac{n}{2c\mu_0} E_o^2$ where E_0 is the peak amplitude of the electric field, n is the refractive index of the medium in which the wave is travelling.

3. Derive the equation for a uniform plane wave at a point 'P' that is at a distance of r from the origin. The wave is travelling along an arbitrary direction $\hat{\mathbf{n}}$ and the phase constant of the wave is β. Make a neat diagram to do your derivation. What is "uniform" here? Give the equation of the plane containing the wave.

4. A 10 MHz uniform plane wave travelling in free space makes an angle of 30^0 and 45^0 with the $x-$ and the $z-$axes respectively. Write down expressions for the electric and the magnetic fields of this wave if the peak amplitude of the electric field is 1 V/m along the axis. Find the field strengths along the and the axes also.

5. A 10 MHz uniform plane wave travelling in free space makes an angle of 30^0 and 60^0 with the and the axes respectively. Does it

have a component for its respective fields? Justify. Sketch this wave and write suitable comments on this.

6. Find the critical angle at the interface between pure silica of refractive index 1.45 and air. What do you mean by this?
7. Repeat P-6 for the interface between doped silica with refractive index 1.46 and pure silica.
8. Consider the case of normal incidence at the interface between pure silica and air. Calculate the energy reflection coefficient.
9. The refractive indices of the core and cladding of a certain fibre are 1.48 and 1.46 respectively. Find out the (i) critical angle at the core-cladding interface, (ii) acceptance angle in air and (iii) the NA of the fibre.
10. The NA of a typical fibre is 0.3. Find out (i) the acceptance angle and (ii) the % age of light collected by the fibre with respect to the incident light? Can all of the incident light be collected by any optical fibre? Justify.
11. A SI fibre has a core diameter of 25 μm, core refractive index of 1.48, the relative refractive index difference of 1.5%. Compute the value of at an operating wavelength of 0.8 μm. Is it a multimode or a monomode fibre?
12. For the above problem, in 11th, it is given that the fibre is a single mode optical fibre. Then find out its maximum core diameter.
13. The refractive indices of the core and cladding of a certain fibre are 1.48 and 1.46 respectively. The operating wavelength is 0.8 μm. Find out the range of for which the wave remains guided by the fibre if it is a monomode fibre having a radius as calculated from P-12.
14. The refractive indices of the core and cladding of a certain fibre are 1.48 and 1.46 respectively. The operating wavelength is 0.8 μm. Find out the range of β for which the wave remains guided by the fibre if it is a multimode fibre having a radius of 25 μm. Compare your results with those obtained from P-13.
15. A SI fibre has a core diameter of 25 μm, core refractive index of 1.48, the relative refractive index difference of 1.5%. The operating wavelength is 0.8 μm. Evaluate the % age of power through the core and the cladding.
16. A step index multimode fibre has a core diameter of 80 μm, relative refractive index difference of 1.5%, the core refractive index 1.48. It is used to carry light at 0.85 μm. Evaluate (i) the V- number and (ii) the number of guided modes. Can you find anything common to all the modes?

44. Is light allowed to travel in an optical fibre when it is incident at any arbitrary angle? Explain.
45. How does light propagate inside a step index fibre?
46. Compare the transit time of light waves in a multimode step index and a graded index fibre.
47. We have shown rays in multimode step index fibre while waves are shown in graded index fibres in understanding light behaviour. Justify.
48. Explain how scattering results in signal loss in an optical fibre.
49. Explain, from a physical point of view, the existence of multiple modes in a multimode fibre.
50. What is the importance of the number 2.405 in optical fibre based systems?
51. Which mode is typically supported by a monomode fibre? Explain.
52. If you wish a fibre to collect more light from the incident light, which optical source would you prefer? Explain.
53. What is the significance of reverse saturation current in a p-n junction diode?
54. Explain the difficulty in producing a blue LED until recently.
55. Why direct bandgap semiconductor materials are not suitable for producing LEDs?
56. Explain how recombination lifetime affects the internal quantum efficiency in a diode.
57. Is laser an amplifier or an oscillator? Explain.
58. What is population inversion? Why is it called so?
59. Which is desirable in a laser, more transition levels or less transition levels? Justify.
60. What is coherent in laser?
61. How threshold in laser is different from that in an LED.
62. Sketch the electric field across the junction of a semiconductor diode. Does this field follow any of Maxwell's equations?
63. What are the primary noises that affect the performance of an optical detector?
64. What is(are) the figure(s) of merit of an optical detector? Why do they serve as the figure(s) of merit?
65. What is(are) the source(s) of Gaussian noise in an optical receiver?
66. Why do you get shot noise in an optical detector?
67. Why does the shot noise in an optical detector follow a Poisson distribution?

68. Is an optical detector the same as an optical receiver? Explain.
69. What do you mean by power spectral density? What is its significance?
70. Distinguish macro bending from microbending with specific reference to their applications.
71. Explain, from a physical point of view, how a bend in an optical fibre affects the intensity of light waves passing through it.
72. Give a physical interpretation of the change in polarization of light.
73. What are the applications of polarization change of a given light wave?
74. State a few applications of phase modulation as far as light waves are concerned.
75. State the fundamental differences between a coherent and a noncoherent optical receiver.
76. Bring out the differences between a physical topology and a logical topology.
77. What are the fundamental objectives of a network?
78. State a few attributes of any network.
79. What is all optical switch?

Subjective Long Questions

1. Explain how optical transmitters are different from their electrical counterparts.
2. What, in your opinion, are the major challenges in an opto electronic system? Explain with reasons and case studies.
3. What do you understand by $\mathbf{k}.\mathbf{r} = \text{a constant}$? What does it represent? Develop an expression that has contribution from all three axes in an orthogonal coordinate system given as x_1, x_2 and x_3. Draw a neat sketch to derive your answer.
4. Redo P-3 in terms of wavelength λ of the wave. Comment on this. Hence, derive the expressions for wave velocity.
5. Prove that $\cos^2\phi_x + \cos^2\phi_y + \cos^2\phi_z = 1$ where $\cos\phi_x, \cos\phi_y$ and $\cos\phi_z$ represent the direction cosines along the three mutually perpendicular axes x,y and Z respectively.
6. The expression for an LP wave is given as $E_x = E_0 \cos(\omega t - \beta z + \phi_0)\hat{x}$? What is the direction of polarization? What is the direction of wave propagation? Comment on the argument of the cosine function.
7. What are meridional and skew rays? Illustrate by means of proper sketches. State the role of these rays in an optical fibre.
8. Why the HE_{11} mode is associated with the number 2.405?
9. Suggest possible method(s) to design a single mode fibre if the relative refractive index between the core and the cladding varies from 0.1% to 0.2% practically. You may assume any suitable operating wavelength.
10. Derive the cut off condition for the HE_{nm} mode.
11. From a theoretical point of view discuss the range of core diameters that are possible for a multimode fibre if it is given that the relative

refractive index difference between the core and the cladding varies from 1% to 2%.

12. Explain the application of boundary conditions in writing down field components at the core-cladding interface of an optical fibre.
13. Discuss the meaning of t_{nm} as applied to the Bessel's functions. Write down the first nine values in the form of a table.
14. Enumerate a procedure for determining the number of modes that can be supported by a multimode fibre.
15. Establish (3.20) from (3.19).
16. Evaluate the average power density as given by (3.91). How does it relate to the incident fields' average power density?
17. Establish (4.16).
18. Establish (4.19).
19. Establish (4.26).
20. Bring out the essential differences in wave propagation for multimode step index and graded index fibres.
21. Explain weakly guided modes with proper illustrations and mathematical expressions. What is 'weak' here?
22. How many total internal reflections take place for a ray traversing a length 'L' of an optical fibre.
23. Find out the condition(s) for which ray optics remains valid as far as propagation of light waves inside an optical fibre is concerned.
24. Discuss the fundamental differences between wave propagation in a metallic waveguide and an optical guide.
25. Sketch a guided mode in a step index fibre in which two plane waves propagate at discrete angles $\pm\theta = \cos^{-}\left(\beta / n_1 k_0\right)$.
26. Verify that $k_1^2 \approx k_2^2 \approx \beta^2$ when $\Delta << 1$ where k_1 and k_2 are the propagation constants in the core and the cladding respectively.
27. Show that the phase change occurring a totally internally reflected ray corresponding to phase shift of the wave component normal to the plane of incidence is $\tan\frac{\delta_N}{2} = \frac{\sqrt{n^2 \cos\theta_1 - 1}}{n \sin\theta_1}$ where θ_1 is the angle of incidence and satisfies $\theta_1 < \theta_c$ and $n = n_1/n_2$.
28. Consider the figure shown below (Fig. P-1). Show that the phase of the twice reflected ray at points A and B must satisfy the following

condition in order for a lightwave to propagate down the length of the fibre $\frac{4n_1ka}{\sin\theta_1}+2\delta_1=2N\pi$ where δ is the phase change of the lightwave at each reflection and N is an integer. What do you infer from this?

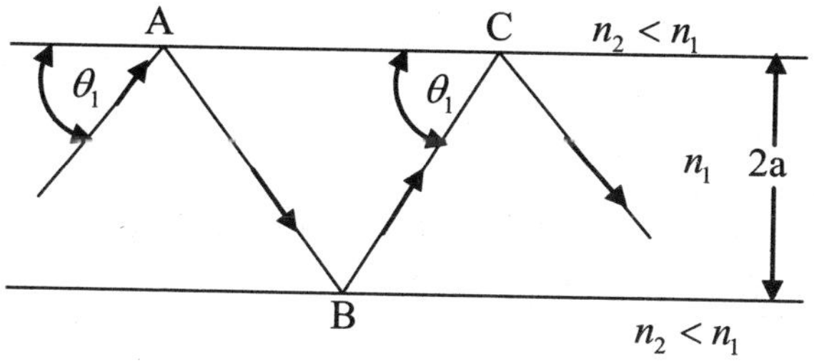

Fig. P-1: Lightwave propagating along an optical fibre of diameter 2a

29. Explain how pulse dispersion phenomena are different in step index and graded index fibres with specific reference to information carrying capacity.
30. What is transit time? Discuss its effect on the maximum information carrying capacity of an optical fibre.
31. Explain the contour of wave propagation in single mode and multimode step index fibres. Derive the appropriate mathematical expressions to support your answer.
32. Explain the relationship between core diameter, operating wavelength, refractive index profile of a fibre and the guided modes that can be supported by such a fibre.
33. A normalized propagation constant is defined as $b=\frac{a^2w^2}{V^2}$. Show that it is equal to $b=\frac{(\beta/k)^2-n_2^2}{n_1^2-n_2^2}$. Hence prepare a sketch between (β/k) and V. Plot V along the $x-$axis. Hence give the limits of b.
34. Make a sketch of $J_1(x)$.
35. Derive the eigenvalue equation of the HE modes. Sketch any such mode.
36. Differentiate between a TE wave and a TE_{mn} mode. Place your comparisons in the form of a table.
37. Derive field expressions for a guided lightwave assuming the optical fibre to be a planer waveguide. Consider the cross section to be in the $xy-$plane.

38. Explain why a plus sign in (5.103) represents an EH mode.
39. Establish (5.84).
40. Sketch the electric fields across the two dimensions of a planer optical fibre when light waves propagate through it.
41. Show that the group delay per unit length in the direction of wave propagation is given as

$$\frac{\tau_g}{L} = \frac{1}{V_g} = \frac{1}{c}\frac{d\beta}{dk} = -\frac{\lambda^2}{2\pi c}\frac{d\beta}{d\lambda}$$

Here, V_g is known as the group velocity of a given mode, L is the distance travelled by the optical pulse, β is the propagation constant along the fibre axis and $k = \frac{2\pi}{\lambda}$.

42. This problem deals with the derivation of material dispersion. Material dispersion, as we have seen is due to the variation of the refractive index as a function of the optical wavelength. The group velocity V_g of a given mode, therefore, becomes a function of the refractive index. The various spectral components of a given mode travel at different velocities, depending on their wavelength. If the RMS spectral width of an optical source this is σ_λ, then show that pulse spreading given as

$$\sigma_g = \sigma_\lambda \frac{d\tau_g}{d\lambda} = -\frac{\sigma_\lambda}{2\pi c}\left(2\lambda\frac{d\beta}{d\lambda} + \lambda^2\frac{d^2\beta}{d\lambda^2}\right)L$$

Hint: Make use of the result derived in P.41.

43. Show that material dispersion can also be expressed as

$\sigma_m = \frac{\sigma_\lambda L}{c}\left|\lambda\frac{d^2 n_1}{d\lambda^2}\right|$ where σ_λ is the RMS spectral width of the source with a mean wavelength λ, L is the length of the optical fibre and is core refractive index.

44. Consider a GaAlAs LED having a spectral width of 40 nm at 800 nm peak output. Find out the pulse spreading due to material dispersion.
45. Explain how scattering loss is different from bending losses.
46. Explain how transit time affects the information carrying capacity of an optical fibre.

47. Derive the mean and variance of a Gaussian noise which has a probability density function expressed as $p(n) = \frac{1}{\sqrt{2\pi\sigma_n^2}} \exp\left[-\frac{(n-\mu)^2}{2\sigma_n^2}\right]$.
48. Derive the mean and variance of a shot noise which has a probability density function given as a Poisson distribution.
49. Compare STM and ATM. Which one allows efficient resource utilization? Justify. Draw appropriate diagrams to support your answer.
50. Explain the difference(s) between physical topology and logical topology with suitable examples.

Subject Index

E

F

G

H

I

L

Y